The Zimányi School and Analytic Hydrodynamics in High Energy Physics

The Zimányi School and Analytic Hydrodynamics in High Energy Physics

Editors

Máté Csanád
Tamás Csörgő

Basel • Beijing • Wuhan • Barcelona • Belgrade • Novi Sad • Cluj • Manchester

Editors
Máté Csanád
Department of Atomic Physics
ELTE Eötvös Loránd University
Budapest
Hungary

Tamás Csörgő
Theoretical Physics Department
HUN-REN Wigner Research Center
Budapest
Hungary

Editorial Office
MDPI
St. Alban-Anlage 66
4052 Basel, Switzerland

This is a reprint of articles from the Special Issue published online in the open access journal *Universe* (ISSN 2218-1997) (available at: https://www.mdpi.com/journal/universe/special_issues/zimanyi_school_analytic_hydro).

For citation purposes, cite each article independently as indicated on the article page online and as indicated below:

Lastname, A.A.; Lastname, B.B. Article Title. *Journal Name* **Year**, *Volume Number*, Page Range.

ISBN 978-3-7258-0053-7 (Hbk)
ISBN 978-3-7258-0054-4 (PDF)
doi.org/10.3390/books978-3-7258-0054-4

© 2024 by the authors. Articles in this book are Open Access and distributed under the Creative Commons Attribution (CC BY) license. The book as a whole is distributed by MDPI under the terms and conditions of the Creative Commons Attribution-NonCommercial-NoDerivs (CC BY-NC-ND) license.

Contents

About the Editors ix

Olivér Surányi on behalf of the CMS Collaboration
Study of Very Forward Neutrons with the CMS Zero Degree Calorimeter
Reprinted from: *Universe* **2019**, *5*, 210, doi:10.3390/universe5100210 1

Davor Horvatić, Dalibor Kekez and Dubravko Klabučar
Temperature Dependence of the Axion Mass in a Scenario Where the Restoration of Chiral Symmetry Drives the Restoration of the $U_A(1)$ Symmetry
Reprinted from: *Universe* **2019**, *5*, 208, doi:10.3390/universe5100208 9

Jan Vanek for the STAR Collaboration
Open-Charm Hadron Measurements in Au+Au Collisions at $\sqrt{s_{NN}}$ = 200 GeV by the STAR Experiment
Reprinted from: *Universe* **2019**, *5*, 196, doi:10.3390/universe5090196 32

Bálint Kurgyis and Máté Csanád
A Particle Emitting Source From an Accelerating, Perturbative Solution of Relativistic Hydrodynamics
Reprinted from: *Universe* **2019**, *5*, 194, doi:10.3390/universe5090194 41

János Takátsy, Péter Kovács, Zsolt Szép and György Wolf
Compact Star Properties from an Extended Linear Sigma Model
Reprinted from: *Universe* **2019**, *5*, 174, doi:10.3390/universe5070174 50

Barnabás Pórfy
Lévy HBT Results at NA61/SHINE
Reprinted from: *Universe* **2019**, *5*, 154, doi:10.3390/universe5060154 61

Péter Pósfay, Gergely Gábor Barnaföldi and Antal Jakovác
Estimating the Variation of Neutron Star Observables by Dense Symmetric Nuclear Matter Properties
Reprinted from: *Universe* **2019**, *5*, 153, doi:10.3390/universe5060153 69

Boris Tomasik, Jakub Cimerman and Christopher Plumberg
Averaging and the Shape of the Correlation Function
Reprinted from: *Universe* **2019**, *5*, 148, doi:10.3390/universe5060148 79

Gábor Bíró, Gergely Gábor Barnaföldi, Gábor Papp and Tamás Sándor Biró
Multiplicity Dependence in the Non-Extensive Hadronization Model Calculated by the HIJING++ Framework
Reprinted from: *Universe* **2019**, *5*, 134, doi:10.3390/universe5060134 90

Máté Csanád, Sándor Lökös and Márton Nagy
Coulomb Final State Interaction in Heavy Ion Collisions for Lévy Sources
Reprinted from: *Universe* **2019**, *5*, 133, doi:10.3390/universe5060133 99

Zoltán Varga, Róbert Vértesi and Gergely Gábor Barnaföldi
Jet Structure Studies in Small Systems
Reprinted from: *Universe* **2019**, *5*, 132, doi:10.3390/universe5050132 110

Róbert Vértesi and on behalf of the ALICE Collaboration
Heavy-Flavor Measurements with the ALICE Experiment at the LHC
Reprinted from: *Universe* **2019**, *5*, 130, doi:10.3390/universe5050130 118

Monika Varga-Kofarago on Behalf of the Bergen pCT Collaboration
Medical Applications of the ALPIDE Detector
Reprinted from: *Universe* **2019**, 5, 128, doi:10.3390/universe5050128 125

Andrey Seryakov
Influence of Backside Energy Leakages from Hadronic Calorimeters on Fluctuation Measures in Relativistic Heavy-Ion Collisions
Reprinted from: *Universe* **2019**, 5, 126, doi:10.3390/universe5050126 130

Filip Krizek on Behalf of the ALICE Collaboration
Correlations of High-p_T Hadrons and Jets in ALICE
Reprinted from: *Universe* **2019**, 5, 124, doi:10.3390/universe5050124 138

Keming Shen, Gergely Gábor Barnaföldi and Tamás Sándor Biró
Hadron Spectra Parameters within the Non-Extensive Approach
Reprinted from: *Universe* **2019**, 5, 122, doi:10.3390/universe5050122 145

Marzio De Napoli
Production and Detection of Light Dark Matter at Jefferson Lab: The BDX Experiment
Reprinted from: *Universe* **2019**, 5, 120, doi:/10.3390/universe5050120 154

David Lafferty and Alexander Rothkopf
Quarkonium Phenomenology from a Generalised Gauss Law
Reprinted from: *Universe* **2019**, 5, 119, doi:10.3390/universe5050119 163

Eszter Frajna and Róbert Vértesi
Correlation of Heavy and Light Flavors in Simulations
Reprinted from: *Universe* **2019**, 5, 118, doi:10.3390/universe5050118 172

Alexander Rothkopf
Quarkonium Production in the QGP
Reprinted from: *Universe* **2019**, 5, 117, doi:10.3390/universe5050117 179

Veronika Agafonova
Study of Jet Shape Observables in Au+Au Collisionsat $\sqrt{s_{NN}}$ = 200 GeV with JEWEL
Reprinted from: *Universe* **2019**, 5, 114, doi:10.3390/universe5050114 188

Xiong-Tao Gong, Ze-Fang Jiang, Duan She and C. B. Yang
Viscous Hydrodynamic Description of the Pseudorapidity Density and Energy Density Estimation for Pb+Pb and Xe+Xe Collisions at the LHC
Reprinted from: *Universe* **2019**, 5, 112, doi:10.3390/universe5050112 195

Ferenc Siklér
Another Approach to Track Reconstruction: Cluster Analysis
Reprinted from: *Universe* **2019**, 5, 105, doi:10.3390/universe5050105 202

Daria Prokhorova, Nikolaos Davis and on behalf of NA61/SHINE Collaboration
Highlights from NA61/SHINE: Proton Intermittency Analysis
Reprinted from: *Universe* **2019**, 5, 103, doi:10.3390/universe5050103 208

Bálint Boldizsár, Márton I. Nagy and Máté Csanád
Polarized Baryon Production in Heavy Ion Collisions: An Analytic Hydrodynamical Study
Reprinted from: *Universe* **2019**, 5, 101, doi:10.3390/universe5050101 216

Astrid Morreale
Nuclear Physics at the Energy Frontier: Recent Heavy Ion Results from the Perspective of the Electron Ion Collider
Reprinted from: *Universe* **2019**, 5, 98, doi:10.3390/universe5050098 229

Balázs Endre Szigeti and Monika Varga-Kofarago
Study of Angular Correlations in Monte Carlo Simulations in Pb-Pb Collisions
Reprinted from: *Universe* **2019**, 5, 97, doi:10.3390/universe5050097 . . . . **244**

Bartosz Malecki
Bose–Einstein Correlations in pp and pPb Collisions at LHCb
Reprinted from: *Universe* **2019**, 5, 95, doi:10.3390/universe5040095 . . . . **250**

Valentina Raskina and Filip Křížek
Characterization of Highly Irradiated ALPIDE Silicon Sensors
Reprinted from: *Universe* **2019**, 5, 91, doi:10.3390/universe5040091 . . . . **258**

Dezső Horváth
Higgs and BSM Studies at the LHC
Reprinted from: *Universe* **2019**, 5, 160, doi:10.3390/universe5070160 . . . . **266**

Michael J. Tannenbaum
Latest Results from RHIC + Progress on Determining $\hat{q}L$ in RHI Collisions Using Di-Hadron Correlations
Reprinted from: *Universe* **2019**, 5, 140, doi:10.3390/universe5060140 . . . . **282**

About the Editors

Máté Csanád

Máté Csanád, Fellow of the Young Academy of Europe, is a professor of physics at the Department of Atomic Physics and the director of the Environmental Science Center, both at Eötvös Loránd University. His research interests focus on femtoscopy and hydrodynamics in high-energy physics and the time evolution of the strongly interacting matter created in heavy-ion collisions.

Tamás Csörgő

Tamás Csörgő, Member of the Academia Europaea, is a research professor at the Femtoscopy Knowledge Center of the Károly Róbert Campus, Hungarian University of Agriculture and Life Sciences (MATE), and the HUN-REN Wigner Research Center. His research interests range from femtoscopy through to hydrodynamics and forward processes in high-energy physics.

Communication

Study of Very Forward Neutrons with the CMS Zero Degree Calorimeter

Olivér Surányi on Behalf of the CMS Collaboration

MTA-ELTE Lendület CMS Particle and Nuclear Physics Group, Eötvös Loránd University, 1053 Budapest, Hungary; oliver.suranyi@cern.ch

Received: 15 August 2019; Accepted: 5 October 2019; Published: 9 October 2019

Abstract: Forward neutrons are studied in proton-lead collisions at the CMS experiment at the CERN LHC. They provide information on the centrality and event plane of collisions and provide an opportunity to study nuclear breakup. At the CMS experiment they are detected by the Zero Degree Calorimeters (ZDCs) in the $|\eta| > 8.5$ pseudorapidity range. The ZDCs are quartz fiber Cherenkov calorimeters using tungsten as absorber. Test beam data and events with a single spectator neutron are used for the calibration of these detectors. A Fourier-based method is used correct for the effect of multiple pPb collisions. The corrected ZDC energy distribution is used to calculate centrality percentiles and unfold the neutron multiplicity distribution.

Keywords: forward neutrons; zero degree calorimetry; centrality; heavy ions

1. Introduction

Very forward ($|\eta| > 8.5$) neutrons are produced in hadron-nucleus and heavy ion collisions. The main physics processes involved are intranuclear cascades [1], nuclear evaporation [2], and nuclear resonances, like the giant dipole resonance [3]. The information gathered from the observation of these neutrons can be used to tag ultraperipheral collisions, calculate the event plane and estimate centrality in heavy ion and hadron-nucleus collisions. Cascade and evaporation nucleons were observed and studied by a wide range of fix target experiments [4]. Ultraperipheral collisions accompanied by nuclear resonances were studied by the STAR Collaboration at RHIC [5,6]. In the ALICE experiments at LHC cascade and evaporation neutrons are used for the estimation of centrality [7,8]. The energy spectrum of very forward neutrons produced in proton-proton collisions was measured by the LHCf experiment [9,10]. In the CMS experiment these neutrons can be observed by the Zero Degree Calorimeters (ZDC). The most important models, which include forward neutron production are DPMJet [2], URQMD [11], Geant4 [12,13], SMASH [14], JAM [15], and PHSD [16].

2. The Zero Degree Calorimeter of CMS Experiment

The two ZDCs are located in the neutral particle absorber (TAN), roughly ±140 m away from the CMS interaction point, between the two beampipes. They measure neutral particles at pseudorapidity values $|\eta| > 8.5$, as the charged products are removed by dipole magnets located between the central CMS detectors and the ZDCs. They are Cherenkov sampling calorimeters consisted of cladded quartz fibers and tungsten plates. They have three different sections (Figure 1): the electromagnetic (EM) section, the hadron (HAD) section, and the reaction plane detector (RPD). The EM section is 19 radiation lengths long, which is equal to one interaction length and has five transverse segments. The RPD is a 4×4 array of quartz tiles and is used to determine the event plane for flow measurements in heavy ion collisions. The HAD section is 5.6 hadronic interaction length and has four longitudinal segments. The plates in the HAD section are tilted by 45° to maximize the light yield of the fibers. This paper presents results using the EM and HAD sections of the ZDC detector on the lead going side ($z < 0$)

measured in pPb collisions at the CMS experiment in 2016. A more detailed description of the ZDC detectors can be found in references [17–20].

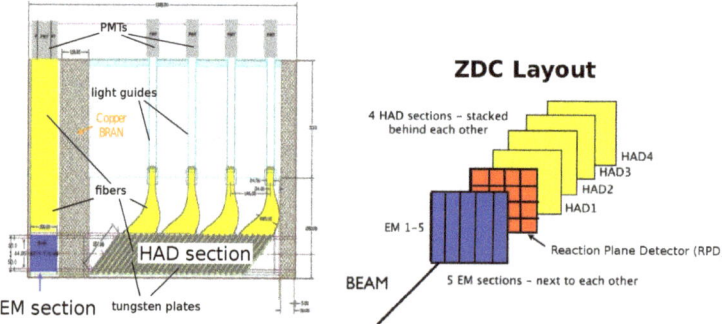

Figure 1. The side view of the CMS Zero Degree Calorimeter (**left**) and the segmentation of each sections (**right**) [21].

Every channel of the detector is read out in 10 timeslices (denoted with TS0 – TS9), each 25 ns long. The main signal arrives in TS4 as shown in the left panel of Figure 2. The bunch spacing in the 2016 pPb datataking was 100 ns, so additional collisions may happen four timeslice before or after the primary signal (out-of-time pileup). The signal value Q_i for a certain i channel is extracted as:

$$Q_i = Q_{i,\text{TS3}} - \frac{1}{2}(Q_{i,\text{TS2}} + Q_{i,\text{TS6}}),\qquad(1)$$

where $Q_{i,\text{TSX}}$ is the signal value in the Xth timeslice for channel i. The second term is used to subtract contribution from pedestal and the tail of out-of-time pileup events.

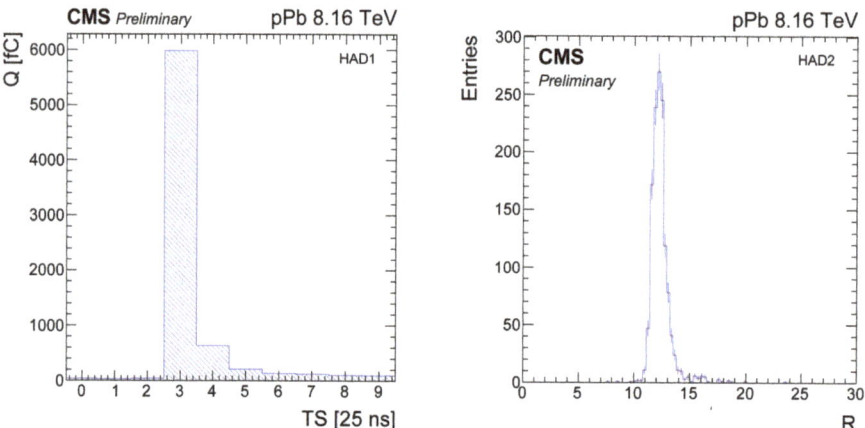

Figure 2. The time dependence of the Zero Degree Calorimeters (ZDC) signal (**left**) and the distribution of TS4 and TS5 signal ratio (**right**) [21].

In order to increase the dynamic range of the detector, the largest signals are let to slightly saturate and their signal values are calculated by scaling up the sum of TS4 and TS5 by a scale-factor $\langle R \rangle$. The scale factor is calculated separately for each channel from the distribution of the values:

$$R = \frac{Q_{i,\text{TS3}} - \frac{1}{2}(Q_{i,\text{TS2}} + Q_{i,\text{TS6}})}{Q_{i,\text{TS4}} - \frac{1}{2}(Q_{i,\text{TS2}} + Q_{i,\text{TS6}})}, \quad (2)$$

which is shown in the right panel of Figure 2. The value of $\langle R \rangle$ is calculated from the mean of this distribution. This method ensures that both the resolution remains good for the low-energy signals and the whole energy range of the detector can be used.

3. Calibration

The relative gains of the different channels are matched (intercalibrated) by using distributions of signal ratios between different channels. These are compared to the same distributions from the 2010 PbPb data collection period, following the test beam calibration of ZDCs. An example of two such distributions is shown in Figure 3.

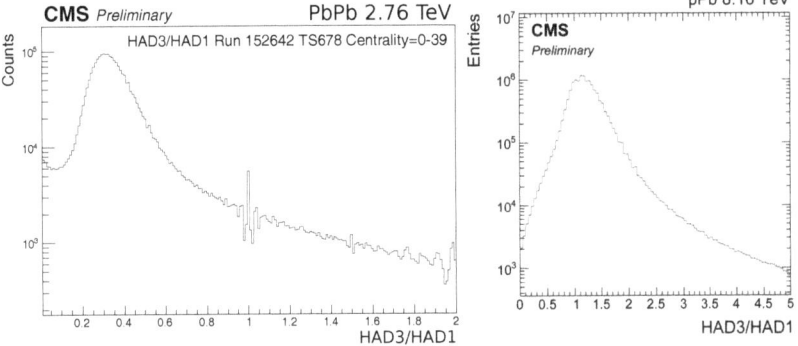

Figure 3. Comparison of signal ratio distributions between 2.76 TeV PbPb (**left**) and 8.16 TeV pPb collisions (**right**) [21].

The low energy part of the total ZDC signal distribution is shown in Figure 4. Since the slow neutrons are close to being monoenergetic, the peaks correspond to events with single, double, and triple neutrons. The widths of the peaks are due to the slight variation in the neutron energy and the finite resolution of the detector. The spectrum is fitted with the sum of Gaussian shapes, with the n-neutron peak position μ_n and width σ_n constrained as:

$$\mu_n = n\mu_1, \quad (3)$$
$$\sigma_n^2 = n\sigma_1^2, \quad (4)$$

where μ_1 and σ_1 are the position and width of the single neutron peak respectively. The results from the fit are used for the absolute energy calibration of the detector, since the position of the first peak corresponds to 2.56 TeV, the energy of nucleons in the Pb ion in pPb collisions with 8.16 TeV collision energy [22]. The detector resolution is found to be 24%, which is slightly larger than the 15% resolution extrapolated from the test beam measurement [20].

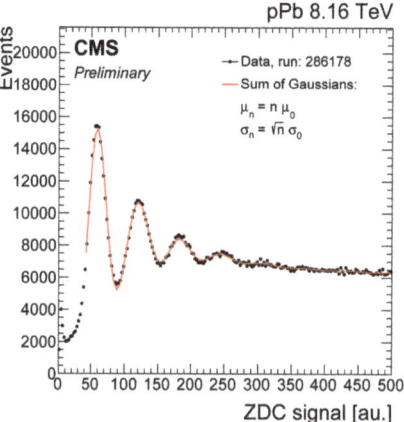

Figure 4. The low end of ZDC signal distribution. The peaks correspond to events with one, two and three neutrons detected by the ZDC respectively. The position of the first peak corresponds to 2.56 TeV, the energy of nucleons in the Pb ion for 8.16 TeV pPb collisions [21].

4. Pileup Correction

The ZDC energy spectrum is shown in Figure 5 for two different data collection periods with different average number of collisions in a single event (pileup). An important feature of these spectra is a shoulder at high energies, which is larger in the data set with a larger pileup. It is caused by pileup collisions. This effect is corrected by the following mathematical procedure, based on the compound Poisson distribution.

Assuming that n is the number of simultaneous pPb collisions in a single event and it follows a Poisson distribution:

$$p_n = \frac{\mu^n}{n!} \frac{e^{-\mu}}{1 - e^{-\mu}}, \quad (5)$$

where only the $n > 0$ case is considered, therefore $1 - e^{-\mu}$ is included in the denominator to ensure the normalization of the above expression. μ is the average number of collisions in an event, that provide at least one neutron in the ZDC. The ZDC energy deposit is described by a random variable X:

$$X = \sum_{i=1}^{n} Y_i, \quad (6)$$

where Y_i is the random variable describing ZDC energy deposit for an event with single collision. The probability density functions of Y_i and X are denoted by $g(x)$ and $f(x)$ respectively. The $f(x)$ function can be expressed as:

$$f(x) = g(x)\, p_1 + (g * g)(x)\, p_2 + (g * g * g)(x)\, p_3 + \ldots, \quad (7)$$

where the $*$ operation stands for convolution. Then using Fourier transform and the convolution theorem one may write:

$$\mathcal{F}f = (\mathcal{F}g)\, p_1 + (\mathcal{F}g)^2\, p_2 + (\mathcal{F}g)^3\, p_3 + \ldots \quad (8)$$

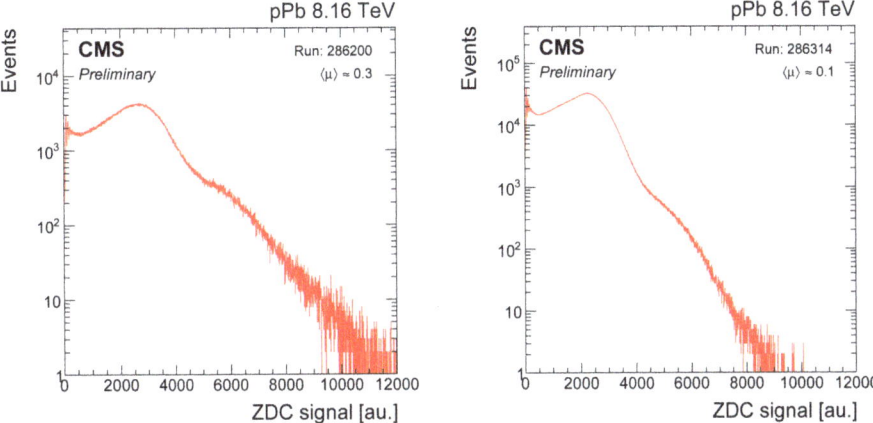

Figure 5. ZDC energy distribution in data sets with average number of collisions 0.3 (**left**) and 0.1 (**right**) [21].

After substituting the definition of p_n into the equation above, the infinite series becomes a geometrical series and can be summed up to:

$$\mathcal{F}f = \frac{e^{-\mu}}{1-e^{-\mu}}\left(e^{\mu \mathcal{F}g} - 1\right). \tag{9}$$

By rearranging this equation, it is possible to express $g(x)$ as:

$$g(x) = \mathcal{F}^{-1}\left[\frac{1}{\mu}\log\left[(e^{\mu}-1)\mathcal{F}f + 1\right]\right], \tag{10}$$

providing a formula to calculate the pileup corrected distribution.

The result of this calculation is shown at three different μ values in Figure 6. As it is expected the shoulder at high energies disappeared. The method gives a similar result when the value of μ is varied; there is only a moderate variation in the tail of the distributions.

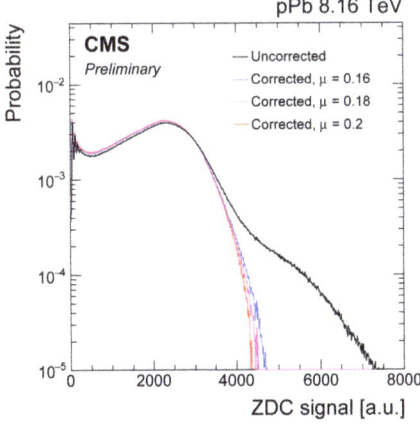

Figure 6. Pileup correction results with Fourier deconvolution method with different μ values assumed (**left**) and the percentiles of the ZDC signal distribution, used for centrality determination (**right**) [21].

5. Using ZDC as a Centrality Estimator in pPb Collisions

The typical quantities used to estimate centrality in heavy ion collisions, such as the multiplicity of charged particles, cannot be used in hadron-nucleus collisions, since they are only loosely correlated with the quantities N_{coll} and N_{part}. A good alternative is to use the very forward energy to estimate centrality, since it is correlated with the number of spectator nucleons. The centrality percentiles, calculated from the corrected energy distribution, are shown in Figure 7. For a proper usage of this zero degree energy as a centrality estimator in a physics measurement, a model of spectator neutron production is needed, which connects the zero degree energy with N_{coll} and N_{part}. The current models are not valid in the LHC collision energies [1,2,4,12], therefore results from ZDC detectors give a useful input for the development of these models.

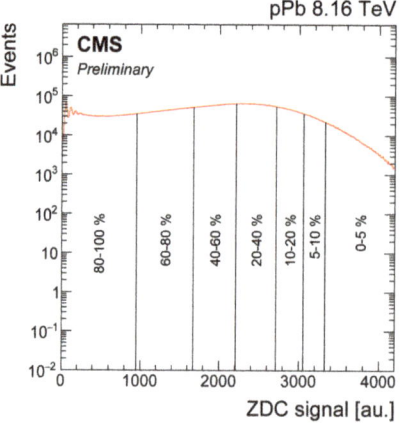

Figure 7. Centrality classes based on ZDC energy [21].

6. Unfolding Neutron Multiplicity Distribution

From the fits in Section 3, neutron multiplicity distribution can be calculated for events where only a few neutrons are produced. But the fit cannot be extended to the larger ZDC energy range, as the Gaussian distributions would overlap causing rapid oscillation of the amplitudes. This problem can be circumvented by constraining the amplitude values to change smoothly with the neutron number. This property is achieved by considering the calculation as an unfolding problem. The ZDC response is calculated for N-neutron events and a response matrix **R** is constructed (left panel of Figure 8), assuming the linearity of the detector. Using the response matrix, the original fit can be performed by minimizing a χ^2 term. The smoothness of the neutron number distribution is ensured by a linear regularization term (Tikhonov regularization [23]), which requires the first derivative of the distribution to be small. The full χ^2 term to minimalize is:

$$\chi^2 = (\mathbf{R} \cdot \mathbf{n} - \mathbf{e})^T \mathbf{V}^{-1} (\mathbf{R} \cdot \mathbf{n} - \mathbf{e}) + \lambda (\mathbf{D} \cdot \mathbf{n})^2, \tag{11}$$

where **n** and **e** are vectors, whose elements represent the unknown neutron distribution and the measured ZDC spectrum respectively, **V** is the covariance matrix of **e**, **D** is the first derivative matrix and λ is the regularization strength.

The optimal solution is calculated by taking the derivative of the above equation, with respect to **n**. The derivative can be rearranged into the form of linear system of equations:

$$(\mathbf{R}^T \mathbf{V}^{-1} \mathbf{R} + \lambda \mathbf{D}^T \mathbf{D}) \mathbf{n} = \mathbf{R}^T \mathbf{V}^{-1} \mathbf{e}. \tag{12}$$

The optimal **n** is calculated from this expression using the LU decomposition method.

The result of this calculation is shown in the right panel of Figure 8. This result serves as a strong constraint and may challenge the models of hadron-nucleus collisions and spectator neutron production [2,11–16].

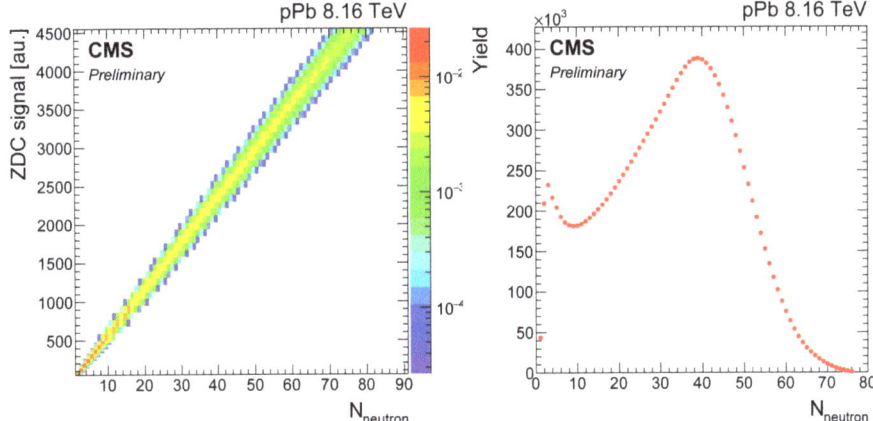

Figure 8. Response matrix (**left**) and the unfolded neutron multiplicity distribution (**right**). [21]

7. Conclusions

Very forward neutrons are produced in hadron-nucleus and heavy ion collisions via intranuclear cascades, evaporation and nuclear resonances. In the CMS experiment they are observed by the ZDCs. The individual channels of ZDCs in 2016 were gain matched using data collected in 2010. Peaks are observed in the ZDC energy spectrum, which correspond to events with one, two and three nearly monoenergetic spectator neutrons. This provides an opportunity to calibrate the detector and measure the resolution, which is 24%. The pileup effects are corrected using a Fourier deconvolution method. As the very forward energy is suitable to be a centrality estimator, the corrected ZDC energy distribution is used to calculate centrality percentiles. The neutron multiplicity distribution can be unfolded using Tikhonov regularization and the results serve as a constraint for the theoretical nuclear break-up models.

Funding: The CMS Zero Degree Calorimeter detector is supported by the Office of Science, US Department of Energy. This research is supported by the ÚNKP-19-3 New National Excellence Program of the Ministry for Innovation and Technology, the National Research, Development and Innovation Office of Hungary (K 124845, K 128713, and K 128786) and the Hungarian Academy of Sciences "Lendület" (Momentum) Program (LP 2015-7/2015).

Conflicts of Interest: The author declares no conflict of interest. The funders had no role in the design of the study; in the collection, analyses, or interpretation of data; in the writing of the manuscript, or in the decision to publish the results.

References

1. Cugnon, J.; Volant, C.; Vuillier, S. Improved intranuclear cascade model for nucleon-nucleus interactions. *Nucl. Phys. A* **1997**, *620*, 475. [CrossRef]
2. Ferrari, A.; Sala, P.R.; Ranft, J.; Roesler, S. Cascade particles, nuclear evaporation, and residual nuclei in high-energy hadron-nucleus interactions. *Z. Phys. C Part. Fields* **1996**, *C70*, 413. [CrossRef]
3. Berman, B.L.; Fultz, S.C. Measurements of the giant dipole resonance with monoenergetic photons. *Rev. Mod. Phys.* **1975**, *47*, 713. [CrossRef]

4. Sikler, F. Centrality Control of Hadron Nucleus Interactions by Detection of Slow Nucleons. Available online: https://arxiv.org/abs/hep-ph/0304065 (accessed on 5 October 2019).
5. Adler, C.; et al. [STAR Collaboration]. Coherent ρ^0 production in ultraperipheral heavy ion collisions. *Phys. Rev. Lett.* **2002**, *89*, 272302. [CrossRef] [PubMed]
6. Adams, J.; et al. [STAR Collaboration]. Production of e^+e^- pairs accompanied by nuclear dissociation in ultra-peripheral heavy ion collision. *Phys. Rev. C* **2004**, *70*, 031902. [CrossRef]
7. Abelev, B.; et al. [STAR Collaboration]. Centrality determination of Pb-Pb collisions at $\sqrt{s_{NN}}$ = 2.76 TeV with ALICE. *Phys. Rev. C* **2013**, *88*, 044909. [CrossRef]
8. Adam, J.; et al. [ALICE Collaboration]. Centrality dependence of particle production in p-Pb collisions at $\sqrt{s_{NN}}$ = 5.02 TeV. *Phys. Rev. C* **2015**, *91*, 064905. [CrossRef]
9. Adriani, O.; Berti, E.; Bonechi, L.; Bongi, M.; Castellini, G.; D'Alessandro, R.; Del Prete, M.; Haguenauer, M.; Itow, Y.; Kasahara, K.; et al. Measurement of very forward neutron energy spectra for 7 TeV proton–proton collisions at the Large Hadron Collider. *Phys. Lett. B* **2015**, *750*, 360. [CrossRef]
10. Ueno, M. Zero degree neutron energy spectra measured by the LHCf at $\sqrt{s}$= 13 TeV proton-proton collision. *PoS* **2018**, *ICRC2017*, 322.
11. Bass, S.A.; Belkacem, M.; Bleicher, M.; Brandstetter, M.; Bravina, L.; Ernst, C.; Gerland, L.; Hofmann, M.; Hofmann, S.; Konopka, J.; et al. Microscopic models for ultrarelativistic heavy ion collisions. *Prog. Part. Nucl. Phys.* **1998**, *41*, 255. [CrossRef]
12. Galoyan, A.; Ribon, A.; Uzhinsky, V. Simulation of neutron production in hadron-nucleus and nucleus-nucleus interactions in Geant4. In Proceedings of the 24th International Baldin Seminar on High Energy Physics Problems: Relativistic Nuclear Physics and Quantum Chromodynamics (ISHEPP 2018) Dubna, Russia, 17–22 September 2018.
13. Allison, J.; Amakoca, K.; Apostolakisd, J.; Arcee, P.; Asaif, M.; Asog, T.; Baglih, E.; Bagulyai, A.; Banerjeej, S.; Barrand, G.; et al. Recent developments in Geant4. *Nucl. Instrum. Meth. A* **2016**, *835*, 186. [CrossRef]
14. Weil, J.; Steinberg, V.; Staudenmaier, J.; Pang, L.G.; Oliinychenko, D.; Mohs, J.; Kretz, M.; Kehrenberg, T.; Goldschmidt, A.; Bäuchle, B.; et al. Particle production and equilibrium properties within a new hadron transport approach for heavy-ion collisions. *Phys. Rev.* **2016**, *C94*, 054905. [CrossRef]
15. Nara, Y.; Otuka, N.; Ohnishi, A.; Niita, K.; Chiba, S. Study of relativistic nuclear collisions at AGS energies from p + Be to Au + Au with hadronic cascade model. *Phys. Rev. C* **2000**, *61*, 024901. [CrossRef]
16. Cassing, W.; Bratkovskaya, E.L. Parton transport and hadronization from the dynamical quasiparticle point of view. *Phys. Rev.* **2008**, *C78*, 034919. [CrossRef]
17. Grachov, O.A.; Murray, M.J.; Ayan, A.S.; Debbins, P.; Norbeck, E.; Onel, Y.; d'Enterria, D.G. Status of zero degree calorimeter for CMS experiment. *AIP Conf. Proc.* **2006**, *867*, 258.
18. Grachov, O.A.; Grachov, O.A.; Metzler, B.; Murray, M.; Snyder, J.; Stiles, L.; Wood, J.; Zhukova, V.; Beaumont, W.; Ochesanu, S.; et al. Measuring photons and neutrons at zero degrees in CMS. *Int. J. Mod. Phys. E* **2007**, *16*, 2137. [CrossRef]
19. Grachov, O.A.; Murray, M.; Snyder, S.; Wood, J.; Zhukova, V.; Ayan, A.S.; Debbins, P.; Ingram,D.F.; Norbeck, E.; Onel, Y.; et al. Performance of the combined zero degree calorimeter for CMS. *J. Phys. Conf. Ser.* **2009**, *160*, 012059. [CrossRef]
20. Grachov, O.; Murray, M.; Wood, J.; Onel, Y.; Sen, S.; Yetkin, T. Commissioning of the CMS zero degree calorimeter using LHC beam. *J. Phys. Conf. Ser.* **2011**, *293*, 012040. [CrossRef]
21. CMS Collaboration. *The Performance of CMS ZDC Detector in 2016*; CMS Detector Performance Summary; CMS-DP-2018-025; CMS Collaboration: 2017. Available online: https://cds.cern.ch/record/2621978?ln=en (accessed on 9 October 2019).
22. Jowett, J.M. The LHC as a Nucleus-Nucleus Collider. *J. Phys. G* **2008**, *35*, 104028. [CrossRef]
23. Tikhonov, A.N. Solution of incorrectly formulated problems and the regularization method. *Soviet Math.* **1963**, *4*, 1035.

© 2019 by the authors. Licensee MDPI, Basel, Switzerland. This article is an open access article distributed under the terms and conditions of the Creative Commons Attribution (CC BY) license (http://creativecommons.org/licenses/by/4.0/).

Article

Temperature Dependence of the Axion Mass in a Scenario Where the Restoration of Chiral Symmetry Drives the Restoration of the $U_A(1)$ Symmetry

Davor Horvatić [1], Dalibor Kekez [2] and Dubravko Klabučar [1,*]

[1] Physics Department, Faculty of Science-PMF, University of Zagreb, Bijenička cesta 32, 10000 Zagreb, Croatia; davorh@phy.hr
[2] Ruđer Bošković Institute, Bijenička cesta 54, 10000 Zagreb, Croatia; kekez@irb.hr
* Correspondence: klabucar@phy.hr

Received: 2 August 2019; Accepted: 1 October 2019; Published: 8 October 2019

Abstract: The temperature (T) dependence of the axion mass is predicted for T's up to ~2.3× the chiral restoration temperature of QCD. The axion is related to the $U_A(1)$ anomaly. The squared axion mass $m_a(T)^2$ is, modulo the presently undetermined scale of spontaneous breaking of Peccei–Quinn symmetry f_a (squared), equal to QCD topological susceptibility $\chi(T)$ for all T. We obtain $\chi(T)$ by using quark condensates calculated in two effective Dyson–Schwinger models of nonperturbative QCD. They exhibit the correct chiral behavior, including the dynamical breaking of chiral symmetry and its restoration at high T. This is reflected in the $U_A(1)$ symmetry breaking and restoration through $\chi(T)$. In our previous studies, such $\chi(T)$ yields the T-dependence of the $U_A(1)$-anomaly-influenced masses of η' and η mesons consistent with experiment. This in turn supports our prediction for the T-dependence of the axion mass. Another support is a rather good agreement with the pertinent lattice results. This agreement is not spoiled by our varying u and d quark mass parameters out of the isospin limit.

Keywords: axion; QCD; non-Abelian axial anomaly; $U_A(1)$ symmetry breaking; chiral restoration

1. Introduction

The axion, one of the oldest hypothetical particles beyond the Standard Model, intensely sought for by many experimentalists already for 40 years now, still escapes detection [1]. It was introduced theoretically [2–5] to solve the so-called Strong CP problem of QCD. The problem is that no experimental evidence of CP-symmetry violation has been found in strong interactions, although the QCD Lagrangian $\mathcal{L}_{QCD}(x)$ can include the so-called θ-term $\mathcal{L}_\theta(x) = \theta Q(x)$ where gluon field strengths $F^b_{\mu\nu}(x)$ form the CP-violating combination $Q(x)$ named the topological charge density:

$$Q(x) = \frac{g^2}{32\pi^2} F^b_{\mu\nu}(x)\widetilde{F}^{b\mu\nu}, \quad \text{where } \widetilde{F}^{b\mu\nu} \equiv \frac{1}{2}\epsilon^{\mu\nu\rho\sigma} F^b_{\rho\sigma}(x). \tag{1}$$

Whereas $Q(x)$ can be re-cast in the form of a total divergence $\partial_\mu K^\mu$, discarding $\mathcal{L}_\theta$ is not justified even if $F^b_{\mu\nu}(x)$ vanish sufficiently fast as $|x| \to \infty$. Specifically, $F\widetilde{F} = \partial_\mu K^\mu$ can anyway contribute to the action integral, since in QCD there are topologically nontrivial field configurations such as instantons. They are important for, e.g., obtaining the anomalously large mass of the η' meson. Also, precisely the form (1) from the θ-term appears in the axial anomaly, breaking the $U_A(1)$ symmetry of QCD - see Equation (2).

For these reasons, one needs $\mathcal{L}_\theta(x) = \theta Q(x)$ in the QCD Lagrangian, as reviewed briefly in Section 1 of Ref. [6]. Moreover, the Strong CP problem cannot be removed by requiring that the coefficient $\theta = 0$, since QCD is an integral part of the Standard Model, where weak interactions break the CP symmetry. This CP violation comes from the complex Yukawa couplings, yielding the complex CKM matrix [7,8] and the quark-mass matrix M which is complex in general. To go to the mass–eigenstate basis, one diagonalizes the mass matrix, and the corresponding chiral transformation changes θ by $\arg \det M$. Hence, in the Standard Model the coefficient of the $Q \propto F\widetilde{F}$ term is in fact $\bar\theta = \theta + \arg \det M$ [6]. Therefore, to be precise, we change our notation to $\bar\theta$-term, $\mathcal{L}_\theta \to \mathcal{L}_{\bar\theta}$.

Since CP is not a symmetry of the Standard Model, there is no *a priori* reason $\bar\theta$, which results from the contributions from *both* the strong and weak interactions, should vanish. And yet, the experimental bound on it is extremely low, $|\bar\theta| < 10^{-10}$ [9], and in fact consistent with zero. Therefore, the mystery of the vanishing strong CP violation is: *why is $\bar\theta$ so small?*

The most satisfactory answer till this very day has been provided by axions, even though the original variant has been ruled out [1]. In the meantime, they turned out to be very important also for cosmology, as promising candidates for dark matter—see from relatively recent references such as [10,11] to the earliest papers [12–14]. (For an example of a broader review of axion physics, see [15].) It is thus no wonder that ever since the original proposal of the axion mechanism [2–5] in 1977–1978, many theorists kept developing various ideas on this theoretically much needed object, trying to pinpoint the properties of this elusive particle and increase chances of finding it.

However, to no avail. There have even been some speculations that the axion is hidden in plain sight, by being experimentally found, paradoxically, already years before it was conjectured theoretically: namely, that the axion should in fact be identified with the well-known η' meson with a minuscule admixture of a pseudoscalar composite of neutrinos [16]. Nevertheless, while an intimate relation between the axion and η' doubtlessly exists, reformulations of the axion theory, let alone so drastic ones, are in fact not needed to exploit this axion-η' relationship: thanks to the fact that both of their masses stem from the axial anomaly and are determined by the topological susceptibility of QCD, in the present paper we show how our previous study [17] of the temperature (T) dependence of the η' and η mesons give us a spin-off in the form of the T-dependence of the axion mass, $m_a(T)$. It is given essentially by the QCD topological susceptibility $\chi(T)$, which is rather sensitive to changes of the lightest quark masses m_q: Equation (9) vanishes linearly when $m_q \to 0$ even for just one flavor q. We thus examine the effect of their values on $\chi(T)$ also out of the isosymmetric limit, and find that such a variation can be accommodated well. The agreement with lattice results on $\chi(T)$ is reasonably good.

2. Connection with the Complex of the η' and η Mesons

2.1. Some Generalities on the Influence of the Anomaly on η' and η

In this paper, we neglect contributions of quark flavors heavier than $q = s$ and take $N_f = 3$ as the number of active flavors.

At vanishing and small temperatures, $T \approx 0$, the physical η' meson is predominantly[1] η_0, the singlet state of the flavor $SU(3)$ group, just like its physical partner, the lighter isospin-zero mass eigenstate η is predominantly the octet state η_8. Unlike the $SU(3)$ octet states π, K and η_8, the singlet η_0 is precluded from being a light (almost-)Goldstone boson of the dynamical breaking of the (only approximate) chiral symmetry of QCD (abbreviated as DChSB). Namely η_0 receives a relatively large anomalous mass contribution from the non-Abelian axial ABJ[2] anomaly, or gluon anomaly for short. An even better name for it is the $U_A(1)$ anomaly, since it breaks explicitly the $U_A(1)$ symmetry of QCD on the quantum level.

---

[1] The mass eigenstate η' is approximated only roughly by the pure $SU(3)$ singlet state η_0, due to the relatively large explicit breaking of the flavor $SU(3)$ symmetry by much heavier s-quark: $2m_s/(m_u + m_d) = 27.3 \pm 0.7$ [1].
[2] ABJ anomaly stands for names of Adler, Bell, and Jackiw, as a reminder of their pioneering work on anomalies [18,19] exactly half a century ago this year.

The breaking of $U_A(1)$ by the anomaly makes the flavor singlet ($a = 0$) axial current of quarks, $A_0^\mu(x) = \sum_{q=u,d,s} \bar{q}(x) \gamma^\mu \gamma_5 q(x)$, not conserved even in the chiral limit:

$$\partial_\mu A_0^\mu(x) = i \sum_{q=u,d,s} 2 m_q \bar{q}(x) \gamma_5 q(x) + 2 N_f Q(x), \qquad (N_f = 3), \qquad (2)$$

unlike the corresponding octet currents $A_a^\mu(x)$, $a = 1, 2, ..., N_f^2 - 1$. In the chiral limit, the current masses of non-heavy quarks all vanish, $m_q \to 0$ ($q = u, d, s$), but the divergence of the singlet current A_0^μ is not vanishing due to the $U_A(1)$ anomaly contributing no other but the topological charge density operator $Q(x)$ (1)—that is, precisely the quantity responsible for the strong CP problem.

The quantity related to the $U_A(1)$-anomalous mass in the η'-η complex is the QCD topological susceptibility χ,

$$\chi = \int d^4x \, \langle 0 | \mathcal{T} Q(x) Q(0) | 0 \rangle, \qquad (3)$$

where $\mathcal{T}$ denotes the time-ordered product.

Figure 1 shows how anomaly contributes to the mass matrix (in the basis of quark–antiquark ($q\bar{q}$) pseudoscalar bound states P) by depicting how hidden-flavor $q\bar{q}$ pseudoscalars mix, transiting through anomaly-dominated gluonic intermediate states.

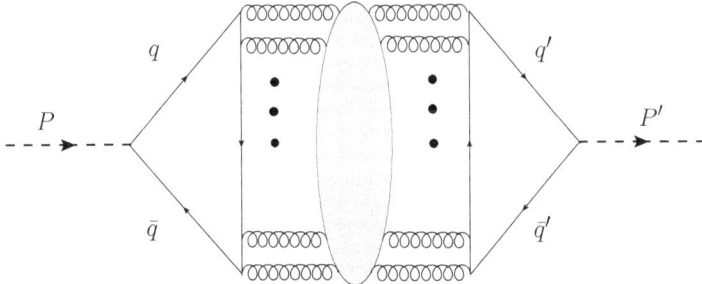

Figure 1. $U_A(1)$ anomaly-induced, hidden-flavor-mixing transitions from pseudoscalar quark–antiquark states $P = q\bar{q}$ to $P' = q'\bar{q}'$ include both possibilities $q = q'$ and $q \neq q'$. Springs symbolize gluons. All lines and vertices are dressed in accord with the nonperturbative QCD. Nonperturbative configurations are essential for nonvanishing anomalous mass [20] contribution to $\eta_0 \sim \eta'$, since $Q(x)$ is a total divergence. The gray blob symbolizes the infinity of all intermediate gluon states enabling such transitions, so that the three bold dots represent any even [21] number of additional gluons. Just one of infinitely many, but certainly the simplest realization thereof, is when such a transition is mediated by just two gluons (and no additional intermediate states), whereby the above figure reduces to the so-called "diamond graph".

2.2. On Some Possibilities of Modeling the $U_A(1)$ Anomaly Influence

Light pseudoscalar mesons can be studied by various methods. We have preferred using [17,21–32] the relativistic bound-state approach to modeling nonperturbative QCD through Dyson–Schwinger equations (DSE), where, if approximations are consistently formulated, model DSE calculations also reproduce the correct chiral behavior of QCD. This is of paramount importance for descriptions of the light pseudoscalar mesons, which are quark–antiquark bound states but simultaneously also the (almost-)Goldstone bosons of DChSB of QCD. (For general reviews of the DSE approach, see, e.g., Refs. [33–36]. About our model choice at $T = 0$ and $T > 0$, in the further text see especially Appendix A).

Figure 1 illustrates how hard computing would be "in full glory" the $U_A(1)$-anomalous mass and related quantities, such as the presently all-important topological susceptibility (3) in the DSE approach with realistically modeled QCD interactions—especially if the calculation should be performed in a consistent approximation with the calculation of the light pseudoscalar bound states, to preserve their correct chiral behavior. (For this reason, they have most often been studied in the rainbow-ladder approximation of DSE, which is inadequate for the anomalous contributions [33–36], as also Figure 1 shows.)

However, our DSE studies of pseudoscalar mesons have been able to address not only pions and kaons, but also η' and η mesons, for which it is essential to include the anomalous $U_A(1)$ symmetry breaking at least at the level of the masses. This was done as described in Refs. [21,24,29,30,37], namely exploiting the fact that the $U_A(1)$ anomaly is suppressed in the limit of large number of QCD colors N_c [38,39]. This allows treating the anomaly contribution formally as a perturbation with respect to the non-anomalous contributions to the η and η' masses [21,24,29]. This way we avoid the need to compute the anomalous mass contribution together, and consistently, with the non-anomalous, chiral-limit-vanishing parts of the masses. The latter must be evaluated by some appropriate, chirally correct method, and our preferred tool—the relativistic bound-state DSE approach [33–36]—is just one such possibility. The point is that they comprise the non-anomalous part of the η'-η mass matrix, to which one can add, as a first-order perturbation, the $U_A(1)$-anomalous mass contribution $M_{U_A(1)}$—and it does not have to be modeled, but taken from lattice QCD [29]. Specifically, at $T = 0$, $M_{U_A(1)}$ can be obtained from χ_{YM}, the topological susceptibility of the (pure-gauge) Yang-Mills theory, for which reliable lattice results have already existed for a long time[3] [40–42].

This can be seen from the remarkable Witten–Veneziano relation (WVR) [38,39] which in a very good approximation relates the full-QCD quantities (η', η and K-meson masses $M_{\eta'}$, M_η and M_K respectively, and the pion decay constant f_π), to the pure-gauge quantity χ_{YM}:

$$M_{\eta'}^2 + M_\eta^2 - 2 M_K^2 = 2N_f \frac{\chi_{YM}}{f_\pi^2} \equiv M_{U_A(1)}^2. \qquad (4)$$

The right-hand-side must be the total $U_A(1)$-anomalous mass contribution in the η'-η complex, since in the combination on the left-hand-side everything else cancels at least to the second order, $\mathcal{O}(m_q^2)$, in the current quark masses of the three light flavors $q = u, d, s$. This is because the non-anomalous, chiral-limit-vanishing parts $M_{q\bar{q}'}$ of the masses of pseudoscalar mesons[4] $P \sim q\bar{q}'$ composed of sufficiently light quarks, satisfy the Gell–Mann–Oakes–Renner (GMOR) relation with their decay constants $f_{q\bar{q}'}$ and the quark–antiquark ($q\bar{q}$) condensate signaling DChSB:

$$M_{q\bar{q}'}^2 = \frac{-\langle \bar{q}q \rangle_0}{(f_{q\bar{q}'}^{\text{ch.lim}})^2}(m_q + m_{q'}) + \mathcal{O}(m_{q'}^2, m_q^2) \qquad (q, q' = u, d, s). \qquad (5)$$

Here $f_{q\bar{q}'}^{\text{ch.lim}} = f_{q\bar{q}'}(m_q, m_{q'} \to 0)$, and $\langle \bar{q}q \rangle_0$ denotes the massless-quark condensate, i.e., the $q\bar{q}$ chiral-limit condensate, or "massless" condensate for short. (In the absence of electroweak interactions, the "massless" condensates have equal values for all flavors: $\langle \bar{q}q \rangle_0 = \langle \bar{q}'q' \rangle_0$.) It turns out that even s-flavor is sufficiently light for Equation (5) to provide reasonable approximations.

Using WVR and χ_{YM} to get the anomalous part of the η' and η masses is successful [29] only for $T \sim 0$, or at any rate, T's well below T_c, the pseudocritical temperature of the chiral transition. In the absence of a systematic re-derivation of WVR (4) at $T > 0$, its straightforward extension (simply replacing all quantities by their T-dependent versions) is tempting, but was found [31] unreliable and with predictions in a drastic conflict with experiment when T starts approaching T_c. This is because the full-QCD quantities $M_{\eta'}(T), M_\eta(T), M_K(T)$ and $f_\pi(T)$ have very different T-dependences from

---

[3] This is in contrast with $\chi = \chi_{QCD}$, the full-QCD topological susceptibility (3), which is much harder to find on the lattice because of the light quark flavors. $\chi = \chi_{QCD}$ approaches χ_{YM} only if one takes the quenched limit of infinitely massive quarks, $\chi_{YM} = \chi_{quench}$, since quarks then disappear from the loops of Equation (3).

[4] The combinations $P \sim q\bar{q}'$ need not always pertain to physical mesons. The pseudoscalar hidden-flavor states $u\bar{u}, d\bar{d}, s\bar{s}$ are not physical as long as the $U_A(1)$ symmetry is not restored (i.e., the anomaly effectively turned off, see around Equation (2.6) in Ref. [43] for example), but build the $SU(3)$ states η_0, η_8 and π^0.

the remaining quantity $\chi_{YM}(T)$, which is pure-gauge and thus much more resilient to increasing temperature: the critical temperature of the pure-gauge, Yang-Mills theory, T_{YM}, is more than 100 MeV higher than QCD's $T_c = (154 \pm 9)$ MeV [44,45]. The early lattice result $T_{YM} \approx 260$ MeV [40,41] is still accepted today [46], and lattice groups finding a different T_{YM} claim only it is even higher, for example $T_{YM} = (300 \pm 3)$ MeV of Gattringer et al. [42]. (There are even some claims about experimentally established $T_{YM} = 270$ MeV [47].)

We thus proposed in 2011. [48] that the above mismatch of the T-dependences in WVR (4) can be removed if one invokes another relation between χ_{YM} and full-QCD quantities, to eliminate χ_{YM}, i.e., substitute pertinent full-QCD quantities instead of χ_{YM} at $T > 0$. This is the Leutwyler–Smilga (LS) relation, Equation (11.16) of Ref. [49], which we used [17,37,48] in the inverted form (and in our notation):

$$\chi_{YM} = \frac{\chi}{1 + \chi \left(\frac{1}{m_u} + \frac{1}{m_d} + \frac{1}{m_s} \right) \frac{1}{\langle \bar{q}q \rangle_0}} \quad (\equiv \widetilde{\chi}) , \tag{6}$$

to express (at $T = 0$) pure-gauge χ_{YM} in terms of the full-QCD topological susceptibility $\chi \equiv \chi_{QCD}$, the current quark masses m_q, and $\langle \bar{q}q \rangle_0$, the condensate of massless, chiral-limit quarks. The combination which these full-QCD quantities comprise, i.e., the right-hand-side of the LS relation (6), we denote (for all T) by the new symbol $\widetilde{\chi}$ for later convenience - that is, for usage at high T, where the equality (6) with χ_{YM} does not hold.

The remarkable LS relation (6) holds for all values of the current quark masses. In the limit of very heavy quarks, it correctly yields $\chi \to \chi_{\text{quenched}} = \chi_{YM}$ for $m_q \to \infty$, but it also holds for the light m_q. In the light-quark sector, the QCD topological susceptibility χ can be expressed as [49–51]:

$$\chi = \frac{-\langle \bar{q}q \rangle_0}{\frac{1}{m_u} + \frac{1}{m_d} + \frac{1}{m_s}} + \mathcal{C}_m , \tag{7}$$

where $\mathcal{C}_m$ represents corrections of higher orders in light-quark masses m_q. Thus, it is small and often neglected, leaving just the leading term as the widely used [52] expression for χ in the light-quark, $N_f = 3$ sector. However, setting $\mathcal{C}_m = 0$ in the light-quark χ (7) returns us $\chi_{YM} = \infty$ through Equation (6) [48]. Or conversely, setting $\chi_{YM} = \infty$ in the LS relation (6), gives the leading term of χ (7) (see also Ref. [43]). This can be a reasonable, useful limit considering that in reality $\chi_{YM}/\chi \gtrsim 40$. Nevertheless, in our previous works on the η'-η complex at $T > 0$ [17,48] we had to fit $\mathcal{C}_m$ (and parameterize it with Ansätze at $T > 0$), since we needed the realistic value of $\chi_{YM}(T = 0) = \widetilde{\chi}(T = 0)$ from lattice to reproduce the well-known masses of η and η' at $T = 0$. (However, just for $\chi(T)$ this is not necessary.)

Replacing [48] $\chi_{YM}(T)$ by the full-QCD quantity $\widetilde{\chi}(T)$ obviously keeps WVR at $T = 0$, but avoids the 'YM vs. QCD' T-dependence mismatch with $f_\pi(T)$ and the LHS of Equation (4), so it is much more plausible to assume the straightforward extension of T-dependences. The T-dependences of $\chi(T)$ (7) and $\widetilde{\chi}(T)$, and thus also of the anomalous parts of the η and η' masses, are then obviously dictated by $\langle \bar{q}q \rangle_0(T)$, the "massless" condensate.

General renormalization group arguments suggest [53] that QCD with three degenerate light-quark flavors has a first-order phase transition in the chiral limit, whereas in QCD with (2+1) flavors (where s-quark is kept significantly more massive) a second-order chiral-limit transition[5] is also possible and even more likely [62–64]. What is important here, is that in any case the chiral-limit condensate $\langle \bar{q}q \rangle_0(T)$ drops sharply to zero at $T = T_c$. (The dotted curve in Figure 2 is just a special example thereof, namely $\langle \bar{q}q \rangle_0(T)$ calculated in Ref. [48] using the same model as in [17] and here.) This causes a similarly sharp drop of $\widetilde{\chi}(T)$ and $\chi(T)$. (We may be permitted to preview similar dotted

---

[5] This is a feature exhibited by DSE models, or at least by most of them, through the characteristic drop of their chiral-limit, massless $q\bar{q}$ condensates [54–61].

curves in Figures 5 and 7 in the next section and anticipate that in this case one would get a massless axion at $T = T_c$.) This was also the reason, besides the expected [65,66] drop of the η' mass, Ref. [48] also predicted so drastic drop of the η mass at $T = T_c$, that it would become degenerate with the pion. However, no experimental indication whatsoever for a decreasing behavior of the η mass, and much less for such a conspicuous sharp mass drop, has been noticed to this day, which seems to favor theoretical descriptions with a smooth crossover. Also, recent lattice QCD results (see [44,67,68] and their references) show that the chiral symmetry restoration is a crossover around the pseudocritical transition temperature T_c.

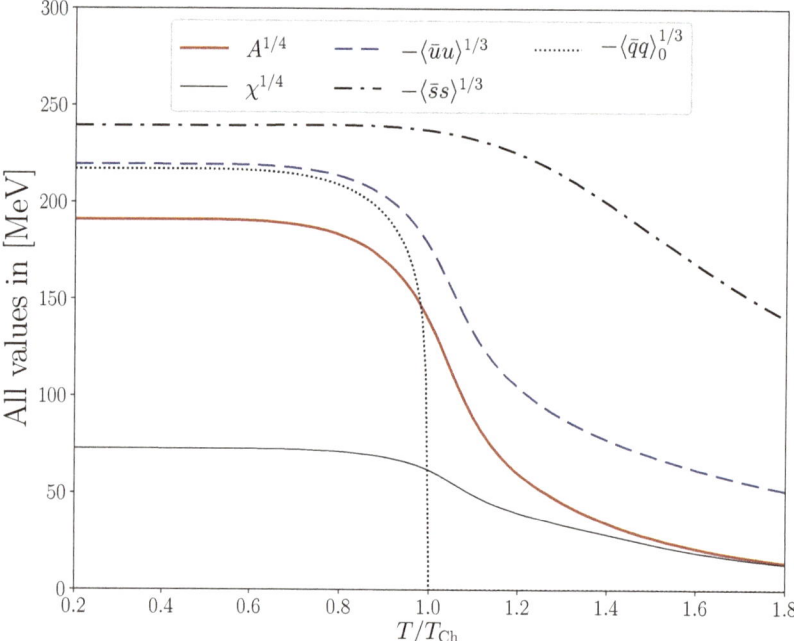

Figure 2. The relative-temperature T/T_c dependences (where $T_{\text{Ch}} \equiv T_c$) of (the 3rd root of the absolute value of) the $q\bar{q}$ condensates, and of (the 4th root of) the topological susceptibility $\chi(T)$ and the full QCD topological charge parameter $A(T)$. Everything was calculated in the isosymmetric limit using the separable rank-2 DSE model which we had already used in Refs. [17,31,48]. (See Appendix A for the model interaction form and parameters, and the first line of Table 1 for the numerical values of the condensates and χ at $T = 0$.) Only the chiral-limit condensate $\langle\bar{q}q\rangle_0(T)$ falls steeply to zero at $T = T_{\text{Ch}}$, indicative of the second-order phase transition. This sharp transition would through (7) and (6) be transmitted to, respectively, the chiral-limit $\chi(T)$ and $\widetilde{\chi}(T)$, and ultimately to the η and η' masses in Ref. [48]. The highest curve (dash-dotted) and the second one from above (dashed) are (3rd roots of the absolute values of) the condensates $\langle\bar{s}s\rangle(T)$ and $\langle\bar{u}u\rangle(T)$, respectively. Their smooth crossover behaviors carry over to $\chi(T)$ and $A(T)$ through Equations (9) and (8), respectively, leading to the empirically acceptable predictions [17] for the T-dependence of the η and η' masses. It turns out that $\chi(T)$ (9) also gives the smooth crossover behavior also to the T-dependence of the axion mass.

To describe such a crossover behavior of the chiral transition, we incorporated [17,37] into our approach Shore's generalization [69] of the Witten–Veneziano relation. To be precise, we studied it at $T = 0$ already in 2008 [30] and adapted it to our DSE bound-state context by applying some very plausible simplifications [20]. The more recent reference [37] presented the analytic, closed-form solutions to Shore's equations for the pseudoscalar meson masses. These solutions showed that Shore's approach is then actually quite similar to the original WVR, leading to a similar η-η' mass matrix [37].

Presently, the most important advantage is that Shore's generalization leads to the crossover T-dependence. As obtained in Section 3 for two specific interactions modeling the nonperturbative QCD interaction, these condensates exhibit a smooth, crossover chiral symmetry transition around T_c. Here Figure 2 illustrates this generic behavior by displaying the results obtained in Section 3 for a specific DSE model: the higher current quark mass, the smoother the crossover behavior, which then results in the crossover behavior also of other quantities, like the presently all-important quantity, the QCD topological susceptibility $\chi(T)$.

This comes about as follows: the quantity which in Shore's mass relations [69] has the role of χ_{YM} in the Witten–Veneziano relation, is called the full-QCD topological charge parameter A. (Shore basically took over this quantity from Di Vecchia and Veneziano [50].) At $T = 0$, it is approximately equal to χ_{YM} in the sense of $1/N_c$ expansion. Shore uses A to express the QCD susceptibility χ through a relation similar to the Leutwyler–Smilga relation (see Equations (2.11) and (2.12) in Ref. [69]), but using the condensates $\langle \bar{u}u \rangle$, $\langle \bar{d}d \rangle$, $\langle \bar{s}s \rangle$ of realistically massive u, d, s quarks. The inverse relation, yielding A (with the opposite sign convention), is the most illustrative for us:

$$A = \frac{\chi}{1 + \chi \left(\frac{1}{m_u \langle \bar{u}u \rangle} + \frac{1}{m_d \langle \bar{d}d \rangle} + \frac{1}{m_s \langle \bar{s}s \rangle} \right)} \qquad (A = \chi_{YM} + \mathcal{O}(\frac{1}{N_c}) \text{ at } T = 0). \qquad (8)$$

Obviously, it is analogous to the inverted LS relation (6) defining $\tilde{\chi}$, except that A is expressed through "massive" condensates. (If they are all replaced by $\langle \bar{q}q \rangle_0$, then $A \to \tilde{\chi}$.) They are in principle different for each flavor, but in the limit of usually excellent isospin symmetry, $\langle \bar{u}u \rangle = \langle \bar{d}d \rangle$.

One can examine the limiting assumption $A = \infty$ in analogy with taking the limit $\chi_{YM} = \infty$ compared to $\chi = \chi_{QCD}$. Then, for $A = \infty$ (be it in our Equation (8) or Shore's Equations (2.11) and (2.12) for χ), one recovers the leading term of the QCD topological susceptibility expressed by the "massive" condensates. However, if one needs a finite A, as in η-η' calculations [17] where one needs to reproduce $A \approx \chi_{YM}$, one also needs the appropriate correction term C'_m, just as C_m in Equation (7), so that:

$$\chi(T) = \frac{-1}{\frac{1}{m_u \langle \bar{u}u \rangle(T)} + \frac{1}{m_d \langle \bar{d}d \rangle(T)} + \frac{1}{m_s \langle \bar{s}s \rangle(T)}} + C'_m. \qquad (9)$$

Again, C'_m is a very small correction term of higher orders in the small current quark masses m_q ($q = u, d, s$), and we can neglect it in the present context, where we actually have a simpler task than finding T-dependence of the η and η' masses in Ref. [17]. Since it turns out that for determining the T-dependence of the mass of the QCD axion we do not need to find A, we set $C'_m = 0$ in this paper throughout. One needs just the topological susceptibility $\chi(T)$ for that, and just the leading term of (9) will suffice to yield the crossover behavior found on lattice (e.g., in Refs. [70–72]).

3. The Axion Mass from the Non-Abelian Axial Anomaly of QCD

Peccei and Quinn (PQ) introduced [2,3] a new global symmetry $U(1)_{PQ}$ which is broken spontaneously at some very large, but otherwise still unknown scale $f_a > 10^8$ GeV [1,73], which determines the absolute value of the axion mass m_a. Nevertheless, this constant cancels from ratios such as $m_a(T)/m_a(0)$, where T is temperature. Thus, useful insights and applications, such as those involving the nontrivial part of axion T-dependence, are possible in spite of f_a being presently unknown.

The factor in the axion mass which carries the nontrivial T-dependence, is the QCD topological susceptibility $\chi(T)$. This quantity is also essential for our description of the η'-η complex at $T > 0$, since it relates the T-dependence of the anomalous breaking of $U_A(1)$ symmetry.

3.1. The Axion as the Almost-Goldstone Boson of the Peccei–Quinn Symmetry

The pseudoscalar axion field $a(x)$ arises as the (would-be massless) Goldstone boson of the spontaneous breaking of the PQ symmetry $U(1)_{PQ}$ [4,5]. The axion contributes to the total Lagrangian its kinetic term and its interaction with fermions of the Standard model. Nevertheless, what is important for the resolution of the strong CP problem, is that the axion also couples to the topological charge density operator $Q(x)$ defined in Equation (1) and generating the $U_A(1)$-anomalous term in Equation (2). The $\bar\theta$-term in $\mathcal{L}_{QCD}$ thus changes into

$$\mathcal{L}_{\bar\theta} \to \mathcal{L}_{axion}^{\bar\theta+} = \left(\bar\theta + \frac{a}{f_a}\right)\frac{g^2}{64\pi^2}\epsilon^{\mu\nu\rho\sigma}F^b_{\mu\nu}F^b_{\rho\sigma}. \tag{10}$$

Because of this axion-gluon coupling, the $U(1)_{PQ}$ symmetry is also broken *explicitly* by the $U_A(1)$ anomaly (gluon axial anomaly). This gives the axion a nonvanishing mass, $m_a \neq 0$ [4,5].

Gluons generate an effective axion potential, and its minimization leads to the axion expectation value $\langle a \rangle$ which makes the modified coefficient of $Q(x)$ in Equation (10) vanish: $\bar\theta + \langle a \rangle/f_a \equiv \bar\theta' = 0$.

Obviously, the experiments excluding the strong CP violation, such as [9], have in fact been finding that consistent with zero is $\bar\theta'$, the coefficient of $Q(x)$ in the QCD Lagrangian when the axion exists. The strong CP problem is thereby solved, irrespective of the initial value of $\bar\theta$. (Relaxation from any $\bar\theta$-value in the early Universe towards the minimum at $\bar\theta = -\langle a \rangle/f_a$ is called misalignment production. The resulting axion oscillation energy is a good candidate for cold dark matter [10–15].)

3.2. Axion Mass from the Topological Susceptibility from Condensates of Massive Quarks

Modulo the (squared) Peccei–Quinn scale f_a^2, the axion mass squared is at all temperatures T given by the QCD topological susceptibility [11,43,70–72,74] very accurately [75,76] (up to negligible corrections of the order $(pion\ mass)^2/f_a^2$):

$$m_a^2(T) = \frac{1}{f_a^2}\chi(T), \tag{11}$$

as revealed by the quadratic term of the expansion of the effective axion potential.

We explained in Section 2 how the $U_A(1)$ symmetry-breaking quantity $\chi(T)$ can be obtained through Equation (9) as a prediction of any method which can provide the quark condensates $\langle \bar q q \rangle(T)$ ($q = u, d, s$). Thus, one can get the T-dependence of the axion mass (11) through the mechanism where DChSB drives the $U_A(1)$ symmetry breaking. (And conversely, of course: the chiral restoration then drives the restoration of $U_A(1)$ symmetry.)

An excellent tool to study DChSB, and in fact "produce" it in the theoretical sense, is one of the basic equations of the DSE approach - the gap equation. The most interesting thing it does for nonperturbative QCD is explaining the notion of the constituent quark mass around $\frac{1}{3}$ of the nucleon mass M_N by generating them via DChSB, *in the same process which produces the $q\bar q$ condensates*. Thanks to this, $q\bar q$ condensates can be evaluated from dressed quark propagators. Specifically, hadronic-scale large ($\sim M_N/3$ at small momenta p) dressed quark-mass functions $M_q(p^2) \equiv B_q(p^2)/A_q(p^2)$ are generated despite two orders of magnitude lighter current quark masses m_q, and in fact even in the chiral limit, when $m_q = 0$! This happens in low-energy QCD thanks to *nonperturbative dressing* via strong dynamics, making strongly dressed quark propagators $S_q(p)$ out of the free quark propagators S_q^{free}:

$$S_q^{free}(p) = \frac{1}{i\gamma \cdot p + m_q} \longrightarrow S_q(p) = \frac{1}{i\gamma \cdot p\, A_q(p^2) + B_q(p^2)} \quad \text{(Euclidean space expressions)}. \tag{12}$$

The solution for the dressed quark propagator $S_q(p)$ of the flavor q, i.e., the dressing functions $A_q(p^2)$ and $B_q(p^2)$, are found by solving the gap equation

$$S_q^{-1}(p) = S_q^{\text{free}}(p)^{-1} - \Sigma_q(p), \qquad (q = u, d, s), \tag{13}$$

where $\Sigma_q(p)$ is the corresponding DChSB-generated self-energy, for example, Equation (14) if the rainbow-ladder truncation is adopted.

In the present work, all we want to model of nonperturbative QCD are the condensates $\langle \bar{q}q \rangle$ at all temperatures T, and for that the solutions of the quark-propagator gap Equation (13) are sufficient, i.e., we do not need the Bethe–Salpeter equation (BSE) for the $q\bar{q}'$ pseudoscalar bound states. However, we want the same condensates, and basically the same (leading term of) the topological susceptibility as we had in our related η-η' paper [17], and in a number of earlier papers, such as Refs. [31,48]. Thus, we now use the same model interaction we have been using then in the consistent rainbow-ladder (RL) truncation of DSE's to produce chirally correctly behaving pseudoscalar mesons - that is, with the non-anomalous parts of their masses given by GMOR (5).

Thus, the quark self-energy in the gap Equation (13) in the RL truncation is

$$\Sigma_q(p) = -\int \frac{d^4\ell}{(2\pi)^4} g^2 D_{\mu\nu}^{ab}(p-\ell)_{\text{eff}} \frac{\lambda^a}{2} \gamma^\mu S_q(\ell) \frac{\lambda^b}{2} \gamma^\nu, \tag{14}$$

where $D_{\mu\nu}^{ab}(k)_{\text{eff}}$ is an effective gluon propagator, which should be chosen to model the nonperturbative, low-energy domain of QCD. This can be done in varying degrees of DSE modeling, depending on the variety of problems one wants to treat [33–36,77]. For example, in the context of low-energy meson phenomenology, if one does not aim to address problems of perturbative QCD, it is better *not to include* the perturbative part of the QCD interaction. Otherwise, in the words of very authoritative DSE practitioners, "the logarithmic tail and its associated renormalization represent an unnecessary obfuscation" [78].

In medium, the original $O(4)$ symmetry is broken to $O(3)$ symmetry. The most general form of the dressed quark propagator then has four independent tensor structures and four corresponding dressing functions. At nonvanishing temperature, $T > 0$, we use the Matsubara formalism, where four-momenta decompose into three-momenta and Matsubara frequencies: $p = (p^0, \vec{p}) \to p_n = (\omega_n, \vec{p})$. Therefore, the (inverted) dressed quark propagator $S_q(p)$ (13) becomes

$$S_q^{-1}(p_n) = S_q^{-1}(\vec{p}, \omega_n) = i\vec{\gamma} \cdot \vec{p}\, A_q(\vec{p}^2, \omega_n) + i\omega_n \gamma_4 \, C_q(\vec{p}^2, \omega_n) + B_q(\vec{p}^2, \omega_n) + i\omega_n \gamma_4 \vec{\gamma} \cdot \vec{p}\, \mathcal{D}_q(\vec{p}^2, \omega_n). \tag{15}$$

(The T-dependence of the propagator dressing functions is understood and, to save space, is not indicated explicitly, except in Appendix A).

Nevertheless, the last dressing function $\mathcal{D}_q(\vec{p}^2, \omega_n)$ is so very small that it is quite safe and customary to neglect it—e.g., see Refs. [34,79]. Thus, also we set $\mathcal{D}_q \equiv 0$, leaving only A_q, C_q and B_q.

For applications in involved contexts, such as calculations at $T > 0$, appropriate simplifications are very welcome for tractability. This is why in Refs. [17,31,48] and presently, we adopted relatively simple, but phenomenologically successful [30–32,77,80] separable approximation [77]. The details on the functional form and parameters of the presently used model interaction can be found in Appendix A.

As already pointed out in the original Ref. [77], the model *Ansätze* for the nonperturbative low-energy interaction ("interaction form factors") are such that they provide sufficient ultraviolet suppression. Therefore, as noted already in Ref. [77], no renormalization is needed and the multiplicative renormalization constants, which would otherwise be needed in the gap Equations (13) with (14), are 1. The usual expression for the condensate of the flavor q then becomes

$$\langle \bar{q}q \rangle = -N_c \oint_p \text{Tr}[S_q(p)] \equiv -N_c T \sum_{n \in \mathbb{Z}} \int \frac{d^3p}{(2\pi)^3} \text{Tr}[S_q(\vec{p}, \omega_n)], \tag{16}$$

where Tr is the trace in Dirac space, and the combined integral-sum symbol indicates that when the calculation is at $T > 0$, the four-momentum integration decomposes into the three-momentum integration and summation over fermionic Matsubara frequencies $\omega_n = (2n+1)\pi T$, $n \in \mathbb{Z}$.

As is well known, the condensates (16) are finite only for massless quarks, $m_q = 0$, i.e., only $\langle \bar{q}q \rangle_0$ is finite, while the "massive" condensates are badly divergent, and must be regularized, i.e., divergences must be subtracted. Since the subtraction procedure is not uniquely defined, the chiral condensate at nonvanishing quark mass is also not uniquely defined. However, the arbitrariness is in practice slight and should rather be classified as fuzziness. It should not be given too large importance in the light of small differences between the results of various sensible procedures.

Our regularization procedure is subtracting the divergence-causing m_q (~ several MeV) from the scalar quark-dressing function $B_q(p^2)$ (~ several hundred MeV) whenever it is found in the numerator of the condensate integrand. To justify our particular regularization of massive condensates as physically meaningful and sensible, we have examined its consistency with two different subtractions used on lattice [81–83] and in a recent DSE-approach paper [84].

We shall now test our massive condensates obtained from the separable rank-2 DSE model (see Appendix A), whose regularized versions have already been shown in Figure 2.

Let us first consider the subtraction on lattice (normalized to 1 for $T = 0$) first proposed in Ref. [81] in their Equation (17), rewritten in our notation and applied to our condensate of u-quarks (and of course d-quarks in the isospin limit):

$$R_{\langle \bar{\psi}\psi \rangle}(T) = R_{\langle \bar{u}u \rangle}(T) = \frac{\langle \bar{u}u \rangle(T) - \langle \bar{u}u \rangle(0) + \langle \bar{q}q \rangle_0(0)}{\langle \bar{q}q \rangle_0(0)}. \tag{17}$$

In Figure 3, the upper, red curve shows (normalized) u-quark condensate $\langle \bar{u}u \rangle(T)/\langle \bar{u}u \rangle(0)$ when regularized in the usual way, by subtracting m_q from $B_q(p^2)$ in the numerator of the condensate integrand. It agrees very well with the lattice regularization $R_{\langle \bar{u}u \rangle}$ (17) of our condensate $\langle \bar{u}u \rangle(T)$, represented by the green curve. The agreement with the lattice data points taken (if pertinent) from Table 6 of Ref. [82] is also rather good.

Next, we examine the consistency of our subtraction with the most usual condensate subtraction on the lattice, which combines the light and strange quark condensates and their masses like this:

$$\tilde{\Delta}_{l,s}(T) = \langle \bar{l}l \rangle_l(T) - \frac{m_l}{m_s} \langle \bar{s}s \rangle_s(T). \tag{18}$$

Following Isserstedt et al. [84], in Figure 4 we make comparison of the normalized version thereof

$$\Delta_{l,s}(T) = \frac{\langle \bar{l}l \rangle(T) - \frac{m_l}{m_s}\langle \bar{s}s \rangle(T)}{\langle \bar{l}l \rangle(0) - \frac{m_l}{m_s}\langle \bar{s}s \rangle(0)} \quad (l = u \text{ or } d \text{ in the isospin symmetric limit}) \tag{19}$$

with the lattice data of Ref. [83]. The agreement is very good, which implies also the agreement with the subtracted condensates in the recent DSE paper [84], which made this successful comparison first (in its Figure 3).

To conclude: results shown in Figures 3 and 4 demonstrate that certain arbitrariness in the choice of regularization does not disqualify our massive condensates from useful applications, such as using them in Equation (9) to make predictions on the topological susceptibility.

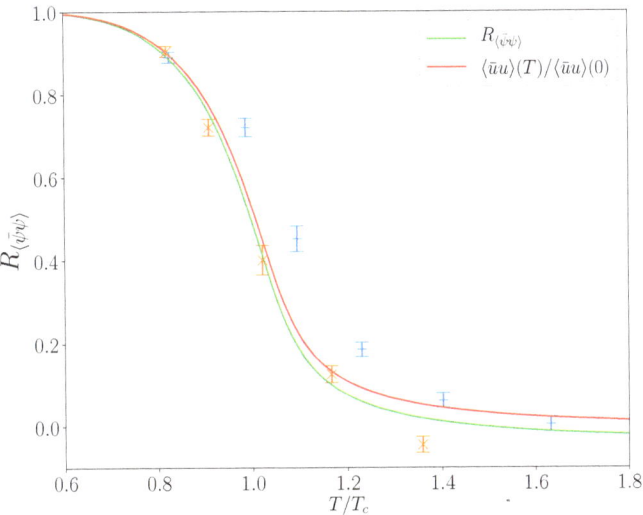

Figure 3. The relative-temperature T/T_c dependence of the subtracted (and normalized) condensate $R_{\langle\bar\psi\psi\rangle}$ defined by Equation (17) and introduced by Ref. [81]. The lattice data points are from Figure 6 of Ref. [82], but scaled for the critical temperatures T_χ from their Table 2, which is different for the "crosses" (data points [82] for $m_\pi \approx 370$ MeV) and "bars" (data points [82] for $m_\pi \approx 210$ MeV). The lower, green curve results from the $R_{\langle\bar\psi\psi\rangle}$ (17) subtraction of our u-quark condensate. The upper, red curve is the T-dependence of our u-quark condensate when regularized in the usual way (see text).

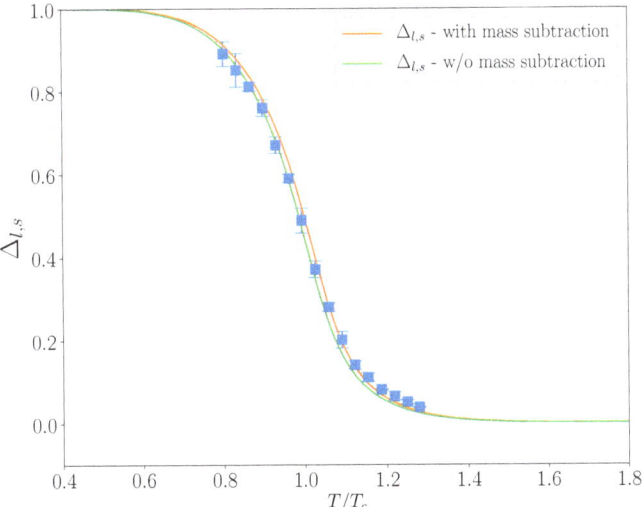

Figure 4. The relative-temperature T/T_c dependence of the (normalized) subtracted quark condensate (19) from the lattice [83] (blue squares) and from our condensates. Slightly lower, green curve results from our unsubtracted condensates plugged in Equation (19), while the very slightly higher, red curve is from our already subtracted condensates.

3.3. Axion Mass and Topological Susceptibility—Results from the Rank-2 Separable Model in the Isosymmetric Limit

Our result for $\chi(T)^{1/4} = \sqrt{m_a(T) f_a}$ is presented in Figure 5 as a solid curve and compared, up to $T \approx 2.3 T_c$, with the corresponding results of two lattice groups [70,71], rescaled to the relative-temperature T/T_c. (Table 1 gives numerical values of our results at $T = 0$.)

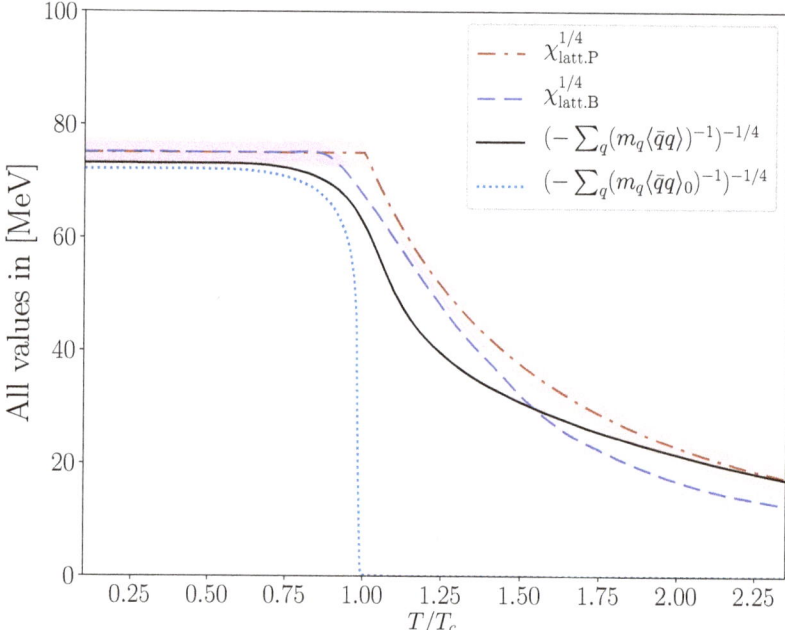

Figure 5. The relative-temperature T/T_c dependence of (the leading term of) $\chi(T)^{1/4}$ from our often adopted [17,30–32,48] isosymmetric DSE rank-2 separable model: solid curve for Equation (9) with massive-quark condensates, while the dotted curve results from using $\langle \bar{q}q \rangle_0$ instead. $\chi(T)^{1/4}$ (with uncertainties) from lattice: dash-dotted curve extracted from Petreczky et al. [70] and long-dashed curve, from Borsany et al. [71]. (Colors online).

In our case, the results for $\chi(T)$ and condensates $\langle \bar{u}u \rangle(T)$, $\langle \bar{d}d \rangle(T)$ and $\langle \bar{s}s \rangle(T)$ needed to obtain it, are predictions of the dynamical DSE model used in the $T > 0$ study of η'-η [17]. This is the same modeling of the low-energy, nonperturbative QCD interactions as we have already employed in our earlier studies of light pseudoscalar mesons at $T \geq 0$ [30–32,48]: the separable model interaction—see, e.g., [77,85], and references therein. We have adopted the so-called rank-2 variant from Ref. [77]. The adopted model with our choice of parameters is defined in detail in the Appendix A of the present work, after the subsection II.A of Refs. [31,86]. It employs the model current-quark-mass parameters $m_u = m_d \equiv m_l = 5.49$ MeV and $m_s = 115$ MeV. The model prediction for condensates at $T = 0$ are $\langle \bar{s}s \rangle = (-238.81$ MeV$)^3$ for the heaviest quark, while isosymmetric condensates of the lightest flavors, $\langle \bar{u}u \rangle = \langle \bar{d}d \rangle \equiv \langle \bar{l}l \rangle = (-218.69$ MeV$)^3$ are quite close to the "massless" one, $\langle \bar{q}q \rangle_0 = (-216.25$ MeV$)^3$.

Contrary to, e.g., Ref. [48], where the condensate of massless quarks $\langle \bar{q}q \rangle_0(T)$ was used, in Ref. [17] and here we follow Shore [69] in using condensates of light quarks with nonvanishing current masses. The smooth, crossover behavior around the pseudocritical temperature T_c for the chiral transition (now confirmed at vanishing baryon density by lattice studies such as [44,67,68,70,71]), is obtained thanks to the DChSB condensates of realistically massive light quarks—i.e., the quarks with realistic explicit chiral symmetry breaking [17].

Table 1. For the both variants of the DSE separable model (with the rank-2 and rank-1 interaction Ansatz) used in the present paper, various sets of values of the model quark-mass parameters m_q ($q = u, d, s$) are related to the model results for the topological susceptibility χ and the "massive" condensates $\langle \bar{q}q \rangle$ at $T = 0$. The topological susceptibility χ varies because of varying m_q and (to a much lesser extent) because of the changes of "massive" condensates induced by changes of the quark-mass parameters m_q. The massless-quark condensate, $\langle \bar{q}q \rangle_0$, depends only on the dynamical DSE model: always $\langle \bar{q}q \rangle_0 = -216.25^3$ MeV^3 for the rank-2 model, and $\langle \bar{q}q \rangle_0 = -248.47^3$ MeV^3 for the rank-1 model. Thus, the topological susceptibility χ_0 calculated with the chiral-limit condensate, varies for a given model only because of varying values of m_q. All values are in MeV (or the indicated 3rd or 4th powers of MeV).

$T = 0$	m_u	m_d	m_s	χ_0 (with $\langle q\bar{q} \rangle_0$)	$\langle \bar{u}u \rangle$	$\langle \bar{d}d \rangle$	$\langle \bar{s}s \rangle$	χ
rank-2								
$m_u = m_d$ [17]	5.49	5.49	115	72.18^4	-218.69^3	-218.69^3	-238.81^3	72.73^4
with constraint $m_u = 0.48\, m_d$ [1], fitted m_u & m_d	4.66	9.71	115	74.64^4	-218.35^3	-220.33^3	-238.81^3	75.44^4
rank-1								
$m_u = m_d$	6.6	6.6	142	83.87^4	-249.27^3	-249.27^3	-251.49^3	84.08^4
with constraint $m_u = 0.48\, m_d$ [1], fitted m_u & m_d	3.15	6.56	142	75.31^4	-248.87^3	-249.21^3	-251.49^3	75.43^4

In contrast, using in Equation (9) the massless-quark condensate $\langle \bar{q}q \rangle_0$ (which drops sharply to zero at T_c) instead of the "massive" ones, would dictate a sharp transition of the second order at T_c [17,48] also for $\chi(T)$, illustrated in Figure 5 by the dotted curve. This would of course imply that axions are massless [87] for $T > T_c$. It is of academic interest to know what consequences would be thereof for cosmology, but now it is clear that only crossover is realistic [70,71].

The rather good agreement with lattice in Figure 5 resulted without any refitting of this model, either in Ref. [17] for η' and η, or in this subsection. The model is in the isosymmetric limit, $m_u = m_d \equiv m_l$, which is perfectly adequate for most purposes in hadronic physics. Nevertheless, the QCD topological susceptibility χ in its version (9) contains the current quark masses in the form of harmonic averages of $m_q \langle \bar{q}q \rangle$ ($q = u, d, s$). A harmonic average is dominated by its smallest argument, and presently this is the lightest current-quark-mass parameter, motivating us to investigate the changes occurring beyond the isospin symmetric point.

3.4. Axion Mass and Topological Susceptibility from Rank-1 and Rank-2 Models out of the Isosymmetric Limit

The previous isosymmetric case, pertinent also for the η'-η study [17], has the current-quark-mass model parameters $m_u = m_d = 5.49$ MeV. This is above the most recent PDG quark-mass values [1], but anyway yields $\chi(T = 0) = (72.73 \text{ MeV})^4$ already a little below the lattice results [70,71], and below the most recent chiral perturbation theory result $\chi(T = 0) = (75.44 \text{ MeV})^4$ [76].

This seems not to bode well for the attempts out of the isosymmetric limit, because lowering the values of the current masses seems to threaten yielding unacceptably low values of the topological susceptibility. Indeed, taking the central values from the current quark masses $m_u = 2.2^{+0.5}_{-0.4}$ MeV and

$m_d = 4.70^{+0.5}_{-0.3}$ MeV and $m_s = 95^{+9}_{-3}$ MeV recently quoted by PDG [1], yields just $(62.50 \text{ MeV})^4$ for the leading term of Equation (9) at $T = 0$.

However, our model m_u, m_d and m_s are phenomenological current-quark-mass *parameters*, and cannot be quite unambiguously and precisely related to the somewhat lower PDG values of the current quark masses. The better relation is through the *ratios* of quark masses, for which PDG gives $m_u/m_d = 0.48^{+0.07}_{-0.08}$ [1].

We thus require that $m_u^{\text{fit}}/m_d^{\text{fit}} = 0.48$ be satisfied by the new non-isosymmetric mass parameters m_u^{fit} and m_d^{fit} when they are varied to reproduce the recent most precise value $\chi(T = 0) = (75.44 \text{ MeV})^4$ [76]. We get $m_u^{\text{fit}} = 4.66$ MeV, resulting in the condensate $\langle \bar{u}u \rangle(T=0) = (-218.35 \text{ MeV})^3$ and $m_d^{\text{fit}} = 9.71$ MeV, resulting in $\langle \bar{d}d \rangle(T=0) = (-220.33 \text{ MeV})^3$. (The s-mass parameter is not varied, i.e., $m_s \equiv m_s^{\text{fit}}$. The rest of model parameters, namely those in the *Ansatz* functions $\mathcal{F}_0(p^2)$ and $\mathcal{F}_1(p^2)$ modeling the strength of the rank-2 nonperturbative interaction (see Appendix), are also not varied.)

The T-dependence of the resulting $\chi(T)^{1/4}$ is given by the short-dashed black curve in Figure 6. Except its better agreement with the lattice results [70,71] at low T, the new (dashed) $\chi(T)^{1/4}$ curve is very close to the isosymmetric (solid) curve.

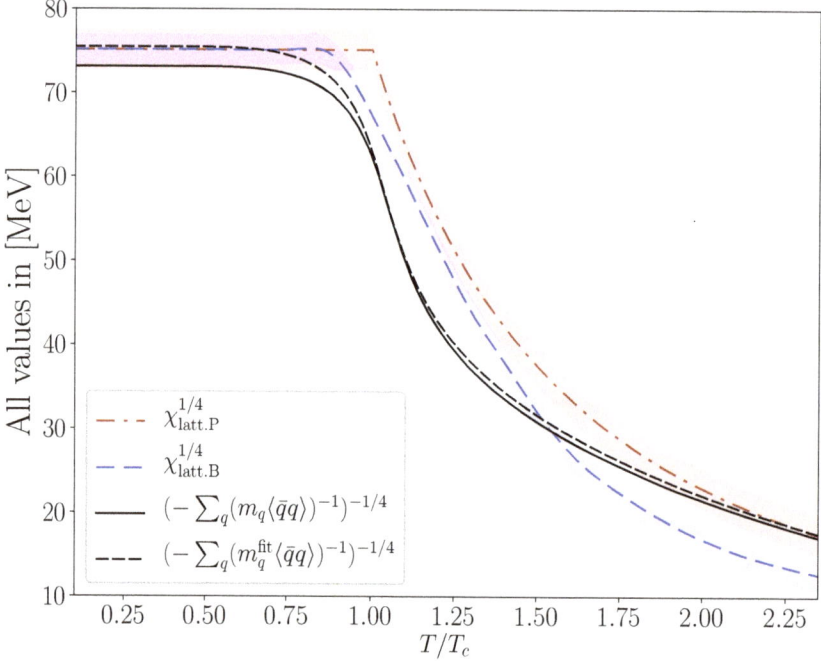

Figure 6. The short-dashed black curve shows the non-isosymmetric case of the leading term of $\chi(T)^{1/4}$, Equation (9), with $m_u^{\text{fit}} = 4.66$ MeV and $m_d^{\text{fit}} = m_u^{\text{fit}}/0.48 = 9.71$ MeV, and appropriately recalculated condensates $\langle \bar{u}u \rangle(T)$ and $\langle \bar{d}d \rangle(T)$. The vertical scale is zoomed with respect to Figure 5 to help resolve the short-dashed curve from the solid curve representing again the isosymmetric case of the same separable rank-2 model. Also, the lattice results [70,71] are again depicted as in Figure 5.

Now we will check the model dependence by comparing our results presented so far (obtained in the rank-2 model) with those we get in the separable rank-1 model of Ref. [77]. It is similar to the previously considered rank-2 one by modeling the low-energy, nonperturbative QCD interaction with an *Ansatz* separating the momenta p_a, p_b of interacting constituents, but is of a simpler form, proportional to just $\mathcal{F}_0(p_a^2)\,\mathcal{F}_0(p_b^2)$. Its presently interesting feature is that for similar quark-mass parameters, it yields significantly larger condensates than those in the separable rank-2 model, and thus also larger χ. (This also holds at low and vanishing T even for $\chi(T)$ calculated using only the "massless" condensate $\langle\bar{q}q\rangle_0(T)$. This case is depicted in Figure 7 as the dotted curve.)

The original rank-1 model employs the light-quark current mass parameters in the isosymmetric limit: $m_u = m_d \equiv m_l = 6.6$ MeV [77]. However, in Equation (9) we also need the s-flavor. The fit to the kaon mass yields $m_s = 142$ MeV. The model prediction for condensates at $T = 0$ are then $\langle\bar{s}s\rangle = (-251.49 \text{ MeV})^3$ for the heaviest quark, while isosymmetric condensates of the lightest flavors, $\langle\bar{u}u\rangle = \langle\bar{d}d\rangle \equiv \langle\bar{l}l\rangle = (-249.27 \text{ MeV})^3$ are quite close to the "massless" one, $\langle\bar{q}q\rangle_0 = (-248.47 \text{ MeV})^3$. This gives too large topological susceptibility at $T = 0$, namely $\chi(0) = (84.08 \text{ MeV})^4$. Nevertheless, for large T, it also falls with T somewhat faster than the rank-2 $\chi(T)$, since rank-1 condensates fall with T somewhat faster than the rank-2 ones.

The isosymmetric rank-1 $\chi(T)$ is depicted by the solid black curve in Figure 7, showing that it actually falls with T faster even than $\chi(T)$'s from lattice [70,71] for practically all T's high enough to induce changes. Then, comparing Figures 6 and 7 shows that the lattice high-T results are in between high-T results of the two separable models.

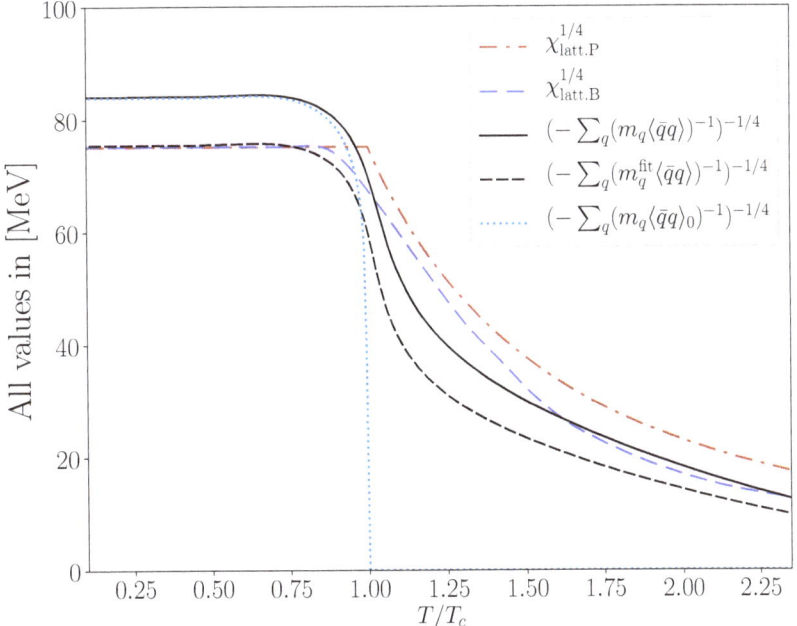

Figure 7. From the calculation in the separable rank-1 DSE model [77], the relative-temperature T/T_c dependence of (the leading term of) $\chi(T)^{1/4}$ is represented by: (i) the solid curve for the isosymmetric case with $m_u = m_d = 6.6$ MeV and $m_s = 142$ MeV, (ii) the dotted curve is for the same mass parameters, but with all condensates approximated by the "massless" condensate $\langle\bar{q}q\rangle_0(T)$, and (iii) the short-dashed curve for the non-isosymmetric case $m_u^{\text{fit}} = 3.15$ MeV, $m_d^{\text{fit}} = 6.56$ MeV, while $m_s^{\text{fit}} \equiv m_s = 142$ MeV. The pertinent lattice results are presented in the same way as in the previous two Figures: the dash-dotted and long-dashed curves extracted, respectively, from Refs. [70,71]. Colors online.

To go out of the isospin limit with the rank-1 model, we again require that the changed parameters m_u^{fit} and m_d^{fit} (together with the condensates resulting from them) fit the currently most precise $T = 0$ value of the topological susceptibility, $\chi = (75.44 \text{ MeV})^4$ [76], while also obeying $m_u^{\text{fit}}/m_d^{\text{fit}} = 0.48$, i.e., the central value of the PDG [1] m_u/m_d ratio. (Again, other model parameters including m_s are not varied: $m_s \equiv m_s^{\text{fit}}$.)

In our rank-1 model, these requirements yield $m_d^{\text{fit}} = 6.56$ MeV, i.e., it practically remained the same as in the originally fitted model [77]. Of course, its condensate $\langle \bar{d}d \rangle$ also remains the same. The lightest flavor has the mass parameter lowered to $m_u^{\text{fit}} = 3.15$ MeV. Now it has only slightly lower condensate $\langle \bar{u}u \rangle(T = 0) = (-248.87 \text{ MeV})^3$, which is even closer to the "massless" $\langle \bar{q}q \rangle_0(T = 0)$. Nevertheless, $\langle \bar{u}u \rangle(T)$ retains the crossover behavior for $T > 0$, although it falls with T steeper than more "massive" condensates.

The resulting non-isosymmetric $\chi(T)^{1/4} = \sqrt{m_a(T) f_a}$ is in Figure 7 shown as the short-dashed black curve, which is everywhere consistently the lowest (among the "massive", crossover curves).

4. Summary and Discussion

In the DSE framework, we have obtained predictions for the nontrivial part of the T-dependence of the axion mass $m_a(T) = \sqrt{\chi(T)}/f_a$, Equation (11), by calculating the QCD topological susceptibility $\chi(T)$, since the unknown Peccei–Quinn scale f_a is just an overall constant. We have used two empirically successful dynamical models of the separable type [77] to model nonperturbative QCD at $T > 0$. We also studied the effects of varying the mass parameters of the lightest flavors out of the isospin limit, and found that our $\chi(T)$, and consequently $m_a(T)$, are robust with respect to the non-isosymmetric refitting of m_u and $m_d = m_u/0.48$.

All these results of ours on $\chi(T)$, and consequently the related axion mass, are in satisfactory agreement with the pertinent lattice results [70,71], and in qualitative agreement with those obtained in the NJL model [88]. Everyone obtains qualitatively similar crossover of $\chi(T)$ around T_c, but it would be interesting to speculate what consequences for cosmology could be if $\chi(T)$, and thus also $m_a(T)$, would abruptly fall to zero at $T = T_c$ due to a sharp phase transition of the "massless" condensate $\langle \bar{q}q \rangle_0(T)$. Of course, dynamical models of QCD can access only much smaller range of temperatures than lattice, where $T \sim 20\, T_c$ has already been reached [71]. (On the other hand, the thermal behavior of the $U_A(1)$ anomaly could not be accessed in chiral perturbation theory [89].)

Since it is now established that (at vanishing and low density) the chiral transition is a crossover, it is important that one can use massive-quark condensates, which exhibit crossover behavior around $T \sim T_c$. In the present work, they give us directly, through Equation (9), the crossover behavior of $\chi(T)$. However, these are regularized condensates, because a nonvanishing current quark mass m_q makes the condensate $\langle \bar{q}q \rangle$ plagued by divergences, which must be subtracted. In Section 3, we have shown that our regularization procedure is reasonable and in good agreement with at least two widely used subtractions on the lattice.

To discuss our approach from a broader perspective, it is useful to recall that JLQCD collaboration [90] has recently pointed out how the chiral symmetry breaking and $U_A(1)$ anomaly are tied, and stressed the importance of the $q\bar{q}$ chiral condensate in that. The axion mass presently provides a simple example thereof: through Equation (11), $m_a(T)$ is at all temperatures directly expressed by the QCD topological susceptibility $\chi(T)$, which is a measure of $U_A(1)$ breaking by the axial anomaly. We calculate $\chi(T)$ through Equation (9) from the quark condensates, which in turn arise from DChSB. In addition, conversely: melting of condensates around $T \sim T_c$ signals the restoration of the chiral symmetry. Therefore, the $U_A(1)$ symmetry breaking and restoration being driven by the chiral ones is straightforward.

The relation of $\chi(T)$ to the η' mass is, however, a little less straightforward [17] because it involves several other elements, but the topological susceptibility remains the main one. Since our present results on the axion are, in a way, a by-product of the framework which was initially formulated to understand better the T-dependence of η' and η masses, we have explained it in detail in Section 2.2.

Therefore, here in the Summary, we should just stress that the topological susceptibility $\chi(T)$ is the strong link between the QCD axion and the η-η' complex. It so also in the case of the present paper regarding our η-η' reference [17]: specifically, we should note that the actual T-dependence of η' and η is rather sensitive to the behavior of $\chi(T)$, and rather accurate $\chi(T)$ is needed to get acceptable $M_\eta(T)$ and $M_{\eta'}(T)$. Thanks to its crossover behavior, our $\chi(T)$ gives in Ref. [17] empirically allowed T-dependence of the masses in the η-η' complex. However, even a crossover, if it were too steep, would lead to the unwanted (experimentally never seen) drop of the η mass, just as a too slow one would not yield the drop of the η' mass required according to some experimental analyses [65,66].

In that sense, our present predictions on $m_a(T)$ are thus supported by the fact that our calculated topological susceptibility $\chi(T)$ gives the T-dependence of the $U_A(1)$ anomaly-influenced masses of η' and η mesons [17] which is consistent with experimental evidence [65,66].

Author Contributions: Conceptualization, D.K. (Dubravko Klabučar); methodology, D.K. (Dubravko Klabučar), D.H. and D.K. (Dalibor Kekez); software, D.H.; validation, D.K. (Dubravko Klabučar), D.H. and D.K. (Dalibor Kekez); formal analysis, D.H., D.K. (Dubravko Klabučar) and D.K. (Dalibor Kekez); investigation, D.H., D.K. (Dubravko Klabučar) and D.K. (Dalibor Kekez); data curation, D.H.; writing–original draft preparation, D.K. (Dubravko Klabučar); writing–review and editing, D.K. (Dubravko Klabučar), D.K. (Dalibor Kekez) and D.H.; visualization, D.H., D.K. (Dalibor Kekez); supervision, D.K. (Dubravko Klabučar).

Funding: This work was supported in part by STSM grants from COST Actions CA15213 THOR and CA16214 PHAROS.

Conflicts of Interest: The authors declare no conflict of interest.

Abbreviations

The following abbreviations are used in this manuscript:

DChSB   Dynamical chiral symmetry breaking
DSE   Dyson–Schwinger equations
QCD   Quantum chromodynamics
ABJ   Adler-Bell-Jackiw
CKM   Cabibbo-Kobayashi-Maskawa
CP   charge conjugation parity

Appendix A. Separable Interaction Models for Usage at $T \geq 0$

At $T = 0$, the Dyson–Schwinger equation (DSE) approach in the rainbow-ladder approximation (RLA) tackles efficiently solving Dyson–Schwinger gap equation and Bethe–Salpeter equations, but extending this to $T > 0$ is technically quite difficult. We thus adopt a simple model for the strong dynamics from Ref. [77], namely the model we already used in Refs. [17,30,31,48]. For the effective gluon propagator in a Feynman-like gauge, we use the *separable Ansatz*:

$$g^2 D^{ab}_{\mu\nu}(p-\ell)_{\text{eff}} = \delta^{ab} g^2 D^{\text{eff}}_{\mu\nu}(p-\ell) \longrightarrow \delta_{\mu\nu} D(p^2, \ell^2, p\cdot\ell) \, \delta^{ab}, \tag{A1}$$

whereby the dressed quark-propagator gap Equations (13) with (14) yields

$$B_q(p^2) = m_q + \frac{16}{3}\int \frac{d^4\ell}{(2\pi)^4} D(p^2,\ell^2,p\cdot\ell) \, \frac{B_q(\ell^2)}{\ell^2 A_q^2(\ell^2) + B_q^2(\ell^2)} \tag{A2}$$

$$[A_q(p^2) - 1]\, p^2 = \frac{8}{3}\int \frac{d^4\ell}{(2\pi)^4} D(p^2,\ell^2,p\cdot\ell)\, \frac{(p\cdot\ell) A_q(\ell^2)}{\ell^2 A_q^2(\ell^2) + B_q^2(\ell^2)}. \tag{A3}$$

More specifically, the so-called *rank-2* separable interaction entails:

$$D(p^2,\ell^2,p\cdot\ell) = D_0\, \mathcal{F}_0(p^2)\, \mathcal{F}_0(\ell^2) + D_1\, \mathcal{F}_1(p^2)\, (p\cdot\ell)\, \mathcal{F}_1(\ell^2). \tag{A4}$$

Then, the solutions of Equations (A2) and (A3) for the dressing functions are of the form

$$B_q(p^2) = m_q + b_q \, \mathcal{F}_0(p^2) \quad \text{and} \quad A_q(p^2) = 1 + a_q \, \mathcal{F}_1(p^2), \tag{A5}$$

reducing Equations (A2) and (A3) to the nonlinear system of equations for the constants b_q and a_q:

$$b_q = \frac{16 D_0}{3} \int \frac{d^4\ell}{(2\pi)^4} \frac{\mathcal{F}_0(p^2) B_q(\ell^2)}{\ell^2 A_q^2(\ell^2) + B_q^2(\ell^2)} \tag{A6}$$

$$a_q = \frac{2 D_1}{3} \int \frac{d^4\ell}{(2\pi)^4} \frac{\ell^2 \mathcal{F}_1(\ell^2) A_q(\ell^2)}{\ell^2 A_q^2(\ell^2) + B_q^2(\ell^2)}. \tag{A7}$$

If one chooses that the second term in the interaction (A4) is vanishing, by simply setting to zero the second strength constant, $D_1 = 0$, one has a still simpler *rank-1 separable Ansatz*, where $A_q(p^2) = 1$.

The analytic properties of these model interactions are defined by the choice of the interaction "form factors" $\mathcal{F}_0(p^2)$ and $\mathcal{F}_1(p^2)$. In the present work we will use the functions [32,86]

$$\mathcal{F}_0(p^2) = \exp(-p^2/\Lambda_0^2) \quad \text{and} \quad \mathcal{F}_1(p^2) = \frac{1 + \exp(-p_0^2/\Lambda_1^2)}{1 + \exp((p^2 - p_0^2)/\Lambda_1^2)}, \tag{A8}$$

which satisfy the constraints $\mathcal{F}_0(0) = \mathcal{F}_1(0) = 1$ and $\mathcal{F}_0(\infty) = \mathcal{F}_1(\infty) = 0$.

For the numerical calculations we fix the free parameters of the model at $T = 0$ as in Refs. [32,86], to reproduce in particular the vacuum masses of the pseudoscalar and vector mesons, $M_\pi = 140$ MeV, $M_K = 495$ MeV, $M_\rho = 770$ MeV, the pion decay constant $f_\pi = 92$ MeV, and decay widths, $\Gamma_{\rho^0 \to e^+ e^-} = 6.77$ keV, $\Gamma_{\rho \to \pi\pi} = 151$ MeV as basic requirements from low-energy QCD phenomenology.

We thus use the same parameter set as in Refs. [32,86], namely $m_u = m_d = m_l = 5.49$ MeV, $m_s = 115$ MeV, $D_0 \Lambda_0^2 = 219$, $D_1 \Lambda_0^4 = 40$, $\Lambda_0 = 0.758$ GeV, $\Lambda_1 = 0.961$ GeV and $p_0 = 0.6$ GeV for the rank-2 model.

For fixing the parameters in the rank-1 model, we use only the masses of pion and kaon, the pion decay constant, and GMOR as one additional constraint. This gives $m_u = m_d = m_l = 6.6$ MeV, $m_s = 142$ MeV, $D_0 \Lambda_0^2 = 113.67$, and $\Lambda_0 = 0.647$ GeV for our values of the rank-1 parameters.

At $T > 0$, $p \to p_n = (\omega_n, \vec{p})$. Presently, pertinent are the fermion Matsubara frequencies $\omega_n = (2n+1)\pi T$. Due to loss of $O(4)$ symmetry in medium, the dressed quark propagator (12) is at $T > 0$ replaced by

$$S_q^{-1}(\vec{p}, \omega_n; T) = i\vec{\gamma} \cdot \vec{p}\, A_q(\vec{p}^2, \omega_n; T) + i\gamma_4 \omega_n\, C_q(\vec{p}^2, \omega_n; T) + B_q(\vec{p}^2, \omega_n; T). \tag{A9}$$

For separable interactions, the dressing functions A_q, C_q and B_q depend only on the sum $p_n^2 = \omega_n^2 + \vec{p}^2$. In the separable models (A4), with their characteristic form (A5) of the propagator solutions at $T = 0$, the dressing functions obtained as solutions of the gap equation at $T > 0$ are:

$$A_q(p_n^2; T) = 1 + a_q(T) \mathcal{F}_1(p_n^2), \quad C_q(p_n^2; T) = 1 + c_q(T) \mathcal{F}_1(p_n^2), \quad B_q(p_n^2; T) = m_q + b_q(T) \mathcal{F}_0(p_n^2). \tag{A10}$$

That is, the former gap constants a_q and b_q become temperature-dependent gap functions $a_f(T), b_f(T)$ and $c_f(T)$ obtained from the nonlinear system of equations:

$$a_q(T) = \frac{8 D_1}{9} T \sum_n \int \frac{d^3 p}{(2\pi)^3} \mathcal{F}_1(p_n^2) \vec{p}^2 A_q(p_n^2, T) d_q^{-1}(p_n^2, T), \tag{A11}$$

$$c_q(T) = \frac{8 D_1}{3} T \sum_n \int \frac{d^3 p}{(2\pi)^3} \mathcal{F}_1(p_n^2) \omega_n^2 C_q(p_n^2, T) d_q^{-1}(p_n^2, T), \tag{A12}$$

$$b_q(T) = \frac{16 D_0}{3} T \sum_n \int \frac{d^3 p}{(2\pi)^3} \mathcal{F}_0(p_n^2) B_q(p_n^2, T) d_q^{-1}(p_n^2, T), \tag{A13}$$

where the denominator function is $d_q(p_n^2, T) = \vec{p}^{\,2} A_q^2(p_n^2, T) + \omega_n^2 C_q^2(p_n^2, T) + B_q^2(p_n^2, T)$.

References

1. Tanabashi, M.; Hagiwara, K.; Hikasa, K.; Nakamura, K.; Sumino, Y.; Takahashi, F.; Tanaka, J.; Agashe, K.; Aielli, G.; Amsler, C.; et al. [Particle Data Group]. Review of Particle Physics. *Phys. Rev.* **2018**, *D98*, 030001. doi:10.1103/PhysRevD.98.030001. [CrossRef]
2. Peccei, R.D.; Quinn, H.R. CP Conservation in the Presence of Instantons. *Phys. Rev. Lett.* **1977**, *38*, 1440–1443. doi:10.1103/PhysRevLett.38.1440. [CrossRef]
3. Peccei, R.D.; Quinn, H.R. Constraints Imposed by CP Conservation in the Presence of Instantons. *Phys. Rev.* **1977**, *D16*, 1791–1797. doi:10.1103/PhysRevD.16.1791. [CrossRef]
4. Weinberg, S. A New Light Boson? *Phys. Rev. Lett.* **1978**, *40*, 223–226. doi:10.1103/PhysRevLett.40.223. [CrossRef]
5. Wilczek, F. Problem of Strong P and T Invariance in the Presence of Instantons. *Phys. Rev. Lett.* **1978**, *40*, 279–282. doi:10.1103/PhysRevLett.40.279. [CrossRef]
6. Peccei, R.D. The Strong CP problem and axions. *Lect. Notes Phys.* **2008**, *741*, 3–17. doi:10.1007/978-3-540-73518-2_1. [CrossRef]
7. Cabibbo, N. Unitary Symmetry and Leptonic Decays. *Phys. Rev. Lett.* **1963**, *10*, 531–533. doi:10.1103/PhysRevLett.10.531. [CrossRef]
8. Kobayashi, M.; Maskawa, T. CP Violation in the Renormalizable Theory of Weak Interaction. *Prog. Theor. Phys.* **1973**, *49*, 652–657. doi:10.1143/PTP.49.652. [CrossRef]
9. Baker, C.A.; Doyle, D.D.; Geltenbort, P.; Green, K.; van der Grinten, M.G.D.; Harris, P.G.; Iaydjiev, P.; Ivanov, S.N.; May, D.J.R.; Pendlebury, J.M.; et al. An Improved experimental limit on the electric dipole moment of the neutron. *Phys. Rev. Lett.* **2006**, *97*, 131801, doi:10.1103/PhysRevLett.97.131801. [CrossRef]
10. Wantz, O.; Shellard, E.P.S. Axion Cosmology Revisited. *Phys. Rev.* **2010**, *D82*, 123508, doi:10.1103/PhysRevD.82.123508. [CrossRef]
11. Berkowitz, E.; Buchoff, M.I.; Rinaldi, E. Lattice QCD input for axion cosmology. *Phys. Rev.* **2015**, *D92*, 034507, doi:10.1103/PhysRevD.92.034507. [CrossRef]
12. Preskill, J.; Wise, M.B.; Wilczek, F. Cosmology of the Invisible Axion. *Phys. Lett.* **1983**, *120B*, 127–132. doi:10.1016/0370-2693(83)90637-8. [CrossRef]
13. Abbott, L.F.; Sikivie, P. A Cosmological Bound on the Invisible Axion. *Phys. Lett.* **1983**, *120B*, 133–136. doi:10.1016/0370-2693(83)90638-X. [CrossRef]
14. Dine, M.; Fischler, W. The Not So Harmless Axion. *Phys. Lett.* **1983**, *120B*, 137–141. doi:10.1016/0370-2693(83)90639-1. [CrossRef]
15. Kim, J.E.; Nam, S.; Semetzidis, Y.K. Fate of global symmetries in the Universe: QCD axion, quintessential axion and trans-Planckian inflaton decay-constant. *Int. J. Mod. Phys.* **2018**, *A33*, 1830002, doi:10.1142/S0217751X18300028. [CrossRef]
16. Dvali, G.; Funcke, L. Domestic Axion. *arXiv* **2016**, arXiv:1608.08969.
17. Horvatić, D.; Kekez, D.; Klabučar, D. η' and η mesons at high T when the $U_A(1)$ and chiral symmetry breaking are tied. *Phys. Rev.* **2019**, *D99*, 014007, doi:10.1103/PhysRevD.99.014007. [CrossRef]
18. Adler, S.L. Axial vector vertex in spinor electrodynamics. *Phys. Rev.* **1969**, *177*, 2426–2438. doi:10.1103/PhysRev.177.2426. [CrossRef]
19. Bell, J.S.; Jackiw, R. A PCAC puzzle: $\pi^0 \to \gamma\gamma$ in the σ model. *Nuovo Cim.* **1969**, *A60*, 47–61. doi:10.1007/BF02823296. [CrossRef]
20. Feldmann, T. Quark structure of pseudoscalar mesons. *Int. J. Mod. Phys.* **2000**, *A15*, 159–207, doi:10.1142/S0217751X00000082. [CrossRef]
21. Kekez, D.; Klabučar, D.; Scadron, M.D. Revisiting the U(A)(1) problems. *J. Phys.* **2000**, *G26*, 1335–1354, doi:10.1088/0954-3899/26/9/305. [CrossRef]
22. Kekez, D.; Klabučar, D. Two photon processes of pseudoscalar mesons in a Bethe-Salpeter approach. *Phys. Lett.* **1996**, *B387*, 14–20, doi:10.1016/0370-2693(96)00990-2. [CrossRef]
23. Kekez, D.; Bistrović, B.; Klabučar, D. Application of Jain and Munczek's bound state approach to gamma gamma processes of Pi0, eta(c) and eta(b). *Int. J. Mod. Phys.* **1999**, *A14*, 161–194, doi:10.1142/S0217751X99000087. [CrossRef]

24. Klabučar, D.; Kekez, D. eta and eta-prime in a coupled Schwinger-Dyson and Bethe-Salpeter approach. *Phys. Rev.* **1998**, *D58*, 096003, doi:10.1103/PhysRevD.58.096003. [CrossRef]
25. Kekez, D.; Klabučar, D. Gamma* gamma —> pi0 transition and asymptotics of gamma* gamma and gamma* gamma* transitions of other unflavored pseudoscalar mesons. *Phys. Lett.* **1999**, *B457*, 359–367, doi:10.1016/S0370-2693(99)00536-5. [CrossRef]
26. Bistrović, B.; Klabučar, D. Anomalous gamma —> 3 pi amplitude in a bound state approach. *Phys. Lett.* **2000**, *B478*, 127–136, doi:10.1016/S0370-2693(00)00241-0. [CrossRef]
27. Kekez, D.; Klabučar, D. Eta and eta-prime in a coupled Schwinger-Dyson and Bethe-Salpeter approach: The gamma* gamma transition form-factors. *Phys. Rev.* **2002**, *D65*, 057901, doi:10.1103/PhysRevD.65.057901. [CrossRef]
28. Kekez, D.; Klabučar, D. Pseudoscalar q anti-q mesons and effective QCD coupling enhanced (A**2) condensate. *Phys. Rev.* **2005**, *D71*, 014004, doi:10.1103/PhysRevD.71.014004. [CrossRef]
29. Kekez, D.; Klabučar, D. Eta and eta' mesons and dimension 2 gluon condensate < A**2 >. *Phys. Rev.* **2006**, *D73*, 036002, doi:10.1103/PhysRevD.73.036002. [CrossRef]
30. Horvatić, D.; Blaschke, D.; Kalinovsky, Y.; Kekez, D.; Klabučar, D. eta and eta-prime mesons in the Dyson-Schwinger approach using a generalization of the Witten-Veneziano relation. *Eur. Phys. J.* **2008**, *A38*, 257–264, doi:10.1140/epja/i2008-10670-x. [CrossRef]
31. Horvatić, D.; Klabučar, D.; Radzhabov, A.E. eta and eta-prime mesons in the Dyson-Schwinger approach at finite temperature. *Phys. Rev.* **2007**, *D76*, 096009, doi:10.1103/PhysRevD.76.096009. [CrossRef]
32. Horvatić, D.; Blaschke, D.; Klabučar, D.; Radzhabov, A.E. Pseudoscalar Meson Nonet at Zero and Finite Temperature. *Phys. Part. Nucl.* **2008**, *39*, 1033–1039, doi:10.1134/S1063779608070095. [CrossRef]
33. Alkofer, R.; von Smekal, L. The Infrared behavior of QCD Green's functions: Confinement dynamical symmetry breaking, and hadrons as relativistic bound states. *Phys. Rept.* **2001**, *353*, 281, doi:10.1016/S0370-1573(01)00010-2. [CrossRef]
34. Roberts, C.D.; Schmidt, S.M. Dyson-Schwinger equations: Density, temperature and continuum strong QCD. *Prog. Part. Nucl. Phys.* **2000**, *45*, S1–S103, doi:10.1016/S0146-6410(00)90011-5. [CrossRef]
35. Holl, A.; Roberts, C.D.; Wright, S.V. Hadron physics and Dyson-Schwinger equations. In Proceedings of the 20th Annual Hampton University Graduate Studies Program (HUGS 2005), Newport News, VA, USA, 31 May–17 June 2005.
36. Fischer, C.S. Infrared properties of QCD from Dyson-Schwinger equations. *J. Phys.* **2006**, *G32*, R253–R291, doi:10.1088/0954-3899/32/8/R02. [CrossRef]
37. Benić, S.; Horvatić, D.; Kekez, D.; Klabučar, D. A $U_A(1)$ symmetry restoration scenario supported by the generalized Witten-Veneziano relation and its analytic solution. *Phys. Lett.* **2014**, *B738*, 113–117, doi:10.1016/j.physletb.2014.09.029. [CrossRef]
38. Witten, E. Current Algebra Theorems for the U(1) Goldstone Boson. *Nucl. Phys.* **1979**, *B156*, 269–283. doi:10.1016/0550-3213(79)90031-2. [CrossRef]
39. Veneziano, G. U(1) Without Instantons. *Nucl. Phys.* **1979**, *B159*, 213–224. doi:10.1016/0550-3213(79)90332-8. [CrossRef]
40. Alles, B.; D'Elia, M.; Di Giacomo, A. Topological susceptibility at zero and finite T in SU(3) Yang-Mills theory. *Nucl. Phys.* **1997**, *B494*, 281–292. [CrossRef]
41. Boyd, G.; Engels, J.; Karsch, F.; Laermann, E.; Legeland, C.; Lutgemeier, M.; Petersson, B. Thermodynamics of SU(3) lattice gauge theory. *Nucl. Phys.* **1996**, *B469*, 419–444, doi:10.1016/0550-3213(96)00170-8. [CrossRef]
42. Gattringer, C.; Hoffmann, R.; Schaefer, S. The Topological susceptibility of SU(3) gauge theory near T(c). *Phys. Lett.* **2002**, *B535*, 358–362, doi:10.1016/S0370-2693(02)01757-4. [CrossRef]
43. Di Vecchia, P.; Rossi, G.; Veneziano, G.; Yankielowicz, S. Spontaneous *CP* breaking in QCD and the axion potential: An effective Lagrangian approach. *J. High Energy Phys.* **2017**, *12*, 104, doi:10.1007/JHEP12(2017)104. [CrossRef]
44. Dick, V.; Karsch, F.; Laermann, E.; Mukherjee, S.; Sharma, S. Microscopic origin of $U_A(1)$ symmetry violation in the high temperature phase of QCD. *Phys. Rev.* **2015**, *D91*, 094504, doi:10.1103/PhysRevD.91.094504. [CrossRef]
45. Bazavov, A.; Ding, H.-T.; Hegde, P.; Kaczmarek, O.; Karsch, F.; Laermann, E.; Maezawa, Y.; Mukherjee, S.; Ohno, H.; Petreczky, P.; et al. The QCD Equation of State to $\mathcal{O}(\mu_B^6)$ from Lattice QCD. *Phys. Rev.* **2017**, *D95*, 054504, doi:10.1103/PhysRevD.95.054504. [CrossRef]

46. Bazavov, A. An overview of (selected) recent results in finite-temperature lattice QCD. *J. Phys. Conf. Ser.* **2013**, *446*, 012011, doi:10.1088/1742-6596/446/1/012011. [CrossRef]
47. Stoecker, H.; Zhou, K.; Schramm, S.; Senzel, F.; Greiner, C.; Beitel, M.; Gallmeister, K.; Gorenstein, M.; Mishustin, I.; Vasak, D.; et al. Glueballs amass at RHIC and LHC Colliders! - The early quarkless 1st order phase transition at T = 270 MeV - from pure Yang-Mills glue plasma to GlueBall-Hagedorn states. *J. Phys.* **2016**, *G43*, 015105, doi:10.1088/0954-3899/43/1/015105. [CrossRef]
48. Benić, S.; Horvatić, D.; Kekez, D.; Klabučar, D. eta-prime Multiplicity and the Witten-Veneziano relation at finite temperature. *Phys. Rev.* **2011**, *D84*, 016006, doi:10.1103/PhysRevD.84.016006. [CrossRef]
49. Leutwyler, H.; Smilga, A.V. Spectrum of Dirac operator and role of winding number in QCD. *Phys. Rev.* **1992**, *D46*, 5607–5632. doi:10.1103/PhysRevD.46.5607. [CrossRef]
50. Di Vecchia, P.; Veneziano, G. Chiral Dynamics in the Large n Limit. *Nucl. Phys.* **1980**, *B171*, 253–272. doi:10.1016/0550-3213(80)90370-3. [CrossRef]
51. Dürr, S. Topological susceptibility in full QCD: Lattice results versus the prediction from the QCD partition function with granularity. *Nucl. Phys.* **2001**, *B611*, 281–310, doi:10.1016/S0550-3213(01)00325-X. [CrossRef]
52. Bernard, V.; Descotes-Genon, S.; Toucas, G. Topological susceptibility on the lattice and the three-flavour quark condensate. *JHEP* **2012**, *06*, 051, doi:10.1007/JHEP06(2012)051. [CrossRef]
53. Pisarski, R.D.; Wilczek, F. Remarks on the Chiral Phase Transition in Chromodynamics. *Phys. Rev.* **1984**, *D29*, 338–341. doi:10.1103/PhysRevD.29.338. [CrossRef]
54. Höll, A.; Maris, P.; Roberts, C.D. Mean field exponents and small quark masses. *Phys. Rev.* **1999**, *C59*, 1751–1755, doi:10.1103/PhysRevC.59.1751. [CrossRef]
55. Kiriyama, O.; Maruyama, M.; Takagi, F. Current quark mass effects on chiral phase transition of QCD in the improved ladder approximation. *Phys. Rev.* **2001**, *D63*, 116009, doi:10.1103/PhysRevD.63.116009. [CrossRef]
56. Ikeda, T. Dressed quark propagator at finite temperature in the Schwinger-Dyson approach with the rainbow approximation: Exact numerical solutions and their physical implication. *Prog. Theor. Phys.* **2002**, *107*, 403–420, doi:10.1143/PTP.107.403. [CrossRef]
57. Blank, M.; Krassnigg, A. The QCD chiral transition temperature in a Dyson-Schwinger-equation context. *Phys. Rev.* **2010**, *D82*, 034006, doi:10.1103/PhysRevD.82.034006. [CrossRef]
58. Fischer, C.S.; Mueller, J.A. Quark condensates and the deconfinement transition. *PoS* **2009**, *CPOD2009*, 023, doi:10.22323/1.071.0023. [CrossRef]
59. Qin, S.x.; Chang, L.; Chen, H.; Liu, Y.x.; Roberts, C.D. Phase diagram and critical endpoint for strongly-interacting quarks. *Phys. Rev. Lett.* **2011**, *106*, 172301, doi:10.1103/PhysRevLett.106.172301. [CrossRef]
60. Gao, F.; Chen, J.; Liu, Y.X.; Qin, S.X.; Roberts, C.D.; Schmidt, S.M. Phase diagram and thermal properties of strong-interaction matter. *Phys. Rev.* **2016**, *D93*, 094019, doi:10.1103/PhysRevD.93.094019. [CrossRef]
61. Fischer, C.S. QCD at finite temperature and chemical potential from Dyson–Schwinger equations. *Prog. Part. Nucl. Phys.* **2019**, *105*, 1–60, doi:10.1016/j.ppnp.2019.01.002. [CrossRef]
62. Ejiri, S.; Karsch, F.; Laermann, E.; Miao, C.; Mukherjee, S.; Petreczky, P.; Schmidt, C.; Soeldner, W.; Unger, W. On the magnetic equation of state in (2+1)-flavor QCD. *Phys. Rev.* **2009**, *D80*, 094505, doi:10.1103/PhysRevD.80.094505. [CrossRef]
63. Ding, H.T.; Hegde, P.; Karsch, F.; Lahiri, A.; Li, S.T.; Mukherjee, S.; Petreczky, P. Chiral phase transition of (2+1)-flavor QCD. *Nucl. Phys.* **2019**, *A982*, 211–214, doi:10.1016/j.nuclphysa.2018.10.032. [CrossRef]
64. Ding, H.T.; Hegde, P.; Kaczmarek, O.; Karsch, F.; Lahiri, A.; Li, S.T.; Mukherjee, S.; Petreczky, P. Chiral phase transition in (2 + 1)-flavor QCD. *PoS* **2019**, *LATTICE2018*, 171, doi:10.22323/1.334.0171. [CrossRef]
65. Csorgo, T.; Vertesi, R.; Sziklai, J. Indirect observation of an in-medium η' mass reduction in $\sqrt{s_{NN}}$ = 200 GeV Au+Au collisions. *Phys. Rev. Lett.* **2010**, *105*, 182301, doi:10.1103/PhysRevLett.105.182301. [CrossRef]
66. Vertesi, R.; Csorgo, T.; Sziklai, J. Significant in-medium η' mass reduction in $\sqrt{s_{NN}}$ = 200 GeV Au+Au collisions at the BNL Relativistic Heavy Ion Collider. *Phys. Rev.* **2011**, *C83*, 054903, doi:10.1103/PhysRevC.83.054903. [CrossRef]
67. Aoki, S.; Fukaya, H.; Taniguchi, Y. Chiral symmetry restoration, eigenvalue density of Dirac operator and axial U(1) anomaly at finite temperature. *Phys. Rev.* **2012**, *D86*, 114512, doi:10.1103/PhysRevD.86.114512. [CrossRef]

68. Buchoff, M.I.; Cheng, M.; Christ, N.; Ding, H.-T.; Jung, C.; Karsch, F.; Lin, Z.; Mawhinney, R.D.; Mukherjee, S.; Petreczky, P.; et al. QCD chiral transition, U(1)A symmetry and the dirac spectrum using domain wall fermions. *Phys. Rev.* **2014**, *D89*, 054514, doi:10.1103/PhysRevD.89.054514. [CrossRef]
69. Shore, G.M. Pseudoscalar meson decay constants and couplings, the Witten-Veneziano formula beyond large N(c), and the topological susceptibility. *Nucl. Phys.* **2006**, *B744*, 34–58, doi:10.1016/j.nuclphysb.2006.03.011. [CrossRef]
70. Petreczky, P.; Schadler, H.P.; Sharma, S. The topological susceptibility in finite temperature QCD and axion cosmology. *Phys. Lett.* **2016**, *B762*, 498–505, doi:10.1016/j.physletb.2016.09.063. [CrossRef]
71. Borsanyi, S.; Fodor, Z.; Kampert, K.H.; Katz, S.D.; Kawanai, T.; Kovacs, T.G.; Mages, S.W.; Pasztor, A.; Pittler, F.; Redondo, J.; et al. Calculation of the axion mass based on high-temperature lattice quantum chromodynamics. *Nature* **2016**, *539*, 69–71, doi:10.1038/nature20115. [CrossRef]
72. Bonati, C.; D'Elia, M.; Mariti, M.; Martinelli, G.; Mesiti, M.; Negro, F.; Sanfilippo, F.; Villadoro, G. Axion phenomenology and θ-dependence from $N_f = 2 + 1$ lattice QCD. *J. High Energy Phys.* **2016**, *03*, 155, doi:10.1007/JHEP03(2016)155. [CrossRef]
73. Takahashi, F.; Yin, W.; Guth, A.H. QCD axion window and low-scale inflation. *Phys. Rev.* **2018**, *D98*, 015042, doi:10.1103/PhysRevD.98.015042. [CrossRef]
74. Bonati, C.; D'Elia, M.; Martinelli, G.; Negro, F.; Sanfilippo, F.; Todaro, A. Topology in full QCD at high temperature: A multicanonical approach. *J. High Energy Phys.* **2018**, *11*, 170. doi:10.1007/JHEP11(2018)170. [CrossRef]
75. Grilli di Cortona, G.; Hardy, E.; Pardo Vega, J.; Villadoro, G. The QCD axion, precisely. *J. High Energy Phys.* **2016**, *01*, 034, doi:10.1007/JHEP01(2016)034. [CrossRef]
76. Gorghetto, M.; Villadoro, G. Topological Susceptibility and QCD Axion Mass: QED and NNLO corrections. *J. High Energy Phys.* **2019**, *03*, 033, doi:10.1007/JHEP03(2019)033. [CrossRef]
77. Blaschke, D.; Burau, G.; Kalinovsky, Y.L.; Maris, P.; Tandy, P.C. Finite T meson correlations and quark deconfinement. *Int. J. Mod. Phys.* **2001**, *A16*, 2267–2291, doi:10.1142/S0217751X01003457. [CrossRef]
78. Alkofer, R.; Watson, P.; Weigel, H. Mesons in a Poincare covariant Bethe-Salpeter approach. *Phys. Rev.* **2002**, *D65*, 094026, doi:10.1103/PhysRevD.65.094026. [CrossRef]
79. Contant, R.; Huber, M.Q. Phase structure and propagators at nonvanishing temperature for QCD and QCD-like theories. *Phys. Rev.* **2017**, *D96*, 074002, doi:10.1103/PhysRevD.96.074002. [CrossRef]
80. Blaschke, D.; Kalinovsky, Y.L.; Radzhabov, A.E.; Volkov, M.K. Scalar sigma meson at a finite temperature in a nonlocal quark model. *Phys. Part. Nucl. Lett.* **2006**, *3*, 327–330. doi:10.1134/S1547477106050086. [CrossRef]
81. Burger, F.; Ilgenfritz, E.M.; Kirchner, M.; Lombardo, M.P.; Muller-Preussker, M.; Philipsen, O.; Urbach, C.; Zeidlewicz, L. Thermal QCD transition with two flavors of twisted mass fermions. *Phys. Rev.* **2013**, *D87*, 074508, doi:10.1103/PhysRevD.87.074508. [CrossRef]
82. Kotov, A.Y.; Lombardo, M.P.; Trunin, A.M. Fate of the η' in the quark gluon plasma. *Phys. Lett.* **2019**, *B794*, 83–88, doi:10.1016/j.physletb.2019.05.035. [CrossRef]
83. Borsanyi, S.; Fodor, Z.; Hoelbling, C.; Katz, S.D.; Krieg, S.; Ratti, C.; Szabo, K.K. Is there still any Tc mystery in lattice QCD? Results with physical masses in the continuum limit III. *J. High Energy Phys.* **2010**, *09*, 073, doi:10.1007/JHEP09(2010)073. [CrossRef]
84. Isserstedt, P.; Buballa, M.; Fischer, C.S.; Gunkel, P.J. Baryon number fluctuations in the QCD phase diagram from Dyson-Schwinger equations. *arXiv* **2019**, arXiv:1906.11644.
85. Cahill, R.T.; Gunner, S.M. The Global color model of QCD for hadronic processes: A Review. *Fizika* **1998**, *B7*, 171.
86. Blaschke, D.; Horvatić, D.; Klabučar, D.; Radzhabov, A.E. Separable Dyson-Schwinger model at zero and finite T. In Proceedings of the Mini-Workshop Bled 2006 on Progress in Quark Models, Bled, Slovenia, 10–17 July 2006; pp. 20–26.
87. Klabučar, D.; Horvatić, D.; Kekez, D. T-dependence of the axion mass when the $U_A(1)$ and chiral symmetry breaking are tied. In Proceedings of the Workshop on Excited QCD 2019, Schladming, Austria, 30 January–3 February 2019.
88. Lu, Z.Y.; Ruggieri, M. Effect of the chiral phase transition on axion mass and self-coupling. *Phys. Rev.* **2019**, *D100*, 014013, doi:10.1103/PhysRevD.100.014013. [CrossRef]

89. Gu, X.W.; Duan, C.G.; Guo, Z.H. Updated study of the η-η' mixing and the thermal properties of light pseudoscalar mesons at low temperatures. *Phys. Rev.* **2018**, *D98*, 034007, doi:10.1103/PhysRevD.98.034007. [CrossRef]
90. Fukaya, H. Can axial U(1) anomaly disappear at high temperature? *EPJ Web Conf.* **2018**, *175*, 01012, doi:10.1051/epjconf/201817501012. [CrossRef]

© 2019 by the authors. Licensee MDPI, Basel, Switzerland. This article is an open access article distributed under the terms and conditions of the Creative Commons Attribution (CC BY) license (http://creativecommons.org/licenses/by/4.0/).

Communication

Open-Charm Hadron Measurements in Au+Au Collisions at $\sqrt{s_{NN}} = 200\,\text{GeV}$ by the STAR Experiment [†]

Jan Vanek for the STAR Collaboration

Nuclear Physics Institute, Czech Academy of Sciences, 250 68 Řež, Czech Republic; vanek@ujf.cas.cz; Tel.: +420-266-177-206

[†] This paper is based on the talk at the 18th Zimányi School, Budapest, Hungary, 3–7 December 2018.

Received: 14 August 2019; Accepted: 5 September 2019; Published: 7 September 2019

Abstract: Study of the open-charm hadron production in heavy-ion collisions is crucial for understanding the properties of the Quark-Gluon Plasma. In these papers, we report on a selection of recent STAR measurements of open-charm hadrons in Au+Au collisions at $\sqrt{s_{NN}} = 200\,\text{GeV}$, using the Heavy-Flavor Tracker. In particular, the nuclear modification factors of D^0 and $D^\pm$ mesons, elliptic and directed flow of D^0 mesons, D_s/D^0 and Λ_c/D^0 yield ratios are discussed. The observed suppression of D^0 and $D^\pm$ mesons suggests strong interactions of the charm quarks with the QGP. The measured elliptic flow of D^0 mesons is large and follows the NCQ scaling, suggesting that charm quarks may be close to thermal equilibrium with the QGP medium. Both D_s/D^0 and Λ_c/D^0 yield ratios are found to be enhanced in Au+Au collisions. The enhancement can be explained by models incorporating coalescence hadronization of charm quarks. In addition, the directed flow of the D^0 mesons is measured to be negative and larger than that of light-flavor mesons which is in a qualitative agreement with hydrodynamic model predictions with a tilted QGP bulk.

Keywords: Quark-Gluon Plasma; open-charm hadrons; nuclear modification factor; elliptic flow; directed flow

1. Introduction

One of the main goals of the STAR experiment is to study the properties of the Quark-Gluon Plasma (QGP), which can be produced in ultra-relativistic heavy-ion collisions. Charm quarks are an excellent probe of the medium created in these collisions since they are produced predominantly in initial hard partonic scatterings and therefore experience the whole evolution of the medium.

As the charm quark propagates through the QGP, it interacts with the QGP and loses energy. The most common way to access the energy loss is by studying the modification of open-charm hadron yields in heavy-ion collisions with respect to those in p+p collisions using the nuclear modification factor:

$$R_{AA}(p_T) = \frac{dN^{AA}/dp_T}{\langle N_{coll}\rangle dN^{pp}/dp_T}, \qquad (1)$$

where $\langle N_{coll}\rangle$ is the mean number of binary collisions, calculated using the Glauber model [1]. $R_{AA} < 1$ for high-p_T open-charm hadrons is considered a signature connected with the presence of the QGP and the level of the suppression gives access to the strength of the interaction between the charm quark and the medium [2,3].

Another way to obtain information about the charm quark interaction with the QGP is to measure the azimuthal anisotropy of the produced charm hadrons (v_2). The magnitude of the v_2 that the charm quarks develop through the interaction with the surrounding medium carries important information about the transport properties of the medium [2,3].

To have a more complete picture of the open-charm hadron production in heavy-ion collisions, it is also important to understand the charm quark hadronization process. The charm quark hadronization mechanism can be studied through the measurements of the Λ_c/D^0 and D_s/D^0 yield ratios [4,5].

Since the charm quarks are created very early in the heavy-ion collisions, they can be used to probe initial conditions in such collisions. Recent theoretical calculations suggest that measurement of the directed flow v_1 of open-charm mesons can be sensitive to the initial tilt of the QGP bulk and also to the initial electro-magnetic field induced by the passing spectators [6,7].

The following section summarizes recent STAR measurements of open-charm hadrons in the context of the observables and phenomena described above.

2. Open-Charm Measurements with the HFT

All results presented in this summary are from Au+Au collisions at $\sqrt{s_{NN}} = 200$ GeV which were collected by the STAR experiment in years 2014 and 2016. Topological reconstruction of the decays, using an excellent vertex position resolution from the Heavy-Flavor Tracker (HFT) [8], was used to extract the signals of the open-charm hadrons listed in Table 1.

Table 1. List of open-charm hadrons measured using the HFT. The left column contains decay channels used for the reconstruction, $c\tau$ is the proper decay length of a given hadron, and BR is the branching ratio. Charge conjugate particles are measured as well. Values are taken from Ref. [9].

Decay Channel	$c\tau$ [μm]	BR [%]
$D^+ \to K^-\pi^+\pi^+$	311.8 ± 2.1	9.46 ± 0.24
$D^0 \to K^-\pi^+$	122.9 ± 0.4	3.93 ± 0.04
$D_s^+ \to \phi\pi^+ \to K^-K^+\pi^+$	149.9 ± 2.1	2.27 ± 0.08
$\Lambda_c^+ \to K^-\pi^+ p$	59.9 ± 1.8	6.35 ± 0.33

The reconstruction of $D^\pm$ mesons in data from 2016 will be used as an example as the steps of reconstruction of all the aforementioned particles are similar. First, a series of selection criteria is applied to the events and tracks. Specific values of the criteria, used in the analysis of $D^\pm$ mesons, are listed in Table 2.

Table 2. Summary of selection criteria used for extraction of $D^\pm$ candidates from the data. For more details, see the text.

Event selection		$\|V_z\| < 6$ cm
		$\|V_z - V_{z(VPD)}\| < 3$ cm
Track selection		$p_T > 500$ MeV
		$\|\eta\| < 1$
		nHitsFit > 20
		nHitsFit/nHitsMax > 0.52
		HFT tracks = PXL1 + PXL2 + (IST or SSD)
Particle identification	TPC	$\|n\sigma_\pi\| < 3$
		$\|n\sigma_K\| < 2$
	TOF	$\|1/\beta - 1/\beta_\pi\| < 0.03$
		$\|1/\beta - 1/\beta_K\| < 0.03$
Decay topology		$DCA_{\text{pair}} < 80$ μm
		30 μm $< L_{D^\pm} < 2000$ μm
		$\cos(\theta) > 0.998$
		$\Delta_{\max} < 200$ μm
		$DCA_{\pi-PV} > 100$ μm
		$DCA_{K-PV} > 80$ μm

The events are selected so that the position of the primary vertex (PV) along the beam axis (V_z), which is determined using the HFT and Time Projection Chamber (TPC) [10], is no further than 6 cm from the center of the STAR detector. This is necessary due to physical dimensions and acceptance of the HFT. The value of V_z is also compared to that measured by the Vertex Position Detector [11] ($V_{z(VPD)}$) which helps with rejection of pile-up events as the VPD is a fast detector.

From these events, only tracks with sufficiently large transverse momentum ($p_T > 300 \text{ MeV}/c$) are selected to reduce the combinatorial background. The pseudorapidity criterion $|\eta| < 1$ is given by the STAR detector acceptance. All tracks are also required to have sufficient number of hits used for track reconstruction inside the TPC (nHitsFit) and to be properly matched to the HFT to ensure their good quality. In this case, a good HFT track is required to have one hit in each of the inner layers (PXL1 and PXL2) and at least one hit in one of the two outer layers (IST or SSD) [1].

Next, all the selected tracks are identified using the TPC and the Time Of Flight (TOF) [12] detectors. The particle identification (PID) with the TPC is done based on energy loss of charged particles in the TPC gas. The measured energy loss is compared to the expected one, which is calculated with Bichsel formula, using $n\sigma$ variable [13]. The PID using TOF is done by comparing velocity of given particle measured by TOF (β) and that calculated from its momentum and rest mass (β_π or β_K).

When charged pions and kaons are identified they are combined into $K\pi\pi$ triplets within each event. The topology of the triplet is then constrained using variables shown in Figure 1. More specifically they are: the maximum distance of closest approach of track pairs (DCA_{pair}), $D^\pm$ meson decay length $L_{D^\pm}$, cosine of the pointing angle $\cos(\theta)$, maximum distance between reconstructed secondary vertices of track pairs ($\Delta_{\max}$), and the distance of closest approach to the primary vertex of the kaon (DCA_{K-PV}) and each of the pions ($DCA_{\pi-PV}$). Specific values used for $D^\pm$ signal extraction are listed in Table 2. The topological selection criteria used for $D^\pm$ mesons will be optimized using the TMVA [14] in near future, as was done for other open-charm hadron results presented in the following section, in order to improve statistical significance and also to extend the p_T range.

The $D^\pm$ signal is subsequently extracted from the invariant mass spectrum of the $K\pi\pi$ triplets which are divided into two sets. The first consists of only correct-sign charge combinations, which may come from decay of $D^\pm$ mesons (see Table 1) and contains the signal together with a combinatorial and a correlated background. The combinatorial background shape can be determined using the second set which contains only wrong-sign charge combinations which cannot originate from decay of $D^\pm$ mesons [2]. The correct-sign and the scaled [3] wrong-sign invariant mass spectrum of the $K\pi\pi$ triplets near invariant mass of the $D^\pm$ mesons is shown in top panel of Figure 2. The scaled wrong-sign spectrum can be then subtracted from the correct-sign one which leads to the spectrum shown in the bottom panel of Figure 2. The invariant mass peak is fitted with Gaussian function in order to determine its width σ and mean. The raw yield Y_{raw} is calculated using bin counting method in $\pm 3\sigma$ region around the peak mean.

---

[1] The HFT consists of total of four layers of silicon detectors. The two innermost layers are Monolithic Active Pixel Sensors (MAPS), PXL1 and PXL2. The outer layers are strip detectors, the Intermediate Silicon Tracker (IST) and the Silicon Strip Detector (SSD).

[2] This method is sufficient for $D^\pm$ analysis. In case of e.g., D^0 or Λ_c, the correlated background needs to be addressed separately as it is more significant for those analyses.

[3] For combinatorial reasons, there are approximately three times as many wrong-sign charge combinations as the correct-sign ones in this case. The wrong-sign spectrum is therefore scaled so that it matches the correct-sign one in order to estimate the combinatorial background. The scale factor is determined from ratio of integrals of the correct and wrong-sign spectrum outside the $D^\pm$ mass peak region which is set $1.795 \text{ GeV}/c^2 < M_{\text{inv}} < 1.945 \text{ GeV}/c^2$.

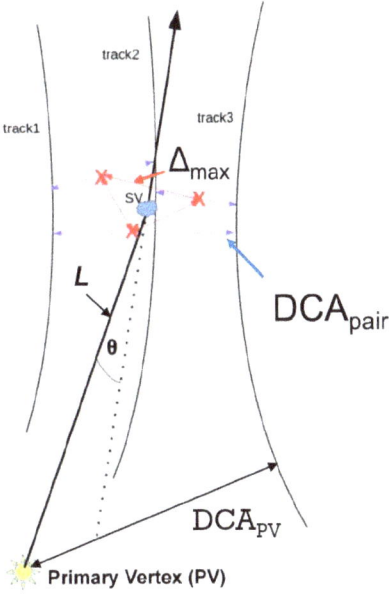

Figure 1. Depiction of a three body decay topology of $D^{\pm}$ mesons. For details about individual variables, see the text.

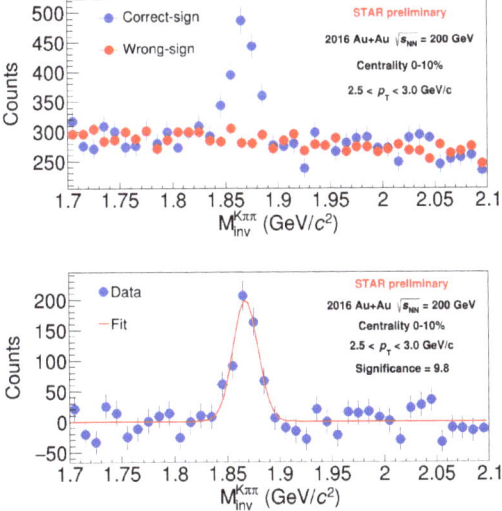

Figure 2. Invariant mass spectrum of $K\pi\pi$ triplets for: (**top**) correct-sign combinations (blue points) and with wrong-sign combinations (red points) and (**bottom**) after background subtraction. The data are fitted with Gaussian function.

The invariant spectrum of the $D^{\pm}$ mesons is then calculated from the raw yield Y_{raw} as:

$$\frac{d^2 N}{2\pi p_T dp_T dy} = \frac{1}{2\pi p_T} \frac{Y_{\text{raw}}}{N_{\text{evt}} BR \Delta p_T \Delta y \varepsilon(p_T)}, \quad (2)$$

where N_{evt} is number of recorded MB events, BR is the branching ratio (see Table 1) and $\varepsilon(p_T)$ is the total reconstruction efficiency calculated using the data-driven fast-simulator. More details about the efficiency calculation can be found in article [15]. An example of reconstruction efficiency of $D^{\pm}$ mesons in 0%–10% central Au+Au collisions extracted with selection criteria from Table 2 is shown in Figure 3.

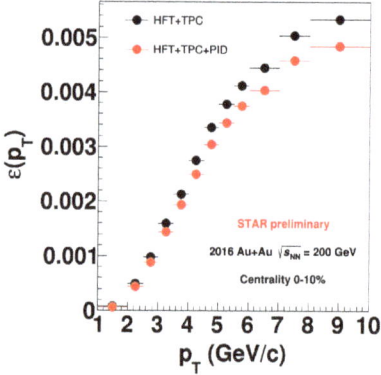

Figure 3. $D^{\pm}$ reconstruction efficiency in 0%–10% central Au+Au collisions calculated using the data-driven fast simulator without (black points) and with the PID efficiency (red points).

3. Results

Figure 4 shows the nuclear modification factor R_{AA} of D^0 [15] and $D^{\pm}$ mesons as a function of p_T in 0%–10% central Au+Au collisions. Both D^0 and $D^{\pm}$ are significantly suppressed in high-p_T region which suggests a significant energy loss of charm quarks in the QGP. The low to intermediate p_T bump structure is consistent with predictions of models incorporating large collective flow of charm quarks [15].

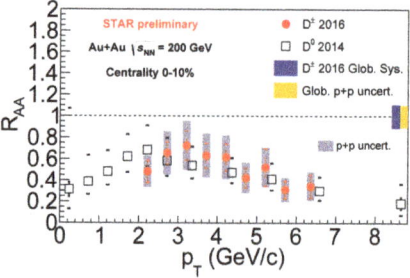

Figure 4. R_{AA} of D^0 [15] and $D^{\pm}$ mesons as a function p_T in 0%–10% central Au+Au collisions at $\sqrt{s_{NN}} = 200$ GeV. The p+p reference is from combined $D^{\star}$ and D^0 measurement by STAR in p+p collisions at $\sqrt{s} = 200$ GeV [16].

STAR has also measured and published the elliptic flow (v_2) of D^0 mesons using 2014 data [17]. Results with improved precision from the combined 2014+2016 data are shown in Figure 5a. The results clearly shows that charm quarks gain significant elliptic flow as they transverse through the medium. It is also of importance to test the Number of Constituent Quarks (NCQ) scaling. In Figure 5b is shown the v_2/n_q as a function of $(m_T - m_0)/n_q$, where n_q is the number of constituent quarks, m_T is the transverse mass and m_0 is the rest mass. In both panels, the D^0 results are compared to similar measurements for light-flavor hadrons [18]. As can be seen in Figure 5b, a similar scaling is observed for all particle species within the uncertainties. The observation of sizable D^0 mesons flow which follows the NCQ scaling, similarly as light-flavor hadrons, suggests that the charm quarks may be in thermal equilibrium with the QGP at RHIC.

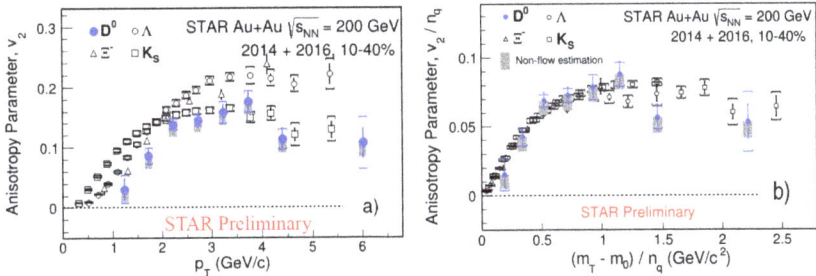

Figure 5. (**a**) The elliptic flow (v_2) of D^0 mesons and light-flavor hadrons [18] as a function of p_T. (**b**) The elliptic flow v_2 divided by the number of constituent quarks n_q as a function of $(m_T - m_0)/n_q$ for the same particle species as shown in panel (**a**). All particle species are on top of each other, which is referred as the NCQ scaling.

To study the charm hadronization and its possible modification in the presence of the QGP, STAR has measured the Λ_c/D^0 yield ratio as a function of p_T and collision centrality, results of which are shown in Figure 6. As can be seen in panel (**a**), the ratio is significantly enhanced compared to PYTHIA model predictions. The data are also compared to models that include coalesence hadronization of charm quarks [4,5] which predict an enhancement of the ratio with a qualitatively similar p_T dependence. The Λ_c/D^0 yield ratio increases from peripheral to central Au+Au collisions, as shown in Figure 6b. This and the qualitative agreement with the coalescence models indicates that the Λ_c enhancement could be a consequence of coalescence hadronization of charm quarks in the medium.

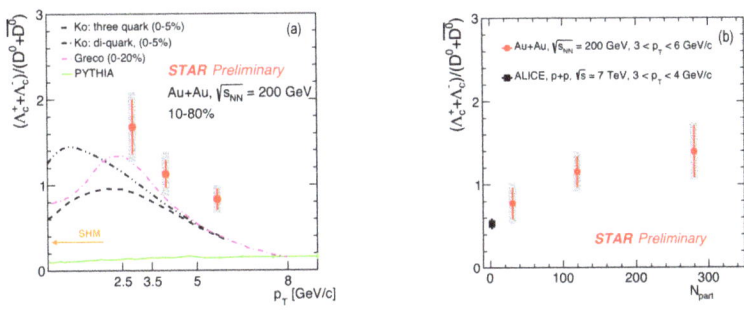

Figure 6. (**a**) The Λ_c/D^0 yield ratio as a function of p_T for 10%–80% central Au+Au collisions at $\sqrt{s_{NN}} = 200$ GeV. The data are compared to coalescence models [4,5], SHM [19] and PYTHIA. (**b**) The Λ_c/D^0 yield ratio as a function of centrality (red circles). The STAR data are compared to ALICE measurement for p+p collisions at $\sqrt{s} = 7$ TeV [20] (black square).

A complementary measurement to the one discussed above is the measurement of the D_s/D^0 ratio. As shown in Figure 7, the D_s is enhanced with respect to the averaged result from elementary collisions [21] as well as PYTHIA model calculations. The TAMU model [22], which includes coalescence hadronization of charm quarks, also shows an enhancement of the ratio, but underpredicts the data. This result also suggests that charm quarks hadronize via coalescence in heavy-ion collisions.

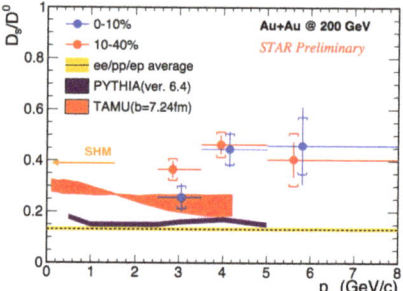

Figure 7. D_s/D^0 ratio as a function of p_T for two centralities. The data are compared to combined e+e, p+p and e+p data [21], PYTHIA, TAMU [22] and SHM [19] models.

The last result presented in this overview is from the measurement of (rapidity odd) directed flow (v_1) of D^0 mesons. There are two main models predicting the origin and magnitude of the v_1 of D^0 mesons. The first one is a hydrodynamical model which predicts larger v_1 slope (dv_1/dy) for heavy-flavor hadrons than for light-flavor hadrons, arising from a difference in the charm quark production profile and the tilted QGP bulk [6]. The second one calculates the v_1 from EM field induced by the passing spectators and predicts opposite v_1 slope for D^0 and $\overline{D^0}$ [7]. When combined, the prediction is that the v_1 slope for both D^0 and $\overline{D^0}$ mesons is negative, larger for D^0 than for $\overline{D^0}$, and much larger than for kaons [23]. As can be seen in Figure 8, the measured slope of v_1 is indeed negative and larger in magnitude for both charmed mesons than for the light-flavor hadrons. On the other hand, the available statistics does not allow firmly concluding on the D^0–$\overline{D^0}$ splitting.

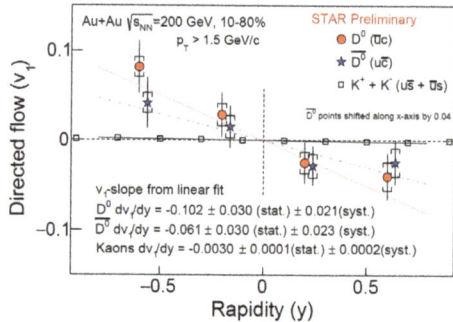

Figure 8. Directed flow of D^0 and $\overline{D^0}$ mesons as a function rapidity y in 10%–80% central Au+Au collisions at $\sqrt{s_{NN}} = 200$ GeV. The data are compared to the similar measurement for charged kaons [24]. The solid black, dashed red and blue lines are linear fits to the data. Parameters of the fits are shown in the figure.

4. Summary

STAR experiment has extensively studied the open-charm hadron production using the excellent vertex position resolution provided by the HFT. In this summary it is shown that D^0 and $D^\pm$ mesons are significantly suppressed in high p_T region which suggests strong interactions of the charm quarks with the QGP. The D^0 mesons also show large elliptic flow v_2 which follows the NCQ scaling, similarly as the light-flavor hadrons, suggesting that the charm quarks may be close to a local thermal equilibrium with the QGP medium at RHIC. Moreover, the Λ_c/D^0 and D_s/D^0 yield ratios are found to be enhanced in Au+Au collisions. Comparison to model predictions suggests that the coalescence plays an important role in charm quark hadronization in heavy-ion collisions at RHIC. The measurement of the D^0 directed flow v_1 shows significantly larger values compared to those from light-flavor hadrons and is in qualitative agreement with hydrodynamic model predictions with a tilted QGP bulk [6]. The v_1 values for D^0 and $\overline{D^0}$ are consistent with each other within the current measurement precision.

Funding: This overview and presentation at Zimányi School 2018 are supported by OPVVV grant CZ.02.1.01/0.0/0.0/16_013/0001569 of the Ministry of Education, Youth and Sports of the Czech Republic.

Acknowledgments: I would like to thank the organizers for giving me the opportunity to present STAR results at the Zimányi School 2018.

Conflicts of Interest: The author declares no conflict of interest.

References

1. Miller, M.L.; Reygers, K.; Sanders, S.J.; Steinberg, P. Glauber modeling in high energy nuclear collisions. *Ann. Rev. Nucl. Part. Sci.* **2007**, *57*, 205–243. [CrossRef]
2. Cao, S.; Luo, T.; Qin, G.Y.; Wang, X.N. Linearized Boltzmann transport model for jet propagation in the quark-gluon plasma: Heavy quark evolution. *Phys. Rev. C* **2016**, *94*, 014909. [CrossRef]
3. Xu, Y.; Bernhard, J.E.; Bass, S.A.; Nahrgang, M.; Cao, S. Data-driven analysis for the temperature and momentum dependence of the heavy-quark diffusion coefficient in relativistic heavy-ion collisions. *Phys. Rev. C* **2018**, *97*, 014907. [CrossRef]
4. Oh, Y.; Ko, C.M.; Lee, S.H.; Yasui, S. Ratios of heavy baryons to heavy mesons in relativistic nucleus-nucleus collisions. *Phys. Rev. C* **2009**, *79*, 044905. [CrossRef]
5. Plumari, S.; Minissale, V.; Das, S.K.; Coci, G.; Greco, V. Charmed hadrons from coalescence plus fragmentation in relativistic nucleus-nucleus collisions at RHIC and LHC. *Eur. Phys. J. C* **2018**, *78*, 348. [CrossRef]
6. Chatterjee, S.; Bożek, P. Large directed flow of open-charm mesons probes the three-dimensional distribution of datter in heavy-ion collisions. *Phys. Rev. Lett.* **2018**, *120*, 192301. [CrossRef]
7. Das, S.K.; Plumari, S.; Chatterjee, S.; Alam, J.; Scardina, F.; Greco, V. Directed flow of charm quarks as a witness of the initial strong magnetic field in ultra-relativistic heavy ion collisions. *Phys. Lett. B* **2017**, *768*, 260. [CrossRef]
8. Beavis, D.; Debbe, R.; Lee, J.H.; LeVine, M.J.; Scheetz, R.A.; Videbaek, F.; Xu, Z.; Bielcik, J.; Krus, M.; Dunkelberger, L.E.; et al. The STAR Heavy-Flavor Tracker, Technical Design Report. 2011. Available online: https://drupal.star.bnl.gov/STAR/starnotes/public/sn0600 (accessed on 6 September 2019).
9. Tanabashi, M.; et al. [Particle Data Group] Review of particle physics. *Phys. Rev. Lett.* **2018**, *98*, 030001.
10. Anderson, M.; Berkovitz, J.; Betts, W.; Bossingham, R.; Bieser, F.; Brown, R.; Burks, M.; Calderón de la Barca Sánchez, M.; Cebra, D.; Cherney, M.; et al. The STAR Time Projection Chamber: A unique tool for studying high multiplicity events at RHIC. *Nucl. Instrum. Meth. A* **2003**, *499*, 659–678. [CrossRef]
11. Llope, W.J.; Zhou, J.; Nussbaum, T.; Hoffmann, G.W.; Asselta, K.; Brandenburg, J.D.; Butterworth, J.; Camarda, T.; Christie, W.; Crawford, H.J.; et al. The STAR Vertex Position Detector. *Nucl. Instrum. Meth. A* **2014**, *759*, 23–28. [CrossRef]
12. STAR TOF Collaboration. Proposal for a Large Area Time of Flight System for STAR. 2004. Available online: https://drupal.star.bnl.gov/STAR/starnotes/public/sn0621 (accessed on 6 September 2019).
13. Shao, M.; Barannikova, O.; Dong, X.; Fisyak, Y.; Ruan, L.; Sorensen, P.; Xu, Z. Extensive particle identification with TPC and TOF at the STAR experiment. *Nucl. Instrum. Meth. A* **2006**, *558*, 419–429. [CrossRef]
14. TMVA Official. Available online: http://tmva.sourceforge.net (accessed on 23 July 2019).
15. Adam, J.; et al. [STAR Collaboration] Centrality and transverse momentum dependence of D^0-meson production at mid-rapidity in Au+Au collisions at $\sqrt{s_{NN}} = 200$ GeV. *Phys. Rev. C* **2019**, *99*, 034908. [CrossRef]
16. Adamczyk, L.; et al. [STAR Collaboration] Measuremets of D^0 and D^* production in p+p collisions at $\sqrt{s} = 200$ GeV. *Phys. Rev. D* **2012**, *86*, 072013. [CrossRef]

17. Adamczyk, L.; et al. [STAR Collaboration] Measurement of D^0 azimuthal anizothropy at midrapidity in Au+Au collisions at $\sqrt{s_{NN}} = 200$ GeV. *Phys. Rev. Lett.* **2017**, *118*, 212301. [CrossRef] [PubMed]
18. Abelev, B.I.; et al. [STAR Collaboration] Centrality dependence of charged hadron and strange hadron elliptic flow from $\sqrt{s_{NN}} = 200$ GeV Au+Au collisions. *Phys. Rev. C* **2008**, *77*, 054901. [CrossRef]
19. Andronic, A.; Braun-Munzinger, P.; Redlich, K.; Stachel, J. Statistical hadronization of charm in heavy-ion collisions at SPS, RHIC and LHC. *Phys. Lett. B* **2003**, *571*, 36–44. [CrossRef]
20. Acharya, S.; et al. [The ALICE Collaboration] Λ_c^+ production in pp collisions at $\sqrt{s} = 7$ TeV and in p-Pb collisions at $\sqrt{s_{NN}} = 5.02$ TeV. *J. High Energ. Phys.* **2018**, *2018*, 108. [CrossRef]
21. Lisovyi, M.; Verbytskyi, A.; Zenaiev, O. Combined analysis of charm-quark fragmentation-function measurements. *Eur. Phys. J. C* **2017**, *76*, 397. [CrossRef]
22. He, M.; Fries, R.J.; Rapp, R. D_s Meson as a quantitative probe of diffusion and hadronization in nuclear collisions. *Phys. Rev. Lett.* **2013**, *110*, 112301. [CrossRef]
23. Chatterjee, S.; Bozek, P. Interplay of drag by hot matter and electromagnetic force on the directed flow of heavy quarks. *arXiv* **2018**, arXiv:1804.04893.
24. Adamczyk, L.; et al. [STAR Collaboration] Beam-energy dependence of directed flow of Λ, $\overline{\Lambda}$, $K^{\pm}$, K_s^0, and ϕ in Au+Au collisions. *Phys. Rev. Lett.* **2018**, *120*, 062301. [CrossRef] [PubMed]

© 2019 by the author. Licensee MDPI, Basel, Switzerland. This article is an open access article distributed under the terms and conditions of the Creative Commons Attribution (CC BY) license (http://creativecommons.org/licenses/by/4.0/).

Article

A Particle Emitting Source From an Accelerating, Perturbative Solution of Relativistic Hydrodynamics

Bálint Kurgyis * and Máté Csanád

Department of Atomic Physics, Eötvös Loránd University, Pázmány P. s. 1/A, H-1117 Budapest, Hungary
* Correspondence: kurgyisb@caesar.elte.hu

Received: 31 July 2019; Accepted: 3 September 2019; Published: 4 September 2019

Abstract: The quark gluon plasma is formed in heavy-ion collisions, and it can be described by solutions of relativistic hydrodynamics. In this paper we utilize perturbative hydrodynamics, where we study first order perturbations on top of a known solution. We investigate the perturbations on top of the Hubble flow. From this perturbative solution we can give the form of the particle emitting source and calculate observables of heavy-ion collisions. We describe the source function and the single-particle momentum spectra for a spherically symmetric solution.

Keywords: hydrodynamics; heavy-ion collisions; Hubble-flow; acceleration

1. Introduction

Our aim is to study the role of acceleration in heavy-ion collisions under an analytic framework. There are many numerical simulations to solve the equations of relativistic hydrodynamics. However, the analytic solutions are also important in understanding the connection between the initial and final state of the matter. The equations of relativistic hydrodynamics can be treated perturbatively to generalize an already known exact solution. We will utilize the known solution Hubble-flow [1] and a perturbative solution, which includes a pressure gradient and acceleration as perturbations on top of the original solution and was given in [2]. From this perturbative solution we can calculate the source function and study the role of the parameters and compare the observables to the ones calculated from the exact solution [3].

2. General Equations

We are using the equations of relativistic perfect fluid hydrodynamics. This can be formulated as the following:

$$\partial_\mu T^{\mu\nu} = 0, \qquad (1)$$

where $T^{\mu\nu}$ is the energy-momentum tensor, which can be expressed with the four-velocity u^μ, pressure p and energy density ϵ; and is the following for perfect fluids:

$$T^{\mu\nu} = (\epsilon + p)u^\mu u^\nu - pg^{\mu\nu}. \qquad (2)$$

We denote the Minkowskian metric tensor by $g^{\mu\nu} = \mathrm{diag}(1, -1, -1, -1)$, and we use $c = 1$ notation. In addition, we use a simple equation of state (EoS), where energy density is proportional to pressure, and κ is constant:

$$\epsilon = \kappa p. \qquad (3)$$

With this EoS the equations of relativistic hydrodynamics can be separated into the following Euler equation and energy equation:

$$\kappa u^\mu \partial_\mu p + (\kappa+1)p\partial_\mu u^\mu = 0, \tag{4}$$
$$(\kappa+1)pu^\mu \partial_\mu u^\nu = (g^{\mu\nu} - u^\mu u^\nu)\partial_\mu p. \tag{5}$$

Finally, we assume that there is a conserved charge density (n), therefore we can formulate a continuity equation for this conserved quantity:

$$\partial_\mu(u^\mu n) = 0. \tag{6}$$

3. Hubble-Flow and Its Perturbations

There are several analytic solutions for the equations of relativistic hydrodynamics. In this paper we investigate the perturbations on top of the Hubble flow.

3.1. Hubble-Flow

The relativistic Hubble-flow is a 1+3D solution without acceleration or pressure gradient [1]. It describes a self-similar expansion. The solution has the following form:

$$u^\mu = \frac{x^\mu}{\tau}, \tag{7}$$
$$n = n_0 \left(\frac{\tau_0}{\tau}\right)^3 \mathcal{N}(S), \tag{8}$$
$$p = p_0 \left(\frac{\tau_0}{\tau}\right)^{3+\frac{3}{\kappa}}. \tag{9}$$

Here we denote the proper time by $\tau = \sqrt{x_\mu x^\mu}$. The self-similarity of the solution is ensured through the scale parameter S:

$$u_\mu \partial^\mu S = 0. \tag{10}$$

3.2. Perturbations on Top of the Hubble-Flow

There are different generalizations of the above mentioned Hubble-flow [4,5]. Next, we would like to include acceleration and a pressure gradient as perturbations. A set of solutions for the first order perturbations on top of the original solution was given in [2]:

$$\delta u^\mu = \delta \cdot F(\tau)g(x_\mu)\partial^\mu S \chi(S), \tag{11}$$
$$\delta p = \delta \cdot p_0 \left(\frac{\tau_0}{\tau}\right)^{3+\frac{3}{\kappa}} \pi(S), \tag{12}$$
$$\delta n = \delta \cdot n_0 \left(\frac{\tau_0}{\tau}\right)^3 h(x_\mu)\nu(S). \tag{13}$$

This is a solution if the following conditions for the functions of the scale parameter and the newly introduced h, F, g functions are satisfied:

$$\frac{\chi'(S)}{\chi(S)} = -\frac{\partial_\mu \partial^\mu S}{\partial_\mu S \partial^\mu S} - \frac{\partial_\mu S \partial^\mu \ln g(x_\mu)}{\partial_\mu S \partial^\mu S}, \tag{14}$$

$$\frac{\pi'(S)}{\chi(S)} = (\kappa + 1)\left[F(\tau)\left(u^\mu \partial_\mu g - \frac{3g(x_\mu)}{\kappa \tau}\right) + F'(\tau)g(x_\mu)\right], \tag{15}$$

$$\frac{\nu(S)}{\chi(S)\mathcal{N}'(S)} = -\frac{F(\tau)g(x_\mu)\partial_\mu S \partial^\mu S}{u^\mu \partial_\mu h(x_\mu)}. \tag{16}$$

3.3. A Concrete Solution

For further studies we chose a simple solution, which is more general than that was investigated in [6]. The scale parameter in this case is:

$$S = r^j/t^j. \tag{17}$$

The perturbations are the following:

$$\delta u^\mu = \delta \cdot \left(\tau + a\tau_0 \left(\frac{\tau}{\tau_0}\right)^{\frac{3}{\kappa}}\right) S^{-\frac{j+1}{j}} \partial^\mu S, \tag{18}$$

$$\delta p = \delta \cdot p_0 \left(\frac{\tau_0}{\tau}\right)^{3+\frac{3}{\kappa}} \frac{(\kappa+1)(\kappa-3)}{\kappa} j S^{-\frac{1}{j}}, \tag{19}$$

$$\delta n = \delta \cdot n_0 \left(\frac{\tau_0}{\tau}\right)^3 \left(\ln\left(\frac{\tau}{\tau_0}\right) + a\frac{\kappa}{3-\kappa}\left(\frac{\tau}{\tau_0}\right)^{\frac{3}{\kappa}-1}\right) j^2 S^{\frac{j-1}{j}} \left(S^{\frac{2}{j}} - 1\right)\left(1 - S^{-\frac{2}{j}}\right) \mathcal{N}'(S). \tag{20}$$

For the scale function of the original charge density we chose a Gaussian shape:

$$\mathcal{N}(S) = e^{-\frac{br^2}{R_0^2 t^2}} = e^{-\frac{b}{R_0^2} S^{2/j}}. \tag{21}$$

This solution has the free parameters τ_0, n_0, p_0, κ and b which are the same as in the original Hubble-flow. In addition to this, for the perturbations there are three new parameters: the perturbation parameter δ, a dimensionless parameter a and the exponent of scale parameter j.

4. Calculation of Observables

In heavy-ion collisions, the velocity field, pressure and energy density can not be measured directly. Let us now investigate the quantities that can be measured in heavy-ion collisions and calculated from hydrodynamical solutions. For this we assume that the particles come from a thermalized medium of quark-gluon plasma and this can be characterized by a source which comes from a relativistic Jüttner-distribution similarly as in [3,6]. Also, we assume a constant freeze-out hypersurface in proper time at τ_0. The temperature of the system is defined through the following equation: $p = nT$. For the pertubative handling we will have to calculate the first order perturbation of this source function.

For the most general set of perturbations of the Hubble-flow described in Equations (11) and (13) the source function has the following form:

$$S(x,p) = N\delta(\tau - \tau_0) \mathrm{d}\tau \mathrm{d}^3 x n_0 \left(\frac{\tau_0}{\tau}\right)^3 \mathcal{N}(S) \exp\left[-\frac{p_\mu u^\mu}{T_0 \left(\frac{\tau_0}{\tau}\right)^{\frac{3}{\kappa}}} \mathcal{N}(S)\right] \left(\frac{\tau p_\mu u^\mu}{t}\right) \cdot$$

$$\cdot \left[1 + \delta\left(-\frac{F(\tau)g(x_\mu)\partial^0 S\chi(S)\tau}{t} + \frac{F(\tau)g(x_\mu)\chi(S)t}{\tau p_\mu u^\mu} p_\mu \partial^\mu S + \frac{F(\tau)g(x_\mu)\chi(S)}{T_0 \left(\frac{\tau_0}{\tau}\right)^{\frac{3}{\kappa}}}\right.\right.$$

$$\left.\left. + \frac{(p_\mu u^\mu)(\mathcal{N}(S)\pi(S) - h(x_\mu)\nu(S))}{T_0 \left(\frac{\tau_0}{\tau}\right)^{\frac{3}{\kappa}}} + \frac{h(x_\mu)\nu(S)}{\mathcal{N}(S)}\right)\right],$$

where N is a normalization factor, and p_μ is the four-momentum of the outgoing particles. For further studies we use a Gaussian approximation of the source. This means that we write the source as the product of a Gaussian peak and some other terms. By performing the proper time integral we can study the spatial dependence of the source. In the case of the concrete solution described in Section 3.3 the source becomes a two component Gaussian:

$$S(x,p)\mathrm{d}^3 x = I_1 + I_2, \text{ where} \tag{22}$$

$$I_1 = Nn_0 \zeta^{(1)} f_0 \left(1 + \epsilon_1 + \epsilon_2 + \epsilon_3\right) \mathrm{d}^3 x, \tag{23}$$

$$I_2 = Nn_0 \zeta^{(2)} f_0 \left(\epsilon_4 + \epsilon_5\right) \mathrm{d}^3 x. \tag{24}$$

With ϵ_i corresponding to the perturbative terms:

$$\epsilon_1 = \delta j \frac{2ab\kappa \tau_0^4}{(\kappa - 3)\dot{R}_0^2 r(\tau_0^2 + r^2)^{3/2}}, \tag{25}$$

$$\epsilon_2 = \delta j \frac{(1+a)\tau_0^2}{r(\tau_0^2 + r^2)^{1/2}}, \tag{26}$$

$$\epsilon_3 = \delta j \frac{(1+a)\tau_0^2 \left((p_x x + p_y y + p_z z)(\tau_0^2 + r^2)^{1/2} - r^2 E\right)}{r^3 \left(E(\tau_0^2 + r^2)^{1/2} - p_x x - p_y y - p_z z\right)}, \tag{27}$$

$$\epsilon_4 = \delta j \frac{(1+a)\tau_0 \left(r^2 E - (p_x x + p_y y + p_z z)(\tau_0^2 + r^2)^{1/2}\right)}{T_0 r^3}, \tag{28}$$

$$\epsilon_5 = \delta j \frac{\left(E(\tau_0^2 + r^2)^{1/2} - p_x x - p_y y - p_z z\right)\left(2ab\kappa^2 \tau_0^2 + \dot{R}_0^2(3-\kappa)^2(\kappa+1)(\tau_0^2 + r^2)^2\right)}{\tau_0 T_0 \dot{R}_0^2 \kappa (3-\kappa) r(\tau_0^2 + r^2)^{3/2}}, \tag{29}$$

with r being the radial distance $r = \sqrt{x^2 + y^2 + z^2}$ and f_0 being the following function:

$$f_0 = \frac{E\sqrt{\tau_0^2 + r^2} - p_x x - p_y y - p_z z}{\sqrt{\tau_0^2 + r^2}}. \tag{30}$$

The $\zeta^{(1)}$, $\zeta^{(2)}$ have the following form in the Gaussian-approximation:

$$\zeta^{(1)} = \exp\left[-\frac{E^2+m^2}{2ET_0} - \frac{p^2}{2ET_{\text{eff}}}\right] \exp\left[-\frac{\left(x-x_s^{(1)}\right)^2}{2R^2} - \frac{\left(y-y_s^{(1)}\right)^2}{2R^2} - \frac{\left(z-z_s^{(1)}\right)^2}{2R^2}\right], \quad (31)$$

$$\zeta^{(2)} = \exp\left[-\frac{E^2+m^2}{2ET_0} - \frac{p^2}{2ET_\delta}\right] \exp\left[-\frac{\left(x-x_s^{(2)}\right)^2}{2R_\delta^2} - \frac{\left(y-y_s^{(2)}\right)^2}{2R_\delta^2} - \frac{\left(z-z_s^{(2)}\right)^2}{2R_\delta^2}\right]. \quad (32)$$

Here, the R and R_δ describe the widths of these Gaussian parts of the source. A visualization of the source can be seen in Figure 1. We can see, that the $\zeta^{(2)}$ term, which has the width R_δ gives a negative contribution to the source with the chosen set of parameters, however the sign of the perturbative peak depends on the choice of parameters and could yield a positive gain.

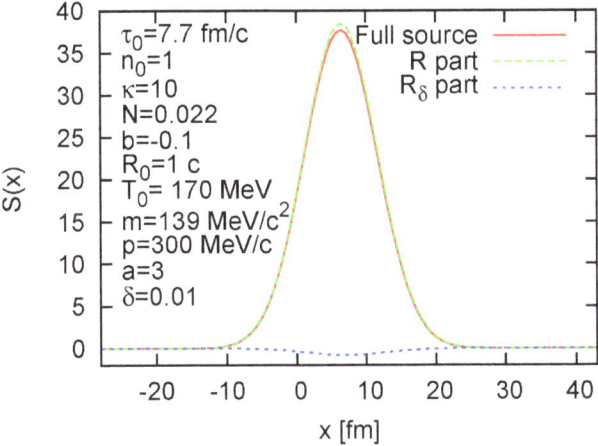

Figure 1. The two component Gaussian source at a given set of parameters denoted on the label.

Furthermore, T_{eff} and T_δ are effective temperatures, corresponding to the inverse logarithmic slope of the Maxwell–Boltzmann like distributions. R and T_{eff} are the same as in the original Hubble-flow, while R_δ and T_δ give the perturbative corrections to the Gaussian width and the effective temperature. The newly introduced notations are the following:

$$T_{\text{eff}} = T_0 + \frac{T_0 E \dot{R}_0^2}{2b(T_0 - E)}, \qquad T_\delta = T_0 + \frac{T_0 E \dot{R}_0^2}{2b(2T_0 - E)}, \quad (33)$$

$$R^2 = \frac{T_0 \tau_0^2 (T_{\text{eff}} - T_0)}{E T_{\text{eff}}}, \qquad R_\delta^2 = \frac{T_0 \tau_0^2 (T_\delta - T_0)}{E T_\delta}, \quad (34)$$

$$x_s^{(1)} = \frac{p_x \tau_0 (T_{\text{eff}} - T_0)}{E T_{\text{eff}}}, \qquad x_s^{(2)} = \frac{p_x \tau_0 (T_\delta - T_0)}{E T_\delta}, \quad (35)$$

$$y_s^{(1)} = \frac{p_y \tau_0 (T_{\text{eff}} - T_0)}{E T_{\text{eff}}}, \qquad y_s^{(2)} = \frac{p_y \tau_0 (T_\delta - T_0)}{E T_\delta}, \quad (36)$$

$$z_s^{(1)} = \frac{p_z \tau_0 (T_{\text{eff}} - T_0)}{E T_{\text{eff}}}, \qquad z_s^{(2)} = \frac{p_z \tau_0 (T_\delta - T_0)}{E T_\delta}. \quad (37)$$

From the source function, the single-particle momentum distribution can be calculated:

$$N_1(p) = \int \mathrm{d}^4 x S(x,p). \tag{38}$$

To perform this integral analytically we use the Gaussian saddlepoint approximation. In general, we have the integrand in the form of $f(x)g(x)$, where $f(x)$ is slowly changing, and $g(x)$ has a sharp, unique peak at x_0:

$$\int f(x)g(x) = f(x_0)g(x_0)\sqrt{\frac{2\pi}{-(\ln(g(x_0)))''}}. \tag{39}$$

From this we can easily get the final form of the single-particle momentum distribution:

$$N(p) = Nn_0 \mathcal{E}_1 \mathcal{V}_1 (1 + \mathcal{P}_1 + \mathcal{P}_2 + \mathcal{P}_3) + Nn_0 \mathcal{E}_2 \mathcal{V}_2 (\mathcal{P}_4 + \mathcal{P}_5). \tag{40}$$

Here, we introduced the following functions:

$$\mathcal{E}_{1,2} = \exp\left[-\frac{E^2 + m^2}{2ET_0} - \frac{p^2}{2ET_{\text{eff},\delta}}\right], \tag{41}$$

$$\mathcal{V}_{1,2} = \sqrt{\frac{2\pi T_0 \tau_0^2}{E}\left(1 - \frac{T_0}{T_{\text{eff},\delta}}\right)^{-3}\left(E - \frac{p^2}{E}\left(1 - \frac{T_0}{T_{\text{eff},\delta}}\right)\right)}. \tag{42}$$

The terms which come from the first order perturbations are denoted with $\mathcal{P}_i$ and are of the following form in this concrete case of the solution with a saddlepoint approximation:

$$\mathcal{P}_i = \begin{cases} \epsilon_i(x = x_s^{(1)}, y = y_s^{(1)}, z = z_s^{(1)}), \text{if } i = 1,2,3, \\ \epsilon_i(x = x_s^{(2)}, y = y_s^{(2)}, z = z_s^{(2)}), \text{if } i = 4,5. \end{cases} \tag{43}$$

Looking at the final form of the momentum distribution we can see that it is spherically symmetric as we have expected from the spherically symmetric solution.

5. Discussion

To understand the role of perturbations on top of the original Hubble-flow we can plot the calculated quantities with given values of parameters. For this we use model parameters of the Hubble-flow from [3] where quantities calculated from the exact solution were fitted to the experimental data. With these parameter values we can study the role of acceleration in this concrete solution and the role of the a, δ and j parameters. We can see from Equations (25) and (29) that the source and the invariant momentum distribution does not depend separately on δ or j, but on their product δj. Also, the form of scale parameter does not affect the observables directly, therefore, we can not study the role of these parameters independently: Their product defines the scale of the perturbations. In Figure 2 we can see the ratio of the original and the perturbated transverse momentum distributions at different values of the a and δj parameters with the Gaussian saddlepoint approximation. It can be seen that with this approach, the perturbations only give small corrections to the low momentum region of the single particle momentum distribution.

However, the saddlepoint approximation might not give back all the properties of the perturbation, as it assumes that the function that multiplies the Gaussian peak is slowly changing. In our case we can see from Equations (25) and (29) that we have terms proportional to τ_0/r that might influence the result, as $r/\tau_0 \ll 1$. Therefore we could make a Laurent-expansion of the terms ϵ_i; as it turns out the series is finite in the negative region with all the terms vanishing below $(r/\tau_0)^{-2}$, which indicates that all the terms are integrable. This approach gives rise to rather complicated integrals and we will not discuss this method further, we simply wanted to note the possibility of such a calculation in the future. For this type of calculation, it is however sufficient to use the saddlepoint calculation, as it provides a good approximation of the results if the requirement $T_0/T_{\text{eff},\delta} \approx 1$ is met, but $p/E \ll 1$ is not.

Let us now turn to study the geometry of the particle emitting source. From femtoscopic measurements, the homogeneity region of the source can be mapped out. The first intensity correlation measuruments were carried out by R. Hanbury Brown and R. Q. Twiss, thus these are often called HBT measurements [7]. The size of the source can be characterized by the HBT-radii, which are often associated with the Gaussian widths of the source [8,9]. However, let us note here that there are more general approaches to characterize the source [10,11]. In this paper, we have used a Gaussian approximation for the analytic calculations, therefore we can associate the Gaussian width of the source with the HBT-radius of the studied, spherically symmetric system. The source is the sum of two terms with different widths. This gives us two different HBT-radii, R and R_δ, where R is the same as it is for the exact solution [3]. The HBT-radius of such a source is some average of the radii R and R_δ.

The values of R and R_δ do not depend on the perturbation parameters δ, j and a, but their averaging does depend on the choice of these. For such model parameters as used for Figure 2 the average HBT-radius is approximately the same as the original R, and only for large δ and a values do we get a significant contribution from R_δ. We can look at the HBT-radius as the function of the transverse mass: $m_t = \sqrt{m^2 + p_t^2}$. Experimentally the HBT-radii usually show a scaling, regardless of particle species, collision energy or centrality [8,9]. The cause of this scaling is the hydrodynamical expansion both in the longitudinal and the radial directions [12]. We can see the $R \propto 1/\sqrt{m_T}$ scaling in Figure 3 as it was already shown in [6].

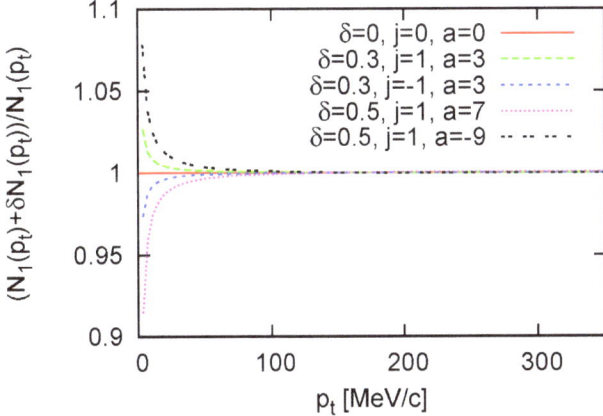

Figure 2. The ratio of the original and the perturbatively corrected single-particle transverse momentum distribution for the investigated solution. The model parameters of the original Hubble-flow come from fits to experimental data [3].

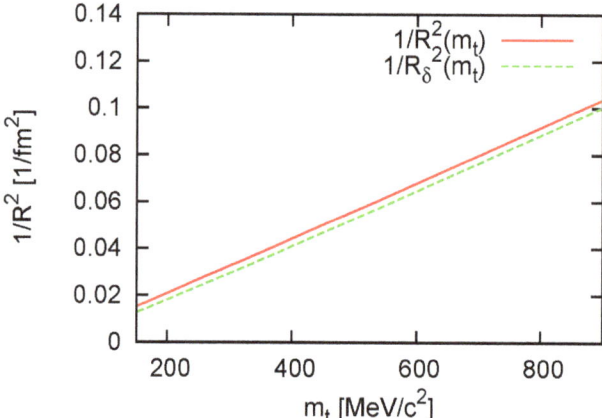

Figure 3. We can see the transverse mass scaling of the calculated HBT-radii, which is usually observed in experimental data.

6. Summary

We have given the perturbated source function for the perturbative, accelerating generalization of the exact Hubble-flow, and calculated the single-particle momentum distribution and the HBT-radius for a spherically symmetric solution. This way the solution includes the acceleration and pressure gradient. For the observables we have found that the perturbations cause only small deviations from the original quantities in the Gaussian saddlepoint approximation. Also, we have seen that the source is a sum of two Gaussians with different widths. Furthermore, we have found that the choice of scale parameter does not affect the calculated observables directly, but results only in a difference in the perturbation scale. For further studies, the elliptical flow could be also calculated, but in a non-spherically symmetric case.

Author Contributions: Conceptualization: M.C.; methodology: M.C.; formal analysis:, B.K. and M.C.; investigation, B.K. and M.C.; writing: M.C. and B.K.; visualization: B.K.; supervision: M.C.

Funding: The authors were supported by Hungarian NKIFH grant No. FK-123842 and the UNKP New National Excellence Program of the Hungarian Ministry of Human Capacities. M. Csanád was supported by the J. Bolyai Research Scholarship of the Hungarian Academy of Sciences.

Conflicts of Interest: The authors declare no conflict of interest.

References

1. Csörgő, T.; Csernai, L.P.; Hama, Y.; Kodama, T. Simple solutions of relativistic hydrodynamics for systems with ellipsoidal symmetry. *Acta Phys. Hung. A* **2004**, *21*, 73–84. [CrossRef]
2. Kurgyis, B.; Csanád, M. Perturbative accelerating solutions of relativistic hydrodynamics. *Universe* **2017**, *3*, 84. [CrossRef]
3. Csanad, M.; Vargyas, M. Observables from a solution of 1 + 3 dimensional relativistic hydrodynamics. *Eur. Phys. J. A* **2010**, *44*, 473–478. [CrossRef]
4. Csanad, M.; Szabo, A. Multipole solution of hydrodynamics and higher order harmonics. *Phys. Rev. C* **2014**, *90*, 054911. [CrossRef]
5. Csanád, M.; Nagy, M.I.; Lökös, S. Exact solutions of relativistic perfect fluid hydrodynamics for a QCD equation of state. *Eur. Phys. J. A* **2012**, *48*, 173. [CrossRef]
6. Kurgyis, B.; Csanád, M. Observables from a perturbative, accelerating solution of relativistic hydrodynamics. arXiv **2019**, arXiv:1810.05402 [hep-th].

7. Hanbury Brown, R.; Twiss, R.Q. A test of a new type of stellar interferometer on Sirius. *Nature* **1956**, *178*, 1046. [CrossRef]
8. Adler, S.S.; Afanasiev, S.; Aidala, C.; Ajitanand, N.N.; Akiba, Y.; Alexander, J.; Amirikas, R.; Aphecetche, L.; Aronson, S.H.; Averbeck, R.; et al. [PHENIX Collaboration], Bose-Einstein correlations of charged pion pairs in Au + Au collisions at $\sqrt{s_{NN}} = 200$ GeV. *Phys. Rev. Lett.* **2004**, *93*, 152302. [CrossRef] [PubMed]
9. Afanasiev, S.; Aidala, C.; Ajitanand, N.N.; Akiba, Y.; Alexander, J.; Al-Jamel, A.; Aoki, K.; Aphecetche, L.; Armendariz, R.; Aronson, S.H.; et al. [PHENIX Collaboration], Kaon interferometric probes of space-time evolution in Au+Au collisions at $\sqrt{s_{NN}} = 200$ GeV. *Phys. Rev. Lett.* **2009**, *103*, 142301. [CrossRef] [PubMed]
10. Adare, A.; Aidala, C.; Ajitanand, N.N.; Akiba, Y.; Akimoto, R.; Alexander, J.; Alfred, M.; Al-Ta'ani, H.; Angerami, A.; Aoki, K.; et al. [PHENIX Collaboration], Lévy-stable two-pion Bose-Einstein correlations in $\sqrt{s_{NN}} = 200$ GeV Au+Au collisions. *Phys. Rev. C* **2018**, *97*, 064911. [CrossRef]
11. Csörgő, T.; Hegyi, S.; Zajc, W.A. Bose-Einstein correlations for Levy stable source distributions. *Eur. Phys. J. C* **2004**, *36*, 67–78. [CrossRef]
12. Csörgő, T.; Lörstad, B. Bose-Einstein correlations for three-dimensionally expanding, cylindrically symmetric, finite systems. *Phys. Rev. C* **1996**, *54*, 1390. [CrossRef] [PubMed]

© 2019 by the authors. Licensee MDPI, Basel, Switzerland. This article is an open access article distributed under the terms and conditions of the Creative Commons Attribution (CC BY) license (http://creativecommons.org/licenses/by/4.0/).

Communication

Compact Star Properties from an Extended Linear Sigma Model

János Takátsy [1,*], Péter Kovács [2], Zsolt Szép [3] and György Wolf [2]

1. Institute of Physics, ELTE Eötvös University, H-1117 Budapest, Hungary
2. Institute for Particle and Nuclear Physics, Wigner Research Centre for Physics, Hungarian Academy of Sciences, H-1525 Budapest, Hungary
3. MTA-ELTE Theoretical Physics Research Group, H-1117 Budapest, Hungary
* Correspondence: takatsyj@caesar.elte.hu
† This paper is based on the talk at the 18th Zimányi School, Budapest, Hungary, 3–7 December 2018.

Received: 2 July 2019; Accepted: 13 July 2019; Published: 15 July 2019

Abstract: The equation of state provided by effective models of strongly interacting matter should comply with the restrictions imposed by current astrophysical observations of compact stars. Using the equation of state given by the (axial-)vector meson extended linear sigma model, we determine the mass–radius relation and study whether these restrictions are satisfied under the assumption that most of the star is filled with quark matter. We also compare the mass–radius sequence with those given by the equations of state of somewhat simpler models.

Keywords: dense baryonic matter; TOV equations; mass–radius relations; chiral effective models

1. Introduction

A lot of theoretical and experimental effort is devoted to study the strong interaction under extreme conditions. The experiments ALICE [1] at CERN, and PHENIX [2] and STAR [3] at RHIC explored the strongly interacting matter at low density and high temperature. In this region the situation is also satisfactory on the theoretical side; however, lattice calculations applicable at low density cannot yet be used at high densities [4]. Hence, effective models are needed in the high density region where the existing experimental data (NA61 [5] at CERN, BES/STAR [3] at RHIC) are scarce and have rather bad statistics. Soon to be finished experimental facilities (NICA [6] at JINR and CBM [7] at FAIR) are designed to explore this region more precisely.

For studying the cold, high density matter, new experimental information emerged in the past decade in a region of the phase diagram that is inaccessible to terrestrial experiments: The properties of neutron stars [8–10]. Since the Tolman–Oppenheimer–Volkoff (TOV) equations [11,12] provide a direct relation between the equation of state (EoS) of the compact star matter and the mass–radius (M–R) relation of the compact star, these data can help to select those effective models, used to describe the strongly interacting matter, whose predictions are consistent with compact star observables. For example, the EoS must support the existence of a two-solar-mass neutron star [13,14]. For the radius, we have less stringent constraints. Bayesian analyses provide some window for the probable values for compact star radii [15–19]. Based on these studies, in this paper we are adopting a radius window of 11.0–12.5 km for compact stars with masses of 2 $M_\odot$. The NICER experiment [9,20] will provide very precise data on the masses and radii of neutron stars simultaneously.

Based on the above considerations, we investigate mass–radius sequences given by the EoS obtained in [21] from the $N_f = 2 + 1$ flavor extended linear sigma model introduced in [22]. The model used here should be undoubtedly regarded in this context as a very crude approximation and the present work has to be considered only as our first attempt to study the problem. This is because the model, which is built on the chiral symmetry of QCD, contains constituent quarks and therefore does not describe a realistic nuclear matter which is expected to form the crust of the compact star. The model is only applicable to some extent under the assumptions that at high densities nucleons dissolve into a sea of quarks and a large part of the compact star is in that state. In other words we investigate here a quark star instead of a neutron or a hybrid star, which would be more realistic. The study in [23] showed that a pure quark star of mass $\sim 2\, M_\odot$ can be achieved in a mean-field treatment of the $N_f = 2 + 1$ linear sigma model if the Yukawa coupling between vector and quark fields is large enough. In the two flavor Nambu–Jona-Lasinio model, inclusion of eight-order quark interaction in the vector coupling channel also resulted in the stiffening of the quark equation of state [24,25]. Recently, the existence of quark-matter cores inside compact stars was investigated also in [15,26]. It was found in [15] within a hybrid star model—in which the quark core was described with a three-flavor Polyakov–Nambu–Jona-Lasinio model—that the current astrophysical constraints can be fulfilled provided the vector interaction is strong enough. While in [26] it is claimed that the existence of quark cores in case of EoSs permitted by observational constraints is a common feature and should not be regarded as a peculiarity.

The paper is organized as follows. In Section 2 we present the model and discuss how its solution, obtained in [21], which reproduced quite well some thermodynamic quantities measured on the lattice, can be used in the presence of a vector meson introduced here to realize the short-range repulsive interaction between quarks in the simplest possible way. In Section 3 we compare our results for the EoS and the M–R relation (star sequences), obtained in the extended linear sigma model (eLSM) for various values of the vector coupling g_v, to results obtained in the two-flavor Walecka model and the three-flavor non-interacting quark model. We draw the conclusions in Section 4 and discuss possible ways to improve the treatment of the model.

2. Methods

The model used in this paper is an $N_f = 2 + 1$ flavor (axial-)vector meson extended linear sigma model (eLSM). The Lagrangian and the detailed description of this model, in which, in addition to the full nonets of (pseudo)scalar mesons, the nonets of (axial-)vector mesons are also included, can be found in [21,22]. The model contains three flavors of constituent quarks, with kinetic terms and Yukawa-type interactions with the (pseudo)scalar mesons. An explicit symmetry breaking of the mesonic potentials is realized by external fields, which results in two scalar expectation values, ϕ_N and ϕ_S.[1]

Compared to [21], the only modification to the model is that we include in the Lagrangian a Yukawa term $-g_v \sqrt{6} \bar{\Psi} \gamma_\mu V_0^\mu \Psi$, which couples the quark field $\Psi^T = (u, d, s)$ to the $U_V(1)$ symmetric vector field, that is $V_0^\mu = \frac{1}{\sqrt{6}} \mathrm{diag}(v_0 + \frac{v_8}{\sqrt{2}}, v_0 + \frac{v_8}{\sqrt{2}}, v_0 - \sqrt{2} v_8)^\mu$. The vector meson field is treated at the mean-field level as in the Walecka model [27], but as a simplification we assign a nonzero expectation value only to v_0^0: $v_0^\mu \to v_0 \delta^{0\mu}$ and $v_8^\mu \to 0$. While this assignment is not physical, in this way the chemical potentials of all three quarks are shifted by the same amount, allowing us to use, as shown below, the result obtained in [21]. With the parameters used in [21], the mass of the vector meson v_0^μ turns out to be $m_v = 871.9$ MeV.

Since a compact star is relatively cold ($T \approx 0.1$ keV), we work at $T = 0$ MeV using the approximation employed in [21]. We have three background fields, ϕ_N, ϕ_S and v_0, and the calculation of the grand potential, Ω, is performed using a mean-field approximation, in which fermionic fluctuations are included at one-loop order, while the mesons are treated at tree-level. Hence, the grand potential can be written in the following form

$$\Omega(\mu_q; \phi_N, \phi_S, v_0) = U_{\mathrm{mes}}(\phi_N, \phi_S) - \frac{1}{2} m_v^2 v_0^2 + \Omega_{q\bar{q}}^{(0)\mathrm{vac}}(\phi_N, \phi_S) + \Omega_{q\bar{q}}^{(0)\mathrm{matter}}(\tilde{\mu}_q; \phi_N, \phi_S), \qquad (1)$$

---

[1] N and S denote the the non-strange and strange condensates, which are coupled to the 3 × 3 matrices $\lambda_N = (\sqrt{2}\lambda_0 + \lambda_8)/\sqrt{3}$ and $\lambda_S = (\lambda_0 - \sqrt{2}\lambda_8)/\sqrt{3}$, with λ_8 being the eighth Gell-Mann matrix and $\lambda_0 = \sqrt{2/3} \mathbb{1}$.

where $\tilde{\mu}_q = \mu_q - g_v v_0$ is the effective chemical potential of the quarks, while $\mu_q = \mu_B/3$ is the physical quark chemical potential, with μ_B being the baryochemical potential. On the right-hand side of the grand potential (1), the terms are (from left to right): The tree-level potential of the scalar mesons, the tree-level contribution of the vector meson, the vacuum and the matter part of the fermionic contribution at vanishing mesonic fluctuating fields. The fermionic part is obtained by integrating out the quark fields in the partition function. The vacuum part was renormalized at the scale $M_0 = 351$ MeV. More details on the derivation can be found in [21].

The background fields ϕ_N, ϕ_S, and v_0 are determined from the stationary conditions

$$\left.\frac{\partial\Omega}{\partial\phi_N}\right|_{\phi_N=\bar{\phi}_N} = \left.\frac{\partial\Omega}{\partial\phi_S}\right|_{\phi_S=\bar{\phi}_S} = 0 \quad \text{and} \quad \left.\frac{\partial\Omega}{\partial v_0}\right|_{v_0=\bar{v}_0} = 0, \qquad (2)$$

where the solution is indicated with a bar. Since $\partial/\partial v_0 = -g_v \partial/\partial \tilde{\mu}_q$, the stationary condition with respect to v_0 reads

$$\bar{v}_0(\phi_N, \phi_S) = \frac{g_v}{m_v^2} \rho_q(\tilde{\mu}_q(\bar{v}_0); \phi_N, \phi_S), \qquad (3)$$

where $\rho_q(x; \phi_N, \phi_S) = -\partial\Omega^{(0)\text{matter}}_{q\bar{q}}(x; \phi_N, \phi_S)/\partial x$.

When solving the model, we give values to g_v in the range $[0, 3)$, while for the remaining 14 parameters of the model Lagrangian we use the values given in Table IV of [21]. These values were determined there by calculating constituent quark masses, (pseudo)scalar curvature masses with fermionic contribution included and decay widths at $T = \mu_q = 0$ and comparing them to their experimental PDG values [28]. Parameter fitting was done using a multiparametric χ^2 minimization procedure [29]. In addition to the vacuum quantities, the pseudocritical temperature T_{pc} at $\mu_q = 0$ was also fitted to the corresponding lattice result [30,31]. We mention that the model also contains the Polyakov-loop degrees of freedom (see [21] for details), but to keep the presentation simple we omitted them from Equation (1), as at $T = 0$ they do not contribute to the EoS directly. Their influence is only through the value of the model parameters taken from [21]: Since they modify the Fermi-Dirac distribution function, they influence the value of T_{pc} used for parameterization, as described above. A parameterization based on vacuum quantities alone could lead to unphysically large values of T_{pc} and, compared to the case when T_{pc} is included in the fit, also to different assignments of scalar nonet states to physical particles, that is χ^2 could become minimal for a different particle assignment.

The solution of the model at $g_v = 0$, obtained in [21], can be used to construct the solution at $g_v \neq 0$ (see, e.g., Chapter 2.1 of [32]): One only has to interpret the solution at $g_v = 0$ as a solution obtained at a given $\tilde{\mu}_q$ and determine μ_q at some $g_v \neq 0$ using Equations (3). To see that the solutions $\bar{\phi}_{N,S}$ for $g_v \neq 0$ can be related to the solution obtained at $g_v = 0$, where $\bar{v}_0 = 0$, consider the grand potential at $g_v = 0$. This potential, denoted as Ω_0, is subject to the stationary Conditions (2) with solutions $\bar{\phi}^0_{N,S}(\mu_q)$. It is then easy to see using Equation (1), that the solution $\bar{\phi}_{N,S}(\mu_q)$ of Conditions (2) satisfies $\bar{\phi}_{N,S}(\mu_q + g_v v_0) = \bar{\phi}^0_{N,S}(\mu_q)$ or, changing the variable μ_q to $\tilde{\mu}_q$, the relation becomes

$$\bar{\phi}_{N,S}(\tilde{\mu}_q + g_v v_0) = \bar{\phi}^0_{N,S}(\tilde{\mu}_q). \qquad (4)$$

The value of the grand potential Ω at the extremum can be given in terms of the value of the grand potential with $g_v = 0$, that is Ω_0, at its extremum. With the extrema of $\Omega_0(\tilde{\mu}_q, \phi_N, \phi_S, v_0 = 0)$ as $\bar{\phi}^0_N$ and $\bar{\phi}^0_S$, one has

$$\Omega(\mu_q; \bar{\phi}_N(\mu_q), \bar{\phi}_S(\mu_q), \bar{v}_0) = \Omega_0(\tilde{\mu}_q, \bar{\phi}^0_N(\tilde{\mu}_q), \bar{\phi}^0_S(\tilde{\mu}_q), v_0 = 0) - \frac{1}{2}m_v^2 \bar{v}_0^2, \qquad (5)$$

where $\bar{v}_0 \equiv \bar{v}_0(\bar{\phi}_N(\mu_q), \bar{\phi}_S(\mu_q)) = \frac{g_v}{m_v^2} \rho_q(\tilde{\mu}_q; \bar{\phi}_N(\mu_q), \bar{\phi}_S(\mu_q)) = \frac{g_v}{m_v^2} \rho_q(\tilde{\mu}_q; \bar{\phi}^0_N(\tilde{\mu}_q), \bar{\phi}^0_S(\tilde{\mu}_q))$ and $\mu_q = \tilde{\mu}_q + g_v \bar{v}_0$.

The pressure p and the energy density ε are calculated from the grand potential. At $v_0 \neq 0$ they can be expressed in terms on the pressure obtained at $g_v = 0$

$$\begin{aligned} p(\mu_q) &= \Omega(\mu_q = 0; \bar{\phi}_N(0), \bar{\phi}_S(0), \bar{v}_0(0)) - \Omega(\mu_q; \bar{\phi}_N, \bar{\phi}_S, \bar{v}_0) \\ &= \Omega_0(\tilde{\mu}_q = 0; \bar{\phi}_N^0(0), \bar{\phi}_S^0(0), v_0 = 0) - \Omega_0(\tilde{\mu}_q; \bar{\phi}_N^0, \bar{\phi}_S^0, v_0 = 0) + \frac{1}{2} m_v^2 \bar{v}_0^2 \\ &= p(\tilde{\mu}_q)|_{g_v = 0} + \frac{1}{2} m_v^2 \bar{v}_0^2, \end{aligned} \qquad (6)$$

where $\bar{v}_0 = \frac{g_v}{m_v^2} \rho_q(\tilde{\mu}_q; \bar{\phi}_N^0(\tilde{\mu}_q), \bar{\phi}_S^0(\tilde{\mu}_q))$, and then $\varepsilon = -p + \mu_q \rho_q$, where $\mu_q = \tilde{\mu}_q + g_v \bar{v}_0$.

With the EoS $p(\varepsilon)$ obtained at $T = 0$ and high densities, we determine the mass–radius relation of non-rotating static compact stars by solving the TOV equation [11,12] using a fourth-order Runge-Kutta differential equation integrator with adaptive stepsize control.

3. Results

Since our eLSM model was fitted to the hadron spectrum and not to the nuclear matter, we compare its results with those obtained in two relativistic models generally used in the description of compact stars, in order to assess the importance of various ingredients involved in these models. For comparison we consider the three-flavor non-interacting constituent quark model (see, e.g., [33,34]) and the Walecka model, which in its simplest form contains the proton and neutron, the scalar-isoscalar meson σ and the isoscalar-vector meson ω [27]. The use of the Walecka model for the description of the neutron stars requires charge neutrality, which calls for the introduction of the ρ meson in order to have a proper description of the nuclear symmetry energy [34].

In the non-interacting constituent quark model the masses are fixed to $m_u = m_d = 75$ MeV, $m_s = 365$ MeV, values obtained from our eLSM at the first order chiral phase transition point, that is at $\mu_{q,c} \approx 323$ MeV, where the potential is degenerate. The calculation of the energy density and pressure was done with the bag constant $B^{1/4} = 163$ MeV. Including electrons in the model, the conditions of β-equilibrium and charge neutrality were taken into account.

We use two mean-field versions of the Walecka model, one that includes the effect of the scalar self-interaction through a classical potential with cubic and quartic terms of the form

$$V_{I,\sigma} = \frac{b}{3} m_N (g_\sigma \sigma)^3 + \frac{c}{4} (g_\sigma \sigma)^4, \qquad (7)$$

and a version where the scalar self-interaction is neglected. Using $m_\sigma = 550$ MeV, $m_\omega = 783$ MeV, and $m_\rho = 775.3$ MeV for the mesons and $m_N = 939$ MeV for the nucleon mass; the parameters are fixed from nuclear matter properties: The value $n_0 = 0.153$ fm^{-3} for the saturation density (where $p = 0$), the nuclear binding energy per nucleon $E_0 = (\varepsilon/n_0 - m_N) = -16.3$ MeV, the symmetry energy coefficient, for which we take the value $a_{\text{sym}} = 31.3$ MeV [35], and in the version with scalar self-interactions there is also the compression modulus $K = 250$ MeV and the Landau mass $m_L = 0.83 m_N$. The values of the parameters used here are basically those of [33]: For the value of the Yukawa couplings of the mesons to the nucleons, one has $g_\sigma^2 = 9.5372/(4\pi)$, $g_\omega^2 = 14.717/(4\pi)$, and $g_\rho = 6.8872$ when the scalar self-interaction is neglected, while in the other case $g_\sigma^2 = 6.003/(4\pi)$, $g_\omega^2 = 5.9484/(4\pi)$, $g_\rho = 8.3235$, $b = 7.95 \cdot 10^{-3}$, and $c = 6.947 \cdot 10^{-4}$. For a recent study of the effect of K, m_L and of the form of the scalar potential on the mass–radius relation, we refer the interested reader to [36].

The EoS of the Walecka model subject to the constraints of β-equilibrium and charge neutrality is applicable only to the core of the compact star. A proper phenomenological description requires the modeling of the stellar matter in the crust and of the crust–core transition. In the present work we only implement, using the tabulated data from Table 5.7 of [34], the BPS EoS [37] for the outer crust of a neutron star whose core is described by the EoS of the Walecka model. This is done by simply replacing, at low energy densities, corresponding to densities below the neutron drip line, $\rho_B \leq 0.01$ fm^{-3}, the EoS of the Walecka model with the BPS EoS, as indicated in the right panel of

Figure 1. More sophisticated procedures for core–crust matching are described in [38] together with their influence on the $M(R)$ relation. A realistic description of astrophysical data would require an additional matching to an EoS for the inner crust that applies for densities above the neutron drip density. This is beyond the scope of our present study and we refer the interested reader to a recent review [39] that provides a detailed discussion of the neutron star crust matter and of the EoS of dense neutron star matter. Convenient analytic parameterizations of unified EoSs derived from a single model and describing the crust and the core of the neutron star are given in [40].

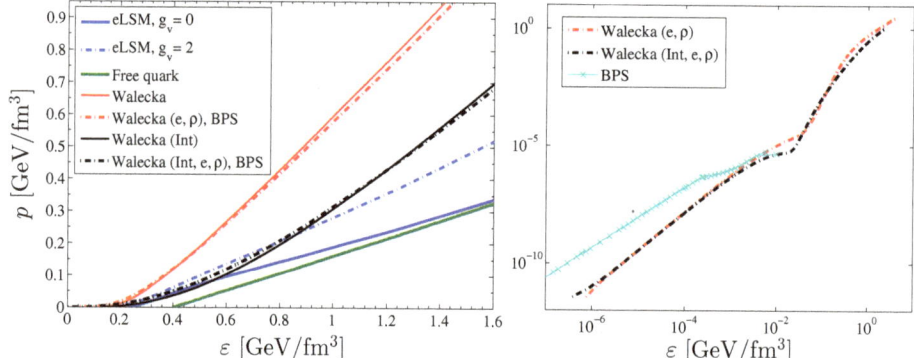

Figure 1. Left panel: The $T = 0$ equation of state (EoS) of the extended linear sigma model (eLSM) (blue solid line for $g_v = 0$ and blue dashed-dotted line for $g_v = 2$) compared to those of the free constituent quark matter with mass values given in the text (green solid line) and of the Walecka model with (black lines) and without (red lines) the scalar self-interaction. For the latter model the dashed–dotted line type indicates that β-equilibrium and charge neutrality are imposed, the ρ meson is included, and that at low energy densities the EoS is replaced by the BPS EoS. Right panel: Matching the EoS of the Walecka model to the BPS EoS (see the text for details). Notice that the consequence of imposing the mentioned compact star constraints (inclusion of electrons and ρ) in the Walecka model is that $p(\varepsilon) > 0$ even at low energy densities.

The zero-temperature EoSs are shown in Figure 1. For the Walecka model, we also consider the case when charge neutrality condition and β-equilibrium with electrons are not imposed and the BPS EoS is not used. We can see in Figure 1 that at small energy densities the pressure in the eLSM with $g_v = 0$ is slightly higher than in the non-interacting quark model (i.e., the EoS is stiffer), but close to the value of the pressure obtained in the Walecka model with scalar self-interaction. This shows that the inclusion of scalar interactions in the Walecka model brings the EoS closer to that of the eLSM, as in case of the Walecka model the higher pressure corresponds to the non-interacting model. At high energy densities the values of the pressure in the eLSM with $g_v = 0$ approach those obtained in the non-interacting quark model. Inclusion of the repulsive interaction between quarks in the eLSM renders the EoS stiffer compared to the $g_v = 0$ case, as expected, and it brings the EoS of the eLSM closer to that obtained in the Walecka model with scalar self-interaction. It is worth noting that relatively small differences in the $p(\varepsilon)$ lead to significant differences in the M–R curves, as we shall see later in Figures 2 and 3.

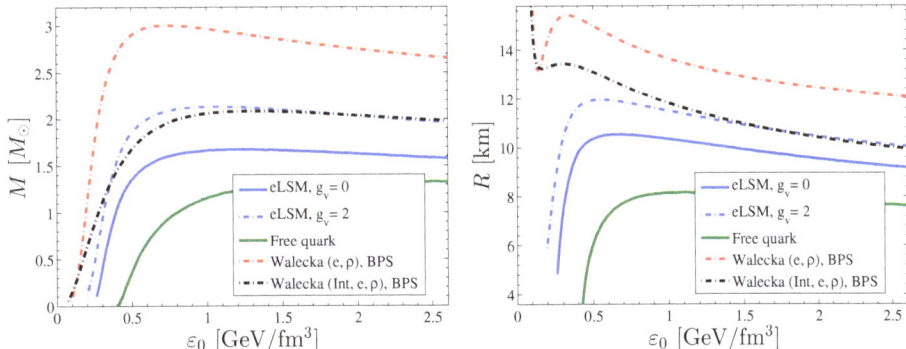

Figure 2. The masses (left panel) and radii (right panel) of the compact stars as functions of the central energy density (ε_0). The line style for the different cases correspond to that of Figure 1.

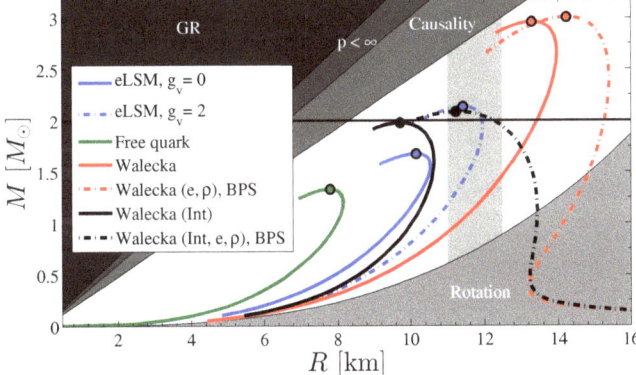

Figure 3. Mass–radius relations for the eLSM (blue solid line for $g_v = 0$ and blue dashed–dotted line for $g_v = 2$), the free constituent quark matter (green solid line), and the Walecka model for the various cases of Figure 1. The ends of the stable sequences of compact stars are marked by blobs. The observational constraint set by observed pulsars with masses of $\sim 2\,M_\odot$ is represented by the black horizontal line, and the applied radius window of 11.0–12.5 km at $2\,M_\odot$ is depicted by the vertical shaded area. The different shaded regions are excluded by the GR constraint $R > 2GM/c^2$, the finite pressure constraint $R > (9/4)GM/c^2$, causality $R > 2.9GM/c^2$, and the rotational constraint based on the 716 Hz pulsar J1748-2446ad, $M/M_\odot > 4.6 \cdot 10^{-4}\,(R/\text{km})^3$ [41]. For a more detailed discussion on these constraints see, e.g., [8].

By solving the TOV equation using a specific EoS, one can obtain the radial dependence of the energy density (and thus of the pressure) for a certain central energy density, ε_0. One can then determine the mass and radius of the compact star for that central energy density. By changing ε_0, one gets a sequence of compact star masses and radii parameterized by the central energy density, as shown in Figure 2 for various models. The sequence of stable compact stars ends when the maximum compact star mass is reached with increasing central energy density.

The mass–radius relations for the four models are shown in Figure 3 together with the physical constraints obtained from observations of binary pulsar systems and X-ray binaries. As expected based on Figure 5.23 of [34], the proper treatment of the neutron star outer crust by the BPS EoS that corresponds to a Coulomb lattice of different nuclei embedded in a gas of electrons has a remarkable influence on both the mass and the radius of the star (see also Figure 4): Without the BPS EoS, the turning point at the smallest radius of that part of the mass–radius diagram which corresponds to large stars with small masses is around 8 km (9 km) in the Walecka model with (without) scalar self-interactions and the minimum mass of the stars with large radii is much smaller. In addition, without the constraints of charge neutrality and β-equilibrium with electrons and without the effect of the ρ meson, even the shape of the $M(R)$ curve obtained in the Walecka model is different.

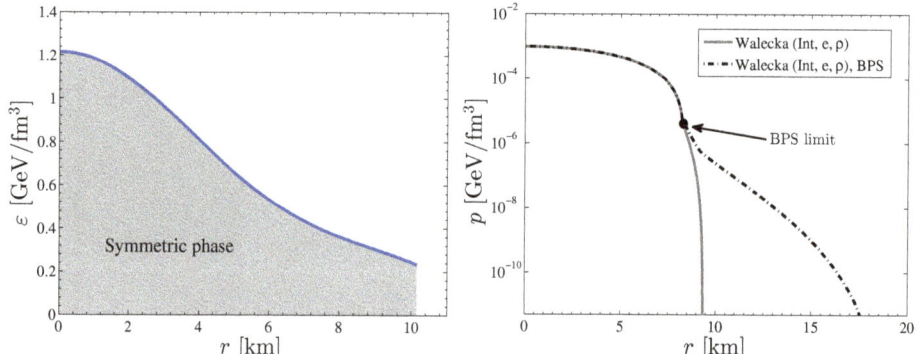

Figure 4. Left panel: The energy density as a function of radial coordinate inside the maximum mass compact star corresponding to the eLSM with $g_v = 0$. The chiral phase transition occurs at the very edge of the star, hence the whole star is basically composed of chirally symmetric quark matter. Right panel: To ilustrate the effect of the BPS EoS in the Walecka model, we show the pressure as a function of the radial distance for a central energy density of $\varepsilon_0 = 7 \cdot 10^8$ MeV^4.

As it can be seen in Figure 3, the maximum possible compact star mass is lower for models with less stiff EoSs. The star sequences corresponding to the non-interacting quark model, the Walecka model with scalar interactions, but without compact star constraints (no electrons and ρ included), and the eLSM model without repulsive interaction in the vector sector do not lie in the desired radius window. The highest mass compact star has a mass of $\sim 1.7\ M_\odot$ for the eLSM with $g_v = 0$, $\sim 1.3\ M_\odot$ for the non-interacting quark model, $\sim 3\ M_\odot$ for the non-interacting Walecka model, and $\sim 2\ M_\odot$ for the interacting Walecka model. It is interesting to note that the star sequence in the eLSM model without repulsive interaction is close to the one of the Walecka model with scalar interaction but without compact star constraints, although the latter contains repulsive interaction as well. As expected, the repulsive interaction makes the EoS stiffer in the eLSM, and for $g_v = 2$ a mass value of $\sim 2.15\ M_\odot$ can be reached with a radius at $M = 2\ M_\odot$ in the permitted radius window. Based on Figure 1, one can observe that, interestingly, it is the stiffer EoS for $\varepsilon < 0.8$ GeV/fm^3, as compared to the Walecka model including scalar interactions and not subject to compact star constraints, that brings the star sequence to the desired range in the eLSM with repulsive interaction.

The energy density as a function of radial position is shown in Figure 4 for the maximum mass compact star obtained with the EoS of the eLSM with $g_v = 0$. Since the chiral phase transition occurs at $\mu_{q,c} \approx 323$ MeV, which essentially corresponds to zero pressure, almost all of the matter in the compact star is in the chirally symmetric phase (i.e., ε corresponds to $\mu_q > \mu_{q,c} \approx 323$ MeV). In the right panel we illustrate in the case of the Walecka model how the BPS EoS, which models the outer crust, influences the solution of the TOV equation.

In Figure 5 we compare $M(R)$ curves obtained in the non-interacting quark model at the three different sets of quark masses listed in the caption (two of them come from the eLSM at the value of μ indicated in the key) and in the interacting eLSM model with $g_v = 0$. For the free quark model, the quark masses increase from right to left, as indicated in the caption, while in case of the eLSM the quark masses change (decreasing with increasing baryochemical potential) and their masses are smaller than or equal to that of the leftmost curve and always larger than that of the rightmost curve obtained in the free quark model. This clearly shows the significant effect of interactions on the M–R curve. The dashed lines show that neglecting the constraints of charge neutrality and β-equilibrium in the non-interacting quark model does not lead to significant changes. Consequently, we also expect these constraints to have a mild effect in the eLSM.

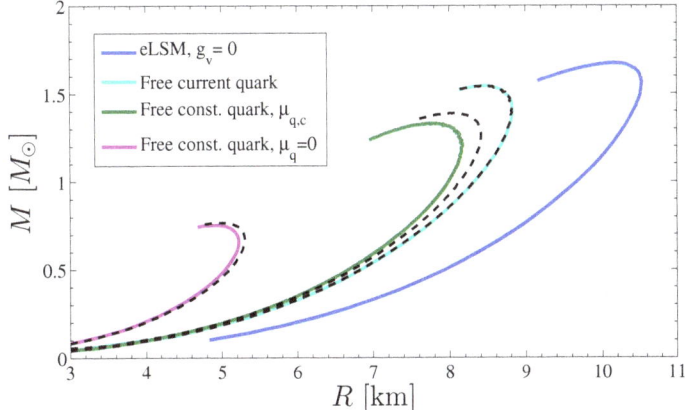

Figure 5. $M(R)$ curves of the non-interacting quark model with three different quark mass setups (left curve: $m_u = m_d = 322$ MeV, $m_s = 458$ MeV; middle curve: $m_u = m_d = 75$ MeV, $m_s = 365$ MeV; and right curve: $m_u = m_d = 0$ MeV, $m_s = 90$ MeV) compared to the $M(R)$ curve of the eLSM model with $g_v = 0$ in which the quark masses change (rightmost curve). The dashed curves are obtained without imposing the constraints of charge neutrality and β-equilibrium.

Finally, in Figure 6 we show the influence of the repulsive interaction on the mass–radius relation obtained in the eLSM. With increasing vector coupling, the EoS becomes stiffer, and more massive and larger stable stars can be attained. For $g_v = 2$ the star sequence is in the permitted radius window at $M = 2 M_\odot$, and the largest mass is $\sim$2.15 $M_\odot$. Beyond a certain value of the coupling, the pressure becomes positive for all positive values of the energy density, which results in star sequences that contain large stars with small masses. Qualitatively similar results were reported in [23].

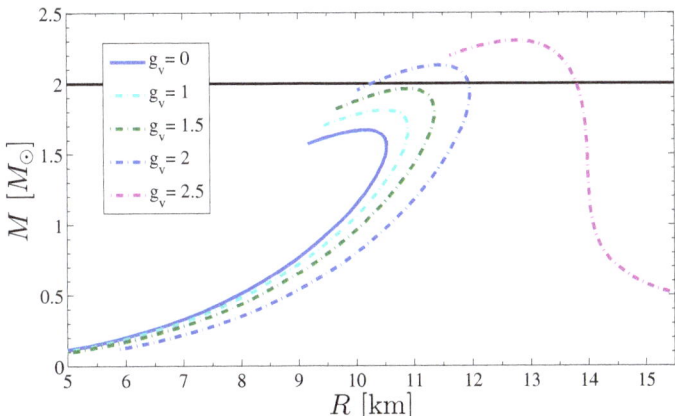

Figure 6. Dependence of the mass–radius relations on the strength of the Yukawa coupling g_v between quarks and the vector meson in the eLSM model.

4. Conclusions

We employed the zero-temperature EoS obtained with some approximations in the eLSM to determine the mass–radius relation of compact stars, assumed to consist of matter described by this model, and compared the resulting mass and radius values to those given by the two-flavor Walecka model and the three-flavor non-interacting quark model. The mass–radius sequence obtained in the eLSM without repulsive interaction mediated by a vector meson is close to that emerging from a Walecka model which includes the self-interaction of scalar mesons, but contrary to the EoS of that model, it can not reach the desired 2 $M_\odot$ mass value. The repulsive interaction in the eLSM model makes the EoS stiff enough to support, in some narrow range of the Yukawa coupling, compact stars with masses larger than 2 $M_\odot$ and in the radius window of 11.0–12.5 km at $M = 2\ M_\odot$, suggested by previous studies.

In the future, we would like to go beyond the mean-field approximation, used for the mesons in the eLSM, in a way that takes into account the effect of fermions in the mesonic fluctuations. At lowest order, this can be done by expanding to quadratic order the fermionic determinant obtained after integrating out the quark fields in the partition function and performing the Gaussian integral over the mesonic fields. In order to have a physically more reliable description, we also plan to include the charge neutrality and β-equilibrium conditions and improve the treatment of the interaction between vector mesons and quarks employed here.

Author Contributions: Authors contributed to the paper equally.

Funding: János Takátsy was supported by the ÚNKP-18-2 New National Excellence Program of the Ministry of Human Capacities. P. Kovács and Zs. Szép acknowledge support by the János Bolyai Research Scholarship of the Hungarian Academy of Sciences.

Acknowledgments: We thank Péter Pósfay for interesting discussions and the academic editor for guidance on the treatment of the neutron star crust in the Walecka model.

Conflicts of Interest: The authors declare no conflict of interest.

References

1. Grelli, A. ALICE Overview. *EPJ Web Conf.* **2018**, *171*, 01005. [CrossRef]
2. Sakaguchi, T. [PHENIX collaboration] Overview of latest results from PHENIX. *PoS* **2019**. *HardProbes2018*, 035. [CrossRef]
3. Tlusty, D. The RHIC Beam Energy Scan Phase II: Physics and Upgrades. In Proceedings of the 13th Conference on the Intersections of Particle and Nuclear Physics (CIPANP 2018), Palm Springs, CA, USA, 29 May–3 June 2018.
4. Borsanyi, S. Frontiers of finite temperature lattice QCD. *EPJ Web Conf.* **2017**, *137*, 01006. [CrossRef]
5. Larsen, D.T. Latest Results from the NA61/SHINE Experiment. *KnE Energy Phys.* **2018**, *3*, 188–194. [CrossRef]
6. Kekelidze, V.D. NICA project at JINR: Status and prospects. *J. Instrum.* **2017**, *12*, C06012. [CrossRef]
7. Ablyazimov, T.; Abuhoza, A.; Adak, R.P.; Adamczyk, M.; Agarwal, K.; Aggarwal, M.M.; Ahammed, Z.; Ahmad, F.; Ahmad, S.; Akindinov, A.; et al. Challenges in QCD matter physics—The scientific programme of the Compressed Baryonic Matter experiment at FAIR. *Eur. Phys. J. A* **2017**, *53*, 60. [CrossRef]
8. Lattimer, J.M.; Prakash, M. Neutron star observations: Prognosis for equation of state constraints. *Phys. Rep.* **2007**, *442*, 109–165. [CrossRef]
9. Watts, A.L.; Andersson, N.; Chakrabarty, D.; Feroci, M.; Hebeler, K.; Israel, G.; Lamb, F.K.; Miller, M.C.; Morsink, S.; Özel, F.; et al. Colloquium: Measuring the neutron star equation of state using X-ray timing. *Rev. Mod. Phys.* **2016**, *88*, 021001. [CrossRef]
10. Tews, I.; Margueron, J.; Reddy, S. Critical examination of constraints on the equation of state of dense matter obtained from GW170817. *Phys. Rev. C* **2018**, *98*, 045804. [CrossRef]
11. Tolman, R.C. Static Solutions of Einstein's Field Equations for Spheres of Fluid. *Phys. Rev.* **1939**, *55*, 364–373. [CrossRef]
12. Oppenheimer, J.R.; Volkoff, G.M. On Massive Neutron Cores. *Phys. Rev.* **1939**, *55*, 374–381. [CrossRef]

13. Demorest, P.B.; Pennucci, T.; Ransom, S.M.; Roberts, M.S.E.; Hessels, J.W.T. A two-solar-mass neutron star measured using Shapiro delay. *Nature* **2010**, *467*, 1081–1083. [CrossRef] [PubMed]
14. Antoniadis, J.; Freire, P.C.C.; Wex, N.; Tauris, T.M.; Lynch, R.S.; van Kerkwijk, M.H.; Kramer, M.; Bassa, C.; Dhillon, V.S.; Driebe, T.; et al. A Massive Pulsar in a Compact Relativistic Binary. *Science* **2013**, *340*, 1233232. [CrossRef] [PubMed]
15. Hell, T.; Weise, W. Dense baryonic matter: Constraints from recent neutron star observations. *Phys. Rev. C* **2014**, *90*, 045801. [CrossRef]
16. Lattimer, J.M.; Steiner, A.W. Neutron star masses and radii from quiescent low-mass X-ray binaries. *Astrophys. J.* **2014**, *784*, 123. [CrossRef]
17. Ayriyan, A.; Alvarez-Castillo, D.; Blaschke, D.; Grigorian, H. Bayesian Analysis for Extracting Properties of the Nuclear Equation of State from Observational Data including Tidal Deformability from GW170817. *Universe* **2019**, *5*, 61. [CrossRef]
18. Most, E.R.; Weih, L.R.; Rezzolla, L.; Schaffner-Bielich, J. New constraints on radii and tidal deformabilities of neutron stars from GW170817. *Phys. Rev. Lett.* **2018**, *120*, 261103. [CrossRef] [PubMed]
19. Abbott, B.P. et al. [The LIGO Scientific Collaboration and the Virgo Collaboration]. GW170817: Measurements of Neutron Star Radii and Equation of State. *Phys. Rev. Lett.* **2018**, *121*, 161101. [CrossRef] [PubMed]
20. Watts, A.L. Constraining the neutron star equation of state using Pulse Profile Modeling. In Proceedings of the Xiamen- CUSTIPEN Workshop on the EOS of Dense Neutron-Rich Matter in the Era of Gravitational Wave Astronomy, Xiamen, China, 3–7 January 2019.
21. Kovács, P.; Szép, Z.; Wolf, G. Existence of the critical endpoint in the vector meson extended linear sigma model. *Phys. Rev. D* **2016**, *93*, 114014. [CrossRef]
22. Parganlija, D.; Kovács, P.; Wolf, G.; Giacosa, F.; Rischke, D.H. Meson vacuum phenomenology in a three-flavor linear sigma model with (axial-)vector mesons. *Phys. Rev. D* **2013**, *87*, 014011. [CrossRef]
23. Zacchi, A.; Stiele, R.; Schaffner-Bielich, J. Compact stars in a SU(3) Quark-Meson Model. *Phys. Rev. D* **2015**, *92*, 045022. [CrossRef]
24. Benic, S. Heavy hybrid stars from multi-quark interactions. *Eur. Phys. J. A* **2014**, *50*, 111. [CrossRef]
25. Benic, S.; Blaschke, D.; Alvarez-Castillo, D.E.; Fischer, T.; Typel, S. A new quark-hadron hybrid equation of state for astrophysics - I. High-mass twin compact stars. *Astron. Astrophys.* **2015**, *577*, A40. [CrossRef]
26. Annala, E.; Gorda, T.; Kurkela, A.; Nättilä, J.; Vuorinen, A. Constraining the properties of neutron-star matter with observations. In Proceedings of the 12th INTEGRAL conference and 1st AHEAD Gamma-ray workshop (INTEGRAL 2019): INTEGRAL looks AHEAD to Multi-Messenger Astrophysics, Geneva, Switzerland, 11–15 February 2019.
27. Walecka, J.D. A theory of highly condensed matter. *Ann. Phys.* **1974**, *83*, 491–529. [CrossRef]
28. Amsler, C.; Doser, M.; Antonelli, M.; Asner, D.M.; Babu, K.S.; Baer, H.; Band, H.R.; Barnett, R.M.; Bergren, E.; Beringer, J.; et al. Review of Particle Physics. *Chin. Phys. C* **2016**, *40*, 100001. [CrossRef]
29. James, F.; Roos, M. Minuit - a system for function minimization and analysis of the parameter errors and correlations. *Comput. Phys. Commun.* **1975**, *10*, 343–367. [CrossRef]
30. Aoki, Y.; Fodor, Z.; Katz, S.; Szabó, K. The QCD transition temperature: Results with physical masses in the continuum limit. *Phys. Lett. B* **2006**, *643*, 46–54. [CrossRef]
31. Bazavov, A.; et al. [HotQCD Collaboration]. The chiral and deconfinement aspects of the QCD transition. *Phys. Rev. D* **2012**, *85*, 054503. [CrossRef]
32. Almási, G. Properties of Hot and Dense Strongly Interacting Matter. Ph.D. Thesis, Technische Universität, Darmstadt, Germany, 2017.
33. Schmitt, A. *Dense Matter in Compact Stars: A Pedagogical Introduction*; Lecture Notes in Physics; Springer: Berlin, Germany, 2010; Volume 811, pp. 1–111.
34. Glendenning, N. *Compact Stars: Nuclear Physics, Particle Physics, and General Relativity*, 2nd ed.; Springer: New York, NY, USA, 2000.
35. Xu, C.; Li, B.A.; Chen, L.W. Symmetry energy, its density slope, and neutron-proton effective mass splitting at normal density extracted from global nucleon optical potentials. *Phys. Rev. C* **2010**, *82*, 054607. [CrossRef]
36. Pósfay, P.; Barnaföldi, G.G.; Jakovác, A. Estimating the variation of neutron star observables by symmetric dense nuclear matter properties. *Universe* **2019**, *5*, 153. [CrossRef]

37. Baym, G.; Pethick, C.; Sutherland, P. The Ground state of matter at high densities: Equation of state and stellar models. *Astrophys. J.* **1971**, *170*, 299–317. [CrossRef]
38. Fortin, M.; Providencia, C.; Raduta, A.R.; Gulminelli, F.; Zdunik, J.L.; Haensel, P.; Bejger, M. Neutron star radii and crusts: Uncertainties and unified equations of state. *Phys. Rev. C* **2016**, *94*, 035804. [CrossRef]
39. Blaschke, D.; Chamel, N. Phases of dense matter in compact stars. *Astrophys. Space Sci. Libr.* **2018**, *457*, 337–400.
40. Potekhin, A.Y.; Fantina, A.F.; Chamel, N.; Pearson, J.M.; Goriely, S. Analytical representations of unified equations of state for neutron-star matter. *Astron. Astrophys.* **2013**, *560*, A48. [CrossRef]
41. Hessels, J.W.T.; Ransom, S.M.; Stairs, I.H.; Freire, P.C.C.; Kaspi, V.M.; Camilo, F. A Radio Pulsar Spinning at 716 Hz. *Science* **2006**, *311*, 1901–1904. [CrossRef] [PubMed]

© 2019 by the authors. Licensee MDPI, Basel, Switzerland. This article is an open access article distributed under the terms and conditions of the Creative Commons Attribution (CC BY) license (http://creativecommons.org/licenses/by/4.0/).

Article
Lévy HBT Results at NA61/SHINE

Barnabás Pórfy [1,2]

[1] Wigner Research Centre for Physics, Konkoly-Thege Miklós út 29-33, H-1121 Budapest, Hungary; bporfy@cern.ch
[2] Department of Atomic Physics, Institute of Physics, Faculty of Science, Eötvös Loránd University, H-1111 Budapest, Hungary

Received: 14 May 2019; Accepted: 14 June 2019; Published: 16 June 2019

Abstract: Bose–Einstein (or Hanbury–Brown and Twiss (HBT)) momentum correlations reveal the space–time structure of the particle emitting source created in high energy nucleus–nucleus collisions. In this paper we present the latest NA61/SHINE measurements of Bose–Einstein correlations of identified pion pairs and their description based on Lévy distributed sources in Be + Be collisions at $150A$ GeV/c. We investigate the transverse mass dependence of the Lévy source parameters and discuss their possible interpretations.

Keywords: quark-gluon plasma; femtoscopy; critical point; small systems

1. Introduction

NA61/SHINE is a fixed target experiment at the CERN Super Proton Synchrotron (SPS). One of its main aims is to study the phase diagram of QCD. In order to accomplish that, different collision systems at multiple energies are investigated. The NA61/SHINE detector is equipped with four large time projection chambers (TPC) [1], these are covering the full forward hemisphere providing excellent tracking down to a transverse momentum of 0 GeV/c. The experiment also features a modular calorimeter, located on the beam axis after the TPCs. This detector is called the projectile spectator detector, and it measures the forward energy which determines the collision centrality of the events. A setup of the NA61/SHINE detector system is shown in Figure 1.

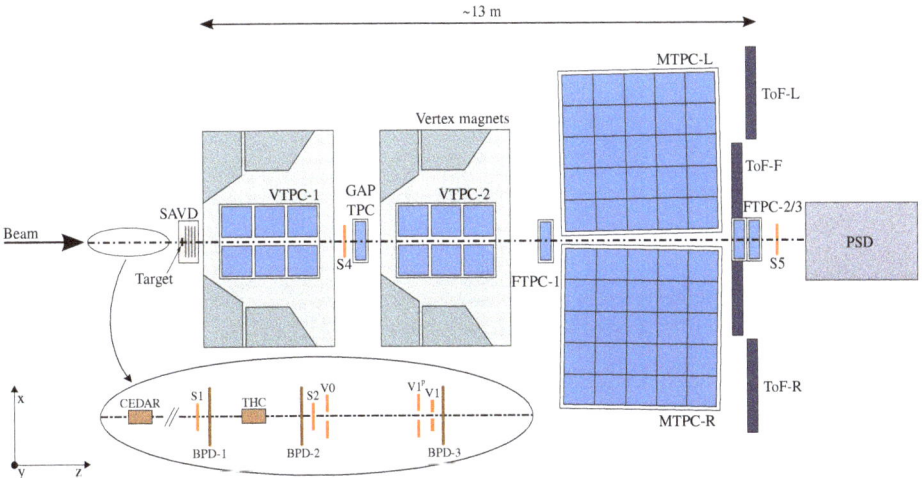

Figure 1. The setup of the NA61/SHINE detector system.

In order to study the QCD phase diagram and search for the critical end point (CEP), vastly different collision systems (p + p, p + Pb, Be + Be, Ar + Sc, Xe + La, Pb + Pb) are investigated at various beam momenta (13A, 20A, 30A, 40A, 75A and 150A GeV/c). There are many observables to accomplish this goal. In the analysis described in this paper we measure Bose–Einstein (or Hanbury–Brown and Twiss (HBT)) correlations of identical pions in Be + Be collisions at 150A GeV/c. These, based on the principles of quantum-statistical correlations, reveal the femtometer scale structure of pion production, hence this field is often called femtoscopy.

2. Femtoscopy, Lévy Sources and the Critical End Point

The method of femtoscopy is based on the work of R. Hanbury Brown and R. Q. Twiss [2] as well as Goldhaber and collaborators [3]. The key relationship of this method shows that the spatial momentum correlations ($C(q)$) are related to the properties of the particle emitting source ($S(x)$, describing the probability density of particle creation) in the following way:

$$C(q) \cong 1 + |\tilde{S}(q)|^2, \tag{1}$$

where $\tilde{S}(q)$ is the Fourier transform of $S(x)$, and q is the momentum difference of the pair (dependence on the average momentum K is suppressed here). See more details e.g., in ref. [4]. The usual assumption for the shape of the source is, based on the central limit theorem, Gaussian. A generalization of this assumption is to assume Lévy distributed sources. A possible reason is that due to the expanding medium, the mean free path may increase and thus anomalous diffusion and Lévy distributed sources may appear [5,6]. Alternatively, due to critical fluctuations and the appearance of large scale spatial correlations, similar power-law tailed sources may be present [7]. Another reason for Lévy distributed sources may be the fractal structure of QCD jets, as discussed in ref. [8]. Here we restrict our investigation to symmetric Lévy distributions, as they have proven to provide a suitable description of Bose–Einstein correlations in nucleus-nucleus collisions [4]. Furthermore, we restrict ourselves to describe the spatial part of the source, and the time dependence is absorbed through the connection of momentum difference q and average momentum K in case of identical particles:

$$\vec{q}\vec{K} = q_0 K_0. \tag{2}$$

Then the symmetric Lévy distribution is characterized by two parameters: Lévy scale parameter R and the Lévy exponent α. The distribution is defined as follows:

$$\mathcal{L}(\alpha, R, r) = \frac{1}{(2\pi)^3} \int d^3q e^{iqr} e^{-\frac{1}{2}|qR|^\alpha}. \tag{3}$$

This distribution can be expressed analytically in two special cases. One is the already mentioned Gaussian distribution for $\alpha = 2$; furthermore, $\alpha = 1$ leads to a Cauchy distribution. An important difference between Lévy distributions and Gaussians is the presence of a power-law tail in case of $\alpha < 2$, i.e., for large distances (r), the following holds:

$$\mathcal{L}(\alpha, R, r) \sim r^{-(d-2+\alpha)}, \tag{4}$$

where d represents the number of spatial dimensions. With Lévy sources, the Bose–Einstein or HBT correlation functions can be expressed in the following way:

$$C(q) = 1 + \lambda \cdot e^{-(qR)^\alpha}, \tag{5}$$

where the λ intercept parameter was introduced, which is defined as

$$\lambda = \lim_{q \to 0} C(q), \tag{6}$$

where the $q \to 0$ extrapolation is done in experimentally available q regions (limited from below by the two-track resolution in the given measurement). The Core-Halo model [9,10] may be utilized to understand this λ intercept parameter. The core–halo model splits the source into two pieces. The core part contains the primordially created pions (directly from hadronic freeze-out) or from short lived (strongly decaying) resonances. The halo consists of pions created from longer lived (compared to the usual source size of a few femtometers) resonances and general background. In this picture, the λ parameter turns out to be connected to the ratio of the Core and the Halo as follows:

$$\lambda = \left(\frac{N_{\text{core}}}{N_{\text{core}} + N_{\text{halo}}} \right)^2. \tag{7}$$

Finally, let us come back to the above mentioned point of connecting Lévy sources to the search for the Critical End Point of QCD. Critical points are characterized by critical exponents, one of which is the exponent of spatial correlations. This appears because due to the second order phase transition at the CEP, the spatial correlation functions becomes a power-law with an exponent of $-(d-2+\eta)$ (where d is the dimension and η is the critical exponent of spatial correlations). We can see that the Lévy exponent α, given in Equation (4), defines a similar power-law, and hence α may be regarded as identical to the critical exponent η. See further discussions in Ref. [7]. Critical exponents are universal in the sense that they take the same values in case of physical systems belonging to the same universality class. It has been shown [11] that the universality class of QCD is that of the 3D Ising model. The value of the critical exponent η has been calculated to be 0.03631(3) [12]. Alternatively, one may rely on the universality class of the 3D Ising model with a random external field, in which case an η value of 0.5 ± 0.05 was calculated [13]. Considering the previous statements, if we "scan" the phase diagram with different energies and systems and measure the values of the α exponent, we might be able to gain more information on the location and characteristics of the CEP.

3. Measurement Details

In this measurement we analyzed the 0–20% most central Be + Be collisions at $150A$ GeV/c. This dataset consists of about three million events, which after various event and track quality selections was reduced to around 300,000 events. The track acceptance in this analysis was as follows. The rapidity region of analyzed particles is $0.85 < \eta < 4.85$ (corresponding to $|\eta| < 2$ in the center-of-mass frame), the azimuthal coverage is 2π; n this track sample, we identified pions based on their deposited energy dE/dx in the TPC gas and charge obtained from the curvature of their trajectories in the magnetic field. We then analyzed negative pion pairs and positive pion pairs, as well as the combination of these two (i.e., created a dataset of identically charged pion pairs). These pairs were sorted into four K_T (average pair transverse momentum) bins in the range of 0–600 MeV/c. In each momentum bin, we measured the pair distribution of pairs from the same event, let us call this the $A(q)$ actual pair distribution. This contains quantumstatistical correlations, as well as many other residual effects related to kinematics and acceptance. To remove this undesirable effects, we created a mixed event for each actual event, by randomly selecting particles from other events of similar parameters, and making sure particles are each selected from a different event. Let us call the pair distribution from this sample $B(q)$, the background distribution. Then the correlation function is calculated as $C(q) = A(q)/B(q)$, provided a proper normalization is done in a q range where quantumstatistical correlations are not expected. Let us mention here, that our analysis was done with a one-dimensional momentum difference variable q, calculated in the longitudinally co-moving system (LCMS), as in this frame, an approximately spherically symmetric source can be expected, furthermore, the extrapolation of $q \to 0$ is equivalent with the three dimensional case.

Final state effects are still present in the $C(q)$ correlation function. Among these, for pion pairs, the most important is the Coulomb effect, responsible for the repulsion of same charged pairs. It is usually handled by a so-called Coulomb correction as follows. The Coulomb correction for Lévy type sources is a complicated numerical integral to calculate and fit, as discussed in Ref. [4]. However it can be observed that the Coulomb correction does not strongly depend on the Lévy exponent α. Hence we can use then the approximate formula published by the CMS Collaboration in ref. [14], valid for α = 1:

$$K_{\text{Coulomb}}(q,R) = \text{Gamow}(q) \cdot \left(1 + \frac{\pi\eta(q)q\frac{R}{\hbar c}}{1.26 + q\frac{R}{\hbar c}}\right), \text{ where} \quad (8)$$

$$\text{Gamow}(q) = \frac{2\pi\eta(q)}{e^{2\pi\eta(q)}-1} \text{ and } \eta(q) = \alpha_{\text{QED}} \cdot \frac{\pi}{q}, \quad (9)$$

where $\eta(q)$ is the so-called Sommerfeld parameter, and α_{QED} is the fine-structure constant (neither should be confused with the above discussed exponents η and α). Utilizing the usual Bowler–Sinyukov method (i.e., Coulomb correcting only for the Core part of the source) as indicated in Refs. [15,16], one obtains:

$$C(q) = N \cdot \left(1 - \lambda + \lambda \cdot \left(1 + e^{-(qR)^\alpha}\right) \cdot K_{\text{Coulomb}}(q)\right), \quad (10)$$

with N being a normalization parameter responsible for the proper normalization of the $A(q)/B(q)$ ratio. This is the final fit function we are using to describe our data.

As also found in ref. [4], the Lévy parameters (α, R, λ) are highly correlated, and especially in a low statistics dataset, it is hard to determine them precisely. We might be able to reduce this correlation and the statistical uncertainty of the parameters, if we fix one of the three parameters to a well motivated value. The resulting statistical uncertainties and free parameter values are modified due to the additional physical assumptions used to fix one of the parameters. From a statistical point of view, a bootstrap type of method may also be used. However, our main aim with this is to see a more clear trend of the m_T dependence of the parameters, with additional physical assumptions. One assumption is that α (i.e., the shape of the pion emitting source) is independent of m_T, with that we may fix α to a weighted average of the four α values obtained in free parameter fits performed in each K_T bin. The other option is fixing R with the following equation motivated by hydrodynamical predictions of the particle emission homogeneity length (essentially the HBT radii) in case of expanding fireballs [10]. In this case, we fit the following equation to the m_T dependence (where $m_T = \sqrt{m^2 + K_T^2}$, i.e., the average transverse mass of the pair) of the R Lévy scale:

$$R(m_T) = \frac{A}{\sqrt{1 + m_T/B}}. \quad (11)$$

Previous results with free parameter fits were shown in Ref. [17], hence here we concentrate on the results of the above mentioned fixed parameter fits. We again note that our aim with fixing one of the parameters to a physically motivated value is to show the trend of the m_T dependence of the parameters.

4. Results

First, the measured correlation functions were fitted with the above mentioned (Equation (10)) function with three free parameters (α, λ and R), as shown in Ref. [17]. Using the results from the free parameter fit, we fitted a constant function to the α values for all m_T bins, as well as the formula of Equation (11) to the R values in each bin. Then we analyzed first with one parameter fixed. All results are shown in Figures 2–4. Let us note here, that all the measurement settings (event selection, track selection, pair cuts, fitting interval) were varied systematically to obtain an estimate of systematic uncertainties. These, along with the statistical uncertainty of the parameters (obtained by the Minos algorithm) are shown in Figures 2–4.

The Lévy stability exponent α determines the source shape, and a value of 2 corresponds to a Gaussian source, a value of 1 to a Cauchy source, and 0.5 is the conjectured value at the critical point. Our results, along with these special cases (dotted yellow lines), are shown in Figure 2. It is clear from the figure that the statistical uncertainties of α from the fixed R fit are reduced by a factor 4–5;

however, the values of α are similar in both cases. These values are far from the Gaussian case as well as the conjectured CEP value, motivating us to perform this measurement in different systems and at different energies as well.

The Lévy scale R determines the correlation length of pion pairs from the given system. From a simple hydrodynamical picture one obtains a $R \propto 1/\sqrt{m_T}$ type of affine linear dependence, as already mentioned above. This, i.e., Equation (11), describes the free parameter fit data points of Figure 3 well. The fixed α fits are within uncertainties compatible with the free parameter results, however, they suggest a more or less constant trend of R versus m_T in the higher m_T bins. This motivated us to perform the same measurement in collision systems with larger multiplicities (where statistical uncertainties are expected to be reduced proportionally to the square of the mean multiplicity).

The last parameter to study was the correlation strength parameter λ, given in Equation (7). The transverse mass dependence of λ is shown in Figure 4. Comparing the three different fits (free parameter fit, fixed α fit, fixed R fit), it was visible that λ in free parameter fit was compatible (within statistical uncertainties) with both other cases. All fits showed a roughly constant $\lambda(m_T)$ trend. This was in contrast to the findings at RHIC, see e.g., the compilation in ref. [18]. This finding was, however, compatible with previous SPS measurements (in different systems), see e.g., ref. [19].

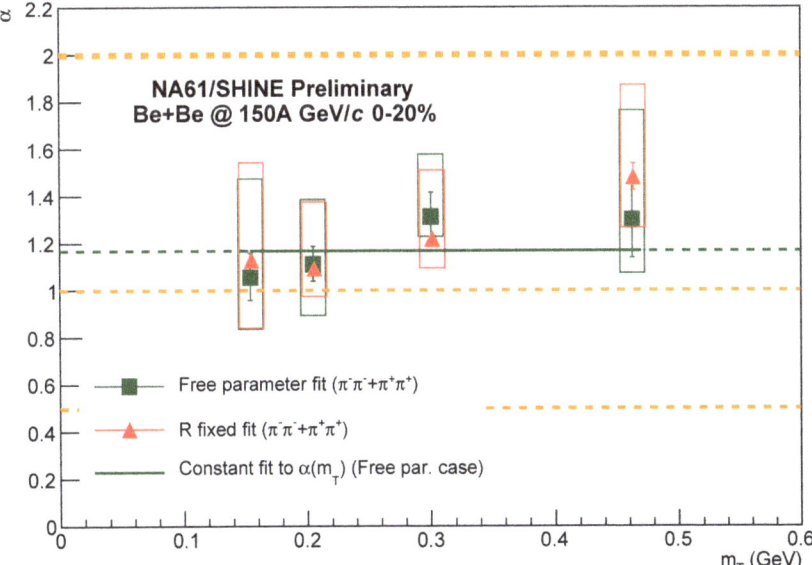

Figure 2. Lévy exponent α versus transverse mass: comparison between free parameter fit, fit with R fixed and fit with α fixed. The boxes represent systematic uncertainties. For each bin, the results are slightly shifted to the right for visibility, but they are in the same bin.

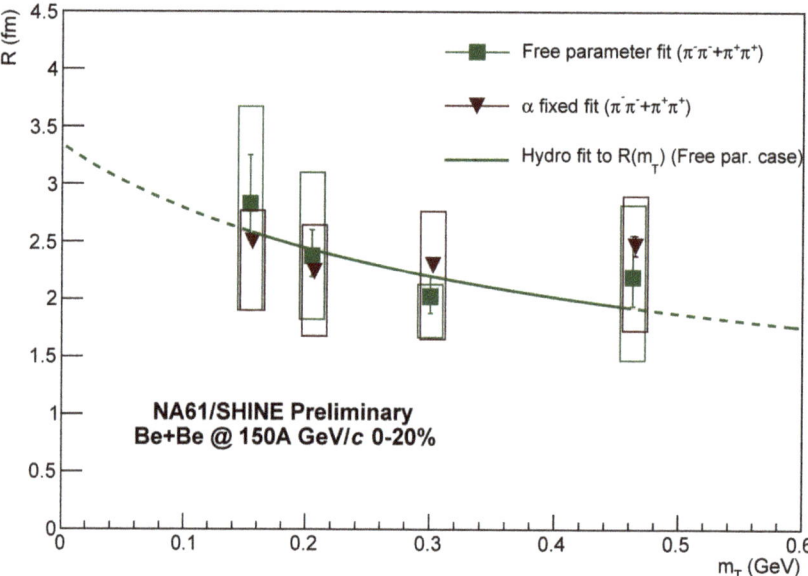

Figure 3. Correlation strength R versus transverse mass: comparison between free parameter fit, fit with α fixed and fit with R fixed. The boxes represent systematic uncertainties. For each bin, the results are slightly shifted to the right for visibility, but they are in the same bin.

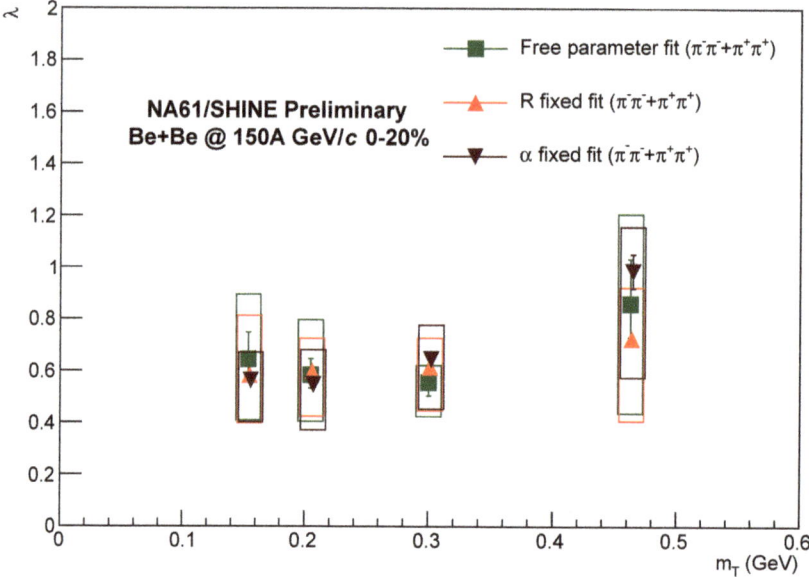

Figure 4. Lévy exponent λ versus transverse mass: comparison between free parameter fit, fit with α fixed and fit with R fixed. The boxes represent systematic uncertainties. For each bin, the results are slightly shifted to the right for visibility, but they are in the same bin.

5. Conclusions

We reported above on the NA61/SHINE measurement of one-dimensional identified two-pion Bose-Einstein correlation functions in the 0–20% most central Be + Be collisions at $150A$ GeV/c. We compared free parameter fits to fixed parameter fits, to reduce statistical uncertainty of the physical parameters (α, λ and R). We found that the results from the free parameter fits and the fixed parameter fits are similar, but the statistical uncertainty of each parameter is reduced by a large factor. The aim of this excercise was to show the trends of the parameters with fixing one parameter to a physically motivated value. Our results confirmed that in this collision system and at this collision energy, the Lévy exponent α is far from the Gaussian case, as well as from the conjectured value at the critical end point. We furthermore found that the $R(m_T)$ dependence is compatible with hydro predictions, and $\lambda(m_T)$ may show different patters at RHIC and SPS energies. These findings will be subsequently investigated in other collision systems as well.

Funding: Barnabás Pórfy was supported by the NKFIH grants FK123842 and FK123959.

Acknowledgments: The author would like to thank the NA61/SHINE collaboration.

Conflicts of Interest: The author declares no conflict of interest.

Abbreviations

The following abbreviations are used in this manuscript:

QCD	Quantum chromodynamics
CERN	European Organization for Nuclear Research
SPS	Super Proton Synchrotron
HBT	Hanbury–Brown and Twiss
CEP	Critical end point
NA61/SHINE	SPS Heavy Ion and Neutrion Experiment

References

1. Abgrall, N.; Andreeva, O.; Aduszkiewicz, A.; Ali, Y.; Anticic, T.; Antoniou, N.; Baatar, B.; Bay, F.; Blondel, A.; Blumer, J.; et al. NA61/SHINE facility at the CERN SPS: Beams and detector system. *J. Instrum.* **2014**, *9*, P06005. [CrossRef]
2. Hanbury Brown, R.; Twiss, R.Q. A Test of a new type of stellar interferometer on Sirius. *Nature* **1956**, *178*, 1046–1048. [CrossRef]
3. Goldhaber, G.; Fowler, W.B.; Goldhaber, S.; Hoang, T.F. Pion-pion correlations in antiproton annihilation events. *Phys. Rev. Lett.* **1959**, *3*, 181–183. [CrossRef]
4. Adare, A.; Aidala, C.; Ajitanand, N.N.; Akiba, Y.; Akimoto, R.; Alexander, J.; Alfred, M.; Al-Ta'ani, H.; Angerami, A.; Aoki, K.; et al. Lévy-stable two-pion Bose-Einstein correlations in $\sqrt{s_{NN}} = 200$ GeV Au+Au collisions. *Phys. Rev.* **2018**, *C97*, 064911. [CrossRef]
5. Metzler, R.; Klafter, J. The random walk's guide to anomalous diffusion: A fractional dynamics approach. *Phys. Rep.* **2000**, *339*, 1–77. [CrossRef]
6. Csanad, M.; Csorgo, T.; Nagy, M. Anomalous diffusion of pions at RHIC. *Braz. J. Phys.* **2007**, *37*, 1002–1013, [CrossRef]
7. Csorgo, T.; Hegyi, S.; Novak, T.; Zajc, W.A. Bose-Einstein or HBT correlation signature of a second order QCD phase transition. *AIP Conf. Proc.* **2006**, *828*, 525–532.
8. Csorgo, T.; Hegyi, S.; Novak, T.; Zajc, W.A. Bose-Einstein or HBT correlations and the anomalous dimension of QCD. *Acta Phys. Polon.* **2005**, *B36*, 329–337.
9. Csorgo, T. Particle interferometry from 40-MeV to 40-TeV. *Acta Phys. Hung.* **2002**, *A15*, 1–80. [CrossRef]
10. Csorgo, T.; Lorstad, B. Bose-Einstein correlations for three-dimensionally expanding, cylindrically symmetric, finite systems. *Phys. Rev.* **1996**, *C54*, 1390–1403.
11. Stephanov, M.A.; Rajagopal, K.; Shuryak, E.V. Signatures of the tricritical point in QCD. *Phys. Rev. Lett.* **1998**, *81*, 4816–4819. [CrossRef]

12. El-Showk, S.; Paulos, M.F.; Poland, D.; Rychkov, S.; Simmons-Duffin, D.; Vichi, A. Solving the 3d Ising Model with the Conformal Bootstrap II. c-Minimization and Precise Critical Exponents. *J. Stat. Phys.* **2014**, *157*, 869–914. [CrossRef]
13. Rieger, H. Critical behavior of the three-dimensional random-field Ising model: Two-exponent scaling and discontinuous transition. *Phys. Rev. B* **1995**, *52*, 6659–6667. [CrossRef] [PubMed]
14. Sirunyan, A.M.; CMS Collaboration. Bose-Einstein correlations in pp, pPb, and PbPb collisions at $\sqrt{s_{NN}} = 0.9 - 7$ TeV. *Phys. Rev.* **2018**, *C97*, 064912. [CrossRef]
15. Sinyukov, Y.; Lednicky, R.; Akkelin, S.V.; Pluta, J.; Erazmus, B. Coulomb corrections for interferometry analysis of expanding hadron systems. *Phys. Lett.* **1998**, *B432*, 248–257. [CrossRef]
16. Bowler, M.G. Coulomb corrections to Bose-Einstein correlations have been greatly exaggerated. *Phys. Lett.* **1991**, *B270*, 69–74. [CrossRef]
17. Porfy, B. NA61/SHINE results on Bose-Einstein correlations. In Proceedings of the 12th International Workshop on Critical Point and Onset of Deconfinement (CPOD 2018), Corfu, Greece, 24–28 September 2018.
18. Vertesi, R.; Csorgo, T.; Sziklai, J. Significant in-medium η' mass reduction in $\sqrt{s_{NN}} = 200$ GeV Au+Au collisions at the BNL Relativistic Heavy Ion Collider. *Phys. Rev.* **2011**, *C83*, 054903.
19. Beker, H.; Boggild, H.; Boissevain, J.; Cherney, M.; Dodd, J.; Esumi, S.; Fabjan, C.W.; Fields, D.E.; Franz, A.; Hansen, K.H.; et al. m(T) dependence of boson interferometry in heavy ion collisions at the CERN SPS. *Phys. Rev. Lett.* **1995**, *74*, 3340–3343. [CrossRef] [PubMed]

© 2019 by the author. Licensee MDPI, Basel, Switzerland. This article is an open access article distributed under the terms and conditions of the Creative Commons Attribution (CC BY) license (http://creativecommons.org/licenses/by/4.0/).

Article

Estimating the Variation of Neutron Star Observables by Dense Symmetric Nuclear Matter Properties

Péter Pósfay [1,*], Gergely Gábor Barnaföldi [1] and Antal Jakovác [2]

1   Department for Theoretical Physics, Wigner Research Centre for Physics of the Hungarian Academy of Sciences, H-1121 Budapest, Hungary; barnafoldi.gergely@wigner.mta.hu
2   Institute of Physics, Eötvös Loránd University, H-1117 Budapest, Hungary; jakovac@caesar.elte.hu
*   Correspondence: posfay.peter@winger.mta.hu

Received: 1 May 2019; Accepted: 12 June 2019; Published: 14 June 2019

Abstract: Recent multi-channel astrophysics observations and the soon-to-be published new measured electromagnetic and gravitation data provide information on the inner structure of the compact stars. These macroscopic observations can significantly increase our knowledge on the neutron star enteriors, providing constraints on the microscopic physical properties. On the other hand, due to the masquarade problem, there are still uncertainties on the various nuclear-matter models and their parameters as well. Calculating the properties of the dense nuclear matter, effective field theories are the most widely-used tools. However, the values of the microscopical parameters need to be set consistently to the nuclear and astrophysical measurements. In this work, we investigate how uncertainties are induced by the variation of the microscopical parameters. We use a symmetric nuclear matter in an extended σ-ω model to see the influence of the nuclear matter parameters. We calculate the dense matter equation of state and give the mass-radius diagram for a simplistic neutron star model. We present that the Landau mass and compressibility modulus of the nuclear matter have definite linear relation to the maximum mass of a Schwarzschild neutron star.

Keywords: dense matter; stars: neutron; equation of state; astro-particle physics

1. Introduction

The investigation of the structure of compact astrophysical objects like neutron stars, magnetars, quark- or hybrid stars, etc. is an active novel research area as a child of astrophysics, gravitational theory and experiment and nuclear physics. Thus far, the extreme dense state of the matter can not be produced in today's Earth-based particle accelerators, thus only celestial objects can be used for tests. Electromagnetic measurements, such as X-ray- and gamma satellites, aim to measure properties of these objects more and more accurately [1–4]. In parallel, radio array data [5] and the newly discovered gravity waves provide a new way to probe their inner structure [6–8]. These observations are particularly important inputs for the theoretical studies of dense nuclear matter [9,10].

From the theoretical point of view, first principle calculations based on lattice field theory are still challenging at high chemical potentials present in compact stars [11–13]. Thus, effective theories play an important role in studying the properties of cold dense nuclear matter [14,15]. Recent studies show the importance of the correct handling of the bosonic sector in effective theories of nuclear matter [16,17]; moreover, applying the functional renormalization group (FRG) method on the simplest non-trivial nuclear matter, the effect of the microscopical parameters on neutron star observables were shown in Refs. [18,19].

We note that we use the simplest nuclear matter for neutron stars without crust. Leptonic fields were not included in the model; therefore, no β-equlibrium was taken into account. During the calculations of the nuclear equation of state, the condition of charge neutrality was not imposed. We note, however, that this does not lead to the violation of the charge neutrality of the neutron star itself, as there are no energy terms related to the electrical charges of the nucleons, and all the hadronic fields considered are identical. The model effectively describes nuclear matter that consists of interacting neutrons and neutral mesons that are parametrized to describe the saturation properties of symmetric nuclear matter as in the case of the original Walecka-model [20]. These assumptions are restricting but led us to investigate the consequences of varying the nuclear matter parameters in a more clear nuclear environment like e.g., in Ref. [21]. We note that an ongoing extended theoretical work is in progress for a more realistic case to compare astrophysical experimental data to our model.

In this paper, we study the connection between the parametrizations of effective nuclear models and measurable properties of compact stars in three differently extended versions of the σ-ω model. All of these include symmetric nuclear matter with various interaction terms in the bosonic sector. After calculating the equation of state (EoS) corresponding to different parametrizations of these models, the mass-radius (M-R) diagrams are calculated by solving the Tollmann–Oppenheimer–Volkoff (TOV) equations. We show how sensitive the mass-radius relation is to differences in the bosonic sector. The dependence of particular properties of compact stars (maximum mass and radius) is presented, influenced by different saturation parameters of the symmetric nuclear matter.

2. The Extended σ-ω in the Mean Field Approximation

Here, we apply the most common mean field model of the dense nuclear matter, formulating the extended σ-ω model [22,23] with the Lagrange-function taken from Refs. [20,24],

$$\mathcal{L} = N_f \overline{\Psi} \left(i\slashed{\partial} - m_N + g_\sigma \sigma - g_\omega \slashed{\omega} \right) \Psi + \frac{1}{2} \sigma \left(\partial^2 - m_\sigma^2 \right) \sigma - U_i(\sigma) - \frac{1}{4} \omega_{\mu\nu} \omega^{\mu\nu} + \frac{1}{2} m_\omega^2 \omega^2, \quad (1)$$

where Ψ is the fermionic nucleon field, $N_f = 2$ is the number of nucleons, and m_N, m_σ, and m_ω are the nucleon, sigma, and omega masses, respectively, for the usual scalar and vector fields. We introduced the $\omega_{\mu\nu} = \partial_\mu \omega_\nu - \partial_\nu \omega_\mu$ and the Yukawa coupling corresponding to the σ–nucleon and ω–nucleon interactions is given by g_σ and g_ω. We denote the general bosonic interaction terms with $U_i(\sigma)$, which can have thee different forms as the considered modified model cases for certain i,

$$\begin{aligned} U_3 &= \lambda_3 \sigma^3, \\ U_4 &= \lambda_4 \sigma^4, \\ U_{34} &= \lambda_3 \sigma^3 + \lambda_4 \sigma^4. \end{aligned} \quad (2)$$

In the mean field (MF) approximation, the kinetic terms are zero for the mesons and only the fermionic path integral has to be calculated at finite chemical potential and temperature. We consider here the symmetric nuclear matter to be in equilibrium, which includes the baryon number conservation. Taking this into account, the standard procedure was applied minimizing the free energy of the infinite symmetric nuclear matter at the zero temperature limit, where, for the proton (n_p) and neutron (n_n), the number of densities are equal, such as the proper chemical potentials, μ_p and μ_n, respectively:

$$n_p = n_n \quad \longrightarrow \quad \mu_p = \mu_n = \mu. \quad (3)$$

After applying this for all three cases in Equation (2) and substituting them into Equation (1), the numerical solution can be obtained after parameter fitting.

3. Parameter Fitting in the Extended σ-ω Model

As the general procedure, all the models' considered cases in Equation (2) need to fit to the nucleon saturation data found in e.g., Refs. [20,25]. In parallel to the effective mass, we introduced the definition of the Landau mass

$$m_L = \frac{k_F}{v_F} \quad \text{with} \quad v_F = \left. \frac{\partial E_k}{\partial k} \right|_{k=k_F}, \quad (4)$$

where $k = k_F$ the Fermi-surface and E_k is the dispersion relation of the nucleons. The Landau mass (m_L) and the effective mass (m^*) are not independent in relativistic mean field theories,

$$m_L = \sqrt{k_F^2 + m^{*2}}. \tag{5}$$

This is the reason why the Landau mass and the effective mass of the nucleons can not be fitted simultaneously in the models we consider [25]. In this paper, we deal with this problem in the following way. We fit all of the models two times: using the effective mass value from Table 1 and one calculated from Equation (5) to reproduce the Landau mass value from Table 1.

Table 1. Nuclear saturation parameter data, from Refs. [20,25].

Parameter	Value	Unit
Binding energy, B	-16.3	MeV
Saturation density, n_0	0.156	fm^{-3}
Nucleon effective mass, m^*	0.6 m_N	MeV
Nucleon Landau mass, m_L	0.83 m_N	MeV
Incompressibility, K	240	MeV

If the models with U_3 and U_4 type interaction terms are used, then there are not enough free parameters to fit the data in Table 1. In these cases, the nucleon effective mass, saturation density, and binding energy are fitted and the compression modulus is a prediction, given by

$$K = k_F^2 \frac{\partial^2}{\partial k_F^2}(\varepsilon/n) = 9n^2 \frac{\partial^2}{\partial n^2}(\varepsilon/n), \tag{6}$$

which has a simple connection to the thermodynamical compressibility at the saturation density n_0.

In the case of U_{34}, all four parameters can be fitted simultaneously, and there is another way to incorporate data regarding both Landau and effective mass. For this model, we consider a third fit, where the value of the effective mass is chosen in a way that minimizes the error coming from not fitting the two types of masses correctly. Technically, this value of the effective mass minimized the χ^2 of the fit, with value

$$m_{opt} = 0.6567\, m_N \approx 616 \text{ MeV}. \tag{7}$$

Since the incompressibility is different for the model cases with different interaction terms, we compared them in Table 2. For model cases with U_3 and U_4, there are two fits, for Landau and effective mass that produce different incompressibility values because they do not have enough free parameters to fit the correct value. However, for U_{34}, there are enough parameters to fit the incompressibility, so it has the same value for all three fits: for the Landau mass, for the effective mass and for the optimal mass. As Table 2 presents incompressibility values for U_3 with Landau fit, it is quite close and the U_{34} results provide the best fit with the saturation nuclear matter parameters in Table 1. These models differ in their predictions for higher densities of nuclear matter, which complicates the description of the compact star interior.

Table 2. The obtained incompressibility values in different model cases and fits.

Models	Calculation Method	K [MeV]
σ-ω model	reference value	563
U_3	effective mass fit	437
U_3	Landau mass fit	247
U_4	effective mass fit	482
U_4	Landau mass fit	334
U_{34}	χ^2 fits for all	240

4. Properties of Nuclear Matter in the Extended σ-ω Model

The nuclear properties of the different model cases were compared with all the possible parameter fits at the equation of state (EOS) level. We used all three types of interaction terms: U_3, U_4, and U_{34}, and each was considered with two parametrizations corresponding to Landau and effective mass fits. All results were cross-checked with the original σ-ω model parameters. In case of the model characterized by U_{34} interactions, we used a fit which reproduces Eequation (7).

The energy density, pressure and density were calculated in all of these models. The equation of states corresponding to these model and fit cases are shown in Figure 1. The results from the modified σ-ω model are compared to other equation of state parametrizations from Refs. [26–28] (*solid lines*). An important feature of Figure 1 is that different model based EoS parametrizations are separated based on whether they are parametrized by the Landau (*full symbols*) or the effective mass (*open symbols*). The models which are fitted to reproduce the correct effective mass of nucleons have smaller energy density at a given pressure. This phenomena becomes more prominent as pressure increases. It is also important to note that, in a given band, the incompressibility corresponding to certain equation of states can be very different. For example, in the group that was fitted for the effective mass in the U_3 type model, (*open rectangles*) has the incompressibility $K = 247$ MeV, but the U_4 type (*open circles*) has almost double the value, $K = 482$ MeV.

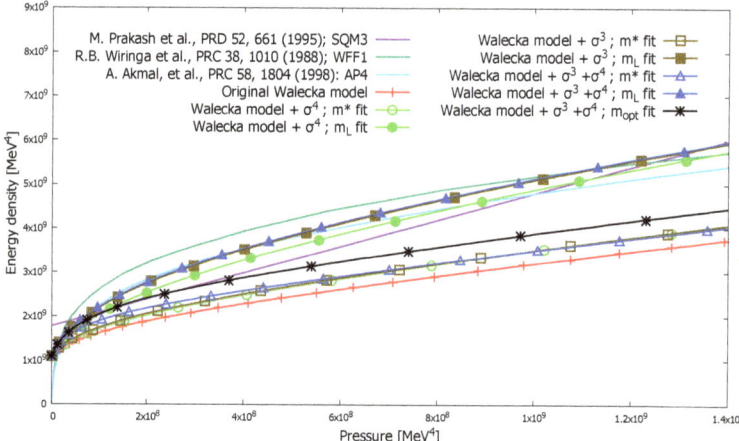

Figure 1. Equation of state in different models in comparison to Refs. [26–28].

Similar grouping can be seen in Figure 2 presenting the binding energy B as the function of the nuclear density n. However, near the minimal B, the value of incompressibility governs the curves because it determines the curvature of the curves around minimum. Effective mass that fits with *open symbols* has a steeper rise with the increasing density, n, while Landau-mass fit curves with *full symbols* are wider. We note that, below the saturation density, curves are getting more independent from the model choice and parameter fits.

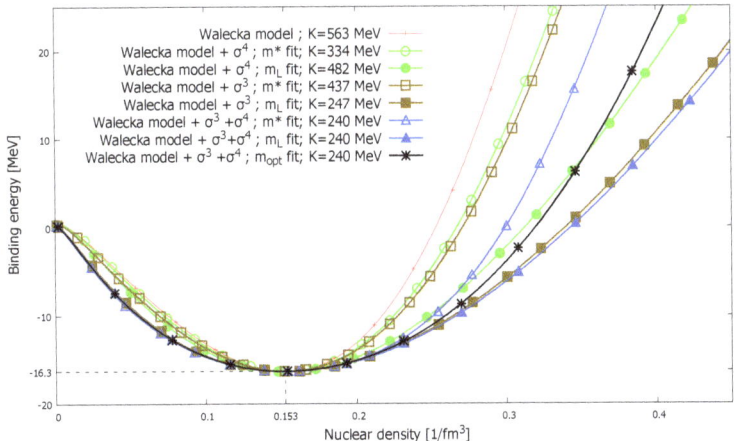

Figure 2. The density dependence of binding energy with the different models and parametrizations.

5. Compact Star Observables from the Effective Models

The Tolman–Oppenheimer–Volkoff equations provide the general relativistic description of the compact stars assuming spherically symmetric and time independent space-time structure [20],

$$\frac{dP(r)}{dr} = -\frac{G\varepsilon(r)m(r)}{r^2}\left[1 + \frac{P(r)}{\varepsilon(r)}\right]\left[1 + \frac{4\pi r^3 P(r)}{m(r)}\right]\left[1 - \frac{2Gm(r)}{r}\right]^{-1},$$

$$\frac{dm(r)}{dr} = 4\pi r^2 \varepsilon(r).$$
(8)

Here, $P(r)$ and $\varepsilon(r)$ are the pressure and energy density as functions of the radius of the star, G is the gravitational constant while $m(r)$ is the mass of star that is included in a shell with radius r. To integrate the equations, one needs a connection between $P(r)$ and $\varepsilon(r)$ at given r which is provided by the nuclear matter equation of state in the form of the relation $P(r) = P(\varepsilon(r))$. To start the integration, one has to choose a central energy density value ε_c for the star as an initial condition.

After solving the Equation (8) using the EoS from the model cases with various fits, the mass (M) and radius (R) of a compact star with a given energy density can be determined. Results corresponding to different energy densities in a given model are summarized on a mass-radius M - R diagram on Figure 3.

The model variants inherited the behaviour as in Figures 1 and 2: the curves are grouped based on whether or not they parametrized to reproduce the effective mass or Landau mass. Models with smaller effective mass (*open symbols*) systematically produce higher maximum mass stars compared to their parametrization with larger effective mass (Landau mass, with *full symbols*). Moreover, all models fitted for the effective mass value in Table 1 produce higher maximum star mass than the ones fitted for the Landau mass. Since the Landau mass and effective mass are not independent as in Equation (4), the above statement is equivalent to saying that higher effective mass produces smaller maximum star mass. This picture is supported by the curve corresponding to the model case parametrized by the optimal mass in Equation (7), which is the best fit of the model. The maximum star mass in this case is between the values produced by parametrizations described by smaller and larger effective nucleon mass. It is interesting to note that parametrizations corresponding to the effective mass and to the optimal mass value may be ruled out by observations, as they produce much larger maximum star mass and radius than the most recent measurements and theoretical predictions suggest [29].

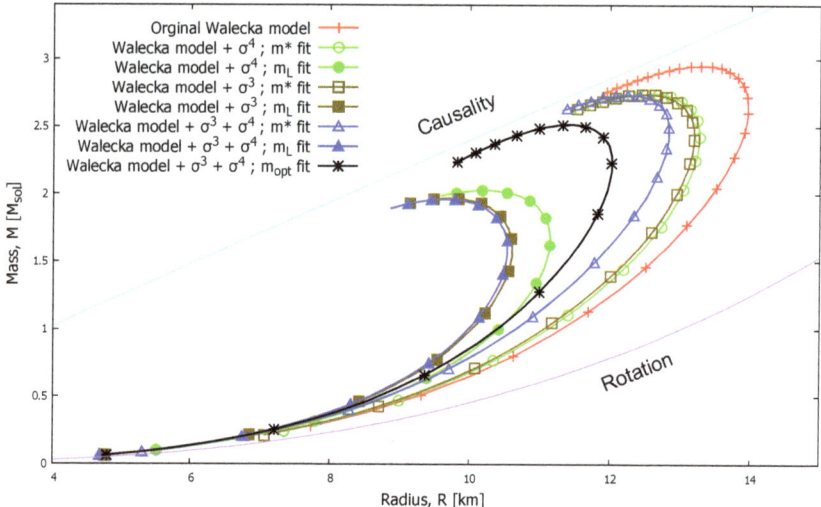

Figure 3. Mass–radius diagram corresponding to the different parametrizations of the modified σ-ω model. The line corresponding to the original σ-ω model is also drawn for comparison with a red color.

The value of incompressibility in these models seems to influence very little the maximum star mass; however, at the same time, it has an effect on the compactness by influencing the radius of the star. Thus, it can be seen in Figure 3 by looking at the curves belonging to the fits with the same effective mass (Landau mass). In these model cases, all of the parameters are the same as listed in Table 1 apart from the incompressibility.

6. Connection between Maximum Star Mass and Nuclear Parameters

Our results present a strong connection between the nuclear effective mass, m_L and maximum star mass, $M_{max\,M}$. To further study this phenomena, the model variant with interaction term U_{34} is used and presented below. The couplings in the model are calculated to reproduce all values listed in Table 1 except for the nuclear effective mass and Landau mass that are kept as external parameters. To study the maximum star mass dependence on the effective nucleon mass, many different fits of this model are considered all with different values of effective nucleon mass. The M–R diagrams corresponding to these parametrizations are calculated, and the maximum star mass is determined in each case. This procedure makes it possible to determine the dependence of maximum star mass on nucleon effective mass in this model, if everything else is kept constant. The results are summarized in Figure 4. The connection between maximum star mass and nucleon effective mass is well described by a linear connection, which gives the best fit of the numerical data,

$$M_{max\,M} = 5.896 - 0.005\,m_L, \tag{9}$$

where $M_{max\,M}$ is given in units of M_{Sun} and m_L is given in MeV.

Although the above equation is derived considering only the model with interaction term U_{34} and taking into account parametrizations that differ only in the value of nucleon effective mass, it generalizes well and it approximates the maximum star mass corresponding to the other model variants we consider with very high accuracy. This seems to indicate that the linear connection is a good approximation regardless of the bosonic interaction term used, and this holds across different model variants.

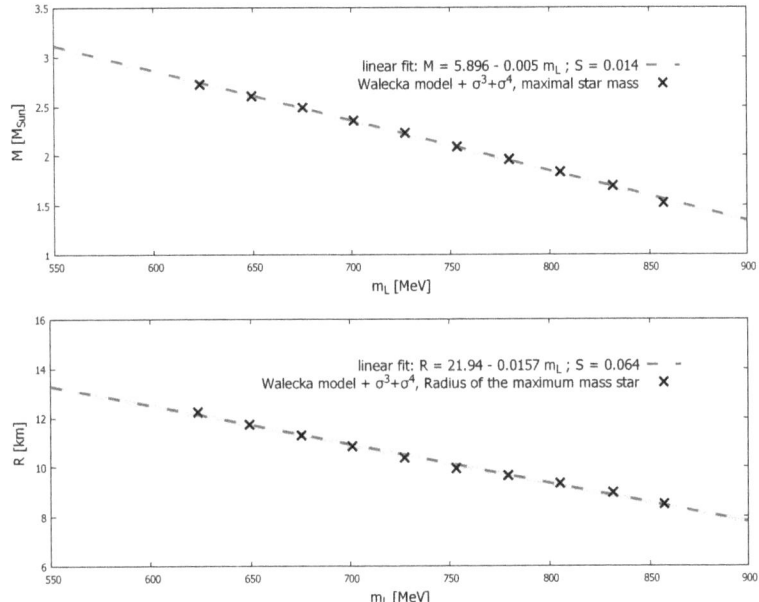

Figure 4. Maximum star mass, $M_{max\,M}$ (*upper panel*) and maximum star size, $R_{max\,M}$ (*lower panel*) as a function of the Landau nucleon mass, m_L.

These observations are further supported by the connection between the radius of the maximum mass star ($R_{max\,M}$) and nuclear Landau mass. This connection also presents strong linear dependence and holds across all model variants we considered in this paper. The results are shown in Figure 4, where the plotted linear relation is described by

$$R_{max\,M} = 21.94 - 0.015\,m_L. \tag{10}$$

Applying R is given in km units and m_L is given in MeV.

Since besides the nucleon Landau (effective) mass the only parameter that is different in our model variants is the incompressibility, it is worth studying the relation between incompressibility and maximum star mass while keeping every other parameter constant. For this end, in the model with U_{34} interaction terms, the Landau mass fixed for the value that is listed in Table 1 and the model is solved to reproduce different values of incompressibility. All these different parametrizations give a different dense matter equation of states that predict different M–R diagrams and different maximum mass star parameters, $M_{Max\,M}$. These results are summarized in the panels of Figure 5.

It can be seen in these plots that the the mass and radius of the maximum mass star are insensitive to the value of the incompressibility. The equations of the best (constant) linear fits are

$$\begin{aligned} M_{max\,M} &= 1.779 + 0.0008\,K, \\ R_{max\,M} &= 7.870 + 0.0070\,K, \end{aligned} \tag{11}$$

where $M_{max\,M}$ is measured in units of M_{Sun}, K in MeV units, and $R_{max\,M}$ is in km.

The slope of both linear functions is tiny and even double the K produces only a few percent of variation in the maximum mass star mass and radius. These results provide a heuristic understanding of the previous results on the dense matter properties: the maximum mass star radius and mass depend strongly on nucleon Landau mass, since its dependence on incompressibility is negligible compared to how strongly it depends on the value of effective mass and all other parameters are kept constant.

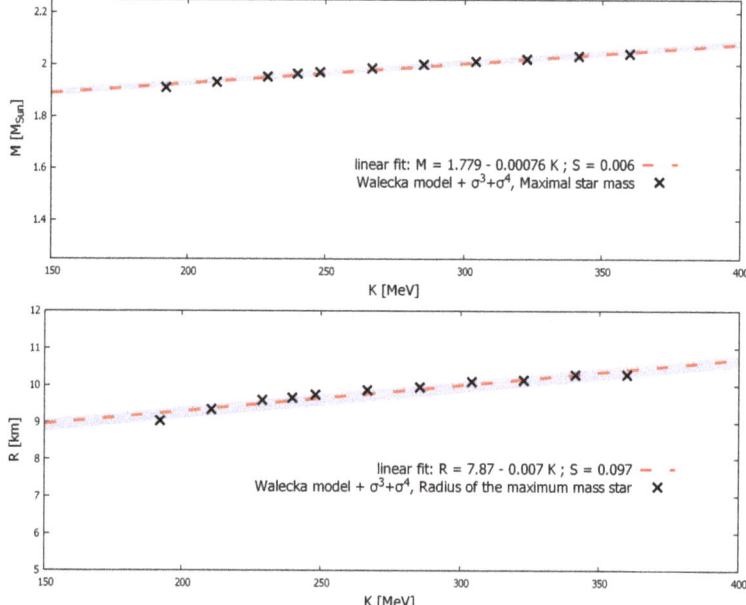

Figure 5. The maximum mass star mass, $M_{Max\,M}$ (*upper panel*) and radius, $R_{max\,M}$ (*lower panel*) as a function of the compression modulus, K. The dashed line is the best linear fit, which is plotted as guidance.

7. Conclusions

We investigate the macroscopical observables of neutron stars with a symmetric dense nuclear and matter in the interior. We compared three cases with different general bosonic interactions, and we used variations of the nuclear parameter fit method. We investigated how these bosonic interactions matter in the nuclear equation of state, and especially how the effective mass and Landau mass parameters play role in the nuclear potentials. The applied model is neglecting leptonic fields, and, therefore, there is β-equilibrium. Although we had this simplified theoretical case, such a clean nuclear environment led us to explore the effects of its parameters on macroscopical observables of a theoretical compact object.

With variation of these microscopical parameters, we explored the mass-radius groups in the case of a Schwarzschild neutron star model. We presented that the Landau mass of the nuclear matter has a definite linear relation to the maximum mass and maximum radius of the neutron star, considered within the best U_{34} model case. We also obtained that the effect of the incompressibility is negligible on these macroscopical parameters, with $M_{max\,M} \approx 1.78$ and $R_{max\,M} = 7.87$ within the regime $K =200–400$, meaning $\lesssim 5\%$ variation for both cases.

These results support the idea that the evolution equation of the strongly interacting matter saves the magnitude of the uncertainty that originated from the microscopical physical parameters, which appears as deterministic variation of the macroscopical observables as suggested in Refs. [18,19].

Finally, we note that an investigation of the asymmetric nuclear matter version of this study is ongoing, like in Ref. [30,31]. Our preliminary results suggest that taking into account β-equilibrium in the asymmetric case does not change much the observed effects, supporting the phenomena summarized in this paper.

Author Contributions: Authors contributed to the paper equally.

Funding: This work is supported by the Hungarian Research Fund (OTKA) under contracts No. K120660, K123815, NKM-81/2016 MTA-UA bilateral mobility program, and PHAROS (CA16214), and THOR (CA15213) COST actions.

Acknowledgments: The authors also acknowledge the computational resources for the Wigner GPU Laboratory.

Conflicts of Interest: The authors declare no conflicts of interest.

References

1. NASA. Nicer. Available online: https://www.nasa.gov/nicer (accessed on 1 May 2019).
2. Merloni, A.; Predehl, P.; Becker, W.; Böhringer, H.; Boller, T.; Brunner, H.; Brusa, M.; Dennerl, K.; Freyberg, M.; Friedrich, P.; et al. eROSITA Science Book: Mapping the Structure of the Energetic Universe 2012. Available online: http://xxx.lanl.gov/abs/1209.3114 (accessed on 1 May 2019).
3. Athena. The Athena X-ray Observatory. Available online: https://www.the-athena-x-ray-observatory.eu/ (accessed on 1 May 2019).
4. Ozel, F.; Psaltis, D.; Arzoumanian, Z.; Morsink, S.; Baubock, M. Measuring Neutron Star Radii via Pulse Profile Modeling with NICER. *Astrophys. J.* **2016**, *832*, 92. [CrossRef]
5. Watts, A.L. Constraining the neutron star equation of state using Pulse Profile Modeling. In Proceedings of the Xiamen-CUSTIPEN Workshop on the EOS of Dense Neutron-Rich Matter in the Era of Gravitational Wave Astronomy, Xiamen, China, 3–7 January 2019.
6. Abbott, B.P.; Abbott, R.; Abbott, T.D.; Acernese, F.; Ackley, K.; Adams, C.; Adams, T.; Addesso, P.; Adhikari, R.X.; Adya, V.B.; et al. GW170817: Measurements of Neutron Star Radii and Equation of State. *Phys. Rev. Lett.* **2018**, *121*, 161101. [CrossRef] [PubMed]
7. Abbott, B.P.; Abbott, R.; Abbott, T.D.; Acernese, F.; Ackley, K.; Adams, C.; Adams, T.; Addesso, P.; Adhikari, R.X.; Adya, V.B.; et al. [The LIGO Scientific Collaboration and the Virgo Collaboration] GW170817: Observation of Gravitational Waves from a Binary Neutron Star Inspiral. *Phys. Rev. Lett.* **2017**, *119*, 161101. [CrossRef] [PubMed]
8. Rezzolla, L.; Takami, K. Gravitational-wave signal from binary neutron stars: A systematic analysis of the spectral properties. *Phys. Rev.* **2016**, *D93*, 124051. [CrossRef]
9. Ozel, F.; Psaltis, D.; Guver, T.; Baym, G.; Heinke, C.; Guillot, S. The Dense Matter Equation of State from Neutron Star Radius and Mass Measurements. *Astrophys. J.* **2016**, *820*, 28. [CrossRef]
10. Raithel, C.A.; Özel, F.; Psaltis, D. From Neutron Star Observables to the Equation of State. II. Bayesian Inference of Equation of State Pressures. *Astrophys. J.* **2017**, *844*, 156. [CrossRef]
11. Guenther, J.; Bellwied, R.; Borsányi, S.; Fodor, Z.; Katz, S.; Pásztor, A.; Ratti, C.; Szabó, K. The QCD equation of state at finite density from analytical continuation. *Nuclear Phys. A* **2017**, *967*, 720–723. [CrossRef]
12. Günther, J.; Bellwied, R.; Borsanyi, S.; Fodor, Z.; Katz, S.D.; Pasztor, A.; Ratti, C. The QCD equation of state at finite density from analytical continuation. *EPJ Web Conf.* **2017**, *137*, 07008. [CrossRef]
13. Endrődi, G.; Fodor, Z.; Katz, S.D.; Sexty, D.; Szabó, K.K.; Török, C. Applying constrained simulations for low temperature lattice QCD at finite baryon chemical potential. *Phys. Rev. D* **2018**, *98*, 074508. [CrossRef]
14. Holt, J.W.; Rho, M.; Weise, W. Chiral symmetry and effective field theories for hadronic, nuclear and stellar matter. *Phys. Rept.* **2016**, *621*, 2–75. [CrossRef]
15. Kojo, T. QCD in stars. In Proceedings of the 10th International Workshop on Critical Point and Onset of Deconfinement (CPOD 2016), Wrocław, Poland, 30 May–4 June 2016.
16. Barnafoldi, G.G.; Jakovac, A.; Posfay, P. Harmonic expansion of the effective potential in a functional renormalization group at finite chemical potential. *Phys. Rev.* **2017**, *D95*, 025004. [CrossRef]
17. Kovács, P.; Szép, Z. Critical surface of the $SU(3)_L \times SU(3)_R$ chiral quark model at nonzero baryon density. *Phys. Rev. D* **2007**, *75*, 025015. [CrossRef]
18. Pósfay, P.; Barnaföldi, G.G.; Jakovác, A. The effect of quantum fluctuations in compact star observables. *Publ. Astron. Soc. Austral.* **2018**, *35*, 19. [CrossRef]
19. Pósfay, P.; Barnaföldi, G.G.; Jakovác, A. Effect of quantum fluctuations in the high-energy cold nuclear equation of state and in compact star observables. *Phys. Rev.* **2018**, *C97*, 025803. [CrossRef]
20. Glendenning, N.K. *Compact Stars: Nuclear Physics, Particle Physics, and General Relativity*; Astronomy and Astrophysics Library; Springer: Berlin, Germany, 1997.

21. Hajizadeh, O.; Maas, A. Constructing a neutron star from the lattice in G_2-QCD. *Eur. Phys. J.* **2017**, *A53*, 207. [CrossRef]
22. Walecka, J. A theory of highly condensed matter. *Ann. Phys.* **1974**, *83*, 491–529. [CrossRef]
23. Johnson, M.H.; Teller, E. Classical Field Theory of Nuclear Forces. *Phys. Rev.* **1955**, *98*, 783–787. [CrossRef]
24. Schmitt, A. Dense matter in compact stars: A pedagogical introduction. *Lect. Notes Phys.* **2010**, *811*, 1–111. [CrossRef]
25. Meng, J. *Relativistic Density Functional for Nuclear Structure*; International Review of Nuclear Physics; World Scientific Publishing Company: Singapore, 2016.
26. Prakash, M.; Cooke, J.R.; Lattimer, J.M. Quark-hadron phase transition in protoneutron stars. *Phys. Rev. D* **1995**, *52*, 661–665. [CrossRef]
27. Wiringa, R.B.; Fiks, V.; Fabrocini, A. Equation of state for dense nucleon matter. *Phys. Rev. C* **1988**, *38*, 1010–1037. [CrossRef]
28. Akmal, A.; Pandharipande, V.R.; Ravenhall, D.G. Equation of state of nucleon matter and neutron star structure. *Phys. Rev. C* **1998**, *58*, 1804–1828. [CrossRef]
29. Özel, F.; Freire, P. Masses, Radii, and the Equation of State of Neutron Stars. *Ann. Rev. Astron. Astrophys.* **2016**, *54*, 401–440. [CrossRef]
30. Pearson, J.M.; Chamel, N.; Potekhin, A.Y.; Fantina, A.F.; Ducoin, C.; Dutta, A.K.; Goriely, S. Unified equations of state for cold non-accreting neutron stars with Brussels-Montreal functionals—I. Role of symmetry energy. *Mon. Not. R. Astron. Soc.* **2018**, *481*, 2994–3026. [CrossRef]
31. Weissenborn, S.; Chatterjee, D.; Schaffner-Bielich, J. Hyperons and massive neutron stars: The role of hyperon potentials. *Nucl. Phys.* **2012**, *A881*, 62–77. [CrossRef]

© 2019 by the authors. Licensee MDPI, Basel, Switzerland. This article is an open access article distributed under the terms and conditions of the Creative Commons Attribution (CC BY) license (http://creativecommons.org/licenses/by/4.0/).

Article

Averaging and the Shape of the Correlation Function

Boris Tomášik [1,2,*], Jakub Cimerman [2] and Christopher Plumberg [3]

1. Faculty of Natural Sciences, Univerzita Mateja Bela, Tajovského 40, 97401 Banská Bystrica, Slovakia
2. FJFI, České vysoké učení technické v Praze, Břehová 7, 11519 Praha 1, Czech Republic; jakub.cimerman@gmail.com
3. Department of Astronomy and Theoretical Physics, Lund University, Sölvegatan 14A, SE-223 62 Lund, Sweden; astrophysicist87@gmail.com
* Correspondence: boris.tomasik@cern.ch

Received: 30 April 2019; Accepted: 11 June 2019; Published: 13 June 2019

Abstract: A brief pedagogical introduction to correlation femtoscopy is given. We then focus on the shape of the correlation function and discuss the possible reasons for its departure from the Gaussian form and better reproduction with a Lévy stable distribution. With the help of Monte Carlo simulations based on asymmetric extension of the Blast-Wave model with resonances we demonstrate possible influence of averaging over many events and integrating over wide momentum bins on the shape of the correlation function. We also show that the shape is strongly influenced by the use of the one-dimensional parametrisation in the q_{inv} variable.

Keywords: correlation femtoscopy; heavy-ion collisions; Lévy stable parametrisation; event-by-event fluctuations

1. Introduction

Correlation femtoscopy is widely used in heavy-ion collisions for the determination of space–time characteristics of hadron-emitting sources. Most commonly, the two-particle correlation functions are fitted by a Gaussian parametrisation augmented with correction terms due to final-state interactions. The widths of this parametrisation are interpreted in terms of space–time (co-)variances of the homogeneity regions [1–5].

Nevertheless, clear indications exist, that the real shape of the correlation function is not Gaussian, as we could also see in a few talks at the 2018 Zimányi School (see e.g., [6–8]). The shape is often better reproduced by a fit with Lévy stable distribution [9]. The choice of this distribution is not random. Stability is a generalisation of the concept of Central Limit Theorems. Lévy stable distributions possess the property that the shape remains unchanged when one more elementary random process is added to the ones which are already accounted for. The excitement about this particular parametrisation is supported by the argument that with the help of such a fit one could access the critical exponents of the strongly interacting matter [10]. We will show in this paper that the observed shape can be caused by more mundane non-critical phenomena. Mostly, we are interested in the role of averaging in influencing the shape of the correlation function. Note that here we shall only be interested in correlation functions from nuclear collisions.

These ideas set out the outline of this contribution. We first review the basic relations of correlation femtoscopy. Then we particularly look at the Lévy stable parametrisations and scrutinise various effects that can lead to such a shape of the correlation function.

2. The Formalism of Correlation Femtoscopy

The two-particle correlation function is constructed in such a way as to reveal the effect of correlations. Since this is a school, we stay on a pedagogical level and introduce the main elements of correlation femtoscopy one by one. For pion pairs one usually uses the correlation stemming from the symmetrisation of the wave function. Nevertheless, the effect of final state interactions is always present, as well. They can be due to electromagnetic or strong interactions. For identical charged pions, the electromagnetic Coulomb final state interactions are important. We shall assume here, that their influence can be factored out from the data with the help of a correction factor [11,12].

The correlation function is then experimentally obtained as

$$C(p_1, p_2) = \frac{P(p_1, p_2)}{P_{\mathrm{mix}}(p_1, p_2)}, \tag{1}$$

where $P(p_1, p_2)$ is the two-particle distribution in the momenta, and $P_{\mathrm{mix}}(p_1, p_2)$ is an analogous distribution in which each particle comes from a different event. Due to wave function symmetrisation for boson pairs, the correlation function exhibits a peak for small momentum differences $p_1 - p_2$, provided that Coulomb repulsion can be filtered out. Thus it is more convenient to study the dependence of the correlation function on the momentum difference and the average momentum

$$q = p_1 - p_2, \qquad K = \frac{1}{2}(p_1 + p_2). \tag{2}$$

The source, which produces particles, can be described with the help of a Wigner density $S(x, p)$. Its classical interpretation is that it is the probability to emit a particle with momentum p from a space–time point x. The correlation function which we express as function of q and K is then given as

$$C(q, K) - 1 \approx \frac{\left| \int d^4x\, S(x, K)\, e^{iqx} \right|^2}{\left(\int d^4x\, S(x, K) \right)^2}. \tag{3}$$

The approximation symbol stands here for two steps: (i) The on-shell approximation which replaces the time-component of $K^0\ (= (p_1^0 + p_2^0)/2)$ with $E_K = \sqrt{\vec{K}^2 + m^2}$, and (ii) the smoothness approximation, which assumes that the denominator can be evaluated at single value of K instead of the two momenta of the particles of the pair: $p_1 = K + q/2$ and $p_2 = K - q/2$. After some manipulations the relation can be rewritten as a simple Fourier transform:

$$C(q, K) \approx 1 + \frac{\int d^4r\, D(r, K)\, e^{iqr}}{\left(\int d^4x\, S(x, K) \right)^2}, \tag{4}$$

where

$$D(r, K) = \int d^4X\, S\left(X + \frac{r}{2}, K\right) S\left(X - \frac{r}{2}, K\right). \tag{5}$$

We see that the correlation function does not measure the distribution of the source itself. Instead, it is a Fourier transform of the distribution of the differences between emission points! This is important! The convolution in Equation (5) often produces a bell-shaped distribution $D(r, K)$ even for emission functions which might possess sharp edges. The Fourier transform in Equation (4) keeps this feature. This is the reason why fitting the correlation function with Gaussian does not seem such a bad idea.

Unfortunately, even measurement of the distribution $D(r, K)$ is not completely possible. Since the momenta of the particles used in the measurement must fulfil the mass-shell constraint, we have

$$q \cdot K = 0 \quad \Rightarrow \quad q^0 = \frac{\vec{q} \cdot \vec{K}}{K^0} = \vec{q} \cdot \vec{\beta} \tag{6}$$

where

$$\vec{\beta} = \frac{\vec{K}}{K^0} \approx \frac{\vec{K}}{E_K}. \tag{7}$$

Hence, only three components of q are independent.

In order to exploit the symmetries of the problem and simplify the interpretation of the measurement, one adopts the *out-side-long* coordinate frame. The longitudinal (or z) direction is parallel to the beam. The outward (or x) direction is identified with the direction of the transverse pair momentum K_T, so that $\vec{K} = (K_T, 0, K_l)$. The sideward (or y) direction is perpendicular to the above two. In this frame the Gaussian parametrisation of the correlation function reads

$$C(q,K) - 1 = \exp\left[-q_o^2 R_o^2 - q_s^2 R_s^2 - q_l^2 R_l^2 - 2q_o q_s R_{os}^2 - 2q_o q_l R_{ol}^2 - 2q_s q_l R_{sl}^2\right]. \tag{8}$$

If this parametrisation is expanded up to second order in q and compared with such an expansion of Equation (3), one recovers the model-independent expressions for the correlation radii [13]

$$\begin{aligned}
R_o^2 &= \left\langle (\tilde{x} - \beta_T \tilde{t})^2 \right\rangle, & R_{os}^2 &= \left\langle (\tilde{x} - \beta_T \tilde{t})\tilde{y} \right\rangle, \\
R_s^2 &= \left\langle \tilde{y}^2 \right\rangle, & R_{ol}^2 &= \left\langle (\tilde{x} - \beta_T \tilde{t})(\tilde{z} - \beta_l \tilde{t}) \right\rangle, \\
R_l^2 &= \left\langle (\tilde{z} - \beta_l \tilde{t})^2 \right\rangle, & R_{sl}^2 &= \left\langle \tilde{y}(\tilde{z} - \beta_l \tilde{t}) \right\rangle.
\end{aligned} \tag{9}$$

Here, the averages are taken with the emission function

$$\langle f(x) \rangle (K) = \frac{\int d^4 x\, f(x)\, S(x, K)}{\int d^4 x\, S(x, K)}, \tag{10}$$

and the coordinates with the tilde are shifted with respect to the means

$$\tilde{x} = x - \langle x \rangle.$$

A caveat must be placed here in connection with the Expressions (9). They are well defined as long as the (co)variances of the emission function can be calculated. If the emission function, however, would be given by a Lévy stable distribution, they would not exist and the interpretation would fail. Moreover—the interpretation is even more complicated: Below we discuss how the measured correlation function results from averaging over different (effective) sources, so strictly speaking no single emission function can be assigned to the measured correlations. Nevertheless, let us consider the Expressions (9) as useful guidelines for the interpretation of measured Gaussian correlation radii.

Sometimes, poor statistics does not allow to sample the q-space densely enough with data so that a decent fit to the histogram can be made. In this case, a one-dimensional parametrisation of the correlation function is sometimes used, which is formulated in terms of the invariant momentum difference

$$q_{inv}^2 = q_o^2 + q_s^2 + q_l^2 - q_0^2 = |\vec{q}|^2 - (\vec{q} \cdot \vec{\beta})^2. \tag{11}$$

In order to calculate the correlation function $C(q_{inv}, K)$ one would need to integrate both the numerator and the denominator of the three-dimensional $C(q, K)$ separately over the hypersurface in q-space

$$C(q_{inv}, K) = 1 + \frac{\int dq_o dq_s dq_l\, \delta(|\vec{q}|^2 - (\vec{q} \cdot \vec{\beta})^2 - q_{inv}^2) \left| \int d^4 x\, S(x, K) e^{iqx} \right|^2}{\int dq_o dq_s dq_l\, \delta(|\vec{q}|^2 - (\vec{q} \cdot \vec{\beta})^2 - q_{inv}^2) \left(\int d^4 x\, S(x, K) \right)^2}. \tag{12}$$

Note that we still assume the validity of the smoothness and the on-shell approximations.

This can be most easily interpreted in the reference frame which co-moves with the particle pair, i.e., with the velocity $\vec{\beta}$. There, all terms which contain $\vec{\beta}$ vanish, and the correlation function becomes

$$C(q_{inv}, K) = 1 + \frac{\int dq_o dq_s dq_l\, \delta(|\vec{q}|^2 - q_{inv}^2) \left| \int d^4 x\, S(x, K) e^{iqx} \right|^2}{\int dq_o dq_s dq_l\, \delta(|\vec{q}|^2 - q_{inv}^2) \left(\int d^4 x\, S(x, K) \right)^2}. \tag{13}$$

If the modification is done on the level of Gaussian parametrisation

$$C(q_{inv}, K) = 1 + \frac{1}{\mathcal{N}} \int dq_o dq_s dq_l \, \delta(|\vec{q}|^2 - q_{inv}^2) \, \exp\left(-q_i q_j \langle x^i x^j \rangle\right), \tag{14}$$

where

$$\mathcal{N} \equiv \int dq_o dq_s dq_l \, \delta(|\vec{q}|^2 - q_{inv}^2) \tag{15}$$

ensures that the normalization of the correlation function is unaffected by integral over the q-surface. Hence, in this frame the averaging runs over the q-surface with constant $|\vec{q}|$. The observed width of $C(q_{inv}, K)$ results from this averaging.

The value of q_{inv} can become 0 (where one would expect the maximum of the correlation function) also for non-vanishing q components (where the maximum is not expected). Therefore, a new variable has been introduced [14]

$$q_{LCMS} = \left(q_o^2 + q_s^2 + q_{l,LCMS}^2\right)^{1/2} \quad \text{where} \quad q_{l,LCMS}^2 = \frac{(p_{1z} E_2 - p_{2z} E_1)^2}{K_0^2 - K_l^2}. \tag{16}$$

Note that q_{LCMS} is invariant under longitudinal boosts. The motivation for this particular variable is that in the longitudinally co-moving frame ($K_l = 0$) it reduces just to $\sqrt{q_o^2 + q_s^2 + q_l^2}$. This variable is thus reasonable for spherically symmetric sources. We have checked that there is no dramatic qualitative difference between the results obtained with q_{inv} and those obtained with q_{LCMS} [15]. Here we shall show results obtained with q_{inv}, while results with q_{LCMS} will be shown elsewhere [15].

We have indicated in Equation (10) that the measured (co-)variances of the source depend on the pair momentum K. Why? The reason is that the particles with specified momentum only come from a part of the whole fireball, the so-called homogeneity region. If we change the momentum K, i.e., we focus on particles with different momentum, then these will be emitted from a different part of the fireball. That part also may have different size. Consequently, the sizes of the (co-)variances in Equation (9) change.

3. Averaging

We shall deal with two kinds of averaging in the discussions of the shape of the correlation function: Averaging over different momenta and averaging over many events.

Averaging over momentum comes through the binning in pair momentum K. Both histograms in q—the numerator and the denominator in Equation (1)—are constructed for $\vec{K}$ within certain interval. The bins always have finite size in the transverse component as well as in the azimuthal angle of $\vec{K}$. The latter is often even integrated over the whole 2π interval. As we pointed out, the correlation function measures the homogeneity lengths corresponding to a given $\vec{K}$. Then by taking an interval of $\vec{K}$ one makes an average over different homogeneity regions. Since the intervals of $\vec{K}$ are integrated on the level of the two histograms in Equation (1), the resulting correlation function is given as

$$C(q, K) \approx 1 + \frac{\int_{bin} dK \left| \int d^4x \, S(x, K) e^{iqx} \right|^2}{\left(\int_{bin} dK \int d^4x \, S(x, K)\right)^2}. \tag{17}$$

Averaging over events results from summing up entries to the histograms from a large number of events. Both the numerator and the denominator fluctuate from event to event, because in each event we have a fireball of different sizes and dynamical state. Averaging must therefore be carried out for the numerator and denominator separately [16,17]. We thus conclude that the correlation function will be

$$C(q, K) \approx 1 + \frac{\int dR \, \rho(R) \left| \int d^4x \, S(x, K; R) e^{iqx} \right|^2}{\int dR \, \rho(R) \left(\int d^4x \, S(x, K; R)\right)^2} \tag{18}$$

where $\rho(R)$ is the distribution of the source sizes. For brevity, we do not write out the averaging over other features of the fluctuating source explicitly; this would be implemented in the same way.

A non-Gaussian shape of the correlation function will be here fitted with the Lévy stable distribution. In one dimension is reads

$$C(q) = 1 + \lambda e^{-(qR)^\alpha}. \tag{19}$$

The three-dimensional generalisation can be formulated as [9]

$$C(q) = 1 + \lambda e^{-(q_o^2 R_o'^2 + q_s^2 R_s'^2 + q_l^2 R_l'^2)^{\alpha/2}}. \tag{20}$$

The Lévy exponent α is one of the fit parameters, together with λ and the R''s. The value $\alpha = 2$ corresponds to Gaussian shape, meaning that a lower value ($\alpha < 2$) implies a non-Gaussian correlation function.

4. The Blast-Wave Model

For the actual calculation of various effects we will generate artificial events with the help of DRAGON Monte Carlo event generator [18,19]. It is based on the Blast-Wave (BW) model [20–24] with resonance decays included. The BW model is described by the emission function

$$S(x,K)\,d^4x = \frac{1}{(2\pi)^3} \left(\exp\left(\frac{u^\mu(x) p_\mu}{T}\right) \pm 1 \right)^{-1} \Theta(r - R(\theta))\, \delta(\tau - \tau_{fo})\, m_t \cosh(\eta - y)\tau\, d\tau\, d\eta\, r\, dr\, d\theta. \tag{21}$$

As spatial coordinates we use here the usual polar coordinates r and θ for the plane transverse to the beam direction, the space–time rapidity η and the longitudinal proper time τ

$$\eta = \frac{1}{2} \ln \frac{t+z}{t-z}, \qquad \tau = \sqrt{t^2 - z^2}. \tag{22}$$

Let us explain the formula representing this particular emission function.

- The factor $(2\pi)^{-3}$ stands for the elementary phase-space cell volume. Recall that $S(x,K)$ represents the distribution in phase-space. (And recall that $\hbar = c = 1$.)
- The thermal distribution—Bose-Einstein or Fermi-Dirac—is formulated with the energy in the rest frame of the fluid, $E^* = u^\mu p_\mu$, where u^μ is the (local) velocity of the fluid.
- The fireball is modelled with a sharp cutoff in the transverse direction: $\Theta(r - R(\theta))$. However, the radius R depends on the azimuthal angle in order to simulate the fireball in non-central collisions.
- Freeze out happens along a hypersurface given by constant $\tau = \tau_{fo}$.
- The fireball is manifestly boost-invariant. There is no limit set on the space–time rapidity. Nevertheless, by choosing the rapidity of particles one effectively selects just a part of the fireball (the relevant homogeneity region) which contributes to the production at that rapidity.
- The factor $m_t \cosh(\eta - y)\tau\, d\tau\, d\eta\, r\, dr\, d\theta$, where y is the rapidity of the emitted particle, comes from the Cooper-Frye [25] factor which stands for the flux of particles across the freeze-out hypersurface Σ: $p_\mu d\Sigma^\mu$.

Our model does not include any corrections to the thermal momentum distribution due to viscosity.

The transverse radius of the fireball depends on the azimuthal angle in order to implement the second-order anisotropy

$$R(\theta) = R_0 \left[1 - a_2 \cos(2(\theta - \theta_2)) \right], \tag{23}$$

where R_0 and a_2 are model parameters, and θ_2 is the angle of the second-order event plane.

The collective expansion velocity field is parametrised with the help of η and the transverse rapidity $\eta_t(r, \theta_b)$

$$u^\mu(x) = (\cosh\eta \cosh\eta_t(r,\theta_b), \cos\theta_b \sinh\eta_t(r,\theta_b), \sin\theta_b \sinh\eta_t(r,\theta_b), \sinh\eta \cosh\eta_t(r,\theta_b)), \quad (24)$$

where the transverse rapidity depends on r and the azimuthal angle

$$\eta_t(r,\theta_b) = \rho_0 \frac{r}{R(\theta)} \left[1 + 2\rho_2 \cos(2(\theta_b - \theta_2))\right]. \quad (25)$$

The model parameters ρ_0 and ρ_2 scale the overall magnitude and the second-order oscillation of the transverse flow, respectively. We have indicated that η_t depends on θ_b, and not directly on θ. The angle θ_b gives the direction perpendicular to the surface of the fireball, and can be obtained from the relation

$$\tan\left(\theta_b - \frac{\pi}{2}\right) = \frac{dx_2}{dx_1} = \frac{\frac{dx_2}{d\theta}}{\frac{dx_1}{d\theta}} = \frac{\frac{dR(\theta)\sin(\theta)}{d\theta}}{\frac{dR(\theta)\cos(\theta)}{d\theta}}, \quad (26)$$

where the functional dependences $x_1(\theta)$, $x_2(\theta)$ refer to the transverse boundary of the fireball [26].

DRAGON also includes resonance decays. Mesonic resonances are included up to masses of 1.5 GeV, baryonic up to 2 GeV. Resonances are produced according to the same emission function as direct pions, with their pole masses. The decay vertex of a given resonance is determined according to an exponential distribution whose width is the lifetime of the resonance in question (in its rest frame): $\rho(\tau_d) \propto e^{-\Gamma_R \tau_d}$. Both two- and three-body decays are included as well as the possibility that one resonance type can decay via various channels according to their branching ratios. Cascades of decays, in which several resonances decay consecutively, are also possible within the model.

5. Results

We begin by considering the effects that the averaging over many fireballs with different shapes may have on the value of the Lévy parameter α. Indeed, in real experiments each fireball is different, with different sizes, eccentricities, and orientations of the event plane. We therefore anticipate that averaging over a distribution of source shapes, as in Equation (18), will cause α to deviate from 2. In this treatment, we limit our focus to second-order anisotropies.

With the help of DRAGON we generated sets of 50,000 events. The basic setting of the parameters includes the freeze-out temperature of 120 MeV, the average transverse radius $R = 7$ fm, freeze-out time $\tau_{fo} = 10$ fm/c, and the strength of the transverse expansion $\rho_0 = 0.8$. To keep the source simple, no resonance decays are included at this point. The correlation function is evaluated in one dimension as a function of q_{inv}.

We first study the effect of averaging over different values of a_2.

Figure 1 (left) compares the Lévy parameter α from two sets of Monte Carlo events. In the first set, all events have spatial eccentricity with $a_2 = 0.05$. In the other set, the eccentricity fluctuates with a_2 between -0.1 and 0.1. We perform the study in a narrow interval of K_T, where the shape of the correlation function changes strongly, even though it cannot be accessed experimentally (although the correlation function has been averaged over the azimuthal angle of $\vec{K}$). We see that the value of α departs from 2 considerably and reaches values between 1.27 and 1.62. The averaging makes up only a small portion of this decrease, at most at a level of 0.05.

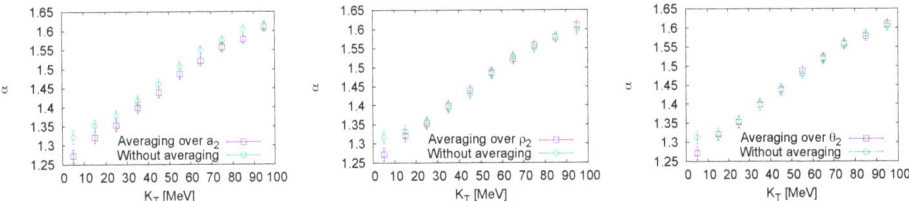

Figure 1. The Lévy parameter of the 1D fits to the correlation function in q_{inv}. Mean transverse momentum K_T in bins of 10 MeV. A green circle shows results calculated with fixed anisotropies. The purple data show results calculated for averaging over a_2 (**left**); averaging over ρ_2 (**middle**), and averaging over θ_2 (**right**). Vertical error bars show the 1σ intervals, resulting from fitting the correlation function with the ansatz of Equation (19).

Almost identical results quantitatively come from the averaging over flow anisotropy (Figure 1, middle) and the event plane orientation (Figure 1, right). In the middle panel we compare Lévy index α obtained from a set with ρ_2 fixed to 0.05 with a set with events for which ρ_2 fluctuates between -0.1 and 0.1. For the event plane averaging we see no change (except in the bin with smallest K_T) if θ_2 fluctuates in comparison to θ_2 fixed to 0. The anisotropy parameters a_2 and ρ_2 in this case fluctuate between -0.1 and 0.1.

We investigate next the influence of resonances on the obtained value of α. We use the source of direct particles with the same basic parameters as in the previous case, and we compare correlation functions obtained with and without resonance decays.

In Figure 2 we show the Lévy indices α obtained from fits to the correlation functions as a function of K_T. We know from the previous Figure already, that the one-dimensional correlation function has quite a non-Gaussian shape. Now we see that the inclusion of the resonance decays pushes down the value of α by another 0.2. The influence of resonance decays is much bigger than that of averaging over different events!

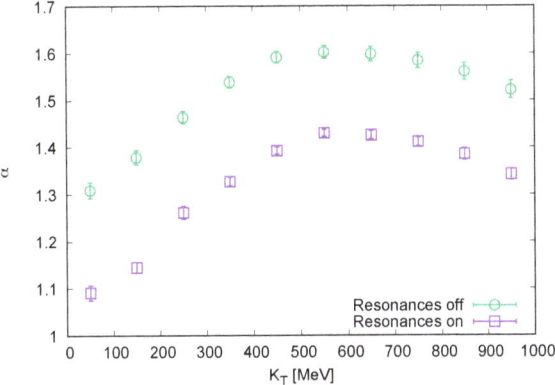

Figure 2. The Lévy parameter α of the 1D fits to the correlation function in q_{inv}. The result from the fits to the correlation function from a source without resonances (green circles) and with resonances (purple squares). Vertical error bars show the 1σ intervals.

We want to perform our analysis more differentially, however. We start by looking individually at each direction of the correlation function and fitting with the Lévy prescription of Equation (19). This is not a three-dimensional analysis, since we do not fit the correlation function in the whole q-space and confine ourselves only to fits along the axes. The aim is to see the differences in its shape along different directions.

Indeed, we observe in Figure 3 that the differences are rather large. If resonance decays are not included, the value of α is around 2 in both transverse directions. However, in the longitudinal direction the Lévy index α is lowered to 1.8 at $K_T = 0$ and increases gradually towards 2 at $K_T = 1$ GeV. In addition, the influence of resonance decays is different for longitudinal and transverse directions. In the longitudinal direction the resonance decays cause a decrease of α by about 0.2. In the transverse directions, however, α drops as low as 1.4 once resonance decays are included.

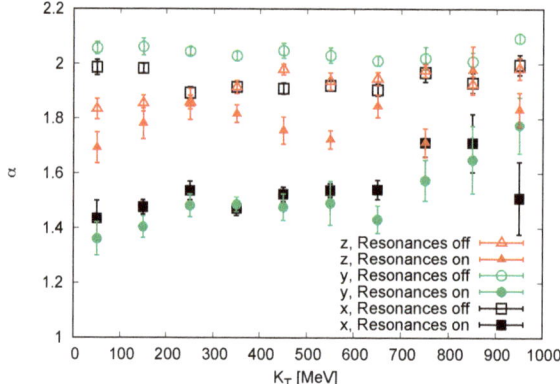

Figure 3. The Lévy parameter α from 1D fits to the correlation function in q_{inv} along the different axes, with or without resonances. Vertical error bars show the 1σ intervals, resulting from fitting the correlation function with the ansatz of Equation (19).

Note the wiggly behaviour of the data points, particularly for larger K_T. Due to the limited statistics of particle pairs, the measured values of the correlation function will tend to fluctuate. The parameter α comes from the fit to the correlation function with Equation (19). The error bars show the $\pm 1\sigma$ intervals for this particular correlation function. They underestimate the real uncertainty of the determination of α. We plan to improve the statistics and the uncertainty intervals in a forthcoming paper [15].

We would like to understand these differences and hence we checked the shape of the source which emits pions.

The profiles of the emission function are plotted in Figure 4, for pions with transverse momentum from the interval (300, 400) MeV. Note that they are produced just from a part of the whole fireball, the so-called homogeneity region. We show the distribution of the production points of pions, with pions from resonance decays included. In order to assess the effect of resonances, we also plot separately both contributions: Directly produced pions as well as those from resonance decays. The upper row shows that there is quite a difference between the longitudinal and the transverse directions. One could argue, however, that due to the on-shell constraint (6) it is not the distribution in x, that is measured, but rather the distribution in $(x - \beta_t t)$. We plot this in the lower left panel of Figure 4.

At this place we would again like to touch upon the discussion concerning the proper choice of the one-dimensional momentum difference variable (q_{inv} vs. q_{LCMS}). We recall that q_{LCMS} was introduced as a reasonable variable for spherically symmetric sources. Figure 4 shows that in this case the symmetry is not present, neither concerning the sizes, nor concerning the shape of the source in different directions.

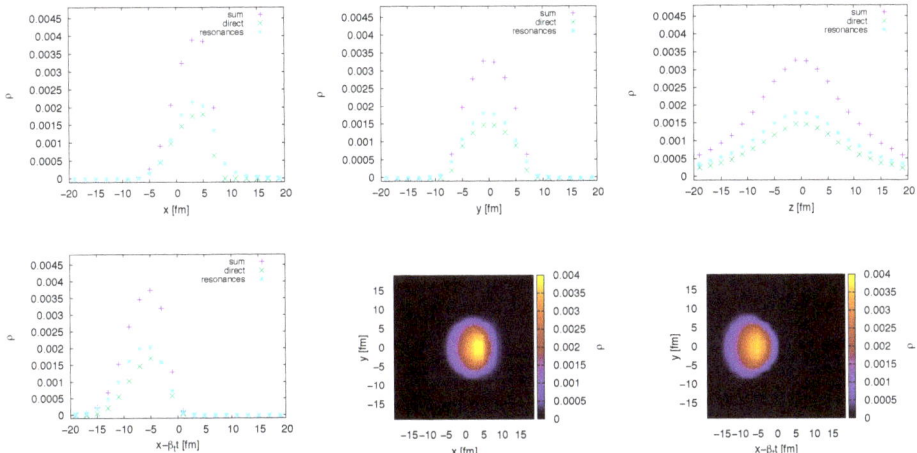

Figure 4. The spatial distribution of the emission points of pions for K_Ts. Upper row: The profiles of the emission points distribution along the x (**left**), y (**middle**), and z-axis (**right**). Lower row: The profile along the variable $(x - \beta_t t)$ (**left**), and two-dimensional distributions in the transverse plane (**middle** and **right**). A green × shows the profile of direct pions, a purple + shows the profile of pions produced by resonances and a blue * shows their sum. All these distributions were calculated as narrow integrals over the remaining coordinates with width 2 fm.

Finally, we extend the fitting to the whole three-dimensional correlation function from the previous simulations. The fit is performed with the three-dimensional Lévy distribution of Equation (20), so it always results in a single value of α.

This is plotted in Figure 5 as a function of K_T. We can see that the obtained α's are closer to 2 than in the case of fitting the one-dimensional correlation functions in q_{inv}, although considerable deviations from 2 are still present. Inclusion of resonance decays lowers α by about 0.1–0.3, depending on K_T.

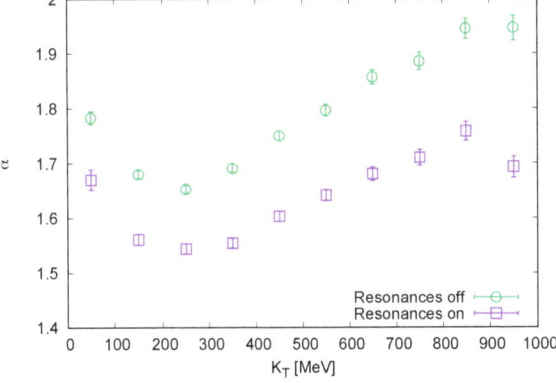

Figure 5. The Lévy parameter of the 3D fits to the correlation function according to Equation (20). Compared are simulations with and without resonances. Vertical error bars show the 1σ intervals, resulting from fitting the correlation function with the ansatz of Equation (20).

6. Conclusions

We explained in the introductory section of this paper that in addition to the generalisation of the concept of Central Limit Theorem [9], an especially important motivation for the use of Lévy stable parametrisation is the search for critical behaviour [10]. The simulations presented here show that the Lévy parameter α may clearly deviate from 2 even for "usual", non-critical sources, like those described by the blast-wave model.

Even without the presence of resonances, the shape of the correlation function looks rather non-Gaussian in the longitudinal direction. Once resonance decays are included, however, the shape deviates even further from Gaussian, especially in the two transverse directions. In this respect it appears interesting to test this with kaons, since a smaller part of them (although not negligible) comes from resonance decays.

In the three-dimensional fit, the value of α is lowered through the influence of resonance decays by about 0.1–0.3, depending on K_T.

It is very important to realise that in our simulations the biggest effect on lowering the value of α was in using only one-dimensional parametrisation of the correlation function in q_{inv}.

Author Contributions: The authors contributed to the paper in this way: conceptualization, B.T.; methodology, B.T., J.C., C.P.; software, J.C.; validation, C.P.; investigation, J.C. and B.T.; writing—original draft preparation, B.T.; writing—review and editing, J.C. and C.P.

Funding: This research was funded by the grant 17-04505S of the Czech Science Foundation (GAČR). B.T. acknowledges support from VEGA via grant No. 1/0348/18. C.P. gratefully acknowledges funding from the CLASH project (KAW 2017-0036).

Conflicts of Interest: The authors declare no conflict of interest. The funders had no role in the design of the study; in the collection, analyses, or interpretation of data; in the writing of the manuscript, or in the decision to publish the results.

References

1. Pratt, S. Pion Interferometry of Quark-Gluon Plasma. *Phys. Rev. D* **1986**, *33*, 1314–1327. [CrossRef]
2. Bertsch, G.; Gong, M.; Tohyama, M. Pion Interferometry in Ultrarelativistic Heavy Ion Collisions. *Phys. Rev. C* **1986**, *37*, 1896–1900. [CrossRef]
3. Makhlin, A.N.; Sinyukov, Y.M. Hydrodynamics of Hadron Matter Under Pion Interferometric Microscope. *Z. Phys. C* **1988**, *39*, 69–73. [CrossRef]
4. Pratt, S.; Csörgő, T.; Zimányi, J. Detailed predictions for two pion correlations in ultrarelativistic heavy ion collisions. *Phys. Rev. C* **1990**, *42*, 2646–2652. [CrossRef]
5. Chapman, S.; Scotto, P.; Heinz, U. A new cross term in the two particle HBT correlation function. *Phys. Rev. Lett.* **1995**, *74*, 4400–4403. [CrossRef] [PubMed]
6. Kincses, D. Lévy Analysis of HBT Correlation Functions in s N N = 62 GeV and 39 GeV Au + Au Collisions at PHENIX. *Universe* **2018**, *4*, 11. [CrossRef]
7. Kurgyis, B. Three-dimensional Lévy HBT Results from PHENIX. *Acta Phys. Pol. B Proc. Suppl.*, **2019**, *12*, 477. [CrossRef]
8. Pórfy, B. Levy HBT results at NA61/SHINE. *Universe*, under review.
9. Csörgő, T.; Hegyi, S.; Zajc, W.A. Bose-Einstein correlations for Lévy stable source distributions. *Eur. Phys. J. C* **2004**, *36*, 67–78. [CrossRef]
10. Csörgő, T. Correlation Probes of a QCD Critical Point. *PoS HIGH-PTLHC* **2008**, *8*, 027.
11. Bowler, M.G. Coulomb corrections to Bose-Einstein correlations have been greatly exaggerated. *Phys. Lett. B* **1991**, *270*, 69–74. [CrossRef]
12. Sinyukov, Y.; Lednický, R.; Akkelin, S.V.; Pluta, J.; Erazmus, B. Coulomb corrections for interferometry analysis of expanding hadron systems. *Phys. Lett. B* **1998**, *432*, 248–257. [CrossRef]
13. Chapman, S.; Scotto, P.; Heinz, U. Model independent features of the two particle correlation function. *Acta Phys. Hung. A* **1995**, *1*, 1–31.

14. Adare, A.; Aidala, C.; Ajitanand, N.N.; Akiba, Y.; Akimoto, R.; Alexander, J.; Alfred, M.; Al-Ta'ani, H.; Angerami, A.; Aoki, K.; et al. [PHENIX Collaboration]. Lévy-stable two-pion Bose-Einstein correlations in $\sqrt{s_{NN}} = 200$ GeV Au+Au collisions. *Phys. Rev. C* **2018**, *97*, 064911. [CrossRef]
15. Cimerman, J.; Plumberg, C.; Tomášik, B. Unpublished work, 2019, in preparation.
16. Plumberg, C.J.; Shen, C.; Heinz, U.W. Hanbury-Brown–Twiss interferometry relative to the triangular flow plane in heavy-ion collisions *Phys. Rev. C* **2013**, *88*, 044914; Erratum: *Phys. Rev. C* **2013**, *88*, 069901.
17. Plumberg, C.; Heinz, U. Probing the properties of event-by-event distributions in Hanbury-Brown–Twiss radii. *Phys. Rev. C* **2015**, *92*, 044906; Addendum: *Phys. Rev. C* **2015**, *92*, 049901. [CrossRef]
18. Tomášik, B. DRAGON: Monte Carlo generator of particle production from a fragmented fireball in ultrarelativistic nuclear collisions. *Comput. Phys. Commun.* **2009**, *180*, 1642–1653. [CrossRef]
19. Tomášik, B. DRoplet and hAdron generator for nuclear collisions: An update. *Comput. Phys. Commun.* **2016**, *207*, 545–546. [CrossRef]
20. Siemens, P.J.; Rasmussen, J.O. Evidence for a blast wave from compress nuclear matter. *Phys. Rev. Lett.* **1979**, *42*, 880–887. [CrossRef]
21. Schnedermann, E.; Sollfrank, J.; Heinz, U. Thermal phenomenology of hadrons from 200-A/GeV S+S collisions. *Phys. Rev. C* **1993**, *48*, 2462–2475. [CrossRef]
22. Csörgő, T.; Lörstad, B. Bose-Einstein correlations for three-dimensionally expanding, cylindrically symmetric, finite systems. *Phys. Rev. C* **1996**, *54*, 1390–1403. [CrossRef]
23. Tomášik, B.; Wiedemann, U.A.; Heinz, U. Reconstructing the freezeout state in Pb + Pb collisions at 158 GeV/c. *Acta Phys. Hung. A* **2003**, *17*, 105–143. [CrossRef]
24. Retiere, F.; Lisa, M.A. Observable implications of geometrical and dynamical aspects of freeze out in heavy ion collisions. *Phys. Rev. C* **2004**, *70*, 044907. [CrossRef]
25. Cooper, F.; Frye, G. Comment on the Single Particle Distribution in the Hydrodynamic and Statistical Thermodynamic Models of Multiparticle Production. *Phys. Rev. D* **1974**, *10*, 186. [CrossRef]
26. Cimerman, J.; Tomášik, B.; Csanád, M.; Lökös, S. Higher-order anisotropies in the Blast-Wave Model - disentangling flow and density field anisotropies. *Eur. Phys. J. A* **2017**, *53*, 161. [CrossRef]

© 2019 by the authors. Licensee MDPI, Basel, Switzerland. This article is an open access article distributed under the terms and conditions of the Creative Commons Attribution (CC BY) license (http://creativecommons.org/licenses/by/4.0/).

Communication

Multiplicity Dependence in the Non-Extensive Hadronization Model Calculated by the HIJING++ Framework

Gábor Bíró [1,2,*,†], Gergely Gábor Barnaföldi [1,*,†], Gábor Papp [2,*,†] and Tamás Sándor Biró [1,*,†]

[1] Department of Theory, Wigner Research Centre for Physics of the Hungarian Academy of Sciences, 29-33 Konkoly-Thege Miklós Str, H-1121 Budapest, Hungary
[2] Institute for Physics, Eötvös Loránd University, 1/A Pázmány P. Sétány, H-1117 Budapest, Hungary
* Correspondence: biro.gabor@wigner.mta.hu (G.B.); barnafoldi.gergely@wigner.mta.hu (G.G.B.); pg@ludens.elte.hu (G.P.); biro.tamas@wigner.mta.hu (T.S.B.)
† These authors contributed equally to this work.

Received: 30 April 2019; Accepted: 23 May 2019; Published: 1 June 2019

Abstract: The non-extensive statistical description of the identified final state particles measured in high energy collisions is well-known by its wide range of applicability. However, there are many open questions that need to be answered, including but not limited to, the question of the observed mass scaling of massive hadrons or the size and multiplicity dependence of the model parameters. This latter is especially relevant, since currently the amount of available experimental data with high multiplicity at small systems is very limited. This contribution has two main goals: On the one hand we provide a status report of the ongoing tuning of the soon-to-be-released HIJING++ Monte Carlo event generator. On the other hand, the role of multiplicity dependence of the parameters in the non-extensive hadronization model is investigated with HIJING++ calculations. We present cross-check comparisons of HIJING++ with existing experimental data to verify its validity in our range of interest as well as calculations at high-multiplicity regions where we have insufficient experimental data.

Keywords: high energy physics; heavy-ion; Monte Carlo; event generator; parallel computing; HIJING; non-extensive; Tsallis

1. Introduction

The transverse momentum (p_T) distribution of identified hadrons stemming from high-energy proton–proton, proton–nucleus, and nucleus–nucleus collisions is one of the most fundamental observables in high-energy physics. In recent years, the Tsallis–Pareto-like distributions, motivated from non-extensive statistical physics, have received close attention because their applicability in this field [1–7]. With the appearance of high precision experimental data spanning from low- to high-p_T, neither the thermal models with a bare Boltzmann–Gibbs exponential distribution nor perturbative Quantum Chromodynamics (pQCD)-motivated power-law distributions are able to describe the whole spectrum. On the other hand, the Tsallis–Pareto distributions combine these two regions perfectly (see, e.g., [1–16] and references therein). During the investigation of the parameters, we showed that they possess non-trivial relations such as mass- and energy scaling [1,8]. There are also implications that for larger systems a *soft-hard* extension is needed [10,16]. These studies indicate that increasing the size of the colliding system (roughly speaking, the volume of the quark-gluon plasma) may also reflect in the parameters. Our goal is therefore to systematically explore the parameter space as the function of the event multiplicity.

The HIJING++ framework is a soon-to-be-published general purpose Monte Carlo event generator, currently in the final phase of development [17–22]. It will serve as the successor of the FORTRAN

HIJING, completely rewritten in modern C++. With the flexibility gained by using modular C++ structures, HIJING++ also utilizes several external packages [23–29]. Currently the internal parameters of HIJING++ are being tuned to main experimental observables using Professor [30,31]. This provides an excellent opportunity to test the capabilities of HIJING++ and calculate high-multiplicity events.

In the next section, we briefly summarize the progress of tuning in HIJING++ and present the current status. In Section 3, a theoretical description of the transverse momentum spectra is given, and the HIJING++ calculations are given in Section 4.

2. Tuning of HIJING++ Parameters

A typical general purpose Monte Carlo event generator, e.g., the HIJING++ framework, developed to be able to simulate high-energy heavy-ion collisions, has parameters that are not determined by theory and need to be tuned to reproduce measured experimental data with the highest possible precision. One of the main features of the HIJING++ framework is that very few input parameters are needed to fully define a run, such as the species of the projectile and target beam and center-of-mass energy. Given this information, all of the other intrinsic parameters are calculated automatically.

Since HIJING++ is based on the convolution of sequential collisions of nucleon–nucleon pairs in each nucleus–nucleus interaction, it is highly important to have a solid proton–proton collisions baseline. In this section, we present the up-to-date result of the tuning process using the following $\sqrt{s} = 7$ TeV proton–proton experimental data:

- p_T spectra of identified $\pi^\pm$, $K^\pm$, and $p(\bar{p})$ hadrons with $INEL > 0$ normalization (at least one charged particle in the $|\eta| < 1.0$ region is required) up to $p_T = 20$ GeV/c [32];
- charged hadron multiplicity distribution in the range of $\langle dN_{ch}/d\eta \rangle = 0 - 70$, where N_{ch} is the number of charged particles [33,34];
- charged hadron $\eta = \frac{1}{2} \ln \frac{p+p_z}{p-p_z}$ pseudorapidity distribution at mid-pseudorapidity $|\eta| < 1.0$ [33].

The tuning process is performed iteratively utilizing the Professor tool [30,31]. In Table 1, we list the main tunable parameters.

Table 1. Main internal parameters in HIJING++.

Parameter	Description
p_0	soft-hard separation scale: minimum p_T transfer of hard or semihard scatterings
σ_{soft}	the inclusive cross section for soft interactions
σ_0	the cross section that characterizes the geometrical size of a nucleon
μ_0	the parameter in the scaled eikonal function of nucleon used to calculate total cross-section
K	K-factor for the differential jet cross sections in the lowest order pQCD calculation
max $p_{T_{cut}}$	p_T cut for classifying the connected-independent type strings at fragmentation
$m_{inv-cut}$	invariant mass cut-off for the dipole radiation of a string system below which soft gluon radiations are terminated
$m_{min-inv-ex.str.}$	minimum value for the invariant mass of the excited string system in a hadron–hadron interaction
$S_{p_{T_1}}$	the parameter that regularizes the singularity at $p_T = 0$ in the distribution of the soft p_T kick
$S_{p_{T_2}}$	the parameter that gives the scale beyond which the p_T kick distribution will be similar to $1/p_T^4$
F	the scale in the form factor to suppress the p_T transfer to diquarks in hard scatterings
v_{q_i}	phenomenological parameters ($i = 1, 2, 3$) of the soft parton distribution function that yield an x distribution of the valence quarks in a soft interaction
s_{q_i}	phenomenological parameters ($i = 1, 2, 3$) of the soft parton distribution function that yield an x distribution of the sea quarks in a soft interaction
StringPT:temperature	the temperature parameter in the Lund fragmentation model as described in [23]
StringPT:tempPreFactor	the temperature prefactor for strange quarks and diquarks in the Lund fragmentation model as described in [23]
StringZ:aExtraSQuark	parameters in the Lund symmetric fragmentation function as described in [23]
StringZ:aExtraDiquark	

The detailed process of tuning and the parameters values will be described in the technical release paper of HIJING++. Here we present only the tuning status regarding the experimental data listed above.

In Figure 1, the multiplicity and the pseudorapidity distribution of charged hadrons are presented, while in Figure 2 the p_T spectrum of identified $\pi^\pm$, $K^\pm$, and $p(\bar{p})$ hadrons calculated and measured at $\sqrt{s} = 7$ TeV proton–proton collisions can be seen.

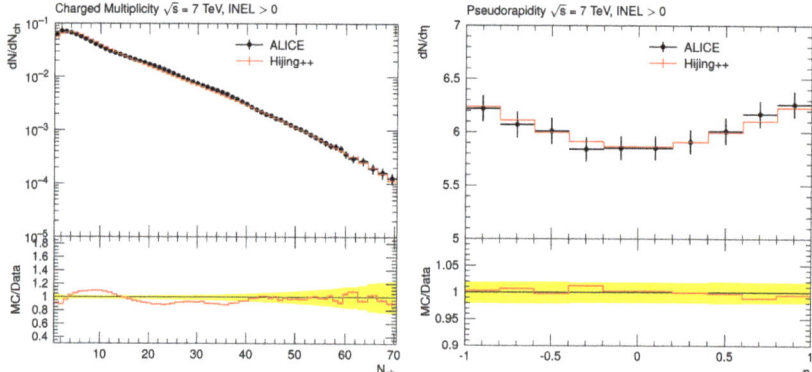

Figure 1. The multiplicity distribution (**left panel**) and pseudorapidity distribution (**right panel**) of charged hadrons stemming from proton–proton collisions at $\sqrt{s} = 7$ TeV calculated with HIJING++ and compared to experimental data [33,34].

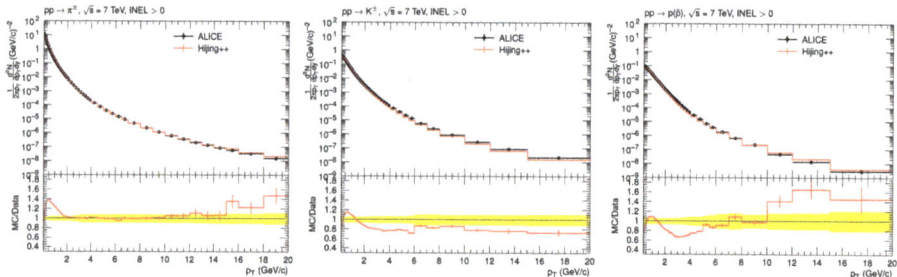

Figure 2. The p_T spectrum of identified $\pi^\pm$ (**left panel**), $K^\pm$ (**middle panel**), and $p(\bar{p})$ (**right panel**) hadrons yield from proton–proton collisions at $\sqrt{s} = 7$ TeV with $INEL > 0$ normalization calculated with HIJING++ and compared to experimental data [32].

The results above show that HIJING++ reproduces the event multiplicity excellently. In Figure 1, the agreement between the HIJING++ results and the experimental data is ∼15% for the multiplicity and ∼1% for the pseudorapidity distribution. The charged pion and kaon spectra also show a good agreement above $p_T = 2$ GeV/c, but the production is slightly overestimated at lower p_T values. The best agreement for the $\pi^\pm$ results is ∼1% between 2 and 15 GeV/c. For kaons, the yield is slightly underestimated above 2 GeV/c, where the agreement is ∼15–20%. On the other hand, the proton yield is overestimated in the large p_T region, the agreement is ∼20–30%.

3. The Non-Extensive Hadronization Model

It is a well-known and an intensively studied phenomenon that the transverse momentum spectra of hadrons stemming from high-energy particle collisions can be described by Tsallis–Pareto type distributions [1–3,6,8–13]. Although this observation itself has further consequences, the theory has even more subtle details because of the observed non-trivial dependence on the center-of-mass energy and hadron mass. In the following sections, we show that the parameters also depend on the event multiplicity, i.e., on the size of the system.

We adopted the usual blast-wave assumptions regarding the system, namely that the fireball is azimuthally symmetric and is expanding with a v radial flow velocity (in units of $c = 1$). Moreover, the freeze-out occurs instantly on a hypersurface according to the Cooper–Frye formulation at a given freeze-out temperature [2,14]. With these assumptions, we used the following simple form of the invariant yield:

$$\left.\frac{d^2N}{p_T dp_T dy}\right|_{INEL>0} = A \cdot m_T \cdot \left(1 + \frac{E}{nT}\right)^{-n} \quad (1)$$

where A is the amplitude incorporating the irrelevant spin degeneracy and constant factors as well as the invariant volume, $m_T = \sqrt{p_T^2 + m^2}$ is the transverse mass, $E = \gamma(m_T - v p_T) - m$ is the one-particle energy in the co-moving coordinate system, $\gamma = 1/\sqrt{1-v^2}$ is the Lorentz factor, T is a parameter with a temperature unit, and finally $n = \frac{1}{q-1}$ is the non-extensivity parameter, characterizing the temperature fluctuations. We note that T is not necessarily the freeze-out temperature and therefore is not necessarily the same for all hadron species [3,15]. The notation $INEL > 0$ means that only those events where there is at least one charged particle in the $|\eta| < 1.0$ region are considered. This choice is in agreement with the experimental definitions described in the previous section.

As a reference, in Table 2 and in Figure 3, we show the parameters and curves fitted on the experimental "minimum bias" (in the sense that there is no event multiplicity classification) data. These results are consistent with our previous observations [8]: the heaviest proton has the largest temperature and the smallest q. We note that, in the lower part of Figure 3, a periodic oscillation is visible. This is an effect in addition to the scaling. This has been investigated, for example, in [35,36].

Table 2. Tsallis parameters extracted from "minimum bias" $INEL > 0$ proton–proton collisions at $\sqrt{s} = 7$ TeV, measured by ALICE [32].

Hadron	n	q	T (GeV)	A	v	χ^2/ndf
$\pi^{\pm}$	7.415 ± 0.033	1.135 ± 0.005	0.089 ± 0.010	73.188 ± 9.700	0.000 ± 0.119	174.225 / 54
$K^{\pm}$	7.539 ± 0.086	1.133 ± 0.013	0.155 ± 0.010	0.915 ± 0.095	0.000 ± 0.066	20.274 / 47
$p(\bar{p})$	8.805 ± 0.184	1.114 ± 0.023	0.191 ± 0.012	0.124 ± 0.013	0.000 ± 0.054	18.462 / 45

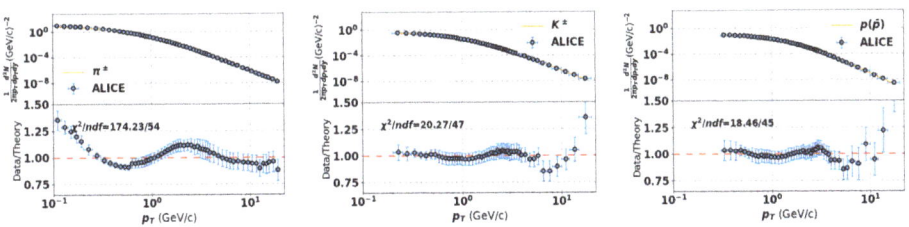

Figure 3. Fits of the Tsallis–Pareto distribution to $\pi^{\pm}$, $K^{\pm}$, and $p(\bar{p})$ hadrons measured by ALICE [32].

4. The Multiplicity Dependence of the Non-Extensive Model

In Section 2, we showed that the HIJING++ framework is able to reproduce the main experimental observables such as event multiplicity distribution and the p_T spectra of various identified hadrons. In Section 3, we briefly summarized the main features of the blast-wave motivated non-extensive hadronization model. In this section, we take advantage of the power of HIJING++ and extract the Tsallis parameters from a wide range of event multiplicity classes. The event classes of the HIJING++ run are classified as

$$\text{Class} = \langle dN_{ch}/d\eta \rangle_{min} < \langle dN_{ch}/d\eta \rangle \le \langle dN_{ch}/d\eta \rangle_{max}. \quad (2)$$

The multiplicity ranges of each class used in this study are listed in Table 3.

Table 3. Multiplicity classes used in HIJING++ runs.

Class	I	II	III	IV	V	VI	VII	VIII	IX	X	XI	XII	XIII	XIV	XV
$\langle dN_{ch}/d\eta \rangle_{min}$	0	10	15	20	25	30	35	40	45	50	55	60	70	80	90
$\langle dN_{ch}/d\eta \rangle_{max}$	10	15	20	25	30	35	40	45	50	55	60	70	80	90	100

Using this event classification, we calculated the mid-rapidity transverse momentum spectra of charge averaged pions, kaons, and protons in $INEL > 0$ events. We generated 200M events. To avoid superfluous overcrowding of the available space, we show only the low, moderate, and high multiplicity spectra along with the fitted Tsallis–Pareto curves defined by Equation (1) in Figure 4.

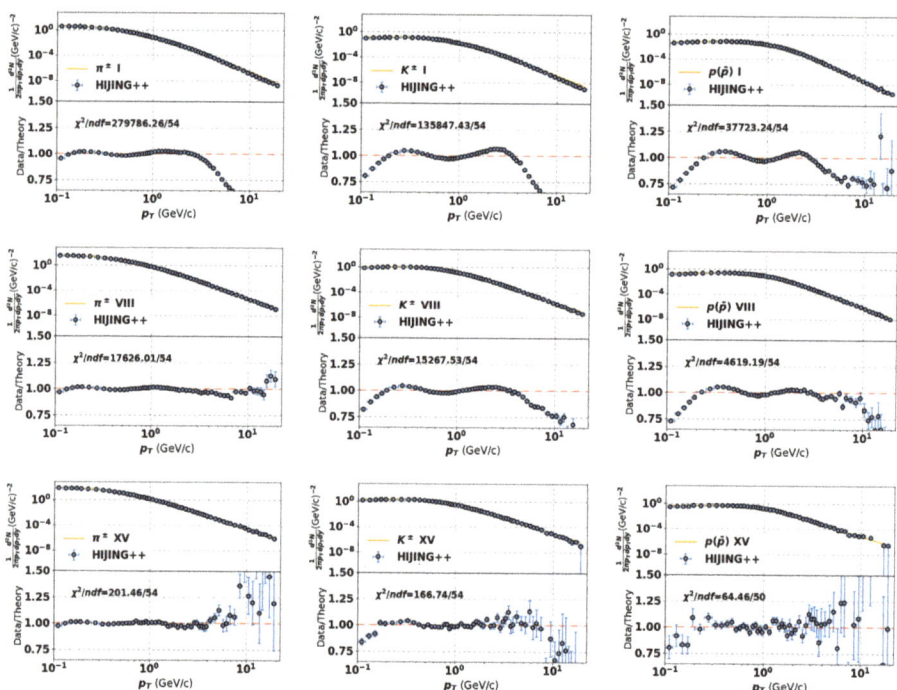

Figure 4. Calculated p_T spectra of charge averaged pions (**left column**), kaons (**middle column**), and protons (**right column**) at low (**top row**), moderate (**middle row**), and high (**bottom row**) multiplicity classes as blue dots and the fitted Equation (1) Tsallis–Pareto curve (orange line). The lower part of each panel shows the *Data/Theory* ratio.

In Figure 4, we can see that the best fits occurred at the high multiplicity events. For the pions, the fitted curves follow the points well at the low-p_T region, while for the kaons and protons with higher mass at the low-p_T region the model overpredicts the yield. We note, however, that for the HIJING++ run we investigated the same 0.1 GeV/c$< p_T <$ 20 GeV/c region for all hadrons, while the low-p_T part for the kaons and protons in the case of the experimental results is missing. At the high-p_T region, the fit breaks down because of the low statistics.

In Figure 5, the fitted parameters in the function of the event multiplicity are shown. Using the distribution form Equation (1), we observe that with increasing multiplicity the q parameter increases for each hadron but with different slopes: the increase of q (or the decrease of n) is the largest for the heaviest hadron. On the other hand, the temperature decreases slowly with the increasing pseudorapidity density. Here, the previously observed $T_{\pi^\pm} < T_{K^\pm} < T_{p(\bar{p})}$ mass hierarchy stays valid with the multiplicity averaged values, as can be seen in Table 4.

Table 4. The multiplicity averaged temperature parameters for charge averaged pions, kaons, and protons.

Hadron	$\langle T_i \rangle$ (GeV)
$\pi^\pm$	0.063 ± 0.003
$K^\pm$	0.092 ± 0.001
$p(\bar{p})$	0.106 ± 0.002

The radial flow velocity also increases with the multiplicity, but also with different rates. While at low multiplicity the lightest pions have the smallest v, it increases rapidly with the increasing multiplicity. On the other hand, the rate of increase in the case of protons and kaons are approximately the same. These observations require further investigation. Finally, the amplitudes are increasing for each hadron species with the multiplicity. The value of pions is much higher than those of the heavier hadrons, which indicates that with increasing multiplicity the number of the produced pions grows faster than the number of kaons and protons.

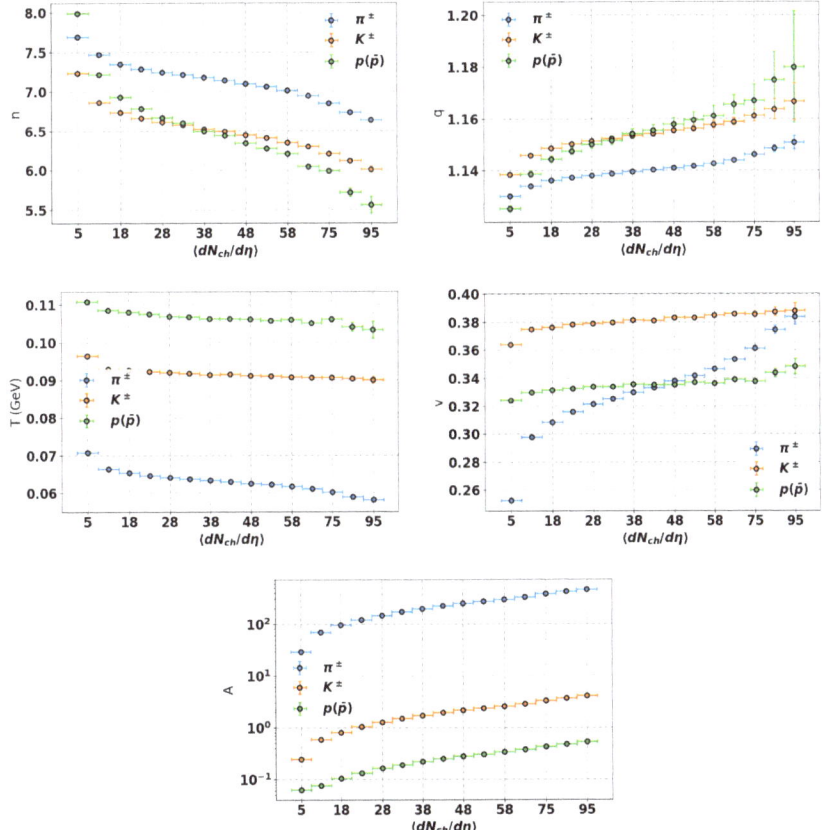

Figure 5. The fitted parameters of the Tsallis–Pareto distribution defined by Equation (1), in the function of the event multiplicity class defined as in Table 3.

5. Conclusions

In this contribution, we investigated the multiplicity dependence of the parameters of the non-extensive hadronization model in proton–proton collisions using HIJING++ calculations. We presented the current status of the tuning process of HIJING++ and showed that it is able to reproduce the main high-energy physics observables such as multiplicity and p_T distributions. We presented the non-extensive hadronization model that we used to describe the transverse momentum distribution of identified hadrons. In accordance with our previous results, we showed that a mass hierarchy emerges in the Tsallis parameters. Utilizing the tuned HIJING++ calculations, we also extracted the parameters from $\sqrt{s} = 7$ TeV proton–proton collisions Monte Carlo calculations with various event multiplicity classifications. Our study showed that the q non-extensivity parameter increases with increasing multiplicity, while the T temperature has only a slight decrease. On the other hand, all hadrons result in a non-zero, increasing radial flow velocity. All parameters show the earlier observed mass hierarchy. Our findings suggest that these parameters are sensitive to the event size and may serve as a thermometer of the collision.

Author Contributions: Software and formal analysis: G.B. and G.P.; investigation: G.B. and G.G.B.; writing—original draft preparation: G.B.; writing—review and editing: G.G.B.; supervision: G.G.B., T.S.B., and G.P.; funding acquisition: G.G.B. and T.S.B.

Funding: This research was funded by Hungarian-Chinese cooperation grant No. MOST 2014DFG02050, the Wigner HAS-OBOR-CCNU grant, and OTKA grants K120660 and K123815. G.B. acknowledges the support of the Wigner Data Center and Wigner GPU Laboratory.

Conflicts of Interest: The authors declare no conflict of interest.

References

1. Bíró, G.; Barnaföldi, G.G.; Biró, T.S.; Ürmössy, K.; Takács, Á. Systematic Analysis of the Non-extensive Statistical Approach in High Energy Particle Collisions—Experiment vs. Theory. *Entropy* **2017**, *19*, 88. [CrossRef]
2. Grigoryan, S. Using the Tsallis distribution for hadron spectra in pp collisions: Pions and quarkonia at $\sqrt{s}$ = 5–13000 GeV. *Phys. Rev. D* **2017**, *95*, 056021. [CrossRef]
3. Zheng, H.; Zhu, L.; Bonasera, A. Systematic analysis of hadron spectra in p+p collisions using Tsallis distributions. *Phys. Rev. D* **2015**, *92*, 074009. [CrossRef]
4. Wong, C.Y.; Wilk, G.; Cirto, L.J.L.; Tsallis, C. Possible Implication of a Single Nonextensive p_T Distribution for Hadron Production in High-Energy pp Collisions. *Eur. Phys. J. Web Conf.* **2015**, *90*, 04002. [CrossRef]
5. Tripathy, S.; Bhattacharyya, T.; Garg, P.; Kumar,P.; Sahoo, R.; Cleymans, J. Nuclear Modification Factor Using Tsallis Non-extensive Statistics. *Eur. Phys. J. A* **2016**, *52*, 289. [CrossRef]
6. Bhattacharyya, T.; Cleymans, J.; Marques, L.; Mogliacci, S.; Paradza, M.W. On the precise determination of the Tsallis parameters in proton–proton collisions at LHC energies. *J. Phys. G Nucl. Part. Phys.* **2018**, *45*, 055001. [CrossRef]
7. Shen, K.; Barnaföldi, G.G.; Biró, T.S. Hadronization within Non-Extensive Approach and the Evolution of the Parameters. *arXiv* **2019**, arXiv:1905.05736.
8. Bíró, G.; Barnaföldi, G.G.; Biró, T.S.; Shen, K. Mass hierarchy and energy scaling of the Tsallis–Pareto parameters in hadron productions at RHIC and LHC energies. *Eur. Phys. J. Web Conf.* **2018**, *171*, 14008. [CrossRef]
9. Bíró, G.; Barnaföldi, G.G.; Biró, T.S.; Ürmössy, K. Application of the Non-extensive Statistical Approach to High Energy Particle Collisions. *AIP Conf. Proc.* **2017**, *1853*, 080001. [CrossRef]
10. Barnaföldi, G.G.; Ürmössy, K.; Bíró, G. A 'soft+hard' model for Pion, Kaon, and Proton Spectra and v_2 measured in PbPb Collisions at $\sqrt{s} = 2.76$ ATeV. *J. Phys. Conf. Ser.* **2015**, *612*, 012048. [CrossRef]
11. Takacs, A.; Barnaföldi, G.G. Non-extensive Motivated Parton Fragmentation Functions. *Proceedings* **2019**, *10*, 12. [CrossRef]
12. Khuntia, A.; Sharma, H.; Kumar Tiwari, S.; Sahoo, R.; Cleymans, J. Radial flow and differential freeze-out in proton–proton collisions at $\sqrt{s} = 7$ TeV at the LHC. *Eur. Phys. J. A* **2019**, *55*, 3. [CrossRef]

13. Wilk, G.; Włodarczyk, Z. Some intriguing aspects of multiparticle production processes. *Int. J. Mod. Phys. A* **2018**, *33*, 1830008. [CrossRef]
14. Urmossy, K.; Biro, T.S. Cooper-Frye Formula and Non-extensive Coalescence at RHIC Energy. *Phys. Lett. B* **2010**, *689*, 14. [CrossRef]
15. Van, P.; Barnafoldi, G.G.; Biro, T.S.; Urmossy, K. Nonadditive thermostatistics and thermodynamics. *J. Phys. Conf. Ser.* **2012**, *394*, 012002. [CrossRef]
16. Shen, K.; Biro,T.S.; Wang, E. Different Non-extensive Models for heavy-ion collisions. *Physica A* **2018**, *492*, 2353–2360. [CrossRef]
17. Wang, X.N.; Gyulassy, M. HIJING: A Monte Carlo model for multiple jet production in pp, pA, and AA collisions. *Phys. Rev. D* **1991**, *44*, 3501–3516. [CrossRef]
18. Deng, W.T.; Wang, X.N.; Xu, R. Hadron production in p+p, p+Pb, and Pb+Pb collisions with the HIJING 2.0 model at energies available at the CERN Large Hadron Collider. *Phys. Rev. C* **2011**, *83*, 014915. [CrossRef]
19. Barnaföldi, G.G.; Bíró, G.; Gyulassy, M.; Haranozó, S.M.; Lévai, P.; Ma, G.; Papp, G.; Wang, X.N.; Zhang, B.W. First Results with HIJING++ in High-Energy Heavy-Ion Collisions. *Nucl. Part. Phys. Proc.* **2017**, *289–290*, 373. [CrossRef]
20. Papp, G.; Barnaföldi, G.G.; Bíró, G.; Gyulassy, M.; Harangozó, S.M.; Ma, G.; Lévai, P.; Wang. X.N.; Zhang, B.W. First Results with HIJING++ on High-energy Heavy Ion Collisions. *arXiv* **2018**, arXiv:1805.02635.
21. Albacete, J.L.; Arleo, F.; Barnaföldi, G.G.; Bíró, G.; d'Enterria, D.; Ducloué, B.; Eskola, K.J.; Ferreiro, E.G.; Gyulassy, M.; Harangozó, S.M.; et al. Predictions for Cold Nuclear Matter Effects in p+Pb Collisions at $\sqrt{s_{NN}}$ = 8.16 TeV. *Nucl. Phys. A* **2018**, *972*, 18–85. [CrossRef]
22. Bíró, G.; Papp, G.; Barnaföldi, G.G.; Nagy, D.; Gyulassy, M.; Lévai, P.; Wang, X.N.; Zhang, B.W. HIJING++, a Heavy Ion Jet INteraction Generator for the High-luminosity Era of the LHC and Beyond. *Proceedings* **2019**, *10*, 4. [CrossRef]
23. Sjöstrand, T.; Ask, S.; Christiansen, J.R.; Corke, R.; Desai, N.; Ilten, P.; Mrenna, S.; Prestel, S.; Rasmussen, C.O.; Skands, P.Z. An Introduction to PYTHIA 8.2. *Comput. Phys. Commun.* **2015**, *191*, 159–177. [CrossRef]
24. Buckley, A.; Ferrando, J.; Lloyd, S.; Nordström, K.; Page, B.; Rüfenacht, M.; Schönherr, M.; Watt, G. LHAPDF6: Parton density access in the LHC precision era. *Eur. Phys. J. C* **2015**, *75*, 132. [CrossRef]
25. Galassi, M.; Davies, J.; Theiler, J.; Gough, B.; Jungman, G.; Alken, P.; Booth, M.; Rossi, F.; Ulerich, R. *GNU Scientific Library Reference Manual*, 3rd ed.; Network Theory Limited: London, UK. ISBN 0954612078
26. Lepage, G.P. A new algorithm for adaptive multidimensional integration. *J. Comp. Phys.* **1978**, *27*, 192–203. [CrossRef]
27. ROOT Data Analysis Framework. Available online: https://root.cern.ch/ (accessed on 30 April 2019).
28. Dulat, S.; Hou, T.J.; Gao, J.; Guzzi, M.; Huston, J.; Nadolsky, P.; Pumplin, J.; Schmidt, C.; Stump, D.; P. Yuan, C. New parton distribution functions from a global analysis of quantum chromodynamics. *Phys. Rev. D* **2016**, *93*, 033006. [CrossRef]
29. Eskola, K.J.; Paakkinen, P.; Paukkunen, H.; Salgado, C.A. EPPS16: Nuclear parton distributions with LHC data. *Eur. Phys. J. C* **2017**, *77*, 163. [CrossRef] [PubMed]
30. Buckley, A.; Hoeth, H.; Lacker, H.; Schulz, H.; von Seggern, J.E. Systematic event generator tuning for the LHC. *Eur. Phys. J. C* **2010**, *65*, 331. [CrossRef]
31. Tange, O. GNU Parallel: The Command-Line Power Tool. Available online: https://www.usenix.org/system/files/login/articles/105438-Tange.pdf (accessed on 26 May 2019).
32. Adam, J.; Adamová, D.; Aggarwal, M.M.; Rinella, G.A.; Agnello, M.; Agrawal, N.; Ahammed, Z.; Ahmad, S.; Ahn, S.U.; Aiola, S.; et al. Multiplicity dependence of charged pion, kaon, and (anti)proton production at large transverse momentum in p-Pb collisions at $\sqrt{s_{NN}}$ = 5.02 TeV. *Phys. Lett. B* **2016**, *760*, 720–735. [CrossRef]
33. Aamodt, K.; Abel, N.; Abeysekara, U.; Quintana, A.A.; Abramyan, A.; Adamova, D.; Aggarwal, M.M.; Rinella, G.A.; Agocs, A.G.; Salazar, S.A; et al. Charged-particle multiplicity measurement in proton–proton collisions at $\sqrt{s}$ = 7 TeV with ALICE at LHC. *Eur. Phys. J. C* **2010**, *68*, 345. [CrossRef]
34. Acharya, S.; Adamová, D.; Adolfsson, J.; Aggarwal, M.M.; AglieriRinella, G.; Agnello, M.; Agrawal, N.; Ahammed, Z.; Ahmad, N.; Ahn, S.U.; et al. Charged-particle multiplicity distributions over a wide pseudorapidity range in proton–proton collisions at $\sqrt{s}$ = 0.9, 7, and 8 TeV. *Eur. Phys. J. C* **2017**, *77*, 852. [CrossRef]

35. Wilk, G.; Wlodarczyk, Z. Tsallis Distribution Decorated With Log-Periodic Oscillation. *Entropy* **2015**, *17*, 384. [CrossRef]
36. Rybczyński, M.; Wilk, G.; Włodarczyk, Z. System size dependence of the log-periodic oscillations of transverse momentum spectra. *Eur. Phys. J. Web Conf.* **2015**, *90*, 01002. [CrossRef]

 © 2019 by the authors. Licensee MDPI, Basel, Switzerland. This article is an open access article distributed under the terms and conditions of the Creative Commons Attribution (CC BY) license (http://creativecommons.org/licenses/by/4.0/).

Article

Coulomb Final State Interaction in Heavy Ion Collisions for Lévy Sources

Máté Csanád [1,*], Sándor Lökös [1,2] and Márton Nagy [1]

[1] Department of Atomics Physics, Eötvös Loránd University, Pázmány Péter sétány 1/A, H-1111 Budapest, Hungary; lokos@caesar.elte.hu (S.L.); nmarci@elte.hu (M.N.)
[2] Department of Applied Informatics, Eszterházy Károly University, Mátrai út 36, H-3200 Gyöngyös, Hungary
* Correspondence: csanad@elte.hu

Received: 17 April 2019; Accepted: 23 May 2019; Published: 28 May 2019

Abstract: Investigation of momentum space correlations of particles produced in high energy reactions requires taking final state interactions into account, a crucial point of any such analysis. Coulomb interaction between charged particles is the most important such effect. In small systems like those created in e^+e^- or $p+p$ collisions, the so-called Gamow factor (valid for a point-like particle source) gives an acceptable description of the Coulomb interaction. However, in larger systems such as central or mid-central heavy ion collisions, more involved approaches are needed. In this paper we investigate the Coulomb final state interaction for Lévy-type source functions that were recently shown to be of much interest for a refined description of the space-time picture of particle production in heavy-ion collisions.

Keywords: Femtoscopy; Bose–Einstein correlations; Coulomb correction; Heavy Ions

1. Introduction

Coulomb repulsion is the most important final state interaction that has to be considered in Bose-Einstein correlation measurements in high-energy physics. In e^+e^- or $p+p$ collisions, where the particle emitting source is much smaller than the wavelength corresponding to the relative momentum of the particle pair, the well-known Gamow factor (essentially the value of the Coulomb interacting pair wave function at the origin) can be used to "correct" for the Coulomb effect. However, for an extended source, the Gamow factor overestimates the correction. A more advanced approach is to take the source-averaged Coulomb wave function (instead of its value at the origin, which may be valid then for a point-like source), see e.g., Refs. [1,2]. In these papers a method (aptly referred to as the Bowler–Sinyukov method) is also described that is widely used to take the effect of long-lived resonances into account.

Traditionally one assumes simple source function shapes (such as exponential, Gaussian ones) for calculating the source averaged Coulomb wave function (as e.g., in the papers referred above); we may mention that more general sources are also considered e.g., in Ref. [3]. Recently, an even more general type of source functions, namely Lévy sources [4,5] has gotten much interest. Lévy-type source functions simplify to Cauchy as well as to Gaussian ones in special cases, and allow for a more refined treatment of the space–time picture of the particle emission. Moreover, a certain parameter of a Lévy distribution (the so-called Lévy exponent) also may carry information about the order of the phase transition between deconfined and hadronic matter [5,6].

Our objective in this paper is to tackle the effect of Coulomb interaction for the case of Lévy-type sources. (Because of the slow, power-law-like decay of Lévy-type sources at large distances, many previously developed methods are unsuitable for them.) We strive for analytical approximate methods that are well suited for use in the actual treatment of experimental Bose–Einstein correlation functions.

2. Coulomb Effect in Bose–Einstein Correlations: Basic Concepts

In this section we briefly review some notions and well-known formulas pertaining to the work presented hereafter.

In a statistical physical (specifically, hydrodynamical) description of particle production in high-energy collisions, a basic ingredient is the (one-particle) *source function* (Wigner function), denoted here by $S(x,p)$. Its physical meaning is essentially that the probability of the production of a particle in the infinitesimal phase-space neighborhood of momentum $\mathbf{p}$ and point $\mathbf{r}$ is proportional to $S(\mathbf{r},\mathbf{p})d^3r d^3p$. Thus it is natural that the one-particle momentum distribution function $N_1(\mathbf{p})$ can be expressed as

$$N_1(\mathbf{p}) = \int d^3 \mathbf{r}\, S(\mathbf{r},\mathbf{p}), \quad \text{with the normalization} \quad \int d^3\mathbf{p}\, N_1(\mathbf{p}) = 1. \tag{1}$$

For a slight convenience we chose the normalization condition of $S(\mathbf{r},\mathbf{p})$, so that $N_1(\mathbf{p})$ is now considered to be the probability distribution of the momentum of the produced particles[1].

According to a simple quantum mechanical treatment of Bose–Einstein correlation effects, the two-particle momentum distribution function $N_2(\mathbf{q},\mathbf{K})$ can be expressed [7] with the source distribution function $S(\mathbf{r},\mathbf{p})$ as an integral over the two-particle final state wave function,

$$N_2(\mathbf{p}_1,\mathbf{p}_2) = \int d^3\mathbf{r}_1 d^3\mathbf{r}_2\, S(\mathbf{r}_1,\mathbf{p}_1) S(\mathbf{r}_2,\mathbf{p}_2) |\psi^{(2)}_{\mathbf{p}_1,\mathbf{p}_2}(\mathbf{r}_1,\mathbf{r}_2)|^2. \tag{2}$$

The two-particle wave function $\psi^{(2)}$ must be symmetric in the space variables (for bosons); this is the main reason for the appearance of quantum statistical (Bose–Einstein) correlations.

With some trivial simplifications, we thus get the correlation function $C_2(\mathbf{q},\mathbf{K})$ as

$$C_2(\mathbf{q},\mathbf{K}) = \int d^3\mathbf{r}\, D(\mathbf{r},\mathbf{p}_1,\mathbf{p}_2) |\psi^{(2)}_\mathbf{q}|^2, \quad \text{where} \quad D(\mathbf{r},\mathbf{p}_1,\mathbf{p}_2) = \int d^3\mathbf{R}\, S\left(\mathbf{R}+\tfrac{\mathbf{r}}{2},\mathbf{p}_1\right) S\left(\mathbf{R}-\tfrac{\mathbf{r}}{2},\mathbf{p}_2\right). \tag{3}$$

The momenta of the two particles were denoted by $\mathbf{p}_1$ and $\mathbf{p}_2$, and we used the combinations of these, the $\mathbf{q}$ relative momentum and the $\mathbf{K}$ average momentum as

$$\mathbf{q} = \mathbf{p}_1 - \mathbf{p}_2, \quad \mathbf{K} = \frac{1}{2}(\mathbf{p}_1+\mathbf{p}_2). \tag{4}$$

The notation $D(\mathbf{r},\mathbf{p}_1,\mathbf{p}_2)$ was introduced for the so-called two-particle source function, obtained as indicated, by integrating over the average spatial position $\mathbf{R}$ of the particle pair (with $\mathbf{r} \equiv \mathbf{r}_1 - \mathbf{r}_2$ thus standing for the relative coordinate).

The (symmetrized) two-particle wave function may depend on all momentum and coordinate components; however, its modulus does not depend on the average momentum $\mathbf{K}$ or the average coordinate $\mathbf{R}$ (owing to translational invariance).

Assuming $\mathbf{p}_1 \approx \mathbf{p}_2$ (in the sense that the pair wave function changes much more rapidly in terms of their difference, $\mathbf{q}$, as does the product of the source functions, we get

$$C_2(\mathbf{q},\mathbf{K}) \approx \frac{\int d^3\mathbf{r}\, D(\mathbf{r},\mathbf{K}) |\psi^{(2)}_\mathbf{q}(\mathbf{r})|^2}{\int d^3\mathbf{r}\, D(\mathbf{r},\mathbf{K})}, \tag{5}$$

where we introduced $D(\mathbf{r},\mathbf{K}) \equiv D(\mathbf{r},\mathbf{K},\mathbf{K}) \approx D(\mathbf{r},\mathbf{p}_1,\mathbf{p}_2)$. In the special case of no final state interactions (i.e., when the $\psi^{(2)}$ wave function is a symmetrized plane wave), we get the well known relation

---

[1] This normalizaton condition is of not much relevance here; one could just as well normalize $N_1(\mathbf{p})$ to $\langle n \rangle$, the mean number of produced particles; $N_1(\mathbf{p})$ would then correspond to the real momentum space distribution function.

$$|\psi^{(2)}_{\text{free}}|^2 = 1 + \cos(\mathbf{q}\mathbf{r}) \quad \Rightarrow \quad C_2^{(0)}(\mathbf{q},\mathbf{K}) \approx 1 + \frac{|\tilde{S}(\mathbf{q},\mathbf{K})|^2}{|\tilde{S}(0,\mathbf{K})|^2}, \quad \text{with} \quad \tilde{S}(\mathbf{q},\mathbf{K}) = \int d^3\mathbf{r}\, S(\mathbf{r},\mathbf{K}) e^{i\mathbf{q}\mathbf{r}}, \quad (6)$$

thus $\tilde{S}$ being the Fourier transform of the source function. In this formula the (0) superscript denotes the neglection of final state interactions.

Returning to the general, interacting case, if one assumes (according to the core-halo model, see Ref. [8]) that a certain fraction of the particle production (denoted by $\sqrt{\lambda}$) happens in a narrow, few fm diameter region ("core"), and the rest from the decay of long-lived resonances (the contribution of which comes from a much wider region), then one can write the source as

$$S(\mathbf{r},\mathbf{p}) = \sqrt{\lambda} S_c(\mathbf{r},\mathbf{p}) + (1-\sqrt{\lambda}) S_h^{R_h}(\mathbf{r},\mathbf{p}), \quad (7)$$

with a normalization that respects the requirement that λ determines the relative weight of the two components:

$$\int d\mathbf{r}\, S(\mathbf{r},\mathbf{p}) = \int d\mathbf{r}\, S_c(\mathbf{r},\mathbf{p}) = \int d\mathbf{r}\, S_h(\mathbf{r},\mathbf{p}) = 1. \quad (8)$$

Here the indices c and h stand for core and halo, respectively. The R_h "radius" parameter (the characteristic size of the halo part) will be assumed to be much higher than the experimentally resolvable distance, $r_{\max} \approx \hbar/Q_{\min}$, where $Q_{\min} \approx 1-2$ MeV, the minimal momentum difference that can be resolved experimentally.

One can also introduce the core–core, core–halo and halo–halo two-particle source functions as

$$D(\mathbf{r},\mathbf{K}) = \lambda D_{cc}(\mathbf{r},\mathbf{K}) + 2\sqrt{\lambda}(1-\sqrt{\lambda}) D_{ch}(\mathbf{r},\mathbf{K}) + (1-\sqrt{\lambda})^2 D_{hh}(\mathbf{r},\mathbf{K}), \quad (9)$$

where the following obvious definitions were used:

$$D_{AB}(\mathbf{r},\mathbf{K}) \equiv \int d^3\mathbf{R}\, S_A\left(\mathbf{R}+\tfrac{\mathbf{r}}{2},\mathbf{K}\right) S_B\left(\mathbf{R}-\tfrac{\mathbf{r}}{2},\mathbf{K}\right), \quad \text{for } A, B = c \text{ or } h. \quad (10)$$

With yet another notation we can write the terms of $D(\mathbf{r},\mathbf{K})$ as

$$D(\mathbf{r},\mathbf{K}) = \lambda D_{cc}(\mathbf{r},\mathbf{K}) + (1-\lambda) D_{(h)}(\mathbf{r},\mathbf{K}), \quad \text{with} \quad D_{(h)} = \frac{2\sqrt{\lambda}(1-\sqrt{\lambda}) D_{ch} + (1-\sqrt{\lambda})^2 D_{hh}}{1-\lambda}, \quad (11)$$

where thus the $D_{(h)}$ term contains all the halo contributions, and D_{cc} is just the core-core component. Perhaps it is useful to explicitly state the (evident) normalization conditions of all these two-particle functions:

$$\int D(\mathbf{r},\mathbf{K}) d^3\mathbf{r} = \int D_{cc}(\mathbf{r},\mathbf{K}) d^3\mathbf{r} = \int D_{ch}(\mathbf{r},\mathbf{K}) d^3\mathbf{r} = \int D_{hh}(\mathbf{r},\mathbf{K}) d^3\mathbf{r} = \int D_{(h)}(\mathbf{r},\mathbf{K}) d^3\mathbf{r} = 1. \quad (12)$$

Using these definitions, the correlation function can be expressed as

$$C_2(\mathbf{q},\mathbf{K}) \approx \lambda \int d^3\mathbf{r}\, D_{cc}(\mathbf{r},\mathbf{K}) |\psi_\mathbf{q}^{(2)}(\mathbf{r})|^2 + (1-\lambda) \int d^3\mathbf{r}\, D_{(h)}(\mathbf{r},\mathbf{K}) |\psi_\mathbf{q}^{(2)}(\mathbf{r})|^2. \quad (13)$$

By taking the $R_h \to \infty$ limit in the second term[2], one arrives at the well-known Bowler–Sinyukov formula [1,2] as

$$C_2(\mathbf{q}, \mathbf{K}) = 1 - \lambda + \lambda \int d^3\mathbf{r}\, D_{cc}(\mathbf{r}, \mathbf{K})|\psi_\mathbf{q}^{(2)}(\mathbf{r})|^2. \tag{14}$$

Specifically, in the free case (with plane-wave wave functions) one arrives at the

$$C_2^{(0)}(\mathbf{q}, \mathbf{K}) = 1 + \lambda \frac{|\tilde{S}_c(\mathbf{q}, \mathbf{K})|^2}{|\tilde{S}_c(\mathbf{0}, \mathbf{K})|^2}, \tag{15}$$

formula (including the normalization term, which is unity in this paper). The experimental observation is that—although the free correlation function defined in Equation (6) takes the value of 2 at 0 relative momentum: $C_2^{(0)}(\mathbf{0}, \mathbf{K}) = 2$—the measured value is $1 + \lambda$. The core-halo model thus naturally explains this fact in terms of the finite momentum resolution of any experiment. In the core-halo model the intercept of the real, measurable correlation function at $\mathbf{q} = 0$ thus tells the fraction of pions coming from the core. In the Coulomb interacting (realistic) case, the interpretation of λ as any intercept parameter is not so simple, however. The Bowler–Sinyukov method, Equation (15) gives a means to take the core-halo model into account when treating the Coulomb effect.

To investigate the λ parameter (which, as it is directly connected to the proportion of resonance decay particles, may have interesting physical consequences, see e.g., Refs. [5,9]) one needs a firm grasp on the effect of the final state interactions in Bose–Einstein correlation functions. For the most important such effect, the Coulomb effect, the $\psi_\mathbf{q}^{(2)}(\mathbf{r})$ wave function (the two-body scattering solution of the Schrödinger equation with Coulomb repulsion) is well known in the center-of-mass system of the outgoing particles (the so-called PCMS system). Its expression is

$$\psi_\mathbf{q}^{(2)}(\mathbf{r}) = \frac{1}{\sqrt{2}} \frac{\Gamma(1+i\eta)}{e^{\pi\eta/2}} \left\{ e^{i\mathbf{k}\mathbf{r}} F(-i\eta, 1, i(kr - \mathbf{k}\mathbf{r})) + [\mathbf{r} \leftrightarrow -\mathbf{r}] \right\}, \quad \text{where} \quad \mathbf{k} = \frac{\mathbf{q}}{2}. \tag{16}$$

Here $F(\cdot, \cdot, \cdot)$ is the confluent hypergeometric function, $\Gamma(\cdot)$ is the Gamma function, and

$$\eta = \frac{q_e^2}{4\pi\varepsilon_0} \frac{1}{\hbar c} \frac{m_\pi c^2}{qc} = \alpha_{\rm EM} \frac{m_\pi c}{q} \tag{17}$$

is the Sommerfeld parameter, with $q_e^2/(4\pi\varepsilon_0)$ being the Coulomb-constant, $\alpha_{\rm EM}$ the fine-stucture constant of the electromagnetic interaction, and m_π the pion mass (as from now on, we restrict this analysis to pion pairs).

For a given source function $S(\mathbf{r}, \mathbf{K})$, the ratio of the (measurable) correlation function $C_2(\mathbf{q})$ and the $C_2^{(0)}(\mathbf{q})$ function is usually called the *Coulomb correction*[3], $K(\mathbf{q})$:

$$K(\mathbf{q}) = \frac{C_2(\mathbf{q})}{C_2^{(0)}(\mathbf{q})} \quad \Rightarrow \quad C_2^{(0)}(\mathbf{q}) = C_2(\mathbf{q}) \cdot \frac{1}{K(\mathbf{q})}. \tag{18}$$

If one focuses on the simple property of the $C_2^{(0)}(\mathbf{q})$ function as being the Fourier transform of the source, then one might want to recover $C_2^{(0)}(\mathbf{q})$ from the measured $C_2(\mathbf{q})$: for this, one uses the Coulomb correction factor. Indeed, many assumptions have been used to estimate the $K(\mathbf{q})$

---

[2] Mathematically, this is the formulation of the condition that the momentum differences corresponding to the halo size, $\hbar/R_h$ are not resolvable by any experimental apparatus. With a re-scaling of the integral by $\mathbf{r} \to R_h \mathbf{r}$, taking advantage of the fact that for large distances, $\psi(\mathbf{r})$ asymptotically becomes the free plane-wave function, one can then use Lebesgue's dominated convergence theorem on the interchangeability of integrals and limits to infer that the second integral indeed gives 1 in the $R_h \to \infty$ limit.

[3] The terminology is not uniform here; it is sometimes this factor, and sometimes its inverse what is called the Coulomb correction.

factor: the simplest case is the so-called Gamow factor that treats the source as a point-like one when calculating $K(\mathbf{q})$:

$$S(\mathbf{r}) = \delta^{(3)}(\mathbf{r}) \quad \Rightarrow \quad K(\mathbf{q}) = K_{\text{Gamow}}(q) = |\psi^{(2)}_{\mathbf{q}}(0)|^2 = \frac{2\pi\eta}{e^{2\pi\eta} - 1}. \tag{19}$$

A method that suits the scope of heavy-ion collisions a little more would be to pre-calculate $K(\mathbf{q})$ for a single specific given assumption for $S(\mathbf{r})$, then apply this correction (with the Bowler–Sinyukov method) and find the $S(\mathbf{r})$ from a fit to the Fourier transform of the recovered $C^{(0)}(\mathbf{q})$. However, it is clear that this process should be done iteratively: after the first "round" of such fits, one would have to re-calculate the Coulomb correction. When this iteration converges, one in principle arrives at the proper $S(\mathbf{r})$.

3. Numerical Table for the Coulomb Correction for Lévy Source

Recent studies have shown that the assumption of a Lévy-type of source function is well suited for the description of two-particle Bose–Einstein correlation functions. The details of the validity of the Lévy-shape assumption is exhaustively expounded in Refs. [4,5]. The (spherically symmetric) Lévy distribution utilized here has two parameters, scale parameter (radius) R and Lévy index α, and is expressed as

$$\mathcal{L}(\alpha, R, \mathbf{r}) := \int \frac{d^3\mathbf{q}}{(2\pi)^3} e^{i\mathbf{q}\mathbf{r}} \exp\left(-\tfrac{1}{2}|\mathbf{q}|^2 R^2|^{\alpha/2}\right). \tag{20}$$

In the $\alpha = 2$ case one gets a Gaussian distribution, in the $\alpha = 1$ case the Cauchy distribution is recovered. For other α values, no simple analytic expression exists for the result of this Fourier transform-like integral. As a remark, we note that the concept of this symmetric Lévy distribution can be generalized without much effort to the non-spherically symmetric case by replacing R^2 with a symmetric 3×3 matrix R^2_{kl}.

In order to apply Lévy-type sources in a self-consistent way, the Coulomb integral defined in Equation (14) has to be calculated. This cannot be carried out in a straightforward analytic manner. In the following we demonstrate two approaches that can be employed to handle the Coulomb final state effect in the presence of a Lévy source.

The integral in Equation (14) cannot be evaluate analytically for a Lévy source so it has to be calculated numerically. For experimental purposes, the results can be loaded to a binary file as a lookup table and can be used in the fitting procedure (thus circumventing the need for an iterative process for the Coulomb correction). Interpolation also should be applied since the correlation function only can be filled into the lookup table for discrete values of the parameters. This interpolation, however, could cause numerical fluctuations in the χ^2 landscape and could mislead the fit algorithm, so an iterative procedure should be applied in the following manner:

1. Fit with the function defined in Equation (14) $\Rightarrow \alpha_0, R_0, \lambda_0$,
2. Fit with $C_2^{(0)}(\lambda, R, \alpha; Q) \frac{C_2(\lambda_0, R_0, \alpha_0; Q)}{C_2^{(0)}(\lambda_0, R_0, \alpha_0; Q)} \Rightarrow \lambda_1, R_1, \alpha_1$,
3. Repeat while λ_1, R_1, α_1 and λ_0, R_0, α_0 differ less then 1%.

In this manner, the fit parameters $\lambda(K), R(K), \alpha(K)$ can be obtained, similarly to Ref. [5].

4. Parametrization of the Coulomb Correction for Lévy Source

In this section, let us review a different approach, where based on the numerical table mentioned above, a parametrization can be formulated. In other words, one can get the Coulomb correction values from the table and parametrize its R and α dependences. This approach was encouraged by the successful parametrization of the $\alpha = 1$ case (the Cauchy case) done by the CMS collaboration (see

Ref. [10], Equation (5) for details). This can be considered as our starting point for the more general, Lévy case (for arbitrary α). The expression used by CMS for the Cauchy distribution, $\alpha = 1$ was

$$K(\mathbf{q})_{\text{Cauchy}} = K_{\text{Gamow}}(\mathbf{q}) \times \left(1 + \frac{\alpha_{\text{EM}} \pi m_\pi R}{1.26\hbar c + qR}\right), \quad \text{where} \quad \alpha_{\text{EM}} = \frac{q_e^2}{4\pi\varepsilon_0} \frac{1}{\hbar c} \approx \frac{1}{137}. \quad (21)$$

Generally, this is a correction of the Gamow correction. This simple formula has the advantage of having only 1 numerical constant parameter (the 1.26 in the denominator). However, it assumes $\alpha = 1$, and we look for a generalization for arbitrary Lévy α values.

A more general correction for the Gamow correction which is able to describe the Coulomb correction for a Lévy source has to fulfill the following requirements:

- It should follow not only the R, but the α dependence.
- In α=1 case, it should reduce to Equation (21).

To fulfill these, we replace R with R/α to introduce the α-dependence and take higher order terms in $\frac{qR}{\alpha\hbar c}$ into consideration. Our trial formula is then assumed to be

$$K_{\text{Lévy}}(q, \alpha, R) = K_{\text{Gamow}}(\mathbf{q}) \times K_{\text{mod}}(\mathbf{q}), \quad \text{with}$$

$$K_{\text{mod}}(\mathbf{q}) = 1 + \frac{A(\alpha, R) \frac{\alpha_{\text{EM}} \pi m_\pi R}{\alpha\hbar c}}{1 + B(\alpha, R) \frac{qR}{\alpha\hbar c} + C(\alpha, R) \left(\frac{qR}{\alpha\hbar c}\right)^2 + D(\alpha, R) \left(\frac{qR}{\alpha\hbar c}\right)^4}, \quad (22)$$

and the task is to find a suitable choice for the $A(\alpha, R)$, $B(\alpha, R)$, $C(\alpha, R)$, $D(\alpha, R)$ functions that yield an acceptable approximation of the results of the numerical integration (contained in our lookup table). The assumed form seems to be sufficient since it simplifies to Equation (21) if $\alpha = 1$ and $C = D = 0$, and could follow the observed weak α dependence of the Coulomb integral (see Figure 1).

We fit the above (22) formula to the numerically calculated results for α parameter values between 0.8 and 1.7 and R parameter values between 3 fm and 12 fm, where the ranges were motivated by the results of Pioneering High Energy Nuclear Interaction eXperiment (PHENIX) [5]. With this we obtained the A, B, C, D values as a function of the given α and R parameters. As a next step, we also parametrized these dependencies empirically, and found that the following expressions give satisfactory agreement with the lookup table:

$$A(\alpha, R) = (a_A \alpha + a_B)^2 + (a_C R + a_D)^2 + a_E(\alpha R + 1)^2 \quad (23)$$

$$B(\alpha, R) = \frac{1 + b_A R^{b_B} - \alpha^{b_C}}{\alpha^2 R(\alpha^{b_D} + b_E R^{b_F})} \quad (24)$$

$$C(\alpha, R) = \frac{c_A + \alpha^{c_B} + c_C R^{c_D}}{c_E} \left(\frac{\alpha}{R}\right)^{c_F} \quad (25)$$

$$D(\alpha, R) = d_A + \frac{R^{d_B} + d_C \alpha^{d_F}}{R^{d_D} \alpha^{d_E}}. \quad (26)$$

The parameters in these functions turn out best to have the values as follows:

$a_A = 0.36060$, $a_B = -0.54508$, $a_C = 0.03475$, $a_D = -1.30389$, $a_E = 0.00378$,
$b_A = 2.04017$, $b_B = 0.55972$, $b_C = 2.47224$, $b_D = -1.26815$, $b_E = -0.11767$, $b_F = 0.52738$,
$c_A = -1.00015$, $c_B = 0.00012$, $c_C = 0.00008$, $c_D = 0.26986$, $c_E = 0.00003$, $c_F = 1.75202$,
$d_A = 0.00263$, $d_B = -0.13124$, $d_C = -0.83149$, $d_D = 1.57528$, $d_E = 0.27568$, $d_F = 0.04937$.

This parametrization describes the R and α dependence of the Coulomb integral in a range where the Coulomb correction deviates from 1 by more than a factor of $\sim 10^{-4}$–10^{-5}. We find that this region is 0 GeV/c < q < 0.2 GeV/c. As an example, for $R = 6$ fm and with different α values, we plotted the results of the parametrization on Figure 1.

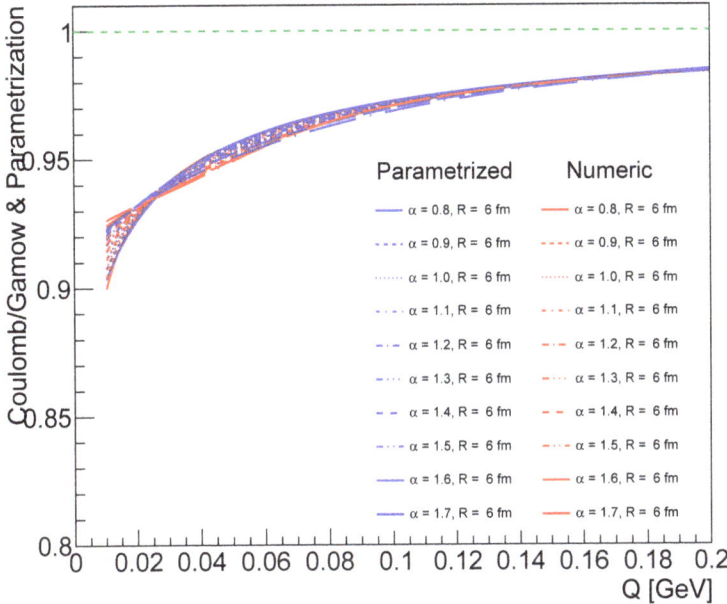

Figure 1. An example for the basic parametrization for a given R and different αs. One can observe that the α dependence is quite weak but still observable and quite complicated.

It turns out that the functional form specified above does yield a satisfactory fit at lower values of q, below 0.1–0.2 GeV/c. However, at higher values, the fit that is acceptable at low q, inevitably starts to deviate from the desired values, i.e., cannot be used to extrapolate beyond the fitted q range. The intermediate q region above and around 0.1 GeV/c can instead be described with an exponential-type function parametrized based on intermediate q fits to the numerical table, with the following functional form:

$$E(q) = 1 + A(\alpha, R) \exp\{-B(\alpha, R)q\}, \tag{27}$$

where the $A(\alpha, R)$ and $B(\alpha, R)$ functions have a form as

$$A(\alpha, R) = A_a + A_b \alpha + A_c R + A_d \alpha R + A_e R^2 + A_f (\alpha R)^2, \tag{28}$$

$$B(\alpha, R) = B_a + B_b \alpha + B_c R + B_d \alpha R + B_e R^2 + B_f (\alpha R)^2. \tag{29}$$

The parameters were chosen based on a fit to numerically calculated Coulomb correction values, and the optimal case was found to be represented by these parameter values:

$A_a = 0.20737,$ $A_b = -0.00999,$ $A_c = -0.02671,$ $A_d = -0.00373,$ $A_e = 0.00119,$ $A_f = 0.00016,$

$B_a = 25.80500,$ $B_b = 4.01674,$ $B_c = 0.00873,$ $B_d = -0.25606,$ $B_e = 0.01077,$ $B_f = -0.00270.$

The exponential damping factor of Equation (27) is "joined" to the proper parametrization valid for the interesting q range by a Wood–Saxon-type of cut-off function:

$$F(q) = \frac{1}{1 + \exp\left(\frac{q - q_0}{D_q}\right)}, \tag{30}$$

where $q_0 = 0.08$ GeV and $D_q = 0.02$ GeV. We investigated different cut-off functions, such as $1/(1 + (q/q_0)^n)$, but found that the results are rather independent from this choice.

Putting all of the above together, our final parametrization, valid for $\alpha = 0.8$–1.7 and $R = 3$–12 fm values, is thus

$$K(q, \alpha, R)^{-1} = F(q) \times K_{\text{Gamow}}^{-1}(q) \times K_{\text{mod}}^{-1}(q; \alpha, R) + (1 - F(q)) \times E(q) \quad (31)$$

and the Coulomb corrected correlation function which could be fitted to data, can be written in the form of

$$C_2(q; \alpha, R) = [1 - \lambda + K(q; \alpha, R)\lambda(1 + \exp[|qR|^\alpha])] \cdot (\text{assumed background}). \quad (32)$$

We used this formula to reproduce PHENIX results Ref. [5] [4]; this can be seen on Figure 2. The two fits are compatible with each other. For an example code calculating the formula of (31), please see Ref. [11]. Example curves resulting from the above (32) formula (with the background being unity) are shown in Figure 3. These clearly show how R changes the scale, and α changes the shape of the correlation functions. Parameter λ provides an overall normalization to the distance of these curves from unity, as described by Equation (32).

We investigated the parametrization by means of its relative deviation from the lookup table. The results can be seen in Figure 4. In the case when $\alpha = 1.2$ with different R values, we present a two-dimensional histogram of the relative differences in in Figure 5. The maximum of these relative differences is around 0.05%.

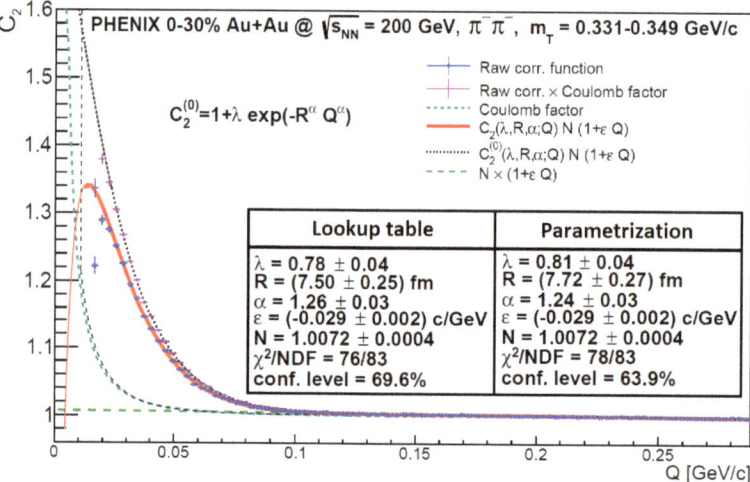

Figure 2. The reproduction of earlier PHENIX results [5] with the parametrization. The original PHENIX fit procedure employed the lookup numerical table, here we show our results from the parameterization.

---

[4] The data of the shown PHENIX correlation function result was retrieved from https://www.phenix.bnl.gov/phenix/WWW/info/data/ppg194_data.html.

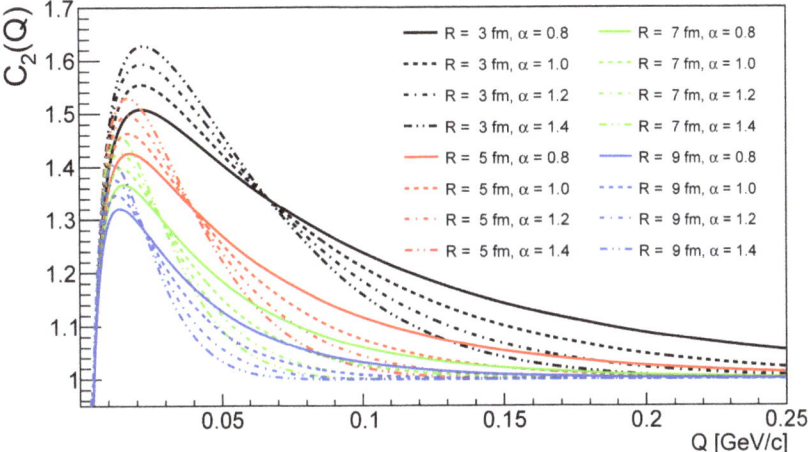

Figure 3. Example correlation functions, based on Equation (32), for different values of parameters R and α.

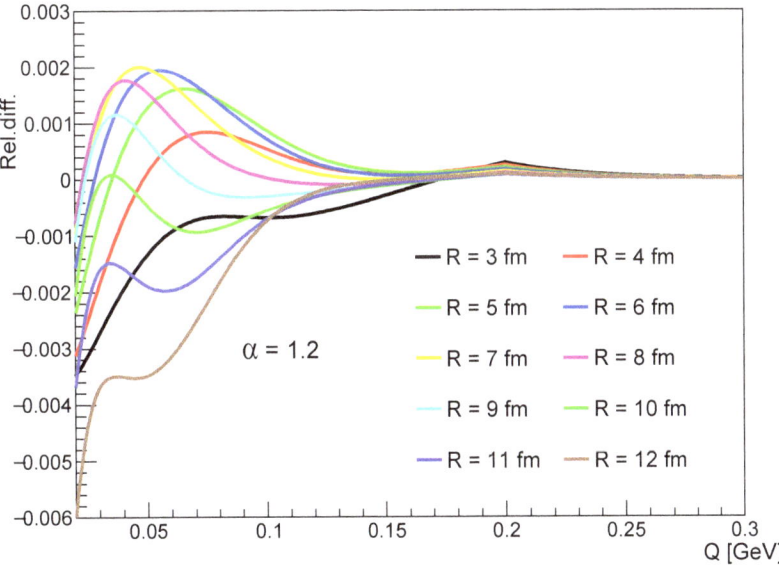

Figure 4. The relative deviation of the parametrization measured in % form the table for a given α with various R values within the domain of the parametrization.

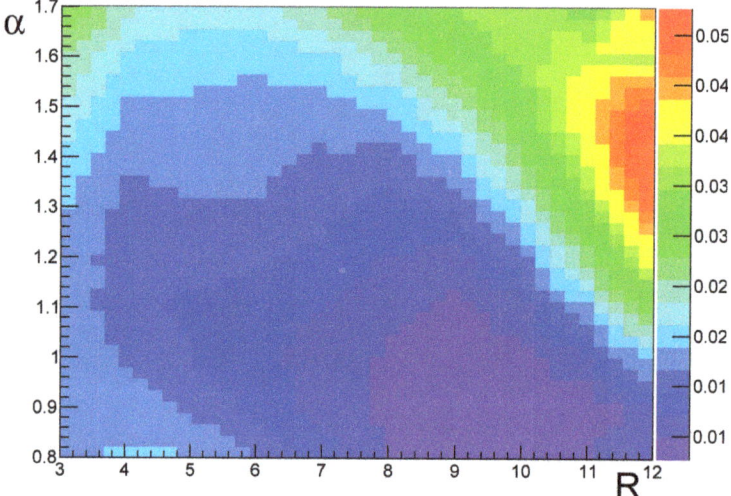

Figure 5. The relative deviation of the parametrization measured in % form the table for the domain of the parameterization in α (0.8–1.7) and in R (3–12 fm), averaged over a q region of 0.01 to 0.1 GeV/c.

5. Conclusions

We investigated the Coulomb correction of Bose–Einstein correlations in high energy heavy ion reactions under the assumption of Lévy source functions. We outlined two equivalent methods that are suited for an experimental analysis. One of them is a numerical lookup table, another one is a parametrization obtained from the former. We investigated the accuracy of the methods and found that a not very complicated ad-hoc parametrization, in the well defined parameter range of $R = 3$–12 fm and $\alpha = 0.8$–1.7, provides an experimentally acceptable description of the results of the numerical integration that is required for the handling of the Coulomb effect. Our parametrization can thus be used effectively in Bose–Einstein correlation analyses that assume Lévy-type source functions.

Author Contributions: Conceptualization: M.C., S.L.; Formal analysis: M.C., S.L. and M.N.; Investigation: M.C., S.L. and M.N.; Software: S.L. and M.N.; Supervision: M.C.; Visualization: S.L.; Writing: M.C., S.L. and M.N.

Funding: This work was supported by the NKFIH grant FK 123842. S.L. is grateful for the support of EFOP 3.6.1-16-2016-00001. N.M. and C.M. were supported by the Hungarian Academy of Sciences through the "Bolyai János" Research Scholarship program as well as the ÚNKP-18-4 New National Excellence Program of the Hungarian Ministry of Human Capacities.

Acknowledgments: We thank D. Kincses for useful discussions on the applicability of the method presented in this paper.

Conflicts of Interest: The authors declare no conflict of interest.

References

1. Bowler, M.G. Coulomb corrections to Bose-Einstein correlations have been greatly exaggerated. *Phys. Lett. B* **1991**, *270*, 69–74. [CrossRef]
2. Sinyukov, Y.; Lednicky, R.; Akkelin, S.V.; Pluta, J.; Erazmus, B. Coulomb corrections for interferometry analysis of expanding hadron systems. *Phys. Lett. B* **1998**, *432*, 248–257. [CrossRef]
3. Biyajima, M.; Mizoguchi, T.; Osada, T.; Wilk, G. Improved Coulomb correction formulae for Bose-Einstein correlations. *Phys. Lett. B* **1995**, *353*, 340–348. [CrossRef]
4. Csörgő, T.; Hegyi, S.; Zajc, W.A. Bose-Einstein correlations for Levy stable source distributions. *Eur. Phys. J. C* **2004**, *36*, 67–78. [CrossRef]

5. Adare, A.; et al. [PHENIX Collaboration]. Lévy-stable two-pion Bose-Einstein correlations in $\sqrt{s_{NN}} = 200$ GeV Au+Au collisions. *Phys. Rev. C* **2018**, *97*, 064911. [CrossRef]
6. Csörgő, T.; Hegyi, S.; Novak, T.; Zajc, W.A. Bose-Einstein or HBT correlation signature of a second order QCD phase transition. *AIP Conf. Proc.* **2006**, *828*, 525–532.
7. Yano, F.B.; Koonin, S.E. Determining Pion Source Parameters in Relativistic Heavy Ion Collisions. *Phys. Lett. B* **1978**, *78*, 556–559. [CrossRef]
8. Csörgő, T.; Lörstad, B.; Zimanyi, J. Bose-Einstein correlations for systems with large halo. *Z. Phys. C* **1996**, *71*, 491–497. [CrossRef]
9. Vértesi, R.; Csörgő, T.; Sziklai, J. Significant in-medium η' mass reduction in $\sqrt{s_{NN}} = 200$ GeV Au+Au collisions at the BNL Relativistic Heavy Ion Collide. *Phys. Rev. C* **2011**, *83*, 054903. [CrossRef]
10. Sirunyan, A.M.; et al. [CMS Collaboration]. Bose-Einstein correlations in pp, pPb, and PbPb collisions at $\sqrt{s_{NN}} = 0.9 - 7$ TeV. *Phys. Rev. C* **2018**, *97*, 064912. [CrossRef]
11. Available online: https://github.com/csanadm/coulcorrlevyparam (accessed on 20 May 2019).

© 2019 by the authors. Licensee MDPI, Basel, Switzerland. This article is an open access article distributed under the terms and conditions of the Creative Commons Attribution (CC BY) license (http://creativecommons.org/licenses/by/4.0/).

Communication
Jet Structure Studies in Small Systems †

Zoltán Varga [1,2,*], **Róbert Vértesi** [1] **and Gergely Gábor Barnaföldi** [1]

1. Wigner Research Centre for Physics, P.O. Box 49, H-1525 Budapest, Hungary; vertesi.robert@wigner.mta.hu (R.V.); barnafoldi.gergely@wigner.mta.hu (G.G.B.)
2. Department of Theoretical Physics, Institute of Physics, Faculty of Science, Eötvös Loránd University, Pázmány Péter sétány 1/A, H-1117 Budapest, Hungary
* Correspondence: varga.zoltan@wigner.mta.hu
† This paper is based on the talk at the 18th Zimányi School, Budapest, Hungary, 3–7 December 2018.

Received: 16 April 2019; Accepted: 22 May 2019; Published: 27 May 2019

Abstract: A study investigating a possible jet shape dependence on the charged event multiplicity was performed on collision samples generated by Monte–Carlo (MC) event generators PYTHIA and HIJING++. We calculated the integral jet shape and found a significant modification caused by multiple-parton interactions. By interchanging and enabling different model ingredients in the simulations and analyzing the results in several p_T bins and event multiplicity classes, we found a characteristic jet size measure that was independent of the chosen tunes, settings, and jet reconstruction algorithms.

Keywords: jet physics; jet structure; jet shapes; color reconnection; multiple-parton interactions; high-energy collisions

PACS: 13.87.-a; 25.75.-q; 25.75.Ag; 25.75.Dw

1. Introduction

The discovery of collective-like behavior in high-multiplicity proton-proton (pp) and proton-nucleus (pA) collisions was one of the major surprises in early LHC results [1,2]. The collective-like behavior previously found in large systems, manifested in long-range correlations and a sizeable azimuthal anisotropy, have traditionally been considered as a signature proving the presence of the quark-gluon plasma (QGP). The creation of the QGP in high-energy nucleus-nucleus (AA) collisions can also be investigated by studying the structure of jets and their modification that leads to the well-known jet quenching phenomenon [3,4]. Whether QGP is created in small systems like pp collisions is still an open question [5]. However, the presence of the QGP is not necessary to explaining collectivity: Relatively soft vacuum-QCD effects such as multiple-parton interactions (MPI) and color reconnection (CR) can also produce a similar behavior [6,7]. These interactions, at least in principle, can modify the jet shapes in even small systems. Although experimental confirmation is not yet available, a recent phenomenology study also suggests the modification of hard processes by soft vacuum-QCD effects in a high-multiplicity environment [8].

MPI is also expected to depend on flavor [9]. Fragmentation of heavy-flavor jets is expected to differ from light-flavor jets because of color charge and mass effects. The internal structures of heavy-flavor jets may therefore provide a deeper insight into the flavor-dependent development of jets and their connection to the underlying event (UE). This paper continues our previous studies [10,11] aimed at the evolution of jet structure patterns and their dependence on simulation components.

2. Analysis

We used PYTHIA 8.226 and HIJING++ Monte–Carlo (MC) generators to simulate pp collision events at $\sqrt{s} = 7$ TeV [12,13]. Three different PYTHIA tunes were investigated, the Monash 2013 tune with the NNPDF2.3LO PDF set [14,15], the Monash* tune with NNPDF2.3LO [16], and tune 4C with the CTEQ6L1 PDF set [17,18]. Collisions with and without multiple-parton interactions and color reconnection were simulated and compared to each other [10,11]. We only considered particles above the transverse momentum threshold $p_T^{track} > 0.15$ GeV/c. We carried out a full jet reconstruction using the anti-k_T, k_T, and Cambridge–Aachen jet reconstruction algorithms, which are part of the FASTJET software package [19]. These choices are typical in jet shape analyses [20]. The multiplicity-integrated jet shape studies of CMS were used as a benchmark for our current multiplicity-differential studies [10,20]. Therefore we chose the jet resolution parameter as $R = 0.7$ and applied a fiducial cut so that the jets were contained in the CMS acceptance $|\eta| < 1$. We did not apply underlying event subtraction in the jet cones. We investigated the jets within the 15 GeV/c $< p_T^{jet} <$ 400 GeV/c transverse momentum range, where multiplicity-differential studies on real data are feasible in the near future.

We chose the following two jet shape measures to study the multiplicity-dependent behavior of the jet structure: The Ψ integral jet shape (or momentum fraction) and the ρ differential jet shape (or momentum density fraction) [21,22]. The former one gives the average fraction of the jet transverse momentum contained inside a sub-cone of radius r around the jet axis, the latter one is the momentum profile of the jet, i.e., the average transverse momentum of the particles contained inside an annulus with a δr width and boundaries r_a and r_b. The exact formulae are given by:

$$\Psi(r) = \frac{1}{p_T^{jet}} \sum_{r_i < r} p_T^i \quad \text{and} \quad \rho(r) = \frac{1}{\delta r} \frac{1}{p_T^{jet}} \sum_{r_a < r_i < r_b} p_T^i \tag{1}$$

respectively, where p_T^i is the transverse momentum of the selected particle and p_T^{jet} is the transverse momentum of the jet. The distance r_i of the given particle from the jet axis is calculated as $r_i = \sqrt{(\phi_i - \phi_{jet})^2 + (\eta_i - \eta_{jet})^2}$, where ϕ is the azimuthal angle and η is the pseudorapidity. For a better understanding, we noted that the aforementioned observables are connected with the equations:

$$\Psi(r) = \int_0^r \rho(r')dr', \quad \text{and} \quad \Psi(R) = \int_0^R \rho(r')dr' = 1, \tag{2}$$

where R is the jet resolution parameter. The differential and integral jet shapes were calculated for the above mentioned p_T^{jet} range. The three tunes reproduced the CMS results within statistical errors [10,11].

3. Results

The multiplicity distributions (multiplicity is defined as the number of charged final state particles in a given collision event) were very similar for the three tunes used in our simulations, as shown in the left panel of Figure 1. However, a significant change is observed when we do not consider the effects of the color reconnection and/or the multiple-parton interactions. Without the CR, the multiplicity distribution becomes wider. If we also switch off the MPI, the multiplicity distribution becomes much narrower compared to the setting where both of them are applied. The width of the multiplicity distribution, however, is not sensitive to the choice of the CR model. In the right panel of Figure 1, we plot the mean values of the multiplicity distributions in events where we reconstructed a jet of a given transverse momentum. We did not exclude non-leading jets in our analysis, but we also investigate the effects of selecting only leading jets later in this paper. As expected, the average event multiplicity grew with the transverse momentum of the selected jet.

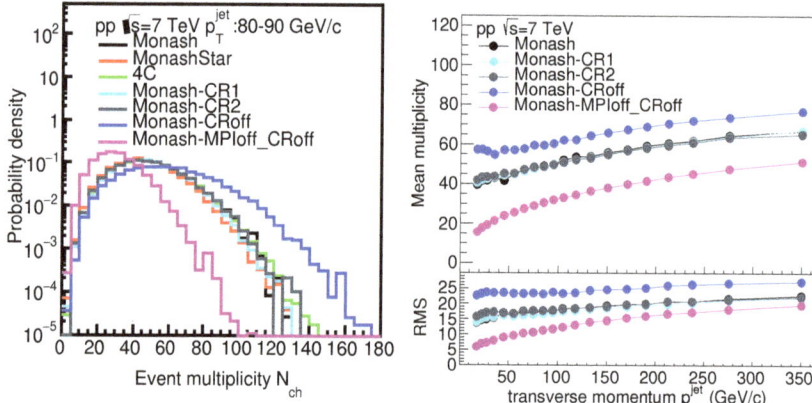

Figure 1. (**Left**) Multiplicity distributions for different tunes and settings in events where the p_T^{jet} of the reconstructed jets falls in a given p_T bin; (**Right**) The mean values of the distributions are shown as a function of the p_T^{jet}. The RMS is the relative mean squared value of the distribution.

The integral jet shape with $r = 0.2$ is shown in Figure 2 to compare the effects of the different tunes and settings. As expected [23], we see similar trends for the multiplicity distributions of the tunes. However, there is a substantial difference between the different MPI and CR settings. The most significant difference in the jet shapes was caused by turning off the MPI, which further supports the current view that the MPI contributions need to be included to correctly describe the jet shapes. We note that we did not subtract the underlying event from the jet. Further investigation is necessary to understand whether the interplay between the UE and the observed hard process is significantly modified by the MPI.

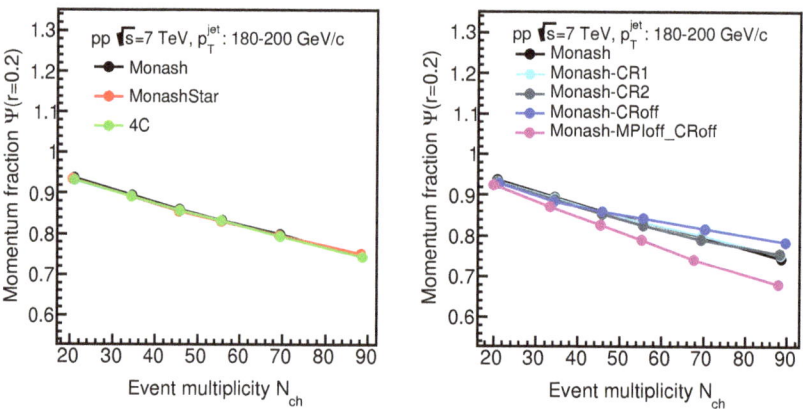

Figure 2. (**Left**) Integrated jet shape for different tunes; (**Right**) Integrated jet shape for different settings.

In order to gain a more comprehensive picture, we use the current differential jet shape to investigate the multiplicity dependence of the jet shapes. On the left side of Figure 3, we categorize the events into a high and low multiplicity class. The differential jet shapes of both categories are compared to the multiplicity-integrated momentum density (ρ_{MI}), computed without any selection in multiplicity. A multiplicity dependence is observed. The jets in the higher multiplicity bin appear to be wider while the jets in the lower multiplicity bin appear to be narrower. This is a trivial multiplicity dependence since the event multiplicity strongly correlates with the jet multiplicity. In our previous work, we canceled this trivial multiplicity dependence by applying a double ratio across the different tunes [10].

In the right panel of Figure 3, we plotted the $\rho(r)/\rho_{MI}$ ratio for the low- and high-multiplicty bins divided by the ρ_{MI} curve. In this way, the difference compared to the ρ_{MI} (black) curve is more visible. We also observed that the $\rho(r)/\rho_{MI}$ curves obtained from the low- and high-multiplicty classes intersect each other at a particular radius inside the jet cone.

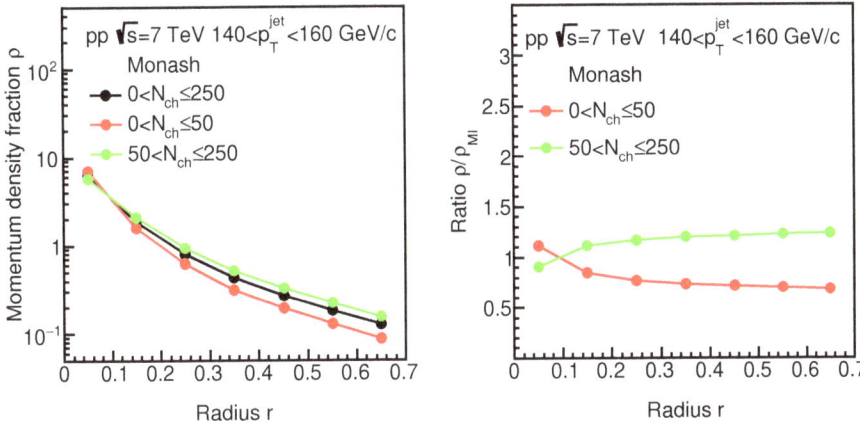

Figure 3. (**Left**) Differential jet shape for low- and high-multiplicity bins; (**Right**) The differential jet shape ρ divided by the ρ_{MI} to emphasize the difference.

We also investigated whether the choice of multiplicity bins affects the location of the intersection point by categorizing the events into several smaller event multiplicity bins. We found that all the ρ/ρ_{MI} curves corresponding to these different multiplicity classes intersect each other at approximately the same radius, as can be seen on the left panel of Figure 4. Therefore we named this specific r value R_{fix}. The R_{fix} depends, however, on the transverse momentum of the jet, as shown in the right panel of Figure 4. The $R_{fix}(p_T^{jet})$ curve goes in a fashion expected by a Lorentz boost and converges to a constant value at higher energies [10].

Since we use a linear interpolation between the points of ρ/ρ_{MI}, the value of R_{fix} will have a slight dependence on the bin width in r (δr), especially at higher p_T values where the value of R_{fix} is smaller. To make sure that the results are robust enough, we repeated the analysis with narrower bins ($\delta r = 0.05$). In the left panel of Figure 5, we show ρ with this finer r binning, while in the right panel of the same figure we plotted the p_T^{jet} dependence of R_{fix}. We can see that the effects by the choice of bin width is rather small and does not change the conclusions, so we do not have to sacrifice the current statistics for finer binnings.

Using different settings in PYTHIA changes the physics enough to expect different jet structures after jet reconstruction. All of the k_T, Cambridge–Aachen and anti-k_T jet clustering algorithms that we investigated reconstructed the jet structures differently since they had different susceptibility to the underlying event. We therefore investigated the effects of varying the physics settings and used three different jet reconstruction algorithms (Figure 6). We conclude that the presence of R_{fix} was very robust for the different physics selections, as its stability was neither an artifact of the particular choice of jet reconstruction algorithms, nor did it depend strongly on the underlying event. We note that R_{fix} is localized to lower r values, therefore we do not expect significant influence from effects that are mostly visible at higher r values, such as the modification of the UE or higher order corrections to the parton shower [24,25].

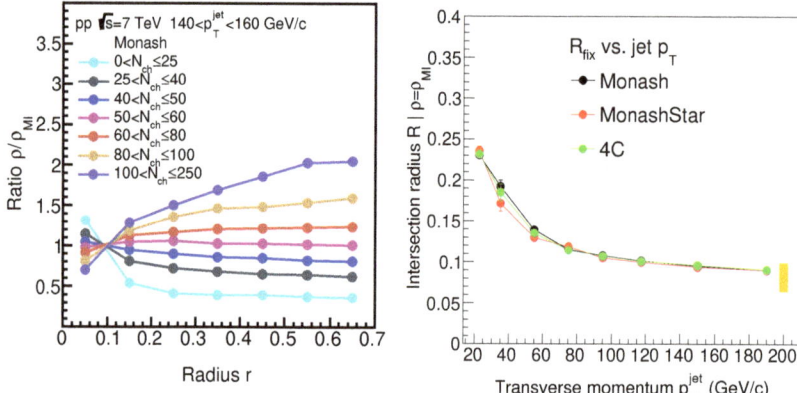

Figure 4. (**Left**) Ratio ρ/ρ_{MI} of the differential jet structure over the multiplicity-integrated curve for many different multiplicity bins; (**Right**) The p_T^{jet} dependence of R_{fix} with $\delta r = 0.1$. The systematic uncertainty from the choice of δr is shown as the yellow band.

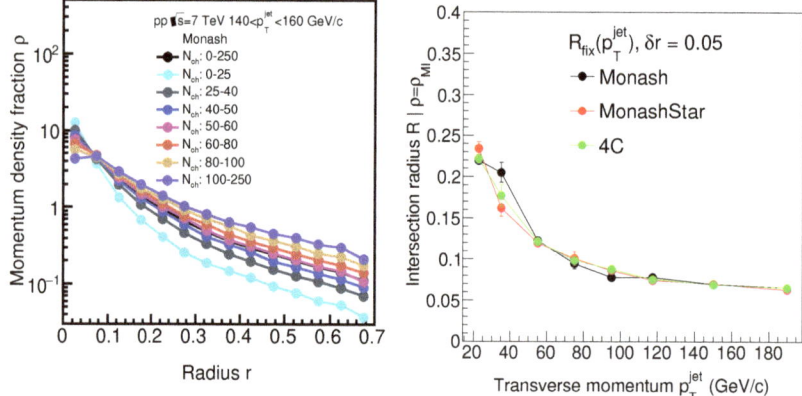

Figure 5. (**Left**) The differential jet shape for many different multiplicity bins with finer ($\delta r = 0.05$) binnings; (**Right**) The p_T^{jet} dependence of R_{fix} with finer ($\delta r = 0.05$) binnings.

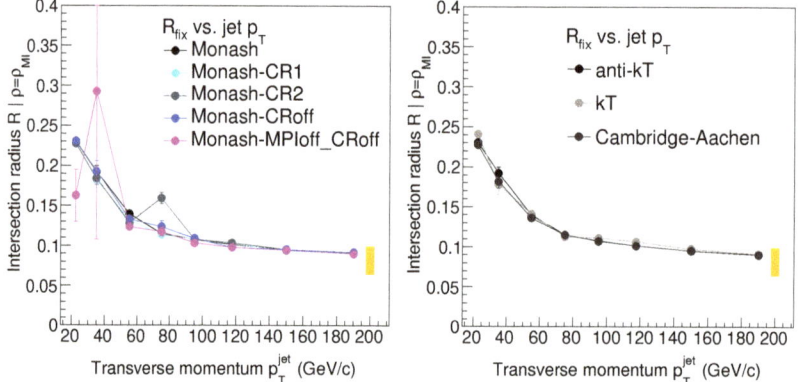

Figure 6. The evolution of R_{fix} with the p_T^{jet} for different settings (**left**); and for different jet reconstruction algorithms (**right**).

We also computed the differential jet shapes using a different MC generator. We selected HIJING++ for this purpose because it implements a mechanism of creating the underlying event and QCD effects on the soft-hard boundary that is different from PYTHIA. HIJING++ uses the PYTHIA jet fragmentation, therefore we do not expect any difference during the later stages. Instead of MPI as implemented in PYTHIA, HIJING++ uses minijet production. Differences at lower momenta may arise below the minijet cutoff. In case of the p_T and multiplicity distributions, these effects do not exceed the variation caused by applying different tunes in PYTHIA.

In the case of HIJING++, we used two different PDF sets. The results show quantitatively the same R_{fix} dependence on p_T^{jet} within systematic errors, as can be seen in the left panel of Figure 7. Jets originating from different flavors undergo different fragmentation due to both the color-charge effect and the dead-cone effect [26]. We compared flavor-inclusive jets to heavy-flavor (beauty and charm) jets in the right panel of Figure 7. We ensured that heavy flavor comes from the initial stages by only enabling leading order processes in PYTHIA. We compared these to leading and subleading flavor-inclusive jets.

It should be noted that the effect of non-leading jets is negligible on the R_{fix}. Although the overall tendency of heavy-flavor R_{fix} is similar to that observed for light flavor, there is also a clear quantitative difference between heavy and light flavors, which points to a different jet structure. The leading b jets differ for higher p_T^{jet} and the leading c jets for lower p_T^{jet}. This suggests that the interplay between the mass and color-charge effects is non-trivial and needs further investigation. One possibility for that would be a parallel study of the UE and the fragmentation region corresponding to a heavy-flavor trigger, in a similar manner to [27]. From all the above, we can assume that R_{fix} is a property of the jets that is associated with the final state.

We also repeated the same analysis by selecting only the leading and sub-leading jets. We observed no significant difference compared to the case where all the jets were used (right panel of Figure 7).

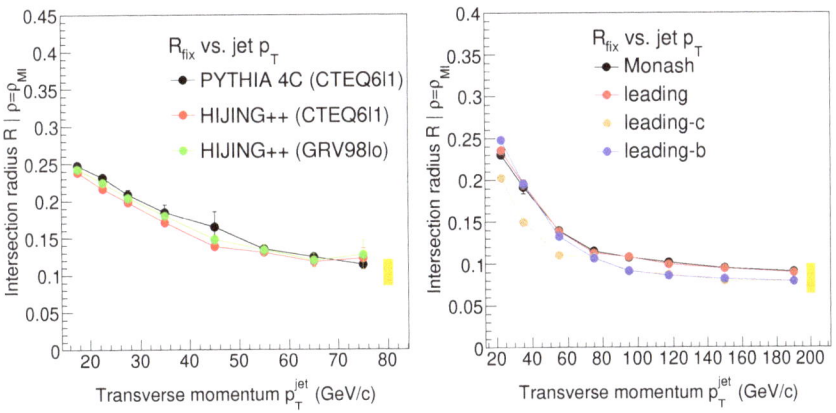

Figure 7. (**Left**) The p_T^{jet} dependence of R_{fix} for the PYTHIA 4C tune is compared with HIJING++; (**Right**) Comparison of different jet selections for the PYTHIA Monash tune.

4. Conclusions

We conducted a systematic study on jet structures for pp collisions using MC generators PYTHIA and HIJING++. We investigated the effects of CR and MPI on jet shapes and showed that not considering the effects of CR and MPI causes significant jet shape modification. We introduced a characteristic jet size (R_{fix}) that depends only on the p_T of the jets but is independent of the jet reconstruction algorithms, MC generators, parton density functions, and even the choice of simulation parameters such as CR and MPI [10,11]. We have also shown that the choice of δr does not change our conclusions about R_{fix}. These observations suggest that R_{fix} is an inherent property of the jets and is characteristic to the space-time evolution of the parton shower at a given momentum.

However, R_{fix} does depend on the flavor of the jet. Flavor-dependent jet structure studies may be a way to access mass versus color charge effects that is complimentary to hadron- or jet-production cross-section measurements. We believe that our findings motivate further phenomenology studies as well as cross-checks with real data to gain a deeper understanding on flavor-dependent jet fragmentation. Another direction for future research could be to investigate the effects of MPI on jets without the underlying event. This could be done either by choosing an observable that depends very weakly on the underlying event [9], or by the parallel understanding of the underlying event and the fragmentation region [27].

Author Contributions: Software and formal analysis, Z.V., R.V.; investigation, R.V., Z.V., G.G.B.; writing—original draft preparation, Z.V.; writing—review and editing, R.V.; supervison, R.V.; funding acquisition, G.G.B.

Funding: This work has been supported by the NKFIH/OTKA K 120660 grant, the János Bolyai scholarship of the Hungarian Academy of Sciences (R.V.), and the MOST-MTA Chinese-Hungarian Research Collaboration. The work has been performed in the framework of COST Action CA15213 THOR.

Acknowledgments: The authors are thankful for the many useful conversations they had with Gábor Bíró, Jana Bielčíková, Miklós Kovács, Filip Křížek, Yaxian Mao, Antonio Ortiz and Guy Paić.

Conflicts of Interest: The authors declare no conflict of interest.

References

1. Yan, L.; Ollitrault, J.Y. Universal fluctuation-driven eccentricities in proton-proton, proton-nucleus and nucleus-nucleus collisions. *Phys. Rev. Lett.* **2014**, *112*, 082301. [CrossRef]
2. Khachatryan, V.; Sirunyan, A.M.; Tumasyan, A.; Adam, W.; Bergauer, T.; Dragicevic, M.; Erö, J.; Fabjan, C.; Friedl, M.; Frühwirth, R.; et al. Observation of Long-Range Near-Side Angular Correlations in Proton-Proton Collisions at the LHC. *J. High Energy Phys.* **2010**, *2010*, 091. [CrossRef]
3. Adcox, K.; Adler, S.S.; Ajitanand, N.N.; Akiba, Y.; Alexander, J.; Aphecetche, L.; Arai, Y.; Aronson, S.H.; Averbeck, R.; Awes, T.C.; et al. Supression of hadrons with large transverse momentum in central Au+Au collisions at $\sqrt{s_{NN}}$ = 130 GeV. *Phys. Rev. Lett.* **2001**, *88*, 022301. [CrossRef] [PubMed]
4. Gyulassy, M.; Levai, P.; Vitev, I. Jet Quenching in Thin Quark-Gluon Plasmas I: Formalism. *Nucl. Phys. B* **2000**, *571*, 197–233. [CrossRef]
5. Dusling, K.; Li, W.; Schenke, B. Novel collective phenomena in high-energy proton–proton and proton–nucleus collisions. *Int. J. Mod. Phys. E* **2016**, *25*, 1630002. [CrossRef]
6. Ortiz, A.; Bencédi, G.; Bello, H. Revealing the source of the radial flow patterns in proton–proton collisions using hard probes. *J. Phys. G* **2017**, *44*, 065001. [CrossRef]
7. Velasquez, A.O.; Christiansen, P.; Flores, E.C.; Cervantes, I.M.; Paić, G. Color Reconnection and Flowlike Patterns in pp Collisions. *Phys. Rev. Lett.* **2013**, *111*, 042001. [CrossRef]
8. Mishra, A. N.; Paić, G. Did we miss the "melting" of partons in pp collisions? *arXiv* **2019**, arXiv:1905.06918.
9. Khachatryan, V.; Sirunyan, A.M.; Tumasyan, A.; Adam, W.; Asilar, E.; Bergauer, T.; Brandstetter, J.; Brondolin, E.; Dragicevic, M.; Erö, J.; et al. Studies of inclusive four-jet production with two b-tagged jets in proton-proton collisions at 7 TeV. *Phys. Rev. D* **2016**, *94*, 112005. [CrossRef]
10. Varga, Z.; Vértesi, R.; Barnaföldi, G.G. Modification of jet structure in high-multiplicity pp collisions due to multiple-parton interactions and observing a multiplicity-independent characteristic jet size. *Adv. High Energy Phys.* **2019**. [CrossRef]
11. Varga, Z.; Vértesi, R.; Barnaföldi, G.G. Multiplicity Dependence of the Jet Structures in pp Collisions at LHC Energies. *arXiv* **2018**, arXiv:1809.10102.
12. Sjöstrand, T.; Ask, S.; Christiansen, J.R.; Corke, R.; Desai, N.; Ilten, P.; Mrenna, S.; Prestel, S.; Rasmussen, C.O.; Skands, P.Z. An Introduction to PYTHIA 8.2. *Comput. Phys. Commun.* **2015**, *191*, 159–177. [CrossRef]
13. Barnaföldi, G.G.; Bíró, G.; Gyulassy, M.; Haranozó, S.M.; Lévai, P.; Ma, G.; Papp, G.; Wang, X.N.; Zhang, B.W. First Results with HIJING++ in High-Energy Heavy-Ion Collisions. *Nucl. Part. Phys. Proc.* **2017**, *373*, 289–290. [CrossRef]

14. Skands, P.; Carrazza, S.; Rojo, J. Tuning PYTHIA 8.1: The Monash 2013 Tune. *Eur. Phys. J. C* **2014**, *74*, 3024. [CrossRef]
15. Ball, R.D.; Bertone, V.; Carrazza, S.; Del Debbio, L.; Forte, S.; Guffanti, A.; Hartland, N.P.; Rojo, J. Parton distributions with QED corrections. *Nucl. Phys. B* **2013**, *877*, 290. [CrossRef]
16. Khachatryan, V.; Sirunyan, A.M.; Tumasyan, A.; Adam, W.; Asilar, E.; Bergauer, T.; Brandstetter, J.; Brondolin, E.; Dragicevic, M.; Erö, J.; et al. Event generator tunes obtained from underlying event and multiparton scattering measurements. *Eur. Phys. J. C* **2016**, *76*, 155. [CrossRef] [PubMed]
17. Buckley, A.; Butterworth, J.; Gieseke, S.; Grellscheid, D.; Höche, S.; Hoeth, H.; Krauss, F.; Lönnblad, L.; Nurse, E.; Richardson, P.; et al. General-purpose event generators for LHC physics. *Phys. Rep.* **2011**, *504*, 145–233. [CrossRef]
18. Pumplin, J.; Stump, D.R.; Huston, J.; Lai, H.L.; Nadolsky, P.M.; Tung, W.K. New generation of parton distributions with uncertainties from global QCD analysis. *J. High Energy Phys.* **2002**, *2002*, 012. [CrossRef]
19. Cacciari, M.; Salam, G.P.; Soyez, G. FastJet user manual. *Eur. Phys. J. C* **2012**, *72*, 1896. [CrossRef]
20. Chatrchyan, S.; Khachatryan, V.; Sirunyan, A.M.; Tumasyan, A.; Adam, W.; Bergauer, T.; Dragicevic, M.; Erö, J.; Fabjan, C.; Friedl, M.; et al. Shape, Transverse Size, and Charged Hadron Multiplicity of Jets in pp Collisions at 7 TeV. *J. High Energy Phys.* **2012**, *2012*, 160. [CrossRef]
21. Ellis, S.D.; Kunszt, Z.; Soper D.E. Large Jets at Hadron Colliders at Order α_s^3: A Look Inside. *Phys. Rev. Lett.* **1992**, *69*, 3615–3618. [CrossRef] [PubMed]
22. Seymour, M.H. Jet Shapes in Hadron Collisions: Higher Orders, Resummation and Hadronization. *Nucl. Phys. B* **1998**, *513*, 269–300. [CrossRef]
23. The ATLAS Collaboration. New ATLAS Event Generator Tunes to 2010 Data. ATL-PHYS-PUB-2011-008, ATL-COM-PHYS-2011-329. Available online: http://inspirehep.net/record/1196773?ln=en (accessed on 27 May 2019).
24. Hoeche, S.; Reichelt, D.; Siegert, F. Momentum conservation and unitarity in parton showers and NLL resummation. *J. High Energy Phys.* **2018**, *2018*, 118. [CrossRef]
25. Cal, P.; Ringer, F.; Waalewijn, W.J. The Jet Shape at NLL'. *arXiv* **2018**, arXiv:1901.06389.
26. Thomas, R.; Kampfer, B.; Soff, G. Gluon Emission of Heavy Quarks: Dead Cone Effect. *Acta Phys. Hung. A* **2005**, *22*, 83–91. [CrossRef]
27. Ortiz, A.; Valencia, P.L. Probing color reconnection with underlying event observables at the LHC energies. *Phys. Rev. D* **2019**, *99*, 034027. [CrossRef]

© 2019 by the authors. Licensee MDPI, Basel, Switzerland. This article is an open access article distributed under the terms and conditions of the Creative Commons Attribution (CC BY) license (http://creativecommons.org/licenses/by/4.0/).

Communication

Heavy-Flavor Measurements with the ALICE Experiment at the LHC

Róbert Vértesi and on behalf of the ALICE Collaboration

Wigner Research Centre for Physics, Konkoly-Thege Miklós út 29-33, 1121 Budapest, Hungary; vertesi.robert@wigner.mta.hu

Received: 19 April 2019; Accepted: 22 May 2019; Published: 25 May 2019

Abstract: Heavy quarks (charm and beauty) are produced early in the nucleus–nucleus collisions, and heavy flavor survives throughout the later stages. Measurements of heavy-flavor quarks thus provide us with means to understand the properties of the Quark–Gluon Plasma, a hot and dense state of matter created in heavy-ion collisions. Production of heavy-flavor in small collision systems, on the other hand, can be used to test Quantum-chromodynamics models. After a successful completion of the Run-I data taking period, the increased luminosity from the LHC and an upgraded ALICE detector system in the Run-II data taking period allows for unprecedented precision in the study of heavy quarks. In this article we give an overview of selected recent results on heavy-flavor measurements with ALICE experiments at the LHC.

Keywords: Large Hadron Collider; heavy flavor; fragmentation; high-energy physics

1. Introduction

Substantial evidence indicates that high-energy heavy-ion collisions can recreate a strongly coupled Quark–Gluon Plasma (QGP) [1,2], an extremely hot, dense, strongly interacting state of matter that was present in the early stages of the Universe. Smaller colliding systems such as pp or p-Pb are useful to study perturbative quantum chromodynamics (QCD) and cold nuclear effects. Heavy quarks are produced early in the reaction and their numbers are almost conserved throughout the reaction. They undergo negligible flavor changing, and there is also very little thermal production or destruction within the QGP or the hadronic nuclear matter. Yet, they are transported through the whole system, thus their kinematics can reveal transport properties such as collisional and radiational energy loss within the hot medium [3]. Since heavy flavor is preserved, it can be used as a penetrating probe down to a very low momentum ($p_T \approx 0$). Observing the hadronization of heavy quarks reveals coalescence mechanisms in the hot medium, as well as fragmentation properties. This latter one, in comparison with light probes, is telltale about color charge and mass/flavor effects.

Measurements of heavy-flavor in pp collisions serve the primary purpose of setting a benchmark on pQCD calculations. As in the case of many other probes, we rely on heavy-flavor production in pp collisions as a reference for measurements in larger collision systems where a substantial effect is expected by the hot or cold nuclear matter. In high-multiplicity events however, we observe signatures of collectivity, such as long-range correlations [4], that resemble those characteristic to QGP production. These are usually attributed to soft and semi-hard vacuum QCD effects such as multiple parton interactions (MPI) [5], although alternative explanations also exist [6]. Heavy-flavor measurements versus event activity play a key role in clarifying the origin of such signatures. Fragmentation of heavy quarks can be studied using jet and correlation observables. Besides color charge, mass and flavor effects such as these can pin down the contribution of late gluon splitting to heavy-flavor production. Production of the heavy-flavored baryons provides additional information on heavy-flavor fragmentation and provides key information for the development of theoretical models.

Some of the most interesting recent heavy-flavor results carried out by the ALICE experiment are summarized in the following sections. The details of the ALICE detector system are described elsewhere [7]. Heavy-flavor is accessed either directly, via the full reconstruction of the decay products, or indirectly by measuring some of the decay products (decay leptons for most of the cases) and statistically disentangling heavy-flavor contribution.

2. Heavy-Flavor Mesons in pp Collisions

Heavy flavor mesons in pp collisions help understand the physics of QCD vacuum. Direct and indirect measurements on their production are carried out in ALICE. While direct reconstruction of charmed mesons such as D^0, D^+ and D^{*+} provide more direct access to decay kinematics, indirect measurements through semi-leptonic decays provide a mixture of beauty and charm contributions. In recent measurements however, secondary vertexing with fine resolution in the Inner Tracking System (ITS) provide means to statistically separate the contributions of charm and beauty decays. Figure 1 shows recent measurements of D^0 mesons as well as heavy-flavor decay electrons. In general, several theoretical models describe the measurements within uncertainties, and heavy-flavor meson measurements already provide restrictive input to them.

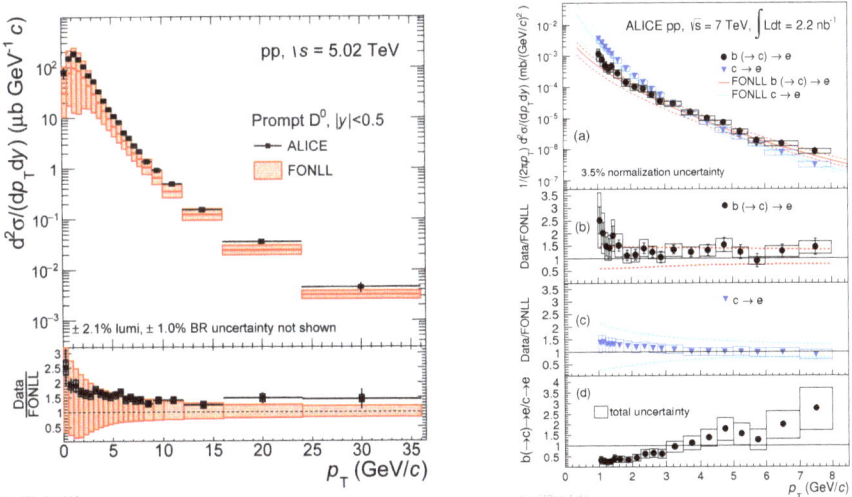

Figure 1. (**Left**) Production cross section of D^0 mesons in pp collisions at $\sqrt{s_{NN}} = 5.02$ TeV, compared to a theoretical calculation [8]. (**Right**) Production cross section and beauty faction of heavy-flavor electrons in pp collisions at $\sqrt{s_{NN}} = 7$ TeV, compared to a theoretical calculation [8].

In pp collisions, the relative yield of D mesons at mid-rapidity depends steeper than linearly on the relative charged multiplicity [9]. This shows that hard processes such as heavy-flavor production; and soft processes like bulk charged hadron production scale differently with event activity. Comparison to theoretical calculations [10–12] in the left panel of Figure 2 shows that the steeper-than-linear trend can be qualitatively described by calculations that include multiple parton interactions (MPI). A recent measurement with heavy-flavor muons indicates a similar trend at forward rapidity, as shown in the right panel of Figure 2.

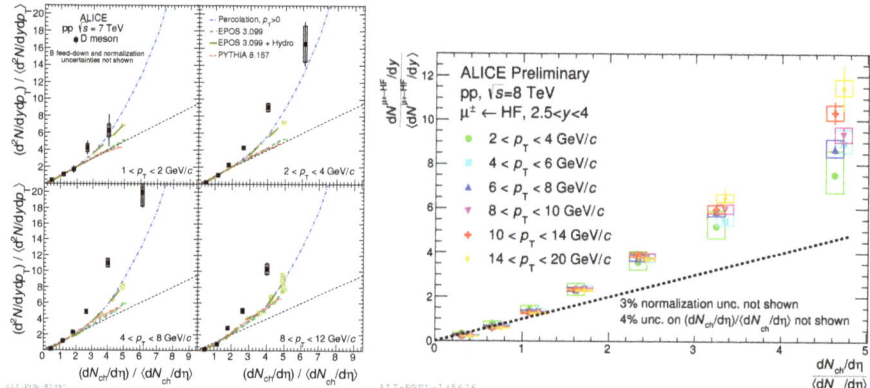

Figure 2. (**Left**) Average self-normalized yields of D mesons in pp collisions at $\sqrt{s_{NN}} = 7$ TeV at mid-rapidity [9], compared to several model calculations [10–12], and (**right**) of muons in pp collisions at $\sqrt{s_{NN}} = 8$ TeV at forward rapidity, for several transverse momentum ranges.

3. The Baryonic Sector

Measurements of baryons with charm content provide valuable input for theories to understand heavy-flavor fragmentation. Recent measurements of Λ_c^+ in pp collisions at $\sqrt{s} = 5$ and 7 TeV and LHC-first Ξ_c^0 measurements in pp collisions at $\sqrt{s} = 7$ TeV [13] show that the production of these mesons are underestimated by widely used theoretical models [8,14,15]. The same is observed in charmed baryon-to-meson ratios with a decreased relative uncertainty, as shown on Figure 3. This shows that our current understanding on heavy-flavor fragmentation in the baryon sector is inadequate.

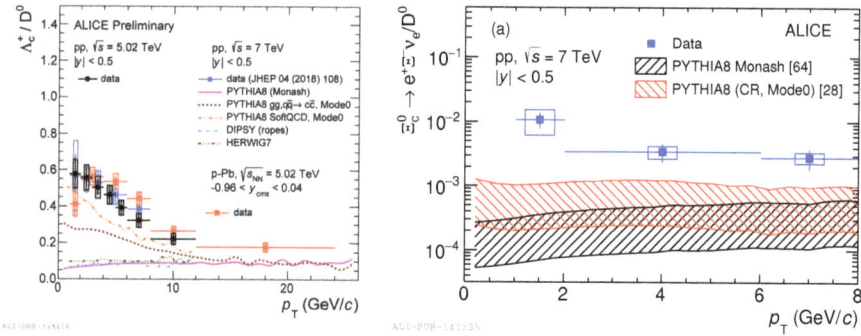

Figure 3. (**Left**) Ratios of Λ_c^+ and (**Right**) Ξ_c^0 p_T-differential cross sections [13] over D^0 in pp collisions at $\sqrt{s} = 7$ TeV, compared to several theoretical calculations [8,14,15].

4. Heavy-Flavor in Cold Nuclear Matter

Cold nuclear matter (CNM) effects are expected to appear both in the initial and in the final state in collisions of protons on heavy ions, in an environment of substantial volume where quarks are still confined into hadrons. Several models predict modification of the nuclear parton distribution functions (nPDF) by (anti)shadowing and gluon saturation. A non-negligible energy loss in the CNM is also expected, as well as the transverse-momentum (k_T) broadening of the initial and final state partons [16–21]. While collectivity has also been observed in p-A collisions, the question whether deconfined matter can be created in p-A collisions has not yet been settled [6].

The left panel of Figure 4 shows the nuclear modification factor R_{pPb} of prompt D mesons as recently measured by the ALICE measurement in p-Pb collisions at $\sqrt{s_{NN}} = 5.02$ TeV.

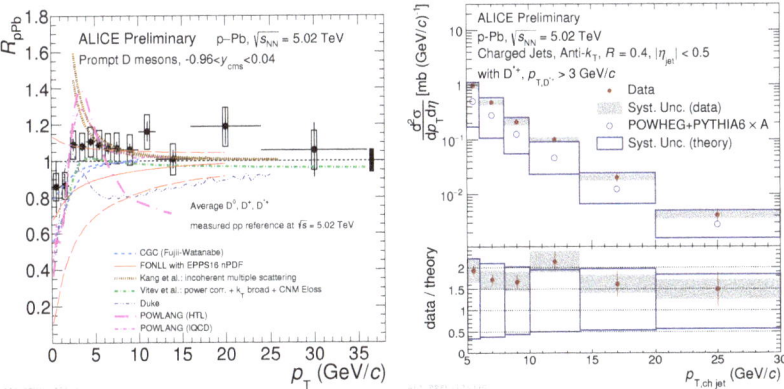

Figure 4. (**Left**) Nuclear modification factor R_{pPb} of prompt D mesons in p-Pb collisions at $\sqrt{s_{NN}} =$ 5.02 TeV, compared to several models [16–18,20,21]. (**Right**) Cross section of D^{*+}-tagged charged jets in p-Pb collisions at $\sqrt{s_{NN}} = 5.02$ TeV, compared to pQCD NLO calculations [22].

The measurement extends down to $p_T \approx 0$ and R_{pPb} is consistent with unity throughout the range. Several models incorporating different CNM mechanisms [16–18,20] adequately describe the weak nuclear modification. The POWLANG model with lattice-QCD calculations, which incorporates QGP formation in a small volume [21], is also able to describe data in a statistically acceptable manner. The right panel of Figure 4 shows charmed jet measurements, defined as jets containing D mesons with $p_T > 3$ GeV/c reconstructed within a jet. POWHEG pQCD NLO calculations with PYTHIA fragmentation [10,22] describe data within uncertainties, indicating the lack of a strong nuclear modification of heavy-flavor jets. However, since the theoretical predictions have large uncertainties, the current measurements provide strong constraints for model development and tuning.

5. Nuclear Modification and Collectivity in Hot Nuclear Matter

The nuclear modification factor R_{AA} of heavy flavor in AA collisions is sensitive to radiative and collisional energy loss processes within the medium and can probe color charge effects as well as flavor-dependent hadronization. At higher momenta, little difference is found between R_{AA} of charmed and light mesons, and both can be described by calculations based on pQCD energy loss [23]. Nuclear modification of heavy flavor at lower momenta, however, shows a significantly weaker suppression pattern than that of light flavor. The left panel of Figure 5 shows the R_{AA} of heavy flavor compared to several model calculations with different ingredients regarding heavy flavor transport [24–28]. Models that contain charm-light coalescence [25–28] typically provide better descriptions of the dataset.

To achieve a stronger discriminative power of data over models, the azimuthal anisotropy parameter v_2 ("elliptic flow") of D_0 mesons in semi-central Pb-Pb collisions is shown in Figure 5. A substantial heavy-flavor anisotropy can be observed. The v_2 of the D_0 mesons is qualitatively similar to that observed for light mesons ($\pi^{\pm}$) in Pb-Pb collisions at $\sqrt{s_{NN}} = 5.02$ TeV.

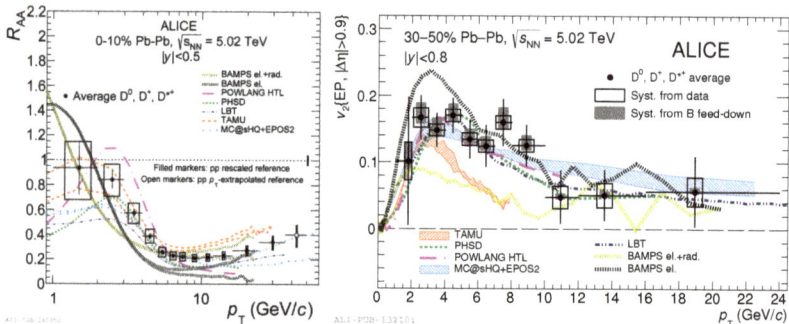

Figure 5. (**Left**) Average nuclear modification factor R_{AA} of D^0, D^+ and D^{*+} mesons in central Pb-Pb collisions at $\sqrt{s_{NN}} = 5.02$ TeV [23], compared to transport model calculations [24–28]. (**Right**) Average azimuthal anisotropy v_2 of D mesons in semi-central Pb-Pb collisions at $\sqrt{s_{NN}} = 5.02$ TeV [29], compared to model calculations [24–28].

6. Summary and Outlook

We gave an overview of selected recent heavy-flavor results from ALICE in pp, p-Pb, and Pb-Pb colliding systems. Transverse momentum differential production of both the charmed and beauty mesons in pp collisions are generally described by pQCD models within uncertainties. The production of charmed baryons is, however, underestimated by theoretical calculations, indicating that models for fragmentation need improvement. The production of heavy flavor increases steeper-than-linearly with event activity, indicating the role of multiple parton interactions. The models, however, fail to describe the data quantitatively. Nuclear modification by cold nuclear matter is weak in the case of both the D mesons and reconstructed D-jets. These recent Run-2 measurements already provide strong restrictions for theoretical calculations. While the suppression of charmed D mesons is similar to that of light hadrons at high-p_T, low-p_T suppression is weaker. A substantial azimuthal anisotropy can be observed for charmed mesons. Although the simultaneous description of R_{AA} and v_2 is a challenge for theory, some transport models that incorporate mechanisms for coalescence between charm quarks and light quarks adequately describe the low-p_T behavior of both observables. Ongoing heavy-flavor measurements at ALICE show unprecedented precision down to very low momenta. The Run-3 phase of LHC with further increased luminosity and detector upgrades [30] will bring about the era of precision beauty measurements.

Funding: This work has been supported by the Hungarian NKFIH/OTKA K 120660 grant and the János Bolyai scholarship of the Hungarian Academy of Sciences.

Conflicts of Interest: The author declares no conflict of interest.

References

1. Adams, J.; et al. [STAR Collaboration]. Experimental and theoretical challenges in the search for the quark gluon plasma: The STAR Collaboration's critical assessment of the evidence from RHIC collisions. *Nucl. Phys. A* **2005**, *757*, 102. [CrossRef]
2. Adare, A.; et al. [PHENIX Collaboration]. Enhanced production of direct photons in Au+Au collisions at $\sqrt{s_{NN}} = 200$ GeV and implications for the initial temperature. *Phys. Rev. Lett.* **2010**, *104*, 132301. [CrossRef]
3. Andronic, A.; Arleo, F.; Arnaldi, R.; Beraudo, E.; Bruna, E.; Caffarri, D.; Conesa del Valle, Z.; Contreras, J.G.; Dahms, T.; Dainese, A.; et al. Heavy-flavour and quarkonium production in the LHC era: From proton–proton to heavy-ion collisions. *Eur. Phys. J. C* **2016**, *76*, 107. [CrossRef]
4. Khachatryan, V.; et al. [CMS Collaboration]. Observation of Long-Range Near-Side Angular Correlations in Proton-Proton Collisions at the LHC. *J. High Energy Phys.* **2010**, *2010*, 91. [CrossRef]
5. Bartalini, P.; Berger, E.L.; Blok, B.; Calucci, G.; Corke, R.; Diehl, M.; Dokshitzer, Y.; Fano, L.; Frankfurt, L.; Gaunt, J.R.; et al. Multi-Parton Interactions at the LHC. *arXiv* **2011**, arXiv:1111.0469.

6. Schenke, B. Origins of collectivity in small systems. *Nucl. Phys. A* **2017**, *967*, 105. [CrossRef]
7. Abelev, B.B.; et al. [ALICE Collaboration]. Performance of the ALICE Experiment at the CERN LHC. *Int. J. Mod. Phys. A* **2014**, *29*, 1430044.
8. Cacciari, M.; Frixione, S.; Houdeau, N.; Mangano, M.L.; Nason, P.; Ridolfi, G. Theoretical predictions for charm and bottom production at the LHC. *J. High Energy Phys.* **2012**, *2012*, 137. [CrossRef]
9. Adam, J.; et al. [ALICE Collaboration]. Measurement of charm and beauty production at central rapidity versus charged-particle multiplicity in proton-proton collisions at $\sqrt{s} = 7$ TeV. *J. High Energy Phys.* **2015**, *2015*, 148. [CrossRef]
10. Sjöstrand, T.; Mrenna, S.; Skands, P.Z. A Brief Introduction to PYTHIA 8.1. *Comput. Phys. Commun.* **2008**, *178*, 852. [CrossRef]
11. Ferreiro, E.G.; Pajares, C. Open charm production in high multiplicity proton-proton events at the LHC. *arXiv* **2015**, arXiv:1501.03381.
12. Werner, K.; Guiot, B.; Karpenko, I.; Pierog, T. Analysing radial flow features in p-Pb and p-p collisions at several TeV by studying identified particle production in EPOS3. *Phys. Rev. C* **2014**, *89*, 064903. [CrossRef]
13. Acharya, S.; et al. [ALICE Collaboration]. First measurement of Ξ_c^0 production in pp collisions at $\sqrt{s} = 7$ TeV. *Phys. Lett. B* **2018**, *781*, 8. [CrossRef]
14. Bahr, M.; Gieseke, S.; Gigg, M.A.; Grellscheid, D.; Hamilton, K.; Latunde-Dada, O.; Platzer, S.; Richardson, P.; Seymour, M.H.; Sherstnev, A.; Webber. B. R. Herwig++ Physics and Manual. *Eur. Phys. J. C* **2008**, *58*, 639. [CrossRef]
15. Bierlich, C.; Christiansen, J.R. Effects of color reconnection on hadron flavor observables. *Phys. Rev. D* **2015**, *92*, 094010. [CrossRef]
16. Fujii, H.; Watanabe, K. Heavy quark pair production in high energy pA collisions: Open heavy flavors. *Nucl. Phys. A* **2013**, *920*, 78. [CrossRef]
17. Mangano, M. L.; Nason, P.; Ridolfi, G. Heavy quark correlations in hadron collisions at next-to-leading order. *Nucl. Phys. B* **1992**, *373*, 295. [CrossRef]
18. Vitev, I. Non-Abelian energy loss in cold nuclear matter. *Phys. Rev. C* **2007**, *75*, 064906. [CrossRef]
19. Cole, B.A.; Barnaföldi, G.G.; Lévai, P.; Papp, G.; Fái, G. EMC effect and jet energy loss in relativistic deuteron-nucleus collisions. *arXiv* **2017**, arXiv:hep-ph/0702101.
20. Kang, Z.B.; Vitev, I.; Wang, E.; Xing, H.; Zhang, C. Multiple scattering effects on heavy meson production in p+A collisions at backward rapidity. *Phys. Lett. B* **2015**, *740*, 23. [CrossRef]
21. Beraudo, A.; De Pace, A.; Monteno, M.; Nardi, M.; Prino, F. Heavy-flavour production in high-energy d-Au and p-Pb collisions. *J. High Energy Phys.* **2016**, *2016*, 123. [CrossRef]
22. Nason, P. A New method for combining NLO QCD with shower Monte Carlo algorithms. *J. High Energy Phys.* **2004**, *2004*, 040. [CrossRef]
23. Acharya, S.; et al. [ALICE Collaboration]. Measurement of D^0, D^+, D^{*+} and D_s^+ production in Pb-Pb collisions at $\sqrt{s_{NN}} = 5.02$ TeV. *J. High Energy Phys.* **2018**, *2018*, 174. [CrossRef]
24. Averbeck, R.; Bastid, N.; Del Valle, Z.C.; Crochet, P.; Dainese, A.; Zhang, X. Reference Heavy Flavour Cross Sections in pp Collisions at $\sqrt{s} = 2.76$ TeV, using a pQCD-Driven $\sqrt{s}$-Scaling of ALICE Measurements at $\sqrt{s} = 7$ TeV. *arXiv* **2011**, arXiv:1107.3243.
25. Cao, S.; Luo, T.; Qin, G.Y.; Wang, X.N. Heavy and light flavor jet quenching at RHIC and LHC energies. *Phys. Lett. B* **2018**, *777*, 255. [CrossRef]
26. He, M.; Fries, R.J.; Rapp, R. Heavy Flavor at the Large Hadron Collider in a Strong Coupling Approach. *Phys. Lett. B* **2014**, *735*, 445. [CrossRef]
27. Nahrgang, M.; Aichelin, J.; Gossiaux, B.P.; Werner, K. Influence of hadronic bound states above T_c on heavy-quark observables in Pb+Pb collisions at at the CERN Large Hadron Collider. *Phys. Rev. C* **2014**, *89*, 014905. [CrossRef]
28. Song, T.; Berrehrah, H.; Cabrera, D.; Cassing, W.; Bratkovskaya, E. Charm production in Pb + Pb collisions at energies available at the CERN Large Hadron Collider. *Phys. Rev. C* **2016**, *93*, 034906. [CrossRef]

29. Acharya, S.; et al. [ALICE Collaboration]. *D*-meson azimuthal anisotropy in midcentral Pb-Pb collisions at $\sqrt{s_{NN}}$ = 5.02 TeV. *Phys. Rev. Lett.* **2018**, *120*, 102301. [CrossRef]
30. Abelev, B.; et al. [ALICE Collaboration]. Upgrade of the ALICE Experiment: Letter Of Intent. *J. Phys. G* **2014**, *41*, 087001. [CrossRef]

Sample Availability: Published ALICE data are available in the HEPData repository.

© 2019 by the author. Licensee MDPI, Basel, Switzerland. This article is an open access article distributed under the terms and conditions of the Creative Commons Attribution (CC BY) license (http://creativecommons.org/licenses/by/4.0/).

Communication

Medical Applications of the ALPIDE Detector [†]

Monika Varga-Kofarago on Behalf of the Bergen pCT Collaboration

HAS Wigner Research Centre for Physics, Konkoly-Thege Miklós út 29-33, H-1121 Budapest, Hungary;
varga-kofarago.monika@wigner.mta.hu
† This paper is based on the talk at the 18th Zimányi School, Budapest, Hungary, 3–7 December 2018.

Received: 18 March 2019; Accepted: 21 May 2019; Published: 24 May 2019

Abstract: The CERN Large Hadron Collider (LHC) ALICE detector is undergoing a major upgrade in the Second Long Shutdown of the LHC in 2019–2020. During this upgrade, the innermost detector, the Inner Tracking System, will be completely replaced by a new detector which is built from the ALPIDE sensor. In the Bergen proton computer tomography (pCT) collaboration, we decided to apply these sensors for medical applications. They can be used for positioning in hadron therapies due to their good position resolution and radiation tolerance. Dose planning of hadron therapy is calculated currently from photon CT measurements, which results in large uncertainties in the planning and therefore in a necessary enlargement of the treatment area. This uncertainty can be reduced by performing the CT scan using protons. The current contribution shows the development of a sampling calorimeter built from the ALPIDE detector for proton CT measurements and describes the state of the project.

Keywords: ALPIDE; silicon sensors; medical applications; proton CT; radiotherapy; computer tomography; cancer; digital sampling calorimeter

1. Introduction

The ALICE detector is undergoing a major upgrade in the second long shutdown of the LHC in 2019–2020. During this upgrade, the innermost detector, the inner tracking system (ITS), will be completely replaced by a new detector [1]. This new detector will be equipped with a MAPS-type silicon detector, the ALPIDE (ALICE pixel detector), which was designed specifically for this upgrade [2]. The ALPIDE is produced in a 50 µm and a 100 µm thick versions and a reverse substrate bias of around -6 V can be applied to it to enlarge its depleted region. The ALPIDE and its prototypes have been thoroughly tested in laboratory and test beam measurements. It was required that the sensor has a position resolution around 5 µm, a detection efficiency above 99% and a noise occupancy below 10^{-6} hits/event/pixel [1]. The ALPIDE fulfills these requirements both before and after irradiating the chips with the radiation doses expected in its lifetime in the ALICE experiment [2].

Due to its low material budget, good position resolution and radiation hardness, the ALPIDE can also be used in other applications where a precise tracking of a high flux of particles is needed. One such application is medical physics, in particular hadron therapy. In these treatments, protons or heavier ions (usually carbon) are used to destroy the DNA of cancer cells. These protons or ions have an energy usually below 200 MeV/u, which is much lower than in the case of the typical particles at the ALICE detector. This means that the ALPIDE has to be tested at these lower energies to see whether the tracking of these particles is possible. The intensity of the beam is also much higher in these medical applications, therefore it needs to be checked maximum how many particles can be distinguished by the ALPIDE in a single frame. This is especially tricky at this low energy, because low-energy particles lose more energy in the silicon, therefore creating more electron–hole pairs, which results in more pixels firing for one passing particle. This makes the distinguishing of particles more difficult at lower energy than at the usual energies of the ALICE detector.

2. Computer Tomography with Protons

The problem of current hadron therapy treatments is that the treatment is planned after acquiring an image of the patient by a photon CT scan. This results in large uncertainties (around 3–4%) in the determination of the stopping power of protons in front of the tumor [3]. This is due to the fact that the relation of the attenuation coefficients of photons and the stopping power of protons is not linear and not one-to-one as it differs depending on the type and geometrical structure of the tissue [4].

This problem can be solved by using protons for the imaging in the CT measurement instead of photons, therefore the measurement will give directly the stopping power for protons. This would reduce the uncertainty by more than a magnitude to 0.3% [3]. Such a measurement would use protons with a higher energy than the ones used for the treatment, such that their Bragg peak would fall outside of the patient and in the detector placed behind the patient. The position of the protons has to be determined before entering the patient and after leaving the patient. Behind the patient the energy of the protons has to be measured as well. Before the patient, the position of the protons can be determined by the measurement of the beam position or by a tracking detector with very low material budget (maximum 50–100 μm of silicon). After the patient, the position and the energy measurement can be achieved by a high resolution sampling calorimeter. The concept of such a detector can be seen in Figure 1. If the measurement is done prior to the treatment, it can be used for the planning of the treatment, while if it is done quasi-simultaneously, it can be used for dose verification, dose optimization or patient alignment. We note that patient alignment is not a trivial task, as the unavoidable movement of the patient has to be taken into account during the treatment. This can be done by monitoring the patient with a quasi-simultaneous proton CT measurement without giving relevant additional dose.

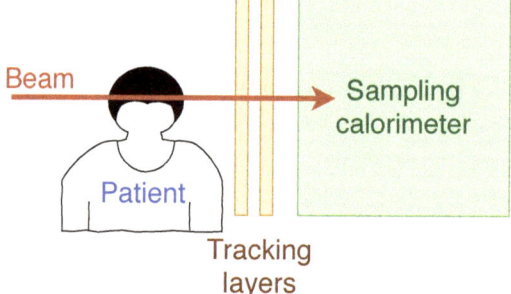

Figure 1. The concept of a proton computer tomography (CT) detector for medical imaging.

3. The Proposed Calorimeter

The calorimeter design has the typical sandwich structure well-known for high-energy experimental setups. The active part of the sampling calorimeter will be the ALPIDE sensor. These ALPIDE layers will alternate with aluminum layers which act as energy degraders for the protons. There will be 41 sensitive layers and 41 degrader layers and each aluminum layer will be 3.5 mm thick. The full front area of the detector will be 27 cm × 15 cm, which is made up of 9 × 9 ALPIDE sensors. The proposed calorimeter design can be seen in Figure 2. In the figure, it can also be seen that there will be no aluminum layers between the first few silicon layers. This is done to allow for a more precise tracking of the particles at their entrance to the detector, because this will result in a more precise determination of the position and angle of the incoming particles.

As the ALPIDE was designed to function in the ALICE detector, it is radiation tolerant up to 1.7×10^{13} 1 MeV n_{eq}/cm^2 non-ionizing dose and up to 2700 krad ionizing radiation. It can be produced in a 50 μm and a 100 μm thick versions, therefore, if needed, it can be used in the tracker in front of the patient on the beam side as well.

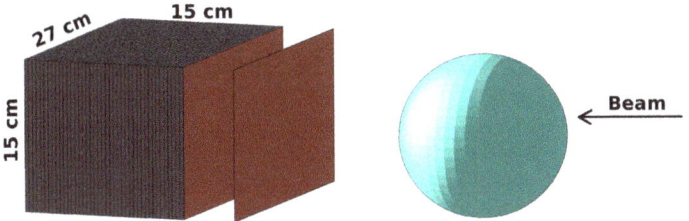

Figure 2. The proposed design of the sampling calorimeter. The red layers are the sensitive layers which alternate with the gray aluminum layers. The blue sphere represents the simplified patient.

4. Results from the First Prototype Tests

The first prototype of the calorimeter was not optimized for detecting protons, but for measuring electromagnetic showers. This prototype used MIMOSA23 sensors [5] and used 3.3 mm tungsten absorbers instead of aluminum as a degrader [6]. It was tested in a proton beam at the KVI-Center for Advanced Radiation Technology in Groningen [7] where the energy was varied from 120 MeV to 188 MeV [8]. The beam energy was changed by introducing an aluminum absorber in the beam line, which introduces a 1.4 MeV energy spread. The beam intensity was set such that one proton per readout frame was delivered to the sensor. The comparison of the results with simulations can be seen in Figure 3 which shows the number of reconstructed protons as a function of their reconstructed range from a 188 MeV proton beam. The protons are reconstructed such that in each layer, the deposited energy is determined from the number of pixels which fire for one passing proton. These deposited energies are then fitted with a Bragg curve, which gives the range of the proton. In both the simulations and the data, the Bragg peak is clearly visible around 230 mm, and the simulation (left panel) describes the test beam data (right panel) well. The lighter green histogram represents nuclear interactions in the detector material, and its structure corresponds to the spacing of the layers. These interactions are not used for the range estimation, as they are not representative for the position of the Bragg peak. Only the dark green points, called accepted tracks, are from reconstructed protons which did not undergo a nuclear interaction and therefore can be used for the estimation of the Bragg peak.

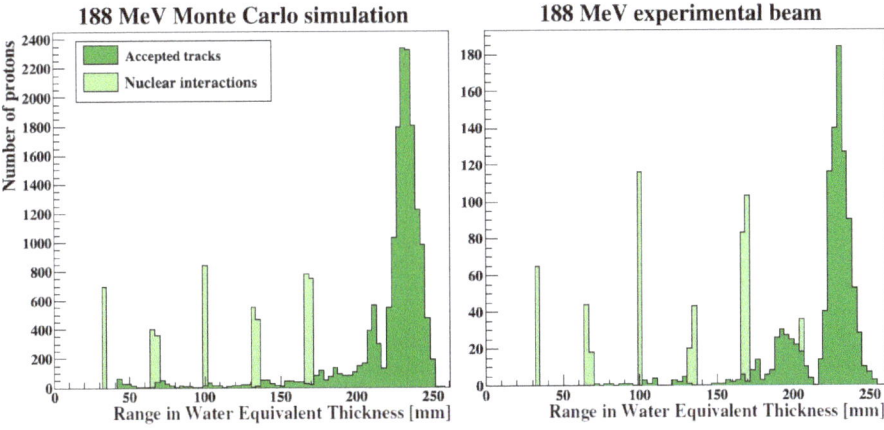

Figure 3. Comparison of the simulations (**left panel**) and the measured results (**right panel**) of the prototype with a 188 MeV proton beam at KVI-Center for Advanced Radiation Technology [8].

In Figure 4, the reconstructed water equivalent thickness (WET) range of the tracks is shown as a function of the energy of the beam from simulations and the energy scan measurement with a proton beam at KVI-Center for Advanced Radiation Technology. The agreement between data and simulation is good here as well, and the observed linear trend shows that the range is a good measurement of the energy of the incoming protons. An oscillation pattern can be seen in the Monte Carlo simulations which arose form a better estimation of the proton range when the Bragg peak is detected in two sensitive layers and a worse estimation when it is detected in only one. This can be reduced by a further optimization of the absorber thickness.

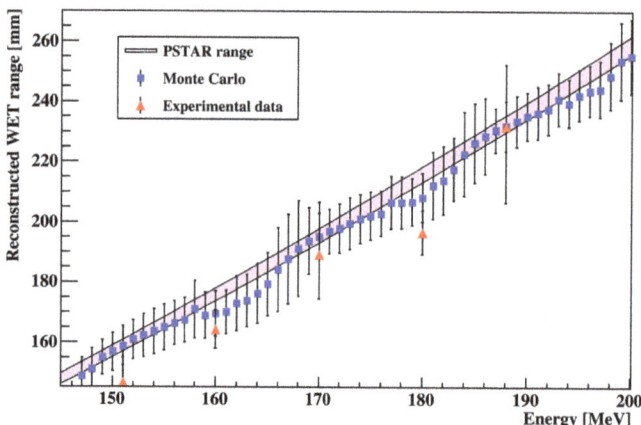

Figure 4. Reconstructed water equivalent thickness range of the protons as a function of the energy of the beam from simulations, from data and from proton stopping power and range (PSTAR) tables that contain numerical integrations of the Bethe equation [8,9].

5. Conclusions

A new detector was developed for the CERN LHC ALICE experiment upgrade to replace the current Inner Tracking System after the Second Long Shutdown of the LHC. This detector will be equipped with the ALPIDE sensor, which can also be used for medical applications. A sampling calorimeter of alternating ALPIDE and aluminum layers is proposed as a proton CT detector. With the help of this detector the dose estimation of hadron therapy will become more accurate, therefore it will have less side effects and can be applied closer to critical organs. The first prototype of such a detector is presented here. This was optimized for electromagnetic showers instead of the energy measurement of protons. It shows a good performance, and its performance can be well described by Monte Carlo simulations. A new prototype, which will be optimized for tracking protons, is being built according to the description given in this paper. First one layer of the new prototype, then later the full detector will be tested in test beam measurements with low energy (50–200 MeV) protons and helium or carbon ions.

Funding: This work has been supported by the Hungarian NKFIH/OTKA K 120660 grant and by the Hungarian—Norwegian Bilateral Cooperation Knowledge Exchange Visits to Norway Grants (NKFIH Grant No: HU01-0030-C1).

Conflicts of Interest: The authors declare no conflict of interest. The funders had no role in the design of the study; in the collection, analyses, or interpretation of data; in the writing of the manuscript, or in the decision to publish the results.

Abbreviations

The following abbreviations are used in this manuscript:

ALICE   A Large Ion Collider Experiment
ALPIDE   ALICE pixel detector
CT   computer tomography
DNA   deoxyribonucleic acid
LHC   Large Hadron Collider
MAPS   Monolithic Active Pixel Sensor
pCT   proton computer tomography
PSTAR   proton stopping power and range
WET   water equivalent thickness

References

1. ALICE Collaboration. Technical Design Report for the Upgrade of the ALICE Inner Tracking System. *J. Phys. G* **2014**, *41*, 087002. [CrossRef]
2. ALICE Collaboration. The ALPIDE pixel sensor chip for the upgrade of the ALICE Inner Tracking System. *Nucl. Instrum. Methods Phys. Res. Sec. A Accel. Spectrom. Detect. Assoc. Equip.* **2017**, *845*, 583–587. [CrossRef]
3. Paganetti, H. Range uncertainties in proton therapy and the role of Monte Carlo simulations. *Phys. Med. Biol.* **2012**, *57*, R99–R117. [CrossRef] [PubMed]
4. Schaffner, B.; Pedroni, E. The precision of proton range calculations in proton radiotherapy treatment planning: Experimental verification of the relation between CT-HU and proton stopping power. *Phys. Med. Biol.* **1998**, *43*, 1579–1592. [CrossRef]
5. Himmi, A.; Bertolone, G.; Brogna, A.; Dulinski, W.; Colledani, C.; Dorokhov, A.; Hu, C.; Morel, F.; Valin, I. *PHASE-1 User Manual*; IPHC: Strasbourg, France, 2008.
6. Rocco, E.; FoCal ALICE Group. Highly granular digital electromagnetic Calorimeter with MAPS. *Nucl. Part. Phys. Proc.* **2016**, *273–275*, 1090–1095. [CrossRef]
7. KVI—Center for Advanced Radiation Technology. Available online: https://www.rug.nl/kvi-cart/?lang=en (accessed on 24 May 2019)
8. Pettersen, H. A Digital Tracking Calorimeter for Proton Computed Tomography. Ph.D. Thesis, The University of Bergen, Bergen, Norway, 2018.
9. Berger, M.J.; Coursey, J.; Zucker, M.; Chang, J. *ESTAR, PSTAR, and ASTAR: Computer Programs for Calculating Stopping-Power and Range Tables for Electrons, Protons, and Helium Ions*; National Bureau of Standards: Gaithersburg, MD, USA, 2005. [CrossRef]

© 2019 by the authors. Licensee MDPI, Basel, Switzerland. This article is an open access article distributed under the terms and conditions of the Creative Commons Attribution (CC BY) license (http://creativecommons.org/licenses/by/4.0/).

Article

Influence of Backside Energy Leakages from Hadronic Calorimeters on Fluctuation Measures in Relativistic Heavy-Ion Collisions

Andrey Seryakov

Laboratory of Ultra-High Energy Physics, St. Petersburg State University, Saint Petersburg 199034, Russia; a.seryakov@spbu.ru or seryakov@yahoo.com

Received: 19 April 2019; Accepted: 17 May 2019; Published: 23 May 2019

Abstract: The phase diagram of the strongly interacting matter is the main research subject for different current and future experiments in high-energy physics. System size and energy scan programs aim to find a possible critical point. One of such programs was accomplished by the fixed-target NA61/SHINE experiment in 2018. It includes six beam energies and six colliding systems: p + p, Be + Be, Ar + Sc, Xe + La, Pb + Pb and p + Pb. In this study, we discuss how the efficiency of centrality selection by forward spectators influences multiplicity and fluctuation measures and how this influence depends on the size of colliding systems. We use SHIELD and EPOS Monte-Carlo (MC) generators along with the wounded nucleon model, introduce a probability to lose a forward spectator and spectator energy loss. We show that for light colliding systems such as Be or Li even a small inefficiency in centrality selection has a dramatic impact on multiplicity scaled variance. Conversely, heavy systems such as Ar + Sc are much less prone to the effect.

Keywords: QGP; critical point; fluctuations; centrality; calorimeters

1. Introduction

Fluctuation measures are considered to be an important tool in the search of the possible critical point of the strongly interacting matter. However, fluctuation quantities are sensitive to various effects along with the critical behavior [1–4] such as volume fluctuations [5,6], resonance decays [7], beam and target material impurities [8] and detector inefficiencies.

Experiments in relativistic heavy ion collisions use different techniques to reduce volume fluctuations by selecting centrality classes. The procedure aims to select events with a restricted number of particle production sources or volume. The centrality selection may be accomplished by measuring produced particle multiplicity in a specific rapidity interval along with energy of non-interacted nucleons-spectators by forward hadronic calorimeters. Although the multiplicity based approach introduces, a bias on any fluctuations due to correlations between multiplicities even in different acceptance windows. However, it is worth noting that this bias can be well reproduced and estimated by using MC generators.

Contrary to this, a solely spectators based centrality selection provides an unbiased method to restrict the collision volume. Technically, it can be accomplished only in fixed target experiments, as it is possible there to place a hadronic calorimeter exactly at the beam line. Nevertheless, such calorimeters suffer from hadronic shower energy leakages from the surface and have much lower resolution capabilities compared to multiplicity detectors. In this paper, we study an influence of the energy leakage from the calorimeter back surface on the average multiplicities and the multiplicity scaled variance and its dependency on the colliding system size.

2. Study with a Geant Calorimeter Model

The main motivation for this work was a study of how Projectile Spectator Detector (PSD) [9] influences the measured quantities in the NA61/SHINE collaboration [10]. PSD is a segmented modular hadronic calorimeter, which is used for triggering, centrality and event plane determination. The detector consists of 44 independent modules and each of them has 60 lead (16 mm) + scintillator (4 mm) layers. The total length of PSD is about 1.2 m, which corresponds to approximately 5.6 interaction lengths. A scheme of the NA61/SHINE setup and a photo of PSD are presented in Figure 1.

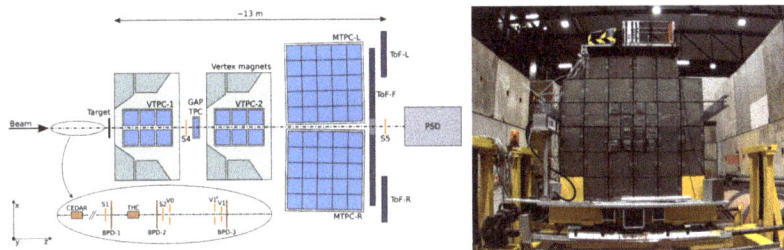

Figure 1. A scheme of the NA61/SHINE experiment [10] and a front view photo of the Projectile Hadronic Calorimeter. PSD is used for triggering, centrality and event plane determination.

Signals from 60 scintillators in each module are grouped by six, therefore, the design of PSD allows collecting information from ten independent areas along the beam axis inside the calorimeter. This makes it possible to study dependencies of different quantities on the calorimeter lengths by selecting centrality by a reduced number of scintillator groups. It is expected that any centrality sensitive measure will saturate at one point with an increase of calorimeter length (see Figure 2). On such plot, the 0 limit corresponds to 0 centrality detector efficiency, an absence of centrality selection and to minimum bias events. The right limit is an absence of any energy leakage from the hadronic calorimeter backside.

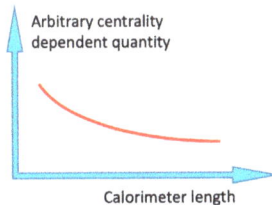

Figure 2. A sketch of how a collision volume depended quantity behaves with the increase of a centrality calorimeter length.

Two MC datasets were generated with the GEANT4 [11] PSD simulation for studying the energy leakage influence on different systems: 100,000 events of $150A$ GeV/c ^{7}Li + ^{9}Be SHIELD MC [12] and 40,000 events of $150A$ GeV/c ^{40}Ar + ^{45}Sc EPOS 1.99 MC [13]. ^{7}Li was chosen instead of experimentally used ^{7}Be as the first one is stable and can be simulated by SHIELD MC. It was possible to compare two completely different MC generators as the studied effect was purely detector based. Moreover, it does not depend on the spectator transverse characteristics as we studied longitudinal shower propagation that is insensitive to a hit position. In each dataset, we selected centrality on the different length of the detector from ≈1.1 to 5.6 interaction lengths. The results for average multiplicities, multiplicity ratios and fluctuation quantities are presented in Figure 3 for ^{7}Li + ^{9}Be and for ^{40}Ar + ^{45}Sc collisions.

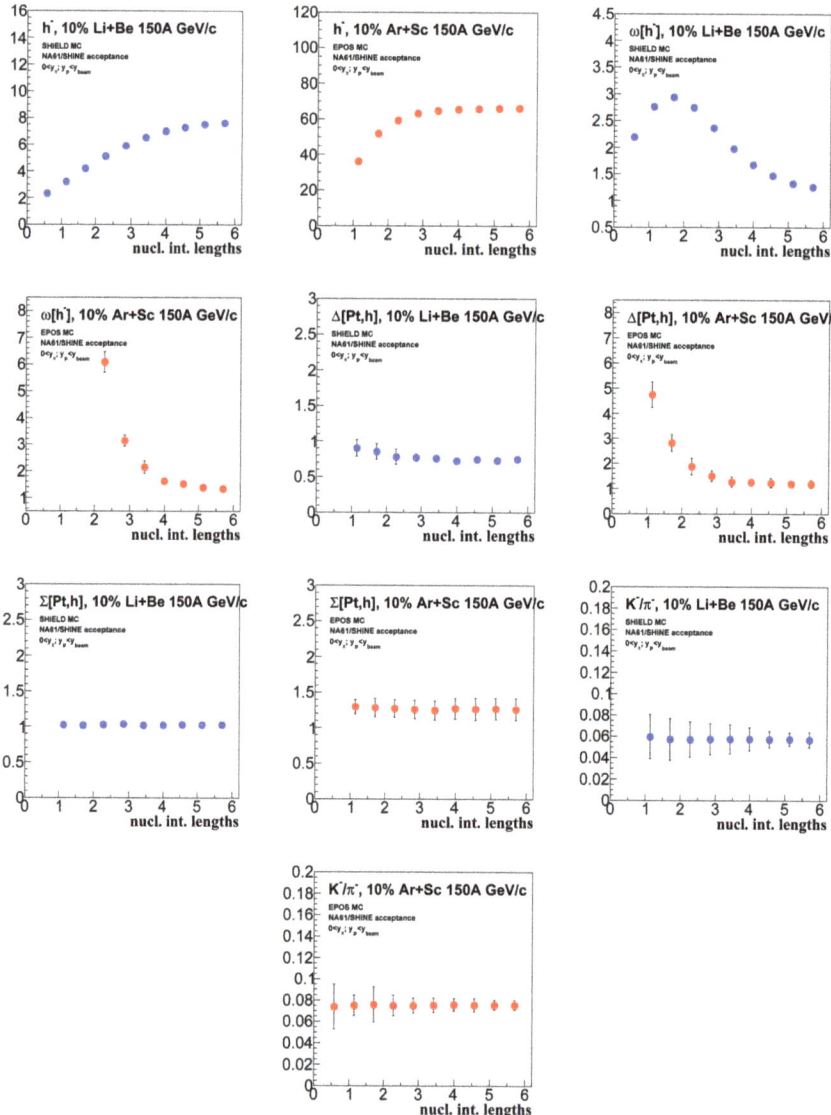

Figure 3. Comparison of different measures behavior versus the centrality calorimeter length for Li + Be (blue dots) and Ar + Sc (red dots) 150A GeV/c collisions. The first two plots present an average number of negatively charged hadrons, the second two show the negative charged hadrons scaled variance, next four show two strongly-intensive quantities $\Delta[Pt,h]$ and $\Sigma[Pt,h]$ [14] and the last two show a ratio of the average number of negative charged kaons to the average number of negative charged pions. All results were calculated in the NA61/SHINE acceptance [15].

Two main conclusions may be drawn from the results:

- The 5.6 interaction lengths were not enough to eliminate the influence of backside energy leakage in light colliding systems (^{7}Li + ^{9}Be) on volume fluctuations as the dependencies did not saturate. The middle size systems as ^{40}Ar + ^{45}Sc were much less prone to the effect. However, Ar + Sc data are more sensitive to energy leakage in case of a short calorimeter.
- Mean multiplicities, scaled variance $\omega[h] = (<h^2> - <h>^2)/<h>$ and strongly intensive $\Delta[Pt,h]$ [14] were sensitive to the effect, while mean multiplicity ratios and another strongly intensive quantity $\Sigma[Pt,h]$ showed steady behavior. The instability of $\Delta[Pt,h]$ contradicted the presumption that such quantities do not depend on the volume fluctuations. Therefore, it was clear that assumptions which lead to the construction of the strongly intensive measure $\Delta[Pt,h]$ [14] are not fulfilled even in MC generators. Investigation of other measures sensitivities transcends the scope of this work.

3. Study within a Wounded Nucleon Model

A simple wounded nucleon model (WNM) was created to understand the unexpected sensitivity of light systems to the energy leak. Three different colliding systems were considered: ^{7}Li + ^{9}Be, ^{35}Cl + ^{40}Ca and ^{208}Pb + ^{208}Pb with 150 A GeV/c beam momentum ($\sqrt{s_{NN}} \approx 17A$ GeV). Nucleon density profiles were taken from [16]. Nucleon core effect was not taken into account. Alpha clustering was not implemented as the goal of the study was to check how the sensitivity to the energy leakage depends on the number of nucleons in colliding systems. Inelastic nucleon–nucleon cross-section was taken equal to 31.75 mb. Multiplicity was introduced based on the number of wounded nucleons; in other words, each wounded nucleon produced a random number of charged particles, which were distributed according to a Poisson with $< N_{ch} > = 3.5$.

In the first version of the model, we introduce dcentrality selection based on the number of forward nucleon-spectators and a probability to lose each of them p. Distributions of forward nucleon spectators for $p = 0\%$ and 10% are shown in Figure 4.

The 10% of events with a lower number of detected forward nucleons spectators were selected as the most central ones. If the boundary between classes di not coincide with the boundary between integer numbers of forward spectators N, then a fraction of events with N + 1 forward spectators was taken to obtain exactly 10% of the whole data sample.

The dependencies of average event multiplicity and multiplicity scaled variance $\omega[N]$ versus the probability to lose a forward spectator showed a striking difference (Figure 5) in the sensitivity to detector efficiency between light system and heavy one (^{7}Li + ^{9}Be and ^{208}Pb + ^{208}Pb). <N> and $\omega[N]$ in Beryllium collisions became sensitive to the spectators lost already, then $p \approx 3$–4% contrary to Pb + Pb collisions where <N> and $\omega[N]$ were steady to the effect until $p \approx 70\%$ and 30% respectively (see Figure 6).

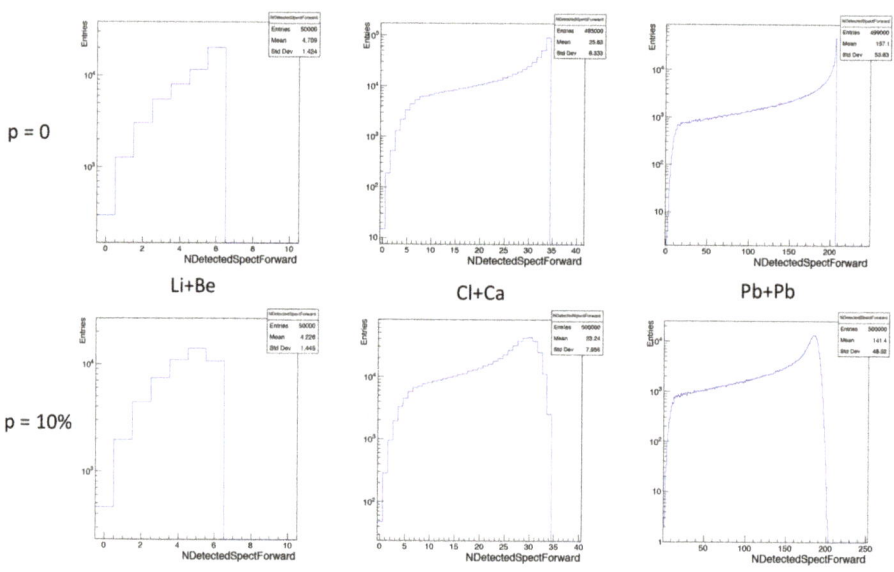

Figure 4. Detected forward spectators distributions for Li + Be, Cl + Ca and Pb + Pb in WNM with a probability to loss a nucleon of 0% and 10%.

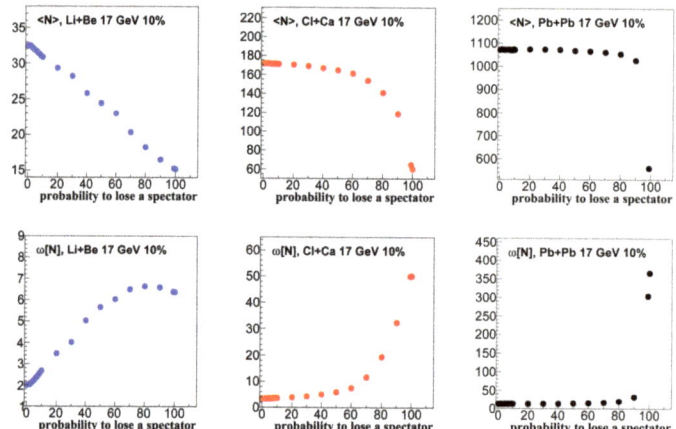

Figure 5. Multiplicity and scaled variance versus the probability to lose a forward spectator in WNM with a centrality selection based on the number of detected forward spectators.

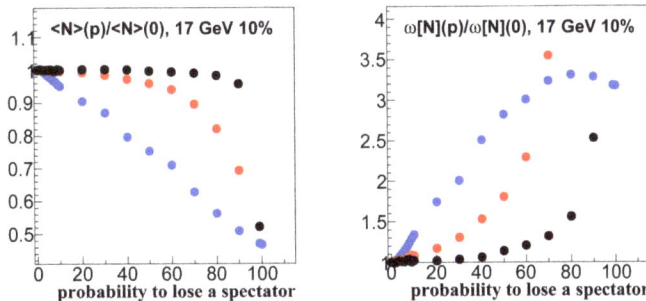

Figure 6. Ratios of multiplicity and scaled variance to values with a zero probability to lose a forward spectator for Li + Be (blue), Cl + Ca (red) and Pb + Pb (black) collisions. WNM with a centrality selection based on the number of detected forward spectators.

Unexpectedly, this simple model reproduced two important features: higher sensitivity of light systems for small energy loss and lower sensitivity for large fraction energy loss. Nevertheless, we are aware of the fact that the probability to lose a spectator is not a realistic model of a hadronic calorimeter. The next step was to introduce a realistic energy losd. For this goal a two-times longer ($\approx$11.2 interaction length) GEANT4 model of PSD was used and a response on a 150 GeV/c proton beam was generated. We calculated and fitted a distribution of ratio between deposited by a proton energy in the first seven sections ($\approx$3.9 int.l.) to the whole calorimeter (20 sections), as shown in Figure 7. We used the obtained function to introduce a random energy loss of each forward spectator in the wounded nucleon model.

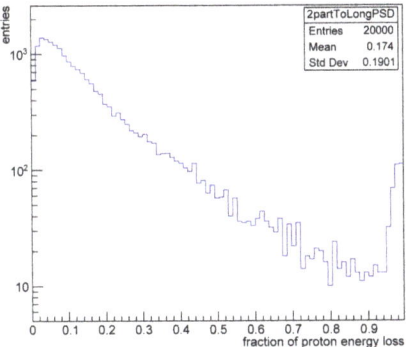

Figure 7. A ratio of deposited energy by a 150 GeV/c proton in the sections from 8 to 20 to the whole long calorimeter PSD (11.2 nuclear int. lengths) in the GEANT4 simulation. The whole calorimeter model has 20 sections. This distribution shows the fraction of a proton energy leak from a calorimeter, which has 3.9 nucl. int. lengths or seven sections in case of PSD.

The 10% of the most central events were selected by the spectator deposited energy (see Figure 8). <N> and ω[N] were calculated and compared with the ideal case (without energy loses). The results are presented in Table 1.

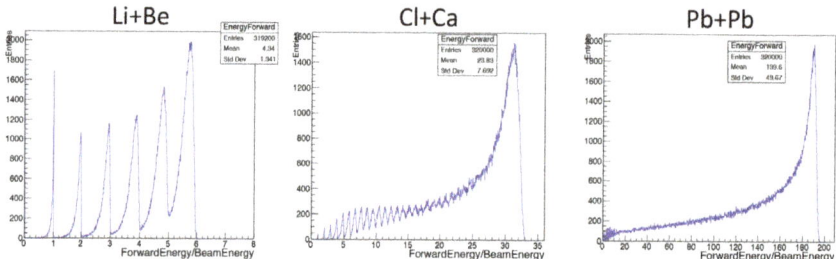

Figure 8. Energy distributions of forward spectators with realistic energy leakage from the calorimeter in WNM.

Table 1. Comparison of results for 10% most central events with realistic energy leakage and without it in a frame of WNM.

	Li + Be	Ca + Cl	Pb + Pb
<N> without energy loss	32.527+/−0.017	172.069+/−0.027	1094.34+/−0.19
<N> with energy loss	32.414+/−0.012	171.996+/−0.034	1094.16+/−0.15
ω[N] without energy loss	2.0192+/−0.0041	3.2594+/−0.0072	13.101+/−0.021
ω[N] with energy loss	2.0625+/−0.0042	3.3111+/−0.0051	13.175+/−0.019
with/without (N)	0.99653+/−0.00057	0.99958+/−0.00032	0.99984+/−0.00052
with/without (ω[N])	1.0214+/−0.0021	1.0159+/−0.0023	1.006+/−0.0017

As shown, the effect is very tiny but nevertheless the lighter the system is the more sensitive it is. The size of the difference is probably a result of an absence of the energy resolution due to the calorimeter sandwich structure, as present in the GEANT4 simulation.

4. Conclusions

It was observed that the light nuclei systems as Li + Be are more sensitive to the energy leakage from the back side of hadronic calorimeters used for centrality determination in fixed target experiments compared to intermediate size systems as Ar + Sc. The probable reason is that in the light systems most of the forward energy is concentrated only in a few nucleons. Therefore, a single nucleon loss produces much bigger volume fluctuations than in a collision of heavy systems, which have a presence of more or less constant energy leakage in each collision. Nevertheless, more investigations are needed to reach the complete understanding of the phenomenon. Even though we succeeded in demonstrating the sensitivity of light systems in the framework of the wounded nucleon model with the probability of a spectator loss, the realistic energy loss simulation in the same model shows only a tiny effect on average multiplicity and scaled variance.

The future fixed target programs, which aim to study light nuclei colliding systems, have to pay attention that a longer calorimeter is needed to control the volume fluctuations for such reactions than for heavier ones.

Funding: This research was funded by the Russian Science Foundation grant number 17-72-20045 in a part of data analysis from MC generators (Section 2) and the Russian Foundation for Basic Research grant number 18-32-01055 mol_a in a part of wounded nucleon model analysis (Section 3).

Acknowledgments: We would like to thank Sergey Morozov, Marina Golubeva and the PSD group for providing the calorimeter and the SHIELD MC simulations; Justyna Monika Cybowska for providing the EPOS MC simulation;

and Evgeny Andronov, Marek Gazdzicki, Peter Seyboth and the NA61/SHINE collaboration for discussions leading to this article.

Conflicts of Interest: The author declares no conflict of interest. The funders had no role in the design of the study; in the collection, analyses, or interpretation of data; in the writing of the manuscript, or in the decision to publish the results.

References

1. Luo, X. Exploring the QCD Phase Structure with Beam Energy Scan in Heavy-ion Collisions. *Nucl. Phys. A* **2016**, *956*, 75–82. [CrossRef]
2. Vovchenko, V.; Anchishkin, D.V.; Gorenstein, M.I.; Poberezhnyuk, R.V.; Stoecker, H. Critical fluctuations in models with van der Waals interactions. *Acta Phys. Pol. Supp.* **2017**, *10*, 753–758. [CrossRef]
3. Luo, X.; Xu, N. Search for the QCD Critical Point with Fluctuations of Conserved Quantities in Relativistic Heavy-Ion Collisions at RHIC: An Overview. *Nucl. Sci. Tech.* **2017**, *28*, 112. [CrossRef]
4. Gazdzicki, M.; Seyboth, P. Search for critical behavior of strongly interacting matter at the CERN Super Proton Synchrotron. *Acta Phys. Pol. B* **2016**, *47*, 1201–1236. [CrossRef]
5. Luo, X.; Xu, J.; Mohanty, B.; Xu, N. Volume Fluctuation and Autocorrelation Effects in the Moment Analysis of Net-proton Multiplicity Distributions in Heavy-Ion Collisions. *J. Phys. G Nucl. Part. Phys.* **2013**, *40*, 105104. [CrossRef]
6. Gorenstein, M.; Gazdzicki, M. Strongly Intensive Quantities. *Phys. Rev. C* **2011**, *84*, 014904. [CrossRef]
7. Bluhm, M.; Nahrgang, M.; Bass, S.A.; Schäfer, T. Impact of resonance decays on critical point signals in net-proton fluctuations. *Eur. Phys. J. C* **2017**, *77*, 210. [CrossRef]
8. Banas, D.; Kubala-Kukuś, A.; Rybczyński, M.; Stabrawa, I.; Stefanek, G. Influence of target material impurities on physical results in relativistic heavy-ion collisions. *Eur. Phys. J. Plus* **2019**, *134*, 44. [CrossRef]
9. Golubeva, M. [NA61/SHINE Collaboration] Hadron Calorimeter (Projectile Spectator Detector—PSD) of NA61/SHINE experiment at CERN. *KnE Energy Phys.* **2018**, *3*, 379–384. [CrossRef]
10. Abgrall, N. [NA61/SHINE Collaboration] NA61/SHINE facility at the CERN SPS: Beams and detector system. *J. Instrum.* **2014**, *9*, P06005. [CrossRef]
11. Agostinellia, S.; Allison, J.; Amako, K.; Apostolakis, J.; Araujo, H.; Arcelmx, P.; Asaig, M.; Axen, D.; Banerjee, S.; Barrand, G.; et al. Geant4—A simulation toolkit. *Nucl. Instrum. Methods Phys. Res. Sect. A* **2003**, *506*, 250–303. [CrossRef]
12. Dementyev, A.; Sobolevsky, N. SHIELD. Available online: www.inr.troitsk.ru/shield/introd-eng.html (accessed on 23 May 2019).
13. Pierog, T.; Ulrich, R. EPOS 1.99 in CRMC. Available online: web.ikp.kit.edu/rulrich/crmc.html (accessed on 23 May 2019).
14. Gazdzicki, M.; Gorenstein, M.I.; Mackowiak-Pawlowska, M. On Normalization of Strongly Intensive Quantities. *Phys. Rev. C* **2013**, *88*, 024907. [CrossRef]
15. NA61/SHINE Collaboration. Acceptance Maps Used in the Paper: Multiplicity and Transverse Momentum Fluctuations in Inelastic Proton-Proton Interactions at the CERN Super Proton Synchrotron. Available online: edms.cern.ch/document/1549298/1 (accessed on 23 May 2019).
16. De Vries, H.; De Jager, C.W.; De Vries, C. Nuclear Charge Density Distributions Parameters from Elastic Electron Scattering. *At. Data Nucl. Data Tables* **1987**, *36*, 495–536. [CrossRef]

© 2019 by the author. Licensee MDPI, Basel, Switzerland. This article is an open access article distributed under the terms and conditions of the Creative Commons Attribution (CC BY) license (http://creativecommons.org/licenses/by/4.0/).

Communication

Correlations of High-p_T Hadrons and Jets in ALICE [†]

Filip Krizek on Behalf of the ALICE Collaboration

Department of Nuclear Spectroscopy, Nuclear Physics Institute of CAS, 25068 Husinec-Rez, Czech Republic; krizek@ujf.cas.cz

[†] This paper is based on the talk at the 18th Zimányi School, Budapest, Hungary, 3–7 December 2018.

Received: 21 March 2019; Accepted: 20 May 2019; Published: 22 May 2019

Abstract: There are two prominent experimental signatures of quark–gluon plasma creation in ultra-relativistic heavy-ion collisions: the jet quenching phenomenon and the azimuthal-momentum space-anisotropy of final-state particle emission. Recently, the latter signature was also observed in lighter collision systems such as p–Pb or pp. This raises a natural question of whether in these systems, the observed collectivity is also accompanied by jet quenching. In this paper, we overview ALICE measurements of the jet quenching phenomenon studied using semi-inclusive distributions of track-based jets recoiling from a high-transverse momentum (p_T) hadron trigger in Pb–Pb and p–Pb collisions at LHC energies. The constructed coincidence observable, the per trigger normalized yield of associated recoil jets, is corrected for the complex uncorrelated jet background, including multi-partonic interactions, using a data-driven statistical subtraction method. In the p–Pb data, the observable was measured in events with different underlying event activity and was utilized to set an upper limit on the average medium-induced out-of-cone energy transport for jets with resolution parameter $R = 0.4$. The associated jet momentum shift was found to be less than $0.4\,\text{GeV}/c$ at 90% confidence.

Keywords: jet quenching; QGP; event activity; small systems

1. Introduction

Ultra-relativistic heavy-ion collisions are used to probe the properties of strongly interacting matter in the regime of extremely high-energy densities and temperatures and vanishing baryochemical potential [1]. Lattice quantum chromodynamics calculations predict that under such conditions, the hadron gas phase undergoes a transition to the state called quark–gluon plasma (QGP) [2]. In this phase, quarks and gluons are released from their confinement in hadrons. The transition has a smooth cross-over character and happens at a temperature of about 150 MeV [3].

A collision of heavy ions creates a rapidly evolving dynamical system where the QGP phase lasts only a short instant. As the collision zone expands and cools, quarks and gluons merge together, giving a rise to a multitude of hadrons that further interact among each other until the kinematic freeze-out is reached [1]. When particles from the collision reach our detector, the QGP does not exist anymore; therefore, properties of the QGP can be investigated indirectly only. Among different experimental observables that are studied in this context, two of them are believed to be directly associated with the production of the QGP in heavy-ion collisions: the large azimuthal-momentum space-anisotropy of produced particles [4,5] and the jet quenching phenomenon [6,7]. The first observable is connected with the time evolution of the initial spatial anisotropy of the collision zone, which results in an azimuthal-momentum space-anisotropy of produced particles. Hydrodynamic calculations, which model this process, show that the magnitude of the observed flow is compatible with QGP behaving like a nearly perfect liquid with a very small shear-viscosity to entropy-density ratio [8].

The jet quenching phenomenon is manifested by a marked reduction of energy of high-p_T hadrons and jets that traversed the QGP medium. Their yield measured in heavy-ion collisions is suppressed when compared to the yield that would be expected from a superposition of the corresponding number of independent pp collisions.

Jets are intuitively understood as collimated sprays of particles that are produced by the fragmentation of highly virtual partons. The exact definition of a jet is done with jet reconstruction algorithms [9]. These algorithms were designed to recover energy of the original parton by summing up momenta of final-state particles. In elementary collisions, the production of jets is well-described with the perturbative quantum chromodynamics [10]. In heavy-ion collisions, highly virtual partons that fragment to jets are produced by hard scatterings that happen before the QGP is formed. The parton shower thus gets modified while these partons traverse the medium [11]. Based on the observed modification, medium properties can be inferred. Jets are thus considered well-suited probes of the produced QGP.

Jets in heavy-ion collisions are accompanied by an intensive underlying event. A jet reconstruction algorithm therefore often clusters together just soft particles from the underlying event and creates a so-called combinatorial background jet. One way to suppress the contribution of these artificial jets is to require the presence of a high-p_T constituent within the reconstructed jet [12]. This condition, however, imposes a fragmentation bias on the reconstructed jets and can essentially affect the selected jet sample. As we will see in the following section, hadron-jet coincidence measurements offer a way to overcome this problem [13]. They allow the removal of the contribution of combinatorial background jets, including the contribution from multi-parton interactions, without imposing the fragmentation bias on analyzed jets. The method is data-driven and uses statistical subtraction. It is suitable also for jets having large R and low p_T.

2. Hadron-Jet Coincidence Measurements in Pb–Pb Collisions at $\sqrt{s_{NN}} = 2.76$ TeV

A principle of hadron-jet coincidence measurements will be explained based on the analysis of Pb–Pb collisions at $\sqrt{s_{NN}} = 2.76$ TeV recorded by ALICE in 2011. The same analysis procedure will be then applied to the data from p–Pb collisions at $\sqrt{s_{NN}} = 5.02$ TeV from 2013. Further details about both data sets and analyses can be found in the original papers [14,15].

ALICE is one of the four big experiments working at the Large Hadron Collider at CERN. The ALICE apparatus is described in detail elsewhere [16]. Let us note that track reconstruction is based on position measurements performed by a six-layer silicon tracker called the Inner Tracking System [17], which surrounds the interaction point, and the Time Projection Chamber [18]. Both detectors have full azimuthal coverage, and tracks can be efficiently reconstructed in the pseudorapidity range $|\eta_{track}| < 0.9$. Both detectors are placed in a 0.5 T solenoidal magnetic field, which ensures reasonable transverse momentum resolution for tracks in the range $0.15 < p_{T,track} < 100$ GeV/c.

Track-based jets are reconstructed from charged tracks with $0.15 < p_{T,track} < 100$ GeV/c using the infrared and collinear safe anti-k_T algorithm [19] as implemented in the FastJet package [20]. Reconstructed jets have a resolution parameter of $R = 0.4$ and are assembled using the boost-invariant p_T-recombination scheme. Pseudorapidity of jets is constrained by a fiducial cut $|\eta_{jet}| < 0.5$ to remove jets whose jet cone overlaps with boarders of the ALICE acceptance. The reconstructed transverse momentum of jets, $p_{T,jet}^{raw,ch}$, is corrected for the mean underlying event contribution on an event-by-event basis by subtracting a product of the mean underlying event density ρ and jet area A_{jet},

$$p_{T,jet}^{reco,ch} = p_{T,jet}^{raw,ch} - \rho \cdot A_{jet}. \quad (1)$$

Here, the mean underlying event density is estimated by the standard area-based approach [21] as

$$\rho = \mathrm{median}_{k_T\ \mathrm{jets}} \frac{p_{T,jet}^{ch}}{A_{jet}}, \quad (2)$$

which is calculated based on a sample of reconstructed k_T track-based jets with $R = 0.4$.

In hadron-jet coincidence measurements, we analyze events that contain a high-p_T track, the so-called trigger track or TT. The presence of a high-p_T particle unambiguously selects events with a hard scattering. The trigger track p_T is required to be in some chosen range $X < p_{T,\text{trig}} < Y\,\text{GeV}/c$, which is denoted TT{$X,Y$} throughout the text. If multiple TT candidates are found, one of them is chosen at random. Jets, which are to be analyzed, are selected to be nearly back-to-back in azimuth w.r.t. to TT,

$$|\varphi_{TT} - \varphi_{\text{jet}}| < \pi - 0.6\,\text{rad} \tag{3}$$

where φ_{TT} and φ_{jet} denote azimuthal angles of TT and recoiling jets. Figure 1 shows the per trigger normalized transverse momentum spectra of recoil jets associated with two exclusive trigger track p_T bins TT{8,9} and TT{20,50} in the 0–10% centrality bin of Pb–Pb collisions at $\sqrt{s_{NN}} = 2.76\,\text{TeV}$. Note that both spectra exhibit a remarkable similarity in the region $p_{T,\text{jet}}^{\text{reco,ch}} \lesssim 0$, where their shape does not depend on the transverse momentum of TT. Jets populating this region are predominantly combinatorial background jets, which are accidentally associated with TT. On the other hand, in the region $p_{T,\text{jet}}^{\text{reco,ch}} > 0$, both spectra differ. Since the presence of a TT{20,50} hadron biases the four-momentum transfer in the associated hard scattering to higher values, the corresponding recoil jet spectrum is harder. The spectrum in the region $p_{T,\text{jet}}^{\text{reco,ch}} > 0$ also has, however, the component coming from combinatorial background jets. Based on the situation in the region $p_{T,\text{jet}}^{\text{reco,ch}} \lesssim 0$, it is assumed that this component is independent of TT p_T. Thus, it will be canceled when both spectra are subtracted.

$$\Delta_{\text{recoil}} = \frac{1}{N_{\text{trig}}} \cdot \frac{d^2 N_{\text{jet}}}{dp_{T,\text{jet}}^{\text{ch}} d\eta_{\text{jet}}}\bigg|_{TT\{20,50\}} - \frac{1}{N_{\text{trig}}} \cdot \frac{d^2 N_{\text{jet}}}{dp_{T,\text{jet}}^{\text{ch}} d\eta_{\text{jet}}}\bigg|_{TT\{8,9\}}. \tag{4}$$

Here N_{trig} denotes the number of TT in a given TT bin. Let us point out that on the theory side, the per trigger normalized yield of recoil jets can be expressed in terms of a cross section to produce a high-p_T hadron and a cross section to produce a high-p_T hadron together with a jet.

$$\frac{1}{N_{\text{trig}}} \cdot \frac{d^2 N_{\text{jet}}}{dp_{T,\text{jet}}^{\text{ch}} d\eta_{\text{jet}}}\bigg|_{p_{T,\text{trig}} \in TT} = \frac{1}{\sigma_{h+X}} \cdot \frac{d^2 \sigma_{h+\text{jet}+X}}{dp_{T,\text{jet}}^{\text{ch}} d\eta_{\text{jet}}}\bigg|_{p_{T,h} \in TT}. \tag{5}$$

The measured Δ_{recoil} spectrum was further corrected for jet reconstruction inefficiency and jet energy scale smearing due to instrumental effects and local underlying event fluctuations. The relation between the measured jet spectrum and the true jet spectrum was assumed to be linear, and the combined effect of the instrumental effects and local background fluctuations on the true jet spectrum was described by means of a response matrix. Regularized inversion of this matrix and the corresponding solution for the true spectrum was found by means of Bayesian unfolding [22,23]. The fully corrected Δ_{recoil} spectrum for anti-k_T jets with $R = 0.2$, 0.4, and 0.5 is shown in the right-hand side panel of Figure 1.

The medium-induced modification of the recoil jet spectrum was quantified by means of a ratio

$$\Delta I_{AA} = \frac{\Delta_{\text{recoil}}^{\text{Pb-Pb}}}{\Delta_{\text{recoil}}^{\text{PYTHIA}}}, \tag{6}$$

where $\Delta_{\text{recoil}}^{\text{Pb-Pb}}$ denotes the fully corrected Δ_{recoil} spectrum measured in Pb–Pb collisions and $\Delta_{\text{recoil}}^{\text{PYTHIA}}$ is a reference Δ_{recoil} spectrum obtained from the PYTHIA 6 Perugia 2010 [24] simulation of pp collisions at the same center-of-mass energy per nucleon–nucleon collision; see Figure 2. Statistics of the measured pp $\sqrt{s} = 2.76\,\text{TeV}$ data by ALICE was found to be insufficient for this analysis. Nevertheless, for pp collisions at $\sqrt{s} = 7\,\text{TeV}$, the PYTHIA Δ_{recoil} spectrum reproduces the measured data very well [14]. The ΔI_{AA} ratio is found to be below unity, which shows that the studied sample of recoil jets was affected by jet quenching. Similar suppression was also found for track-based anti-k_T jets having $R = 0.2$ and $R = 0.5$ [14].

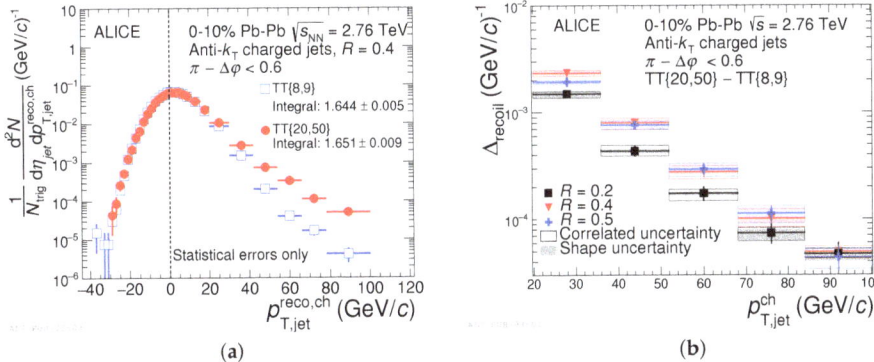

Figure 1. (a) Per trigger normalized transverse momentum spectra of recoil jets associated with trigger track p_T bins TT{8,9} and TT{20,50} measured in the 0–10% centrality bin of Pb–Pb collisions at $\sqrt{s_{NN}} = 2.76$ TeV. (b) Fully corrected Δ_{recoil} distributions for track-based anti-k_T jets with $R = 0.2, 0.4$, and 0.5. Systematic uncertainties in the data are shown by boxes. Taken from [14].

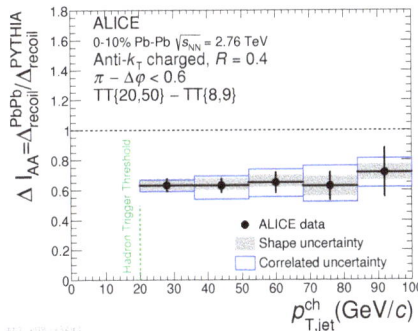

Figure 2. Ratio of Δ_{recoil} spectra obtained from the 0–10% centrality bin of Pb–Pb collisions at $\sqrt{s_{NN}} = 2.76$ TeV and from the PYTHIA 6 Perugia 2010 simulation of pp collisions at $\sqrt{s} = 2.76$ TeV. Systematic uncertainties in the data are shown by boxes. Taken from [14].

3. Searches for Jet Quenching in p–Pb Collisions at $\sqrt{s_{NN}} = 2.76$ TeV

In past, the PHENIX collaboration has searched for jet quenching in d–Au collisions at $\sqrt{s_{NN}} = 200$ GeV [25]. PHENIX measured a nuclear modification factor of inclusive anti-k_T jets with $R = 0.3$. The nuclear modification factor was calculated as

$$R_{\text{dAu}} = \frac{\mathrm{d}N_{\text{jets}}^{\text{d-Au}}/\mathrm{d}p_{T,\text{jet}}}{T_{\text{dAu}} \cdot \mathrm{d}\sigma_{\text{pp}}/\mathrm{d}p_{T,\text{jet}}}, \tag{7}$$

where T_{dAu} is the nuclear overlap function for a given centrality bin of deuteron–gold, $\mathrm{d}\sigma_{\text{pp}}/\mathrm{d}p_{T,\text{jet}}$ is the inclusive cross section for jets in pp collisions at the same nucleon–nucleon center-of-mass energy and $\mathrm{d}N_{\text{jets}}^{\text{d-Au}}/\mathrm{d}p_{T,\text{jet}}$ is the measured spectrum of jets for a given centrality in d–Au. PHENIX found that R_{dAu} is compatible with unity for minimum bias events. However, once they sorted measured events in centrality classes, they found significant enhancement for peripheral events and suppression for central events. The behavior of R_{dAu} for peripheral events was surprising since peripheral events were expected to be similar to pp. The similar ordering of the nuclear suppression factor for jets was seen also by ATLAS in peripheral and central p–Pb collisions at $\sqrt{s_{NN}} = 2.76$ TeV [26]. The interpretation of these results in terms of medium-induced modification of jet production is problematic, since the calculation of the nuclear overlap function does not take into account conservation laws, which play an important role in small systems [27], e.g., momentum conservation. The detection of a high-p_T jet at midrapidity affects measurement of event activity or "centrality" in forward rapidity, and the assigned event geometry can be biased.

Hadron-jet coincidence observables have the advantage that they can be used to identify jet quenching without the need to know the corresponding nuclear overlap function [15]. If there were no medium-induced modifications of jet production in p–Pb, cross sections appearing on the right-hand side of (5) could be expressed in terms of the corresponding pp cross sections and the the nuclear overlap functions T_{pPb},

$$\frac{1}{\sigma_{\text{h+X}}^{\text{p-Pb}}} \cdot \frac{d^2\sigma_{\text{h+jet+X}}^{\text{p-Pb}}}{dp_{\text{T,jet}}^{\text{ch}} d\eta_{\text{jet}}}\bigg|_{p_{\text{T,h}} \in \text{TT}} = \frac{1}{T_{\text{pPb}} \cdot \sigma_{\text{h+X}}^{\text{pp}}} \cdot \frac{T_{\text{pPb}} \cdot d^2\sigma_{\text{h+jet+X}}^{\text{pp}}}{dp_{\text{T,jet}}^{\text{ch}} d\eta_{\text{jet}}}\bigg|_{p_{\text{T,h}} \in \text{TT}}. \tag{8}$$

Since the nuclear overlap function term appears in the numerator and denominator, they cancel. It is thus not necessary to know the relation between the measured event activity and the collision geometry.

Analysis of Δ_{recoil} spectra in p–Pb collisions at $\sqrt{s_{\text{NN}}} = 5.02$ TeV is analogous to what was done in Pb–Pb. In p–Pb, event activity was measured by two forward detectors, the neutron zero degree calorimeter (ZNA) and the V0A scintillator array. Figure 3 shows the per trigger normalized yield of recoil jets associated with the chosen trigger track p_T bins TT{6,7} and TT{12,50} in 20% of p–Pb events with the largest event activity in the ZNA. The resulting raw Δ_{recoil} distribution is shown by open circles.

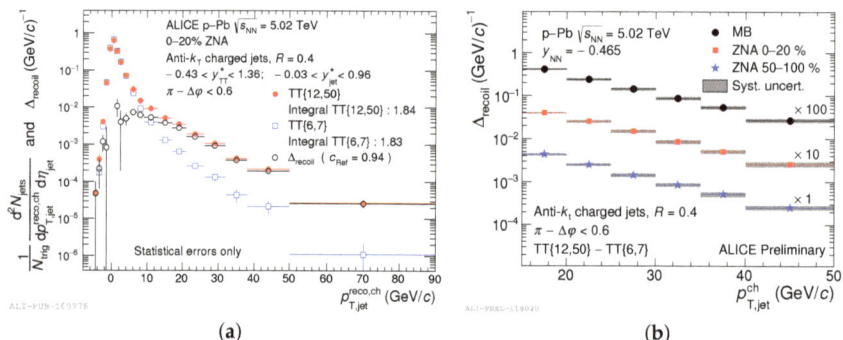

(a) (b)

Figure 3. (a) Per trigger normalized transverse momentum spectra of recoil jets associated with trigger track p_T bins TT{6,7} and TT{12,50} measured in the 0–20% neutron zero degree calorimeter (ZNA) centrality bin of p–Pb collisions at $\sqrt{s_{\text{NN}}} = 5.02$ TeV. (b) Fully corrected Δ_{recoil} distributions for track-based anti-k_T jets with $R = 0.4$. Systematic uncertainties in the data are shown by boxes. Taken from [15].

The right-hand side panel of Figure 3 shows the fully corrected Δ_{recoil} spectra obtained for minimum bias p–Pb events, for 20% of events that had the largest event activity in the ZNA, and for 50% of events that had the lowest event activity in the ZNA. Medium-induced modification of the spectrum was studied by means of the ratio of the Δ_{recoil} spectra measured for the high and low event activity; see Figure 4. The data are compatible with unity and do not exhibit a visible jet-quenching effect. Furthermore, we used these data to provide a limit on the magnitude of medium-induced energy transport to large angles out of the $R = 0.4$ jet cone. The fully corrected Δ_{recoil} spectra have an exponential shape (Figure 3). Under the assumption that higher energy density in events with higher event activity would cause a horizontal shift of the Δ_{recoil} spectrum, we can parameterize the Δ_{recoil} spectra measured for low- and high-event activity as $\Delta_{\text{recoil}}|_{0-20\%} = a \cdot \exp\left(-p_{\text{T,jet}}^{\text{ch}}/b\right)$ and $\Delta_{\text{recoil}}|_{50-100\%} = a \cdot \exp\left((-p_{\text{T,jet}}^{\text{ch}} + \bar{s})/b\right)$, respectively. Here, a and b are constants, and $\bar{s}$ is the spectrum shift. The corresponding ratio is then $\Delta_{\text{recoil}}|_{0-20\%} / \Delta_{\text{recoil}}|_{50-100\%} = \exp(-\bar{s}/b)$. The ratio is independent of jet p_T and can be used to extract $\bar{s}$ provided that the ratio and the slope parameter b are extracted from data. Thus from the data in Figures 3 and 4, the following estimates for the spectrum shift were obtained [15]: $\bar{s} = (-0.12 \pm 0.35_{\text{stat}} \pm 0.03_{\text{syst}})\,\text{GeV}/c$ for events where event activity was measured by ZNA, and $\bar{s} = (-0.06 \pm 0.34_{\text{stat}} \pm 0.02_{\text{syst}})\,\text{GeV}/c$ for events where event activity was measured by V0A. Both values are consistent with zero within uncertainties.

The measured $\bar{s}$ values were further used to set a one-sided 90% confidence upper limit on the medium-induced charged energy transport out of the jet cone of $R = 0.4$ for jets with $15 < p_{\text{T,jet}}^{\text{ch}} < 50\,\text{GeV}/c$. In events with high V0A or high ZNA activity, the medium-induced charged energy transport is less than $0.4\,\text{GeV}/c$ at 90% confidence. This limit is shown by the red line in Figure 4.

In summary, observables that are based on correlations of high-p_T hadrons with jets provide a powerful tool to probe properties of the medium created in heavy-ion collisions. This approach makes it possible to cope with a large underlying event and does not induce fragmentation bias on the studied jet sample. Is is also well-suited for jet quenching studies in small systems since their interpretation does not depend on models that relate collision geometry with event activity.

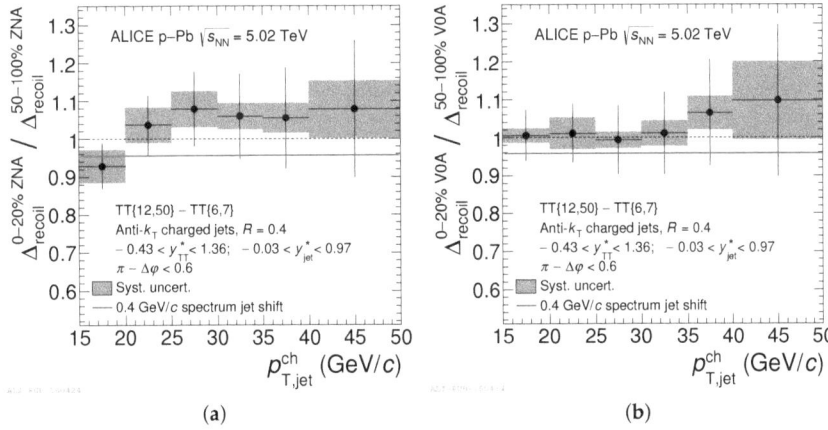

Figure 4. Ratio of fully corrected Δ_{recoil} spectra measured in p–Pb collisions at $\sqrt{s_{\text{NN}}} = 5.02\,\text{TeV}$ in events having different event activity biases. Event activity was measured by ZNA (**a**) and V0A (**b**). Systematic uncertainties in the data are shown by boxes. The red line shows a value of the ratio for the situation when medium-induced charged energy transport out of the $R = 0.4$ jet cone would be $\bar{s} = -0.4\,\text{GeV}/c$. Taken from [15].

Funding: This research was funded by the Ministry of Education, Youth, and Sports of the Czech Republic, grant number LTT17018.

Conflicts of Interest: The author declares no conflict of interest.

References

1. Stock, R. Relativistic Nucleus-Nucleus Collisions and the QCD Matter Phase Diagram. *arXiv* **2008**, arXiv:0807.1610.
2. Stephanov, M.A. QCD phase diagram: An overview. *arXiv* **2006**, arXiv:hep-lat/0701002.

3. Aoki, Y.; Fodor, Z.; Katz, S.D.; Szabo, K.K. The QCD transition temperature: results with physical masses in the continuum limit. *Phys. Lett. B* **2006**, *643*, 46–54. [CrossRef]
4. Ollitraut, J.-Y. Anisotropy as a signature of transverse collective flow. *Phys. Rev. D* **1992**, *43*, 229–245. [CrossRef]
5. Heinz, U.; Snellings, R. Collective flow and viscosity in relativistic heavy-ion collisions. *Annu. Rev. Nucl. Part. Sci.* **2013**, *63*, 123–151. [CrossRef]
6. Mehtar-Tani, Y.; Milhano, J.G.; Tywoniuk, K. Jet physics in heavy-ion collisions. *Int. J. Mod. Phys. A* **2013**, *28*, 1340013. [CrossRef]
7. Majumder, A.; Van Leeuwen, M. The Theory and Phenomenology of Perturbative QCD Based Jet Quenching. *Prog. Part. Nucle Phys. A* **2011**, *66*, 41–92. [CrossRef]
8. Song, H.; Bass, S.A.; Heinz, U.; Hirano, T.; Shen, C. 200 A GeV Au+Au collisions serve a nearly perfect quark-gluon liquid. *Phys. Rev. Lett.* **2011**, *106*, 192301. [CrossRef]
9. Salam, G.P. Towards Jetography. *Eur. Phys. J. C* **2010**, *67*, 637–686. [CrossRef]
10. Chatrchyan, S.; et al. [CMS Collaboration]. Measurement of the Inclusive Jet Cross Section in pp Collisions at $\sqrt{s} = 7$ TeV. *Phys. Rev. Lett.* **2011**, *107*, 132001. [CrossRef]
11. Acharya, S.; et al. [ALICE Collaboration]. Medium modification of the shape of small-radius jets in central Pb–Pb collisions at $\sqrt{s} = 2.76$ TeV. *J. High Energy Phys.* **2018**, *2018*, 139. [CrossRef]
12. Adam, J.; et al. [ALICE Collaboration]. Measurement of jet suppression in central Pb–Pb collisions at $\sqrt{s_{NN}} = 2.76$ TeV. *Phys. Lett. B* **2012**, *712*, 176–197.
13. de Barros, G.O.V.; Fenton-Olsen, B.; Jacobs, P.; Płoskoń, M. Data-driven analysis methods for the measurement of reconstructed jets in heavy ion collisions at RHIC and LHC. *Nucl. Phys. A* **2013**, *910*, 314–318. [CrossRef]
14. Adam, J.; et al. [ALICE Collaboration]. Measurement of jet quenching with semi-inclusive hadron-jet distributions in central Pb–Pb collisions at $\sqrt{s_{NN}} = 2.76$ TeV. *J. High Energy Phys.* **2015**, *2015*, 170. [CrossRef]
15. Acharya, S.; et al. [ALICE Collaboration]. Constraints on jet quenching in p–Pb collisions at $\sqrt{s_{NN}} = 5.02$ TeV measured by the event-activity dependence of semi-inclusive hadron-jet distributions. *Phys. Lett. B* **2018**, *783*, 95. [CrossRef]
16. Abelev, B.; et al. [ALICE Collaboration]. Performance of the ALICE Experiment at the CERN LHC. *Int. J. Mod. Phys. A* **2014**, *29*, 1430044.
17. Aamodt K. et al. [ALICE Collaboration]. Alignment of the ALICE Inner Tracking System with cosmic rays. *J. Instrum.* **2010**, *5*, P03003.
18. Aamodt, K.; et al. [ALICE Collaboration]. The ALICE TPC, a large 3-dimensional tracking device with fast readout for ultra-high multiplicity events. *Nucl. Instrum. Methods Phys. Res. A* **2010**, *622*, 316–367.
19. Cacciari, M.; Salam, G.P.; Soyez, G. The anti-k_t jet clustering algorithm. *J. High Energy Phys.* **2008**, *2008*, 63. [CrossRef]
20. Cacciari, M.; Salam, G.P.; Soyez, G. FastJet User Manual. *Eur. Phys. J. C* **2012**, *72*, 1896. [CrossRef]
21. Cacciari, M.; Salam, G.P. Pileup subtraction using jet areas. *Phys. Lett. B* **2008**, *659*, 119–126. [CrossRef]
22. D'Agostini, G. A multidimensional unfolding method based on Bayes' theorem. *Nucle Instrum. Methods Phys. Res. A* **1995**, *362*, 487–498. [CrossRef]
23. Adye, T. Unfolding algorithms and tests using RooUnfold. *arXiv* **2011**, arXiv:1105.1160.
24. Skands, P.Z. Tuning Monte Carlo Generators: The Perugia Tunes. *Phys. Rev. D* **2010**, *82*, 074018. [CrossRef]
25. Adare, A.; et al. [PHENIX Collaboration]. Centrality categorization for $R_{p(d)+A}$ in high-energy collisions. *Phys. Rev. C* **2014**, *90*, 034902. [CrossRef]
26. Aad, G.; et al. [ATLAS Collaboration]. Centrality and rapidity dependence of inclusive jet production in $\sqrt{s_{NN}} = 5.02$ TeV proton-lead collisions with the ATLAS detector. *Phys. Lett. B* **2015**, *748*, 392–413. [CrossRef]
27. Kordell, M.; Majumder, A. Jets in d(p)–A Collisions: Color Transparency or Energy Conservation. *Phys. Rev. C* **2018**, *97*, 054904. [CrossRef]

© 2019 by the authors. Licensee MDPI, Basel, Switzerland. This article is an open access article distributed under the terms and conditions of the Creative Commons Attribution (CC BY) license (http://creativecommons.org/licenses/by/4.0/).

Communication

Hadron Spectra Parameters within the Non-Extensive Approach [†]

Keming Shen *, Gergely Gábor Barnaföldi [‡] and Tamás Sándor Biró [‡]

Wigner Research Center for Physics of HAS, 29-33 Konkoly-Thege Miklós Street, 1121 Budapest, Hungary; barnafoldi.gergely@wigner.mta.hu (G.G.B.); biro.tamas@wigner.mta.hu (T.S.B.)
* Correspondence: shen.keming@wigner.mta.hu
† This paper is based on my talk at the 18th Zimányi School, Budapest, Hungary, 3–7 December 2018.
‡ These authors contributed equally to this work.

Received: 8 April 2019; Accepted: 20 May 2019; Published: 21 May 2019

Abstract: We investigate how the non-extensive approach works in high-energy physics. Transverse momentum (p_T) spectra of several hadrons are fitted by various non-extensive momentum distributions and by the Boltzmann–Gibbs statistics. It is shown that some non-extensive distributions can be transferred one into another. We find explicit hadron mass and center-of-mass energy scaling both in the temperature and in the non-extensive parameter, q, in proton–proton and heavy-ion collisions. We find that the temperature depends linearly, but the Tsallis q follows a logarithmic dependence on the collision energy in proton–proton collisions. In the nucleus–nucleus collisions, on the other hand, T and q correlate linearly, as was predicted in our previous work.

Keywords: transverse momentum spectra; non-extensive; heavy-ion collisions

1. Introduction

In high-energy nuclear physics, the investigation of transverse momentum (p_T) spectra is a fundamental measure in statistical approaches. The p_T spectrum reveals information on the kinetic properties of the particles produced in high-energy collisions. Strong correlation phenomena were recently observed in proton–proton and heavy-ion collisions [1,2], their statistical and thermodynamical description points beyond the classical Boltzmann–Gibbs (BG) statistics. It has long been realized that data on single inclusive particle distributions show a power-law behavior in the high-p_T region. For these, the Pareto–Hagedorn–Tsallis distribution has been frequently applied [3–5]. Its form coincides with the generalized q-exponential function [6]:

$$e_q(x) := [1 + (1-q)x]^{\frac{1}{1-q}}. \qquad (1)$$

Hadron spectra can be described by the Lorentz-invariant particle spectra. These were successfully fitted by the non-extensive distributions in a wide center-of-mass energy and p_T range [7–20]. In the following, we focus on the most often used formulas from [7–15] for representing identified particle spectra in various collisions. This work explores differences between ($m_T - m$) and m_T-dependent, as well as simple p_T functions:

$$E\frac{d^3 N}{d^3 p} = \frac{d^3 N}{dy\, p_T\, dp_T\, d\phi} = \frac{1}{2\pi p_T} \frac{d^2 N}{dy\, dp_T}. \qquad (2)$$

Different research groups used various kinds of expressions of it in order to describe p_T spectra. We consider functions of $m_T - m$ and p_T in the non-extensive approach, after applying the normalized functions and the thermodynamically motivated ones [21]. Our aim is to find the best-fitting functions among these, while assigning a physical interpretation to their parameters. We investigate the following distribution forms:

$$f_0 = f_{BG} = A_0 \cdot \exp\left(\frac{m_T - m}{T_0}\right),$$

$$f_1 = A_1 \cdot \left(1 + \frac{m_T - m}{n_1 T_1}\right)^{-n_1},$$

$$f_2 = A_2 \cdot \frac{(n_2 - 1)(n_2 - 2)}{2\pi n_2 T_2 [n_2 T_2 + m(n_2 - 2)]} \cdot \left(1 + \frac{m_T - m}{n_2 T_2}\right)^{-n_2},$$

$$f_3 = A_3 \cdot m_T \left(1 + \frac{m_T - m}{n_3 T_3}\right)^{-n_3},$$

$$f_4 = A_4 \cdot \left(1 + \frac{m_T}{n_4 T_4}\right)^{-n_4},$$

$$f_5 = A_5 \cdot \left(1 + \frac{p_T}{n_5 T_5}\right)^{-n_5}. \tag{3}$$

There are relations among the distributions defined above. It is easy to realize that f_1 and f_2 coincide whenever their amplitudes satisfy the relation

$$A_1 = A_2 \cdot \frac{(n_2 - 1)(n_2 - 2)}{2\pi n_2 T_2 [n_2 T_2 + m(n_2 - 2)]} = A_2 \cdot C_q, \quad \text{and} \quad n_1 = n_2. \tag{4}$$

Accounting for the differences between $(m_T - m)$ and m_T dependencies, we re-cast f_1 and f_4 described in Equation (3) as follows:

$$f_1 = A_1 \cdot \left(1 - \frac{m}{n_1 T_1}\right)^{-n_1} \cdot \left(1 + \frac{m_T}{n_1 T_1 - m}\right)^{-n_1}. \tag{5}$$

Comparing this with f_4, we arrive at the relations

$$A_1 \cdot \left(1 - \frac{m}{n_1 T_1}\right)^{-n_1} = A_4, \quad n_1 = n_4, \quad \text{and} \quad n_1 T_1 - m = n_4 T_4. \tag{6}$$

These comments are important for the comparison of different approaches. They also demonstrate that no inconsistency occurs by applying different fit formulas. However, differences arise from the statistical physical motivations behind these formulas [7–12,21,22]. The corresponding results and discussions are investigated next. Note that for all the physical quantities, we use the natural units, $c = 1$, for convenience in this paper.

2. Results and Discussions

In this section, we analyze the transverse momentum distributions of identified pions and kaons stemming from the elementary (pp) and heavy-ion (pPb and $PbPb$) collisions fitted by the functions listed in Equation (3). All the relevant parameters are then analyzed in order to investigate further the non-extensive physics behind these collisions.

2.1. Analysis of the pp Spectra

In high-energy physics, even the smallest hadron–hadron (pp) collisions are rather complicated processes. One usually separates two main regimes of hadron production: one is a soft multiparticle production, dominant at low transverse momenta, where the spectra can also be fitted by an exponential behavior [23], cf. the curve f_{BG} in Figure 1. We realize that f_{BG} describes well this part of the spectra even in pp collisions. As p_T gets higher ($p_T > 3$ GeV), the spectrum displays a power-law tail. They are predicted by perturbative QCD, owing to the hard scattering of current quarks and gluons. In a number of publications [16–20], the Tsallis statistical distribution was successfully applied to describe data for pp collisions over a wide range of the transverse momenta because of its two limits: the exponential shape at small p_T and the power-like distribution at large p_T,

$$e_q(-\frac{p_T}{T}) \longrightarrow \begin{cases} e^{-p_T/T} & p_T \to 0 \\ \left((q-1)\frac{p_T}{T}\right)^{\frac{1}{1-q}} & p_T \to \infty. \end{cases} \quad (7)$$

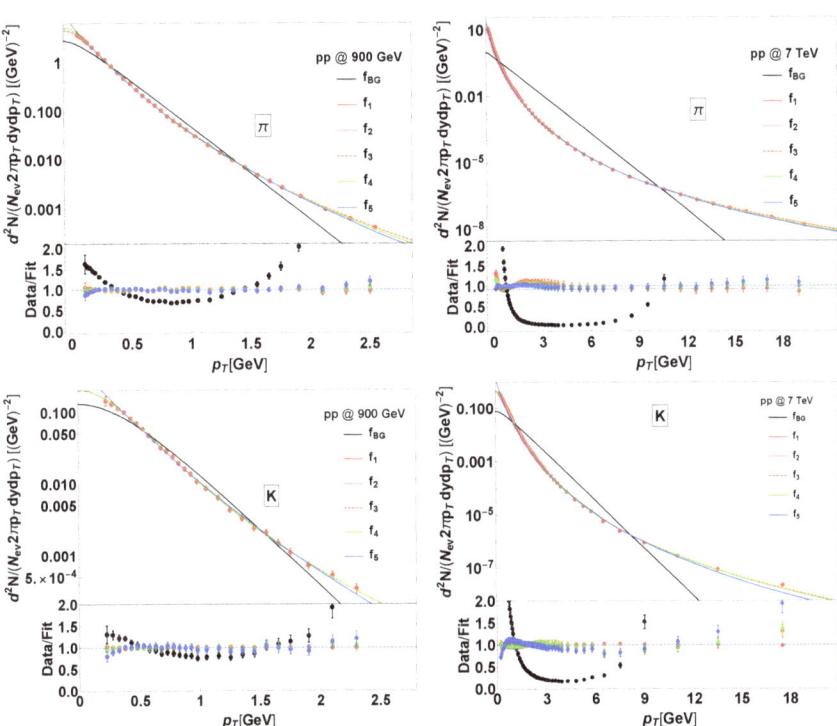

Figure 1. The p_T spectra for pions (upper) and kaons (lower) in pp collisions at $\sqrt{s}$ = 900 GeV and 7 TeV at midrapidity as examples. Data are taken from [16,17]. All are fitted with all of the functions of Equation (3) in the ranges of $0.1 < p_T < 2.6$ GeV at $\sqrt{s}$ =900 GeV and $0.1 < p_T < 20$ GeV at 7 TeV, respectively. Ratios of the net fits to data are also shown in the lower panel. The relevant values of $\chi^2/d.o.f.$ are shown in Table 1.

Table 1. The values of $\chi^2/d.o.f.$ of spectral fits for pions, kaons, and protons in pp collisions at 900 GeV and 7 TeV as examples.

Collision Energy ($\sqrt{s}$)	Produced Hadrons	f_{BG}	f_1	f_2	f_3	f_4	f_5
900 GeV	π	110.8	0.2814	0.2814	0.4697	0.2814	1.456
	K	8.047	0.1748	0.1749	0.1698	0.1749	0.6669
	p	3.491	0.3724	0.3724	0.3735	0.3724	0.4145
7 TeV	π	1316.0	0.9681	0.9681	3.417	0.9681	0.3049
	K	520.2	0.4202	0.4202	0.4313	0.4202	3.100
	p	254.3	0.4481	0.4481	0.4356	0.4481	4.357

We focus on the fittings of the produced charged particle spectra in elementary collisions with the non-extensive functions in Equation (3). Data were taken for pions, kaons, and protons in pp collisions at $\sqrt{s} = 62.4$ GeV, 200 GeV from the PHENIX Collaboration [18] and at 900 GeV [16], 2.76 TeV [19], 5.02 TeV, and 7 TeV [17] from the ALICE Collaboration. We restrict our analysis to the midrapidity region $|y| < 0.5$ within the p_T ranges, as shown in Table 2. Note that in the following, π, K, and p mark the spectra of $\frac{\pi^+ + \pi^-}{2}$, $\frac{K^+ + K^-}{2}$, and $\frac{p + \bar{p}}{2}$, respectively.

Table 2. Fitting p_T ranges of spectra for different charged particles in pp collisions [16–19].

$\sqrt{s}$	π [GeV]	K [GeV]	p [GeV]
62.4 GeV	0.3–2.9	0.4–2	0.6–3.6
200 GeV	0.3–3	0.4–2	0.5–4.6
900 GeV	0.1–2.6	0.2–2.4	0.35–2.4
2.76 TeV	0.1–20	0.2–20	0.3–20
5.02 TeV	0.1–20	0.2–20	0.3–20
7 TeV	0.1–20	0.2–20	0.3–20

Figure 1 shows that all of the different non-extensive functions we used fit the pion and kaon spectra very well for various kinds of beam energies at midrapidity. The ratios of $\chi^2/d.o.f.$ of the relevant fits are given in Table 1. Specifically, the first two distributions (f_1 and f_2) of $m_T - m$ and f_4 of m_T show close-fitting results. The distribution, f_3, derived thermodynamically, does not display large differences in the goodness of fit either. Checking the fitting parameters A, T, and $q = 1 + 1/n$, we observe that, as we expected and introduced in the previous section, all these functions share the same Tsallis parameter n. The two $m_T - m$ functions (f_1 and f_2) lead to fitting values of the temperature T, which are different from the pure m_T fit (f_4). This indicates that the normalization constant does not affect the fitted T and q parameters but the integrated yield dN/dy. Namely, by normalizing the momentum spectrum

$$\frac{1}{2\pi p_T} \frac{d^2 N}{dy\, dp_T} = A_2 \cdot C_q \cdot \left(1 + \frac{m_T - m}{n_2 T_2}\right)^{-n_2} \tag{8}$$

with the C_q normalization constant and the condition of $A_2 = dN/dy$, we obtain the integral over p_T from 0 to its maximal values p_{Tmax}:

$$\int_0^{p_{Tmax}} \frac{1}{2\pi p_T} \frac{d^2 N}{dy\, dp_T} 2\pi p_T\, dp_T = \frac{dN}{dy} . \tag{9}$$

Moving towards physical interpretation issues, we investigate the temperature, T, and the non-extensive parameter, q. Investigations in [18,24] showed that both of them express $\sqrt{s}$ dependence. In this paper, we found that they are also dependent on the hadron mass, m. The $\sqrt{s}/m$ dependence, as a result, is studied in order to analyze hadron spectra parameters within the non-extensive approach. Following the phenomenological observations in [25,26], a QCD-like evolution can be introduced for both the parameters T and q. While analyzing data, we found that the temperature T had a weak logarithmic $\sqrt{s}/m$ dependence. Thus, here we assume a linear $\sqrt{s}/m$ dependence to analyze the temperature T, but the non-extensive parameter q is kept with the stronger logarithmic distribution:

$$T = T_0 + T_1 \cdot \left(\frac{\sqrt{s}}{m}\right), \quad \text{and} \quad q = q_0 + q_1 \cdot \ln\left(\frac{\sqrt{s}}{m}\right) . \tag{10}$$

In summary, our work indicates that the BG distribution is not suitable for describing the hadron spectra over a wide range of p_T. Comparisons of their corresponding fitting errors $\chi^2/d.o.f.$ show that both $m_T - m$ and m_T functions share the same goodness between f_1 and f_2, cf. Equation (3). Together with the thermodynamically derived f_3, all the non-extensive approaches ($f_1 \sim f_4$) follow the experimental data accurately. The fitting temperature, T, is nearly constant when changing the ratio of the collision energy to hadron mass, $\sqrt{s}/m$. Specifically, distributions of f_1, f_2, f_4, and f_5 are described best with such a connection, as shown in the left panel of Figure 2. From Table 3, we also see that the slope parameters in these four cases are almost zero, which means that they are constant around some values. The non-extensive parameter q, on the other hand, follows a logarithmic dependence, agreeing with a pQCD-based motivation, cf. [21]. Note that our results on T and q are different from the work by Cleymans et al. [24]. Those authors parameterized this relation as a power-law.

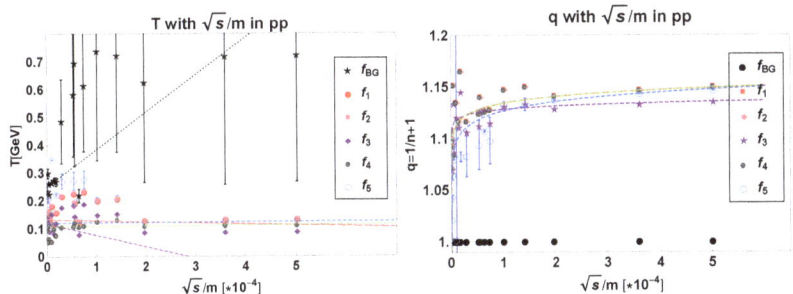

Figure 2. Both the center-of-mass energy $\sqrt{s}$ and hadron mass m distributions of the fitting temperature T and the non-extensive parameter q. In this work, we analyze the results at all given energies with the relationship cf. Equation (10). Here we list the results for $\sqrt{s}$ =62.4 GeV, 200 GeV, 900 GeV, 2.76 TeV, 5.02 TeV, and 7 TeV. and hadron species of pions, kaons, and protons. We have extracted a factor of 10^4 from the values of $\sqrt{s}/m$ for convenience.

Table 3. Fitting parameters of Equation (10) in use within Figure 2.

Fitting Functions	T_0	T_1	q_0	q_1
f_{BG}	0.2515 ± 0.0005	0.1335 ± 0.0002	-	-
f_1	0.1343 ± 0.0003	-0.0041 ± 0.0001	1.135 ± 0.002	0.009 ± 0.001
f_2	0.1343 ± 0.0003	-0.0041 ± 0.0001	1.135 ± 0.002	0.009 ± 0.001
f_3	0.1190 ± 0.0002	-0.0412 ± 0.0002	1.129 ± 0.001	0.004 ± 0.001
f_4	0.1083 ± 0.0003	0.0011 ± 0.0004	1.135 ± 0.002	0.009 ± 0.001
f_5	0.1222 ± 0.0005	0.0007 ± 0.0001	1.127 ± 0.002	0.013 ± 0.002

2.2. Analysis of the pPb and PbPb Results

In pPb [17] collisions at 5.02 TeV and in $PbPb$ [27–30] collisions at 2.76 TeV, more kinds of hadron spectra are analyzed within the formulas of Equation (3). Data are taken from the ALICE Collaboration within wide p_T ranges, as seen in Table 4. We observe that all of them present good fittings over the whole range of p_T for each hadron at various kinds of centrality bins. On the other hand, similar to the pp cases, the BG formula can still perform well just in the low p_T region ($p_T < 3$ GeV).

Table 4. Fitting p_T range of different hadron spectra in heavy-ion collisions in this work [17,27–30].

Particles	Mass [GeV]	pPb [GeV]	$PbPb$ [GeV]
π	0.140	0.11–2.85	0.11–19
K	0.494	0.225–2.45	0.225–19
K_S^0	0.498	0.05–7	0.45–11
K^*	0.896		0.55–4.5
p	0.938	0.325–3.9	0.325–17.5
ϕ	1.019		0.65–4.5
Λ	1.116	0.65–7	0.65–11
Ξ	1.321		0.7–7.5
Ω	1.672		1.3–7.5

In this work, as an example, we analyzed the fitting results of p_T spectra of pions and kaons produced in all kinds of collisions mentioned above. It is instructive to plot the relationship between the fitting temperature T and the Tsallis parameter q for the same hadron spectra for different centralities in the same heavy-ion collisions. The results of pions and kaons in pp collisions are also analyzed as comparisons. In Figure 3, we show the linear correlating appearances for both π and K in pPb at 2.76 TeV [17] and in $PbPb$ at 5.02 TeV [27,28] as well as the pp results in all kinds of collision energies [16–19] in this paper. In fact, whatever kinds of particle we study, all these non-extensive fittings give a similar dependence of T on the parameter q:

$$T \approx T_0 - (q-1)T_1 ,\qquad (11)$$

which agrees with our previous work [21,22] and that of others [31].

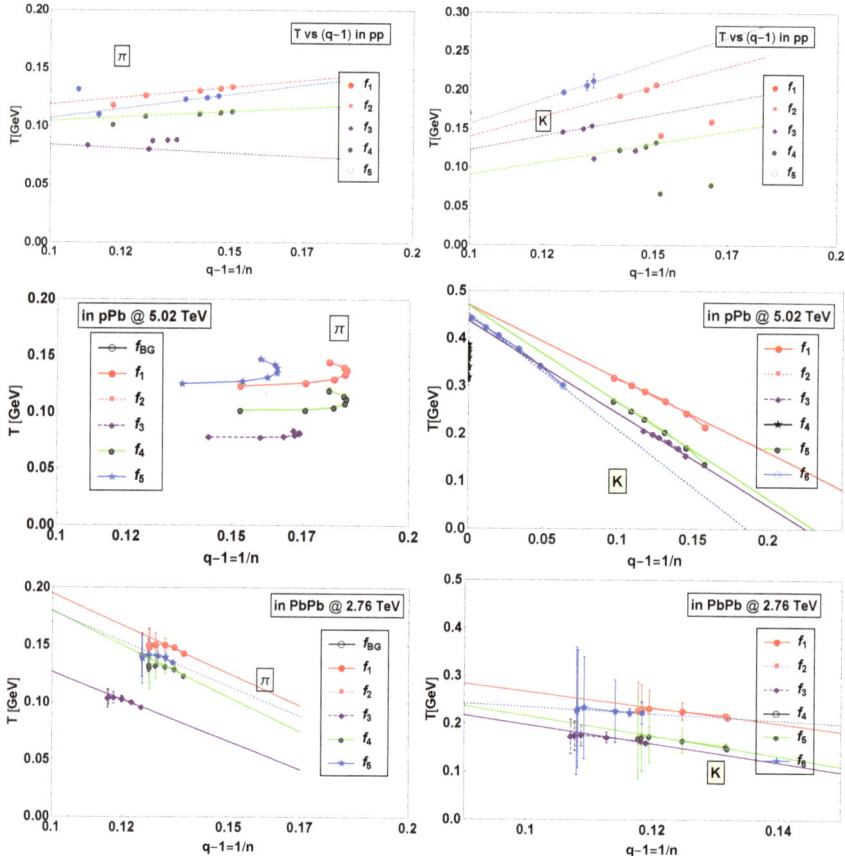

Figure 3. Correlations between T and $q - 1 = 1/n$ for spectra of π (left) and K (right) in pp, pPb, and $PbPb$ collisions. The corresponding p_T range is listed in Table 4, and the values of fitting parameters in Equation (11) are listed in Table 5.

Note that the slope parameter T_1 in Table 5 turns negative and T_0 is nearly zero for the pp case, as discussed in [22]. Results of fittings on pion spectra, typically in pPb collisions at 5.02 TeV, fail in the obvious linear combinations probably due to the small mass of pions and high multiplicities. It is found that all forms of non-extensive distributions feature a similar relation between the temperature

T and non-extensive parameter q. This, in turn, hopefully promotes a better understanding of the meaning of the non-extensive parameter q.

Table 5. Fitting parameters of Equation (11) between T and $q - 1 = 1/n$ for spectra of π (upper) and K (lower) in pp, pPb, and $PbPb$ collisions (note that f_{BG} is not included because $q = 1$ is a constant).

Particles	Fittings	T_1 in pp	T_0 in pp	T_1 in pPb	T_0 in pPb	T_1 in $PbPb$	T_0 in $PbPb$
	f_1	-0.36 ± 0.02	0.08 ± 0.01	-	-	1.40 ± 0.02	0.335 ± 0.004
	f_2	-0.36 ± 0.02	0.08 ± 0.01	-	-	1.40 ± 0.02	0.335 ± 0.004
π	f_3	-0.14 ± 0.04	0.07 ± 0.02	-	-	1.22 ± 0.02	0.249 ± 0.005
	f_4	-0.22 ± 0.01	0.08 ± 0.01	-	-	1.52 ± 0.02	0.333 ± 0.005
	f_5	-0.31 ± 0.03	0.08 ± 0.01	-	-	1.31 ± 0.01	0.311 ± 0.007
	f_1	-1.30 ± 0.02	0.011 ± 0.001	1.55 ± 0.02	0.470 ± 0.001	1.67 ± 0.06	0.434 ± 0.003
	f_2	-1.30 ± 0.02	0.011 ± 0.001	1.55 ± 0.02	0.470 ± 0.001	1.67 ± 0.06	0.434 ± 0.003
K	f_3	-0.90 ± 0.04	0.032 ± 0.005	1.94 ± 0.03	0.436 ± 0.004	1.96 ± 0.03	0.394 ± 0.007
	f_4	-0.81 ± 0.01	0.010 ± 0.004	2.03 ± 0.03	0.470 ± 0.003	2.11 ± 0.05	0.427 ± 0.006
	f_5	-1.59 ± 0.02	-0.001 ± 0.0005	2.43 ± 0.01	0.453 ± 0.002	0.73 ± 0.02	0.309 ± 0.007

3. Summary

In this work, we analyzed various fitting formulas of the hadron spectra in order to explore their sensitivity to different fitting parameters in use within the non-extensive approaches, cf. Equation (3). The hadronization, as well as the p_T distributions in high-energy physics (in proton–proton, proton–nucleus, and nucleus–nucleus collisions) are being studied here. For more details, see [21].

Our results reveal that normalization parameters have no major effect on the shape of these functions. In other words, the fitting formulas of either $m_T - m$ or m_T lead to the same fit quality. As shown in Table 1, they obtained similar fitting values of $\chi^2/d.o.f$. Finally, we investigated the relationship between the fitting parameters, T and q. In pp collisions, the temperature values were fitted by the linear relation of $\sqrt{s}/m$, while the non-extensive parameter q had a logarithmic $\sqrt{s}/m$ dependence, motivated by the QCD-like evolution [25,26]. All kinds of approaches led to linear relations between the temperature, T, and the non-extensive parameter, $q - 1$, in heavy-ion collisions at different centralities. This agrees well with our previous results [21,22] and others in [31].

Summarizing, based on the Tsallis q-exponential, five types of non-extensive formulas in Equation (3) were investigated in parallel to the usual BG distribution. Results showed that the BG statistics failed in describing the hadronization in the whole p_T range. Within the non-extensive approaches, $m_T - m$ functions obtained similar fitting results to the m_T ones. This provides a free choice between the functions $m_T - m$ and m_T when analyzing the hadron spectra. On the other hand, it does not make any differences with regards to the normalization. Nevertheless, the normalized function, f_2, is the best choice since it is also connected to the particle yield per unit rapidity, dN/dy, by its normalization, A_2.

Author Contributions: Authors contributed equally to the contribution.

Funding: This research was funded by the Hungarian National Research, Development and Innovation Office (NKFIH) under the contract numbers K120660 and K123815 and THOR COST CA 15213.

Acknowledgments: Computational resources were provided by the Wigner GPU Laboratory and the Wigner Datacenter.

Conflicts of Interest: The authors declare no conflict of interest. The funders had no role in the design of the study; in the collection, analyses, or interpretation of data; in the writing of the manuscript, or in the decision to publish the results.

Abbreviations

The following abbreviations are used in this manuscript:

p_T transverse momentum
BG   Boltzmann–Gibbs
m_T transverse mass

References

1. ALICE Collaboration. Enhanced production of multi-strange hadrons in high-multiplicity proton-proton collisions. *Nat. Phys.* **2017**, *13*, 535–539.
2. ALICE Collaboration. ALICE measures pA collisions: Collectivity in small systems? *J. Phys. Conf. Ser.* **2017**, *798*, 012068.
3. Pareto, V. *Cours d'Economie Politique*; Droz: Geneva, Switzerland, 1896.
4. Hagedorn, R.; Multiplicities, T. p_T distributions and the expected hadron $\to$ quark-gluon phase transition. *Riv. Nuovo Cimento* **1983**, *6*, 1–50. [CrossRef]
5. Tsallis, C. Possible generalization of Boltzmann-Gibbs statistics. *J. Stat. Phys.* **1988**, *52*, 479–487. [CrossRef]
6. Tsallis, C. Generalizing What We Learnt: Nonextensive Statistical Mechanics. In *Introduction to Nonextensive Statistical Mechanics*; Springer: New York, NY, USA, 2009; p. 382.
7. Osada, T.; Wilk, G. Nonextensive hydrodynamics for relativistic heavy-ion collisions. *Phys. Rev. C* **2008**, *77*, 044903. [CrossRef]
8. STAR Collaboration. Identified Baryon and Meson Distributions at Large Transverse Momenta from $Au + Au$ Collisions at $\sqrt{s_{NN}}$ = 200 GeV. *Phys. Rev. Lett.* **2006**, *97*, 152301.
9. BRAHMS Collaboration. Charged Meson Rapidity Distributions in Central $Au + Au$ Collisions at $\sqrt{s_{NN}}$ = 200 GeV. *Phys. Rev. Lett.* **2005**, *94*, 162301.
10. CMS Collaboration. Transverse momentum and pseudorapidity distributions of charged hadrons in pp collisions at sqrt(s) = 0.9 and 2.36 TeV. *J. High Energy Phys.* **2010**, *2010*, 41.
11. CMS Collaboration. Transverse-Momentum and Pseudorapidity Distributions of Charged Hadrons in pp Collisions at $\sqrt{s}$ = 7 TeV. *Phys. Rev. Lett.* **2010**, *105*, 022002.
12. PHENIX Collaboration. Identified charged particle spectra and yields in $Au + Au$ collisions at $\sqrt{s_{NN}}$ = 200 GeV. *Phys. Rev. C* **2004**, *69*, 034909.
13. STAR Collaboration. Identified hadron spectra at large transverse momentum in $p + p$ and $d + Au$ collisions at $\sqrt{s_{NN}}$ = 200 GeV. *Phys. Lett. B* **2006**, *637*, 161–169.
14. STAR Collaboration. $K(892)^*$ Resonance Production in $Au + Au$ and $p + p$ Collisions at $\sqrt{s_{NN}}$ = 200 GeV at RHIC. *Phys. Rev. C* **2005**, *71*, 064902.
15. Wilk, G.; Wlodarczyk, Z. Interpretation of the Nonextensivity Parameter q in Some Applications of Tsallis Statistics and Lévy Distributions. *Phys. Rev. Lett.* **2000**, *84*, 2770–2773. [CrossRef] [PubMed]
16. ALICE Collaboration. Production of pions, kaons and protons in pp collisions at $\sqrt{s}$ = 900 GeV with ALICE at the LHC. *Eur. Phys. J. C* **2011**, *71*, 1655.
17. ALICE Collaboration. Multiplicity Dependence of Pion, Kaon, Proton and Lambda Production in $p + Pb$ Collisions at $\sqrt{s_{NN}}$ = 5.02 TeV. *Phys. Lett. B* **2014**, *728*, 25–38.
18. PHENIX Collaboration. Identified charged hadron production in $p + p$ collisions at $\sqrt{s}$ = 200 and 62.4 GeV. *Phys. Rev. C* **2011**, *83*, 064903.
19. ALICE Collaboration. Production of charged pions, kaons and protons at large transverse momentum in pp and $Pb - Pb$ collisions at $\sqrt{s_{NN}}$ = 2.76 TeV. *Phys. Lett. B* **2014**, *736*, 196–207.
20. Tang, Z.; Xu, Y.; Ruan, L.; van Buren, G.; Wang, F.; Xu, Z. Spectra and radial flow in relativistic heavy ion collisions with Tsallis statistics in a blast-wave description. *Phys. Rev. C* **2009**, *79*, 051901. [CrossRef]
21. Shen, K.; Barnaföldi, G.G.; Biró, T.S. Hadronization within Non-Extensive Approach and the Evolution of the Parameters. *arXiv* **2019**, arXiv:1905.05736.
22. Shen, K.M.; Biro, T.S.; Wang, E.K. Different non-extensive models for heavy-ion collisions. *Phys. A* **2018**, *492*, 2353–2360. [CrossRef]
23. Hagedorn, R. *Hot and Hadronic Matter: Theory and Experiment*; Plenum Press, Publishing House: New York, NY, USA, 1995; p. 13.
24. Cleymans, J.; Lykasov, G.I.; Parvan, A.S.; Sorin, A.S.; Teryaev, O.V.; Worku, D. Systematic properties of the Tsallis distribution: Energy dependence of parameters in high energy p-p collisions. *Phys. Lett. B* **2013**, *723*, 351–354. [CrossRef]
25. Barnaföldi, G.G.; Ürmössy, K.; Biró, T.S. Tsallis–Pareto–like Distributions in Hadron-Hadron Collisions. *J. Phys. Conf. Ser.* **2011**, *270*, 357–363. [CrossRef]
26. Takacs, A.; Barnaföldi, G.G. Non-Extensive Motivated Parton Fragmentation Functions. *Multidiscipl. Digit. Publ. Inst. Proc.* **2019**, *10*, 12. [CrossRef]

27. ALICE Collaboration. Centrality dependence of the nuclear modification factor of charged pions, kaons, and protons in $Pb - Pb$ collisions at $\sqrt{s_{NN}}$ = 2.76 TeV. *Phys. Rev. C* **2016**, *93*, 034913.
28. ALICE Collaboration. K_S^0 and Λ Production in $Pb - Pb$ Collisions at $\sqrt{s_{NN}}$ = 2.76 TeV. *Phys. Rev. Lett.* **2013**, *111*, 222301.
29. ALICE Collaboration. Multi-strange baryon production at mid-rapidity in $Pb - Pb$ collisions at $\sqrt{s_{NN}}$ = 2.76 TeV. *Phys. Lett. B* **2014**, *728*, 216.
30. ALICE Collaboration. $K^*(892)^0$ and $\phi(1020)$ production in $Pb - Pb$ collisions at $\sqrt{s_{NN}}$ = 2.76 TeV. *Phys. Rev. C* **2015**, *91*, 024609.
31. Wilk, G.; Włodarczyk, Z. On possible origins of power-law distributions. *AIP Conf. Proc.* **2015**, *1558*, 893–896, and its corresponding references.

© 2019 by the authors. Licensee MDPI, Basel, Switzerland. This article is an open access article distributed under the terms and conditions of the Creative Commons Attribution (CC BY) license (http://creativecommons.org/licenses/by/4.0/).

Article

Production and Detection of Light Dark Matter at Jefferson Lab: The BDX Experiment

Marzio De Napoli [†]

Istituto Nazionale di Fisica Nucleare, Sezione di Catania, 95125 Catania, Italy; marzio.denapoli@ct.infn.it; Tel.: +39-095-3785331

† On Behalf of the BDX Collaboration.

Received: 6 April 2019; Accepted: 16 May 2019; Published: 20 May 2019

Abstract: The Beam Dump eXperiment (BDX) is a an electron-beam thick-target experiment aimed to investigate the existence of light Dark Matter particles in the MeV-GeV mass region at Jefferson Lab. The experiment will make use of a 10.6 GeV high-intensity electron-beam impinging on the Hall-A beam-dump to produce the Dark Matter particles (χ) through the Dark Photon portal. The BDX detector located at $\sim$20 m from the dump consists of two main components: an electromagnetic calorimeter to detect the signals produced by the χ-electron scattering and a veto system to reject background. The expected signature of the DM (Dark Matter) interaction in the Ecal (Electromagnetic calorimeter) is a $\sim$GeV electromagnetic shower paired with a null activity in the surrounding active veto counters. Collecting 10^{22} electrons on target in 285 days of parasitic run at 65 µA of beam current, and with an expected background of O(5) counts, in the case of a null discovery, BDX will be able to lower the exclusion limits by one to two orders of magnitude in the parameter space of dark-matter coupling versus mass. This paper describes the experiment and presents a summary of the most significant results achieved thus far, which led to the recent approval of the experiment by JLab-PAC46.

Keywords: Dark Matter; Dark Photon; beam dump experiment

---

1. Introduction

The existence of a copious quantity of Dark Matter (DM) in the Universe is proved by a rich collection of astrophysical and cosmological observations. Nevertheless, its elementary properties remains largely elusive [1], making the search for DM one of the hottest topics in physics today. At the same time, the Standard Model (SM) of particle physics does not explain some experimental facts, such as neutrino masses, the cosmological baryon asymmetry and, of course, the existence of DM, which is, itself, an overwhelming evidence of physics beyond the SM. Various extensions of the SM have the merit to propose also candidates for the role of DM particles, such as the popular Weakly Interacting Massive Particles (WIMPs) ($\sim$10 GeV–10 TeV mass range) expected to weakly interact with SM [1]. Due to the lack of evidence for WIMPs either from LHC or direct DM searches, other well motivated models of DM gained recently the interest of the physics community [2]. Physics beyond the SM might eventually emerge as a whole new sector containing new particles as well as new interactions. These new states do not need to be particularly heavy, with masses below 1 GeV/c^2, and would have easily escaped detection by underground experiments seeking for halo DM. Thus complementary searches attempting to explore these new scenarios are well motivated.

In a popular scenario, light Dark Matter (LDM) with mass in the $\sim$1 MeV–1 GeV range is charged under a new $U(1)_D$ broken symmetry, whose vector boson mediator A' (heavy photon, also called Dark Photon) is massive. The Dark Photon can be kinetically mixed with the SM photon field, resulting in SM–DM interaction through an effective weak coupling of A' to electric charge ϵe [3]. The kinetic mixing parameter ϵ is expected to be in the range of $\sim 10^{-4}$–10^{-2} ($\sim 10^{-6}$–10^{-3}) if the mixing is generated by one-loop (two-loop) interaction [4–6]. The minimal parameter space of vector-mediated LDM is characterized by ϵ, the coupling α_D of the A' to the LDM particle χ and two masses $m_{A'}$ and m_χ.

Depending on the relative mass of the A' and χ, A' can decay only into SM particles (visible decay) or dominantly to LDM states (invisible decay). In particular, if $m_\chi < m_{A'}/2$, and provided that $\alpha_D > \epsilon e$, the latter scenario dominates. This picture is compatible with the well-motivated hypothesis of DM thermal origin, a hypothesis which provides constraints to model parameters from the observed DM density in the Universe [2].

LDM received strong attention in recent years, motivating many theoretical and phenomenological studies. It also stimulated the reanalysis and interpretation of old data and promoted new experimental programs to search both for the A' and LDM states [2,7]. In this context, accelerator-based experiments that make use of a lepton beam of moderate energy (~ 10 GeV) on a thick target or a beam-dump show a sizable sensitivity to a wide area of LDM parameter space [8,9]. Different experimental approaches are possible, each affected by different backgrounds, and with specific sensitivity to model parameters. In particular, high intensity $\sim$ GeV electron-beam fixed-target experiments offer large sensitivity to a broad class of Dark Sector scenarios that feature particles in the elusive MeV-GeV mass range [10].

2. BDX Overview

The Beam Dump experiment (BDX) at Jefferson Lab [11,12] aims to produce and detect LDM, assuming valid the above cited theoretical paradigm. Taking advantage of the high-intensity electron beam available at JLab, BDX has the unique capability of significantly improve the sensitivity to MeV–GeV DM, extending well beyond the reach of existing experiments. BDX will take advantage of the CEBAF (Continuous Electron Beam Accelerator Facility) beam, impinging on the JLab Hall-A beam-dump, which is enclosed in a concrete tunnel at the end of the beam transport line. The Hall-A can receive from CEBAF a 11 GeV electron beam with a current up to 65 µA. Such a beam intensity will allow BDX to collect $\sim 10^{22}$ electron-on-target (EOT) in 285 days, in full parasitic runs. The interaction between the energetic electrons and the atoms of the dump leads to the production of Dark Photons through a Bremsstrahlung-like radiative process (A' – $strahlung$, Figure 1, left) [10] and e^+e^- annihilations [13,14]. The A' could then decay into forward-boosted DM particles (χ) (Figure 1, left). Having a small coupling to ordinary matter, LDM particles propagate through the dump and the shielding region up to the BDX detector.

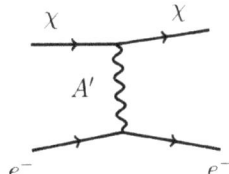

Figure 1. (**Left**). production of Dark Photons through a Bremsstrahlung-like radiative process and decay of A' into a pair of DM particles; (**Right**) the χ-e^- scattering is also mediated by A'.

The detector will be placed along the LDM beam trajectory ~ 8 m underground, i.e., at the beam-dump level, in a new underground facility located ~ 20 m downstream of the Hall-A dump (Figure 2). A specific shielding configuration made by ~ 7 m of iron plus ~ 7 m of concrete and installed between the dump and the detector will be used to suppress the high-energy component of the beam-related background.

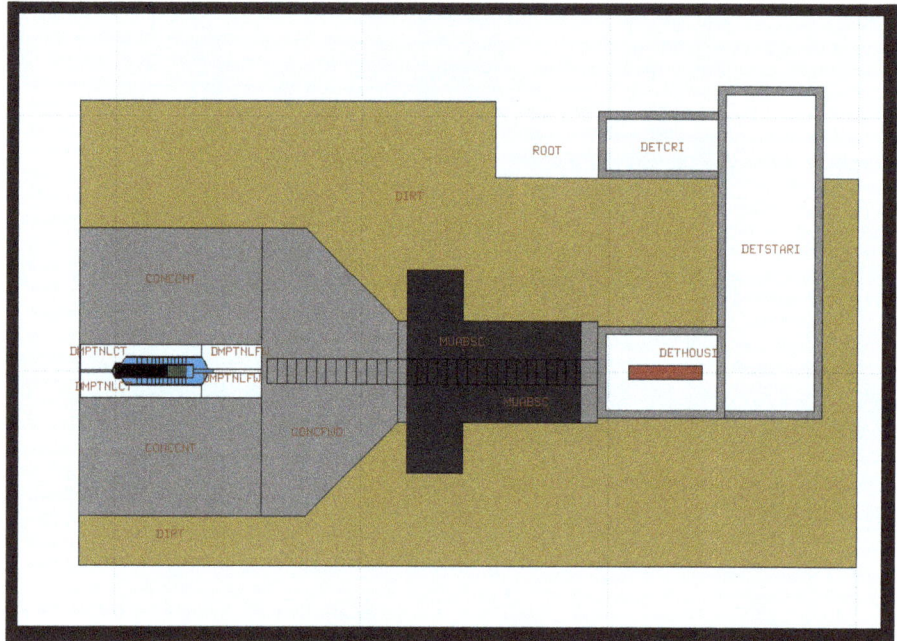

Figure 2. The BDX experimental setup, as implemented in FLUKA [15]. From left to right: the Hall-A beam dump (blue), the concrete (light-gray) and iron (dark-gray) shielding, and the BDX detector (red) located inside the new underground facility.

A fraction of DM particles will then scatter on the electrons of the BDX detector active material (Figure 1, right). For χ-e^- interaction, since $m_e \ll m_\chi$, the typically scattered electron carries GeV-scale energy producing an electromagnetic shower in the GeV energy range, generated by the recoiling electron, that represent an easily detectable signal in the BDX electromagnetic calorimeter. To identify and reduce the SM background that could mimic the expected Dark Matter signals, a combination of passive shielding, active vetos and analysis cuts will be applied.

3. The BDX Experimental Setup

The BDX detector is made of two main components: an electromagnetic calorimeter used to detect signals produced by the interacting DM particles, and an active veto system used to reject the background (see Figure 3 for a sketch of the detector). A signal event in BDX is characterized by the presence of an electromagnetic shower in the Ecal coupled with a null activity in the veto system.

The calorimeter consists of ~800 CsI(Tl) crystals, arranged in eight modules of 10×10 CsI(Tl) crystals each, with the long size along the beam direction. The average size of each crystal is $4.7 \times 5.4 \times 32.5$ cm^3. This arrangement results in a cross section of ~50×55 cm^2 for a total length of ~3 m. Light generated in the crystals will be read-out by Silicon Photomultipliers (SiPM); a rapidly growing technology for the detection of visible photons that is substituting more traditional PMTs and APDs in many physics fields. The good performance of SiPMs as the light readout system of a large-size CsI(Tl) crystal, same size as the BDX Ecal crystals, has been recently proved in Ref. [16]. The BDX Ecal is operated inside two hermetic layers of active veto counters, made of plastic scintillators: the outermost called Outer Veto (OV) and the innermost Inner Veto (IV). Both vetos consists of 1/2 cm-thick plastic scintillators. Due to the relatively large volume to cover, they are divided in paddles. The light from each of them is readout by one or more SiPMs, depending on the paddle size, through wavelength shifting plastic scintillators and scintillating fibers. Between the Ecal and the vetos, a layer of lead

~5 cm thick will reduce the number of events where the EM shower is not entirely contained in the Ecal and a fraction of its energy is deposited in the vetos, increasing, in this way, the detection efficiency to DM signals. Signals fromt the SiPMs will be amplified by custom charge amplifiers and digitized in the framework of a triggerless data acquisition system. For this purpose, a dedicated front-end board has been recently developed [17]. This highly configurable digitizer board includes 12 complete acquisition channels: the analog-to-digital converter components no the board can be chosen to fit the needs of the specific application within the range from 12 bits at 65 MHz to 14 bits at 250 MHz. The board allows time synchronization using various methods including GPS and White Rabbit. The configurability of the board and the various options implemented permit its use in a triggerless data acquisition system. Up to 240 channels can be hosted in a single 6U crate.

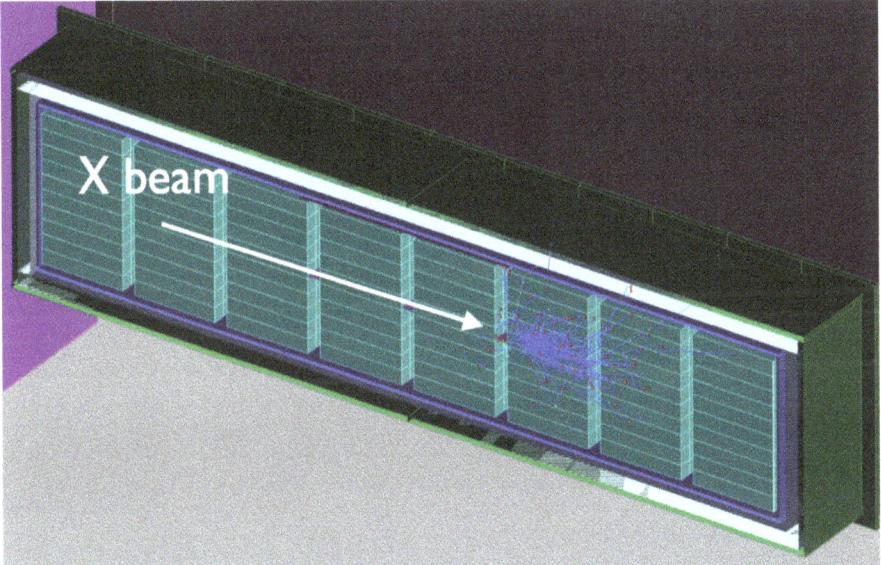

Figure 3. The BDX detector as implemented in GEANT4 [18]. The outer veto is shown in green, the inner veto is gray and the lead vault in blue. Crystals arranged in eight blocks of 10×10 are shown in light blue. A simulated electromagnetic shower from a χ-e^- scattering in the Ecal is also shown.

4. Background

Background is usually the limiting factor in experiments searching for rare events. This is the case for BDX where the low signal rate expected due to the two-step processes involving weak mixing between the SM photon and A' (see Figure 1), makes background rejection a critical issue. Even though BDX will search for electromagnetic showers with energies on the range of hundreds of MeV, thus not requiring the low energy thresholds needed in standard DM direct searches, it is nevertheless mandatory to identify and reject the SM particles that can mimic a DM signal in the Ecal.

4.1. Beam-Related Background

In beam-dump experiments, where a high intensity $O(\text{GeV})$ electron/proton beam is directed into a dump, an overwhelming shower of standard model particles is produced in addition to the rare DM particles of interest. While most of the radiation (gamma, electron/positron and neutron) is contained in the dump or degraded down to harmless energy levels, deep penetrating radiation propagate for long distances before depositing their energy far from the point of origin. In BDX, we used Monte Carlo simulations to find the best combination of shielding and analysis cuts to minimize such background. A summary of the most significant results found is reported in the following. Details on BDX simulations can be found in Refs. [11,19,20].

We simulated an 11 GeV electron-beam interacting with the beam-dump and propagated all particles to the location of interest sampling the flux in different locations. Exploiting biasing techniques available in FLUKA an equivalent statistics of $\sim 0.5 \times 10^{17}$ EOT is obtained. In order to estimate the number of expected background events, the number of particles per EOT was multiplied by 10^{22} EOT. Figure 4 shows the particle rate per EOT at different depths in the shielding.

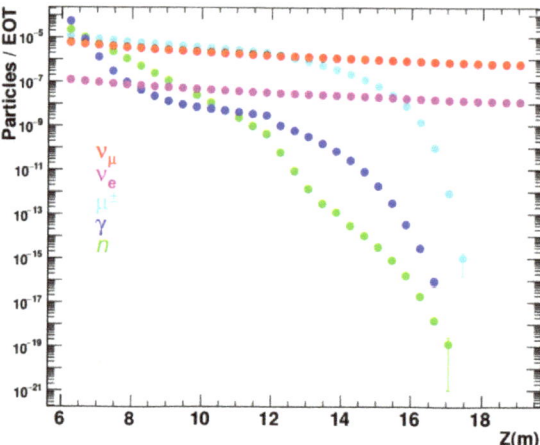

Figure 4. Particle rates per EOT at different depths in the shielding. Particle transport threshold was fixed to 100 MeV (10 MeV for neutrinos).

Results from simulations show that no neutrons or photons above 100 MeV transport threshold hit the detector; muons emitted forward and passing through the shielding are ranged-out; muons emitted at large angles in the dump, propagating in the dirt and then, after a hard interaction, re-scattering in the detector, result in a non-zero background rate. However, they have a kinetic energy lower than 300 MeV and the expected rate is much lower (about a factor 1000) than the rate of cosmic muons that we proved can be efficiently identified and removed with the veto system (see next paragraph) and using an energy threshold in the single crystal of $\sim$350 MeV.

Neutrinos are produced in muon decays and hadronic showers (pion decay). The majority come from pion and muon decay at rest but a non negligible fraction, due to in-flight pion decay, experience a significant boost to several GeV energy. High energy neutrinos interacting with BDX detector by elastic and inelastic scattering may result in a significant energy deposition O(300) MeV that may mimic an EM shower produced by the χ–atomic electron interaction. The $\nu_\mu N \to \mu X$ CC interaction produces a μ in the final state (beside the hadronic state X). This reaction can be identified and used to provide an experimental assessment of the ν_μ background (and therefore estimate the ν_e contribution) by detecting a μ scattering in the detector (a MIP signal inside the calorimeter with or without activity in IV and OV).

The NC $\nu_\mu N \to \nu_\mu X$ and $\nu_e N \to \nu_e X$ interactions produce an hadronic state X that may interact in the detector (while the scattered ν escapes detection). This can mimic an EM shower if π_0 (γs) are produced. However, due to the difference in mass, the scattered ν carries most of the available energy providing a small transfer to the hadronic system and reducing the probability of an over-threshold energy deposition.

The critical background source for the experiment is the $\nu_e N \to eX$ process since the CC interaction could produce a high energy electron/positron into the detector that mimics the signal.

This background can be rejected considering the different kinematics of the ν interaction with respect to the χ-electron scattering. The significant difference in the polar angle of the scattered electron (with respect to the beam direction) allows defining a selection criterion to identify ν_e and separate from the χ. This difference is shown in Figure 5, reporting the angular distribution of scattered e^- from ν_e CC, compared to the characteristics kinematics of the $\chi e^- \to \chi e^-$ kinematics.

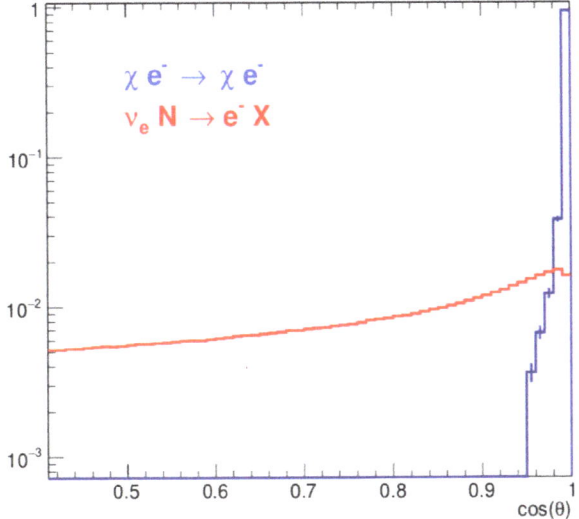

Figure 5. Scattered electron angle distribution for the signal $\chi e^- \to \chi e^-$ and ν_e CC background. The two histograms have been scaled to the same unitary area.

Indeed, such difference in the kinematics has an effect on the shower transverse dimension R, quantity indicating the shower deviation from the beam direction [20], which can consequently be used as an efficient analysis cut to reduce the neutrino background. By using an energy threshold on the single crystal of 350 MeV, the vetoes in anticoincidence and a cut of $R < 0.6$, the number of expected beam-related background (from neutrinos) is ~ 5.

To validate our MC simulations with real data, an on-site experimental campaign was performed to measure the muon flux in the present unshielded configuration at the location of the future BDX detector [21]. The measurement used an electron beam with the proposed energy (10.6 GeV) and one third the current ($\sim 20\,\mu$A) expected in the BDX experiment. We measured the fluence of muons produced by interactions of 10.6 GeV electron beam with the JLab Hall-A beam-dump. Beam-produced muons were measured with a CsI(Tl) crystal sandwiched between a set of segmented plastic scintillators placed at two different distances from the dump: 25.7 m and 28.8 m. At each location, the muon flux was sampled at different vertical positions with respect to the beam height. Data were compared with detailed Monte Carlo simulations using FLUKA for the muon production in the dump and propagation to the detector, and GEANT4 to simulate the detector response. The good agreement in absolute value and shape demonstrates that the simulation framework can safely be used to estimate the beam-related muon background in the BDX experimental set-up.

4.2. Cosmogenic Background

Beam-unrelated background is mainly due to cosmic neutrons, cosmic muons and their decay products. Both direct cosmic flow and secondary particles contribute to the beam-unrelated background rate in the detector.

To validate the BDX detector concept and prove the capability of rejecting high energy cosmic background, we performed an experimental campaign of cosmic-ray measurements at INFN-Sezione di Catania and LNS (Laboratoi Nazioni del SUd (INFN), Catania, Italy), using a prototype of the proposed BDX detector [19]. The BDX-Proto incorporates all the elements of the final detector, built using the same proposed technologies. One of the CsI(Tl) crystals that will be used for the final detector readout by a SiPM was placed inside two layers of plastic scintillator paddles forming the inner and outer vetos and a 5 cm lead vault. Cosmic ray data were taken for about one year inside and outside a similar overburden as the one expected in the BDX experiment. Details of the experimental conditions and data analysis are reported in Ref. [16]. The extrapolation of the expected cosmogenic background was performed by conservatively scaling the experimental rates of a single crystal observed in anticoincidence with the veto systems, to the 800 crystals comprising the full detector. This is certainly an upper limit on the expected rates since this assumes crystal-to-crystal fully uncorrelated counts, which overestimates the case for χ-e^- scattering. The results show that, for energy thresholds high enough, 300–350 MeV, the number of expected cosmogenic background counts in 285 days reduces to zero.

5. Status and Perspectives

The Beam Dump eXperiment (BDX) is an electron-beam thick-target experiment aimed to investigate the existence of light Dark Matter (LDM) particles in the MeV-GeV mass range at Jefferson Lab. The experiment has been approved last year with the maximum scientific grade (A) by JLab PAC46 and is expected to run in a dedicated underground facility located $\sim$ 20 m downstream of the Hall A beam-dump. It will make use of a 10.6 GeV e^- beam collecting up to 10^{22} electrons on target. The detector consists of two main components: a CsI(Tl) electromagnetic calorimeter (Ecal) and a veto system used to reject the background. The expected signature of the DM interaction in the Ecal is a GeV electromagnetic shower paired with a null activity in the surrounding active veto counters. In addition to the veto system, a specific shielding configuration installed between the dump and the detector will be used to suppress the high-energy component of the beam-related background. Indeed, simulations have shown that, provided enough shielding is installed between the beam-dump and the detector, neutrinos are the only source of beam-related background (O(5) background events expected)—considering a detection threshold of O(300) MeV. Using similar energy thresholds coupled with vetos in anticoincidence, the expected cosmogenic background can be considered negligible, as demonstrated by the BDX-Prototype. With 285 days of a parasitic run at 65 µA (corresponding to 10^{22} EOT), the BDX experiment will lower the exclusion limits in the case of no signal by more than one order of magnitude in the parameter space of dark-matter coupling versus mass (Figure 6).

Very recently, a proof of concept measurement has already started at JLAB in the present unshielded configuration. It is using a 2.2 GeV e^- beam and is expected to run parasitically for one year. The compact detector used, called BDX-Mini, is made by a PbWO4 electromagnetic calorimeter, surrounded by a layer of tungsten shielding and two hermetic plastic scintillator veto systems. BDX-Mini is currently lowered in a well, dug downstream of Hall-A at the location of the proposed BDX facility. Although it is an early stage experiment, it represents the first dedicated new-generation beam-dump experiment whose physics reach could almost cover a kinematic region measured by summing up old not-optimized experiments.

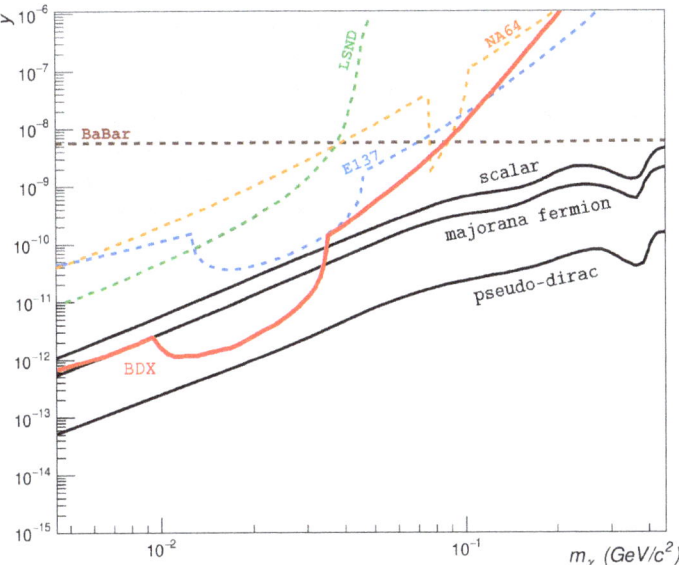

Figure 6. BDX exclusion limits (red line) from Ref. [14]. Limits are given for the parameter $y = \alpha_D \epsilon^2 (m_\chi/m_{A'})^4$ as a function of m_χ, assuming $\alpha_D = 0.5$ and $m_{A'} = 3m_\chi$. Black lines indicate various thermal relic targets.

Funding: This research received no external funding.

Conflicts of Interest: The authors declare no conflict of interest.

References

1. Arcadi, G.; Dutra, M.; Ghosh, P.; Lindner, M.; Mambrini, Y.; Pierre, M.; Profumo, S.; Queiroz, F.S. The waning of the WIMP? A review of models, searches, and constraints. *Eur. Phys. J. C* **2018**, *78*, 203. [CrossRef]
2. Battaglieri, M.; Belloni, A.; Chou, A.; Cushman, P.; Echenard, B.; Essig, R.; Estrada, J.; Feng, J.L.; Flaugher, B.; Fox, P.J.; et al. US Cosmic Visions: New Ideas in Dark Matter 2017: Community Report. *arXiv* **2017**, arXiv:1707.04591.
3. Holdom, B. Two U (1)'s and epsilon charge shifts. *Phys. Lett. B* **1986**, *166*, 196–198. [CrossRef]
4. Essig, R.; Kaplan, J.; Schuster, P.; Toro, N. On the Origin of Light Dark Matter Species. *arXiv* **2010**, arXiv:1004.0691.
5. Aguila, F.D.; Coughlan, G.; Quirs, M. Gauge coupling renormalisation with several U(1) factors. *Nucl. Phys. B* **1988**, *307*, 633–648. [CrossRef]
6. Arkani-Hamed, N.; Weiner, N. LHC signals for a SuperUnified theory of Dark Matter. *J. High Energy Phys.* **2008**, *2008*, 104. [CrossRef]
7. Alexander, J.; Battaglieri, M.; Echenard, B.; Essig, R.; Graham, M.; Izaguirre, E.; Jaros, J.; Krnjaic, G.; Mardon, J.; Morrissey, D.; et al. Dark Sectors 2016 Workshop: Community Report. *arXiv* **2016**, arXiv:1608.08632.
8. Bjorken, J.D.; Essig, R.; Schuster, P.; Toro, N. New fixed-target experiments to search for dark gauge forces. *Phys. Rev. D* **2009**, *80*, 075018. [CrossRef]
9. Andreas, S.; Niebuhr, C.; Ringwald, A. New limits on hidden photons from past electron beam dumps. *Phys. Rev. D* **2012**, *86*, 095019. [CrossRef]
10. Izaguirre, E.; Krnjaic, G.; Schuster, P.; Toro, N. New electron beam-dump experiments to search for MeV to few-GeV dark matter. *Phys. Rev. D* **2013**, *88*, 114015. [CrossRef]

11. Battaglieri, M.; Bersani, A.; Caiffi, B.; Celentano, A.; De Vita, R.; Fanchini, E.; Marsicano, L; Musico, P.; Osipenko, M.; Panza, F.; et al. [BDX Collaboration]. Dark matter search in a Beam-Dump eXperiment (BDX) at Jefferson Lab. *arXiv* **2016**, arXiv:1607.01390.
12. Bondí, M.; [BDX Collaboration]. Light Dark Matter search in a beam-dump experiment: BDX at Jefferson Lab. *Eur. Phys. J. Web Conf.* **2017**, *142*, 01005. [CrossRef]
13. Marsicano, L.; Battaglieri, M.; Bondí, M.; Carvajal, C.D.R.; Celentano, A.; De Napoli, M.; De Vita, R.; Nardi, E.; Raggi, M.; Valente, P. Dark photon production through positron annihilation in beam-dump experiments. *Phys. Rev. D* **2018**, *98*, 015031. [CrossRef]
14. Marsicano, L.; Battaglieri, M.; Bondí, M.; Carvajal, C.D.R.; Celentano, A.; De Napoli, M.; De Vita, R.; Nardi, E.; Raggi, M.; Valente, P. Novel Way to Search for Light Dark Matter in Lepton Beam-Dump Experiments. *Phys. Rev. Lett.* **2018**, *121*, 041802. [CrossRef] [PubMed]
15. Böhlen, T.T.; Cerutt, F.; Chin, M.P.W.; Fassò, A.; Ferrari, A.; Ortega, P.G.; Mairani, A.; Sala, P.R.; Smirnov, G.; Vlachoudis, V. The FLUKA Code: Developments and Challenges for High Energy and Medical Applications. *Nucl. Data Sheets* **2014**, *120*, 211–214. [CrossRef]
16. Bondí, M.; Battaglieri, M.; Carpinelli, M.; Celentano, A.; De Napoli, M.; De Vita, R.; Marsicano, L.; Randazzo, N.; Sipala, V.; Smith, S.S. Large-size CsI(Tl) crystal read-out by SiPM for low-energy charged-particles detection. *Nucl. Instrum. Methods A* **2017**, *867*, 148–153. [CrossRef]
17. Ameli, F.; Battaglieri, M.; Bondì, M.; Capodiferro, M.; Celentano, A.; Chiarusi, T.; Chiodi, G.; De Napoli, M.; Lunadei, R.; Marsicano, L.; et al. A low cost, high speed, multichannel analog to digital converter board. *Nucl. Instrum. Methods A* **2019**, in press. [CrossRef]
18. Agostinelli, S.; Allison, J.; Amako, K.; Apostolakis, J.; Araujo, H.; Arce, P.; Asai, M.; Axen, D.; Banerjee, S.; Barrand, G.; et al. Geant4—A simulation toolkit. *Nucl. Instrum. Methods A* **2003**, *506*, 250–303. [CrossRef]
19. Battaglieri, M.; Bersani, A.; Bracco, G.; Caiffi, B.; Celentano, A.; De Vita, R.; Marsicano, L.; Musico, P.; Osipenko, M.; Panza, F.; et al. [BDX Collaboration]. Dark matter search in a Beam-Dump eXperiment (BDX) at Jefferson Lab: an update on PR12-16-001. *arXiv* **2017**, arXiv:1712.01518v2.
20. Battaglieri, M.; Bersani, A.; Bracco, G.; Caiffi, B.; Celentano, A.; De Vitay, R.; Marsicano, L.; Musico, P.; Panza, F.; Ripani, M.; et al. [BDX Collaboration]. Dark Matter Search in a Beam-Dump eXperiments (BDX) at Jefferson Lab–2018 Update to PR12-16-001. Available online: https://www.jlab.org/exp_prog/proposals/18/C12-16-001.pdf (accessed on 1 June 2018).
21. Battaglieri, M.; Bondí, M.; Celentano, A.; De Napoli, M.; De Vita, R.; Fegan, S.; Marsicano, L.; Ottonello, G.; Parodi, F.; Randazzo, N.; et al. Measurements of the muon flux produced by 10.6 GeV electrons in a beam dump. *Nucl. Instrum. Methods A* **2019**, *925*, 116–122. [CrossRef]

© 2019 by the author. Licensee MDPI, Basel, Switzerland. This article is an open access article distributed under the terms and conditions of the Creative Commons Attribution (CC BY) license (http://creativecommons.org/licenses/by/4.0/).

Article

Quarkonium Phenomenology from a Generalised Gauss Law

David Lafferty [1],* and Alexander Rothkopf [2]

1. Institute for Theoretical Physics, Heidelberg University, Philosophenweg 16, 69120 Heidelberg, Germany
2. Faculty of Science and Technology, University of Stavanger, 4021 Stavanger, Norway; alexander.rothkopf@uis.no
* Correspondence: lafferty@thphys.uni-heidelberg.de

Received: 17 April 2019; Accepted: 13 May 2019; Published: 20 May 2019

Abstract: We present an improved analytic parametrisation of the complex in-medium heavy quark potential derived rigorously from the generalised Gauss law. To this end we combine in a self-consistent manner a non-perturbative vacuum potential with a weak-coupling description of the QCD medium. The resulting Gauss-law parametrisation is able to reproduce full lattice QCD data by using only a single temperature dependent parameter, the Debye mass m_D. Using this parametrisation we model the in-medium potential at finite baryo-chemical potential, which allows us to estimate the $\Psi'/J/\Psi$ ratio in heavy-ion collisions at different beam energies.

Keywords: quarkonium; heavy-quark potential; heavy-ion collisions; quarkonium phenomenology

1. Introduction

The study of heavy-quarkonium—the bound states of a heavy quark anti-quark pair—has become a central tenet in our understanding of strongly interacting matter under extreme conditions in the context of heavy-ion collisions. Experimentally, the decay of heavy quarkonia into di-leptons leaves a clean signal that allows the probing of different stages of the quark gluon plasma (QGP) and ensures the continued importance of heavy quarkonium measurements at future accelerators [1]. On the theory side, the heavy masses of the constituent quarks permits the use of effective field theories (EFTs) to simplify the description of heavy quarkonium behaviour [2]. This powerful framework has led to considerable progress both in direct lattice QCD studies of equilibrated quarkonium as well as in real-time descriptions of their non-equilibrium evolution. The formulation of EFTs relies on a separation of scales inherent to the heavy-quark system, $m_Q \ll m_Q v \ll m_Q v^2$ with m_Q the heavy-quark mass and v its typical velocity, denoted respectively as hard, soft, and ultra-soft. Two additional scales are present, namely the characteristic scale of quantum fluctuations Λ_{QCD} and of thermal fluctuations T. Integrating out the hard scale $\sim m_Q$ from the full Quantum ChromoDynamics (QCD) Lagrangian leaves Non-Relativistic QCD (NRQCD) given in terms of non-relativistic Pauli spinor fields; this can be achieved non-perturbatively. Further integrating out the soft scale $\sim m_Q v$ results in Potential Non-Relativistic QCD (pNRQCD), where the potential governing the quarkonium dynamics enters as a matching coefficient. While the perturbative derivation of pNRQCD has been successfully completed, its non-perturbative definition is still an active field of research.

In the static limit, the EFT-based definition of such a potential has been suggested based on the real-time evolution on the QCD Wilson loop [3]:

$$V(r) = \lim_{t \to \infty} \frac{i \partial_t W_\Box(r,t)}{W_\Box(r,t)}. \qquad (1)$$

The evaluation of Equation (1) in hard thermal loop (HTL) resummed perturbation theory has demonstrated that this potential is a complex quantity [4]. In addition to the well-known Debye screening in the real part, an imaginary part arises owing to Landau damping or gluo-dissociation, depending on the hierarchy of scales present [5]. At high temperatures the former dominates and the potential reads:

$$V_{HTL}(r) = -\tilde{\alpha}_s \left[m_D + \frac{e^{-m_D r}}{r} + iT\phi(m_D r) \right] + \mathcal{O}(g^4), \quad \phi(x) = 2\int_0^\infty dz \frac{z}{(z^2+1)^2}\left(1 - \frac{\sin(xz)}{xz}\right). \quad (2)$$

Here $\tilde{\alpha}_s = C_F g^2/4\pi$ is the rescaled strong coupling constant. It should be emphasised that this potential does not govern the evolution of the bound state wavefunction; instead it evolves the correlator of unequal time wavefunctions. The question of how this potential can be related to the evolution of the wavefunction itself is an active field of research—an open-quantum-systems approach appears to be promising in this regard (see, e.g., [6]).

Significant progress has been made in understanding the equilibrated properties of heavy quarkonium by extracting the heavy quark potential directly from lattice QCD simulations. These works have confirmed that at low temperatures the potential closely resembles the Cornell form [7],

$$V^{\text{vac}}(r) = -\frac{\tilde{\alpha}_s}{r} + \sigma r + c, \quad (3)$$

where σ denotes the string-tension and c an additive constant. Equation (3) already captures the two most prominent features of QCD, namely asymptotic freedom via the running coupling at small distances and confinement via the non-perturbative linear rise. At finite temperature, the same extraction procedure reveals a weakening of the real part as one moves into the deconfined phase, as well as an imaginary part persisting beyond the QCD pseudo-critical temperature. In order to employ these numerical results in computations of quarkonium spectral functions, which inform us of the in-medium properties, we require an accurate analytic parametrisation of the in-medium heavy quark potential—in particular that holds at the lower and more phenomenologically relevant temperatures below the strict validity range of HTL perturbation theory.

To this end, in this contribution we improve upon the work of [8] and utilise the generalised Gauss law to reproduce the in-medium heavy quark potential. The non-perturbative vacuum bound state is described by the Cornell potential in Equation (3) and will be inserted into a weakly coupled deconfined medium characterised by the HTL in-medium permittivity. Taking into account string breaking, we are able to derive expressions for ReV and ImV with a closed and simple functional form. This parametrisation captures the in-medium behaviour of the real and imaginary parts of the lattice-QCD-calculated potential very well, based on a single temperature dependent parameter—the Debye mass m_D. Our new derivation overcomes the main technical limitation of the previous work, namely an ad-hoc assumption about the functional form of the real-space in-medium permittivity.

2. The Gauss Law Potential Model

2.1. A Novel Formulation

The central idea of this approach is to calculate the in-medium modification to the Coulombic and string-like parts of the Cornell potential given in Equation (3). In linear response theory, the electric potential at finite temperature is obtained from its vacuum counterpart via a division in momentum-space by the static dielectric constant [9]:

$$V(\mathbf{p}) = \frac{V^{\text{vac}}(\mathbf{p})}{\varepsilon(\mathbf{p}, m_D)}. \quad (4)$$

The permittivity, defined as an appropriate limit of the real-time in-medium gluon propagator, will encode the medium effects. Equation (4) does not rely on a weak-coupling approximation and remains valid so long as the vacuum field is weak enough to justify the linear response ansatz. The real space equivalent via the convolution theorem is

$$V(\mathbf{r}) = \left(V^{\text{vac}} * \varepsilon^{-1}\right)(\mathbf{r}), \tag{5}$$

where '$*$' represents the convolution. We now consider the other main building block of our approach, the generalised Gauss law,

$$\nabla \cdot \left(\frac{\mathbf{E}^{\text{vac}}}{r^{a+1}}\right) = 4\pi q \delta(\mathbf{r}), \tag{6}$$

which holds for electric fields of the form $\mathbf{E}^{\text{vac}}(r) = -\nabla V^{\text{vac}}(r) = q r^{a-1} \hat{r}$. This reduces to the well-known Coulombic potential for $a = -1, q = \tilde{\alpha}_s$ while the linearly rising string case corresponds to $a = 1, q = \sigma$. For a general a,

$$-\frac{1}{r^{a+1}}\nabla^2 V^{\text{vac}}(r) + \frac{1+a}{r^{a+2}}\nabla V^{\text{vac}}(r) = 4\pi q \delta(\mathbf{r}). \tag{7}$$

Denoting the differential operator on the left-hand-side above as $\mathcal{G}_a$ and applying it to Equation (5), the general integral expressions for each term in the in-medium heavy-quark potential are deduced:

$$\mathcal{G}_a[V(r)] = \mathcal{G}_a \int d^3y \left(V^{\text{vac}}(r-y)\varepsilon^{-1}(y)\right) = 4\pi q \left(\delta * \varepsilon^{-1}\right)(r) = 4\pi q\, \varepsilon^{-1}(r, m_D). \tag{8}$$

Here we have used Equation (7) and that the convolution commutes with $\mathcal{G}_a$. For the Coulombic and string cases respectively, this gives

$$-\nabla^2 V_C(r) = 4\pi \tilde{\alpha}_s\, \varepsilon^{-1}(r, m_D), \quad -\frac{1}{r^2}\frac{d^2 V_S(r)}{dr^2} = 4\pi\sigma\, \varepsilon^{-1}(r, m_D). \tag{9}$$

From the perturbative HTL expression in momentum-space [10],

$$\varepsilon^{-1}(p, m_D) = \frac{p^2}{p^2 + m_D^2} - i\pi T \frac{p m_D^2}{\left(p^2 + m_D^2\right)^2}, \tag{10}$$

the expression for the coordinate space in-medium permittivity is obtained by inverse Fourier transform. Now, using Equation (10) to solve for the in-medium modified Coulombic part of the potential, we find that our ansatz reproduces the HTL result

$$\text{Re}V_C(r) = -\tilde{\alpha}_s \left[m_D + \frac{e^{-m_D r}}{r}\right], \quad \text{Im}V_C(r) = -\tilde{\alpha}_s\left[iT\phi(m_D r)\right], \tag{11}$$

with ϕ as defined in Equation (2). The next step is to turn to the string part, for which the formal solution can be immediately written down as

$$V_S(r) = c_0 + c_1 r - 4\pi\sigma \int_0^r dr' \int_0^{r'} dr''\, r''^2 \varepsilon^{-1}(r'', m_D). \tag{12}$$

The constants c_0 and c_1 will be chosen to ensure the physically motivated boundary conditions $\text{Re}V_S(r)|_{r=0} = 0$, $\text{Im}V_S(r)|_{r=0} = 0$ and $\partial_r \text{Im}V_S(r)|_{r=0} = 0$. This leads to the following analytical form:

$$\text{Re}V_S(r) = \frac{2\sigma}{m_D} - \frac{e^{-m_D r}(2 + m_D r)\sigma}{m_D}, \quad \text{Im}V_S(r) = \frac{\sqrt{\pi}}{4} m_D T \sigma\, r^3\, G_{2,4}^{2,2}\left(\begin{array}{c}-\tfrac{1}{2},-\tfrac{1}{2}\\ \tfrac{1}{2},\tfrac{1}{2},-\tfrac{3}{2},-1\end{array}\Big|\tfrac{1}{4}m_D^2 r^2\right), \tag{13}$$

where G denotes the Meijer-G function. In the real parts the short distance limit $r \to 0$ recovers the Cornell potential as does the zero temperature limit $m_D \to 0$. At large distances $\text{Re}V_C(r)$ displays an exponential decay $\sim e^{-m_D r}$ (i.e., Debye screening) while $\mathrm{Im}V_C(r)$ asymptotes to a constant which is expected for Landau damping. Only the imaginary string part in Equation (13), at first sight appears problematic as it diverges logarithmically at large r. We argue that this is a manifestation of the absence of an explicit string breaking in the original vacuum Cornell potential.

In the preceding computation the explicit expression for $\mathrm{Im}V_S$ can be written, after substituting the imaginary part of Equation (10) into Equation (12) and performing the angular integration of the inverse Fourier transform, as follows:

$$\mathrm{Im}V_S(r) = c_0 + c_1 r + 2T\sigma m_D^2 \int_0^r dr' \int_0^{r'} dr'' \, r''^2 \int_0^{\infty} dp \, p^2 \, \frac{\sin(pr'')}{pr''} \, p^2 \, \frac{1}{p(p^2 + m_D^2)^2}. \quad (14)$$

We have arranged the momentum factors as above to make clear their different origins: the first term (p^2) arises from integrating in spherical coordinates and the second ($\mathrm{sinc}(pr'')$) after completing the polar integration. The last two terms are contributions from the in-medium permittivity. It is the $1/p$ factor here that we identify as causing the weak infrared divergence. In order to regularise, we modify this term as

$$\frac{1}{p(p^2 + m_D^2)^2} \to \frac{1}{\sqrt{p^2 + \Delta^2}(p^2 + m_D^2)^2}, \quad (15)$$

where Δ will be a suitably chosen regularisation scale. In Equation (14) the spatial integrals can be carried out analytically, which combined with the regularisation above gives our new definition of the string imaginary part:

$$\mathrm{Im}V_S(r) = 2T\sigma m_D^2 \int_0^{\infty} dp \, \frac{2 - 2\cos(pr) - pr\sin(pr)}{\sqrt{p^2 + \Delta^2}(p^2 + m_D^2)^2}, \quad (16)$$

after imposing the boundary conditions stated above Equation (13). The only remaining step is to determine the regularisation scale Δ. To do so, note that if we rescale momentum $p \to p/m_D$ and slightly rearrange, Equation (16) takes on a suggestive form:

$$\mathrm{Im}V_S(r) = \frac{\sigma T}{m_D^2} \chi(m_D r), \quad \chi(x) = 2\int_0^{\infty} dp \, \frac{2 - 2\cos(px) - px\sin(px)}{\sqrt{p^2 + \Delta_D^2}(p^2 + 1)^2}, \quad (17)$$

with $\Delta_D = \Delta/m_D$. That is, we can express $\mathrm{Im}V_S(r)$ using a temperature dependent prefactor with dimensions of energy, multiplied by a dimensionless momentum integral. This is very similar to the Coulombic expression, where the integral asymptotes to unity in the limit $r \to \infty$. We thus impose the same condition for the string part. This procedure also recovers the correct behaviour at large T (large m_D), i.e., the string contribution to the imaginary part diminishes until the HTL result is recovered. The value of the regularisation parameter Δ_D can be computed numerically. Furthermore, since it is expressed in terms of the Debye mass it remains constant and the computation need only be performed once. It is found that $\Delta_D = \Delta/m_D \simeq 3.0369$ gives $\chi(\infty) \simeq 1$ and thus Equation (17) represents the final closed form of a physically consistent in-medium string imaginary part.

2.2. Vetting with Lattice QCD Data

The most important benchmark for any description of the in-medium heavy quark potential is its ability to reproduce the non-perturbative lattice QCD results. This vetting process is carried out here against potential values [11] calculated on finite temperature ensembles generated by the HotQCD collaboration on $48^3 \times 12$ lattices with $N_f = 2+1$ flavours of dynamical light quarks discretised with the asqtad action [12]. The pion mass on these lattices is $m_\pi \approx 300$ MeV and the QCD transition temperature is $T_C \approx 175$ MeV.

Following the steps in [13], we first calibrate the vacuum parameters by fitting the Cornell potential to the two low-temperature ensembles included in the lattice dataset. As in that study, the

Cornell ansatz gives an excellent fit. The entire temperature dependence in our parametrisation then enters only via the Debye mass m_D, which will be fit using only the real part. The imaginary data points can be used as a cross-check. Note that since the heavy quark potential is a generic quantity that is unspecific to either of the heavy quark families, this fit need only be performed once.

The results are shown in Figure 1. From the left panel we see that the Gauss law parametrisation provides an excellent fit, capturing the behaviour of the non-perturbative data points from the Coulombic region at small r through the intermediate region and up to the screening regime at high temperature and large distances. Furthermore, the central panel shows a good agreement between the Gauss law predictions and corresponding tentative values of the imaginary part extracted from the lattice. The predicted values lie within the considerable errors of the lattice $\mathrm{Im}V_S(r)$ for all but the lowest temperature. We observe that the imaginary part from the Gauss law rises more steeply with increasing temperature but the asymptotic value at large distances behaves non-monotonously, reflecting the competing Coulombic and string parts. The best fit values of m_D are shown in the right panel. We conclude that our novel parametrisation captures the relevant physics encoded within the non-perturbative in-medium potential.

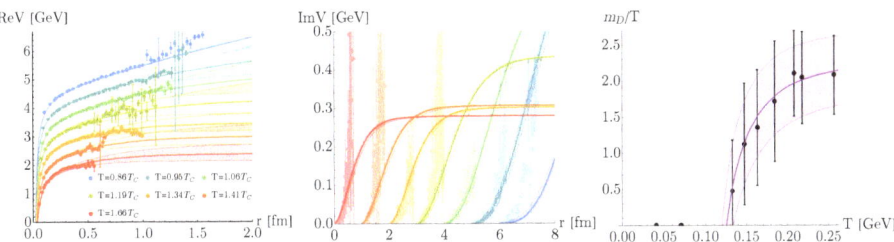

Figure 1. Gauss-law parametrisation and the lattice QCD potential. (**left**) Real part (symbols) and best fit results (solid lines). (**centre**) Tentative imaginary part (symbols) and the Gauss-law prediction (solid lines). Errorbands from uncertainty in both the $T > 0$ fit and the vacuum parameters. (**right**) Best fit values of the Debye mass and interpolation.

3. Phenomenology

3.1. Spectral Functions at Finite Temperature

The next natural step is to employ our validated Gauss law potential model in a realistic investigation of heavy quarkonium in-medium behaviour. As we have calibrated the Debye mass temperature dependence against lattice data with an unphysical pion mass, we first must carry out a continuum extrapolation. Since this has not been rigorously achieved so far we resort to using continuum corrections as outlined in detail in [13]. The outcome is a set of phenomenological vacuum parameters for the Cornell potential, which in our case read

$$\tilde{\alpha}_s = 0.513 \pm 0.0024 \text{ GeV}, \quad \sqrt{\sigma} = 0.412 \pm 0.0041 \text{ GeV}, \quad c = -0.161 \pm 0.0025 \text{ GeV}, \qquad (18)$$

to be used in conjunction with a "fit" of the charm mass $m_c^{\text{fit}} = 1.4692$ GeV. The continuum corrected values for the Debye mass parameter are interpolated via the HTL inspired ansatz

$$m_D(T) = Tg(\Lambda)\sqrt{\frac{N_c}{3} + \frac{N_f}{6}} + \frac{N_c Tg(\Lambda)^2}{4\pi}\log\left(\frac{1}{g(\Lambda)}\sqrt{\frac{N_c}{3} + \frac{N_f}{6}}\right) + \kappa_1 Tg(\Lambda)^2 + \kappa_2 Tg(\Lambda)^3. \qquad (19)$$

Here, the first and second term respectively are the leading order perturbative result plus logarithmic correction in $SU(N_c)$ with N_f fermions, $m_{u,d} = 0$, and at zero baryon chemical potential. κ_1 and κ_2 absorb the non-perturbative corrections, which in our case take the values $\kappa_1 = 0.686 \pm 0.221$

and $\kappa_2 = -0.317 \pm 0.052$. The resulting interpolation for m_D is shown as the purple band in the right panel of Figure 1.

With these corrections in place, we may now calculate realistic quarkonium spectral functions at finite temperature by solving the appropriate Schrödinger equation using the Fourier space method as described in [14].

In Figure 2 we show the results for S-wave charmonium states, which exhibit the characteristic broadening of in-medium peaks and their shifts to lower frequencies. This corresponds to the in-medium state being lighter than the vacuum state, while at the same time being less strongly bound. The in-medium modification is shown quantitatively in Figure 3. In the following section we look at phenomenological extensions and will focus on charmonium where it is expected that our model will be most applicable.

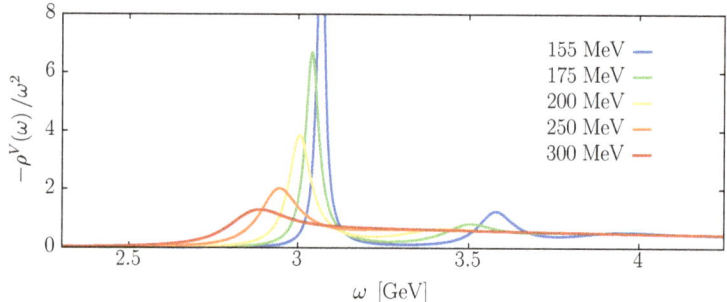

Figure 2. Illustrative spectral functions for S-wave Charmonium.

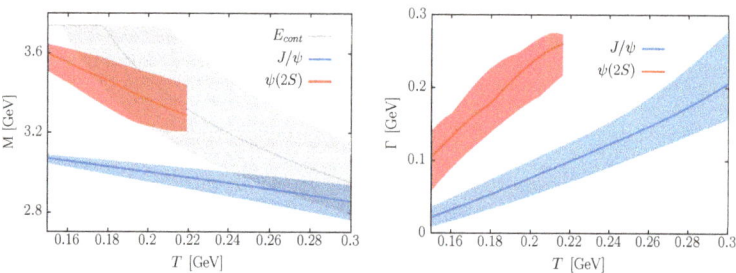

Figure 3. Thermal mass (**left**) and spectral width (**right**) of charmonium as a function of temperature. The error bands denote the Debye mass uncertainty arising from the fitting procedure. The continuum threshold energy on the left figure is defined as $\mathrm{Re}V(r \to \infty)$.

3.2. Applications to Heavy Ion Collisions

An observable of current interest at RHIC and LHC is the production ratio of Ψ' to J/Ψ particles. The reason is that it is expected to be highly discriminatory among different phenomenological models. Using thermal in-medium quarkonium spectral functions this ratio has already been estimated at vanishing baryo-chemical potential in [13], showing good agreement with prediciton from the statistical model of hadronisation. Here we wish to extend the computation of the ratio to different (lower) beam energies, relevant for future collider facilities such as FAIR and NICA.

We require a prescription to evaluate our Gauss law potential model at a given centre-of-mass energy. The strategy here is two-fold. Firstly, we note that the statistical hadronisation model already provides a well-established scheme with which to estimate the thermal parameters (temperature and baryo-chemical potential μ_B) of the produced bulk medium at chemical freeze-out with a given $\sqrt{s_{NN}}$. The most recent results [15] are:

$$T(\sqrt{s_{NN}}) = \frac{158 \text{ MeV}}{1 + \exp(2.60 - \ln(\sqrt{s_{NN}})/0.45)}, \qquad \mu_B(\sqrt{s_{NN}}) = \frac{1307.5 \text{ MeV}}{1 + 0.288\sqrt{s_{NN}}}, \qquad (20)$$

where $\sqrt{s_{NN}}$ is the dimensionless numerical value of the centre-of-mass energy measured in GeV.

Secondly, since the physical information within our potential model is captured entirely by the dependence on the Debye mass m_D, we need only modify m_D to include the effects on finite baryo-chemical potential. At leading order, the Debye mass can be calculated perturbatively at finite baryo-chemical potential [16]. As a first step, we propose to add this μ_B-term to the temperature dependence of the Debye mass in Equation (19). The result is:

$$m_D(T, \mu_B) = \sqrt{m_D(T, 0)^2 + T^2 g(\Lambda)^2 \frac{N_f}{18\pi^2} \frac{\mu_B^2}{T^2}}. \qquad (21)$$

Here, the renormalisation scale is now $\Lambda = 2\pi\sqrt{T^2 + \mu_B^2/\pi^2}$. At high μ_B the chemical potential itself becomes the only relevant scale and a similar (linear) dependence of m_D is expected. This leads us to adopt Equation (21) over the entire finite baryo-chemical potential regime. In the absence of reliable lattice data at finite chemical potential, we hold the non-perturbative constants κ_1 and κ_2 in Equation (19) the same.

With all ingredients now in place, we may now compute the compute the $\Psi'/J/\Psi$ ratio over a range of centre-of-mass energies. Through Equations (21) and (20) we scan the $\sqrt{s_{NN}}$ range and update the Debye mass that encodes the physics of our potential model. The in-medium spectral functions are calculated in the same manner as Section 3.1 and finally, the number ratio is estimated via the procedure in [13]—assuming an instantaneous freeze-out scenario where all in-medium bound states are projected onto the corresponding vacuum state. The final ratio is expressed as

$$\left.\frac{N_{\Psi'}}{N_{J/\Psi}}\right|_{\sqrt{s_{NN}}} = \left.\frac{R_{\ell\ell}^{\Psi'}}{R_{\ell\ell}^{J/\Psi}}\right|_{\sqrt{s_{NN}}} \times \frac{M_{\Psi'}^2 |\psi_{J/\Psi}(0)|^2}{M_{J/\Psi}^2 |\psi_{\Psi'}(0)|^2}, \quad R_{\ell\ell}^{\Psi_n} \propto A_n \int d^3\mathbf{p} \, n_B\left(\sqrt{M_n^2 + \mathbf{p}^2}\right) \frac{M_n}{\sqrt{M_n + \mathbf{p}^2}}. \qquad (22)$$

Here, M_n is the thermal mass of the state, i.e., the frequency at which the corresponding spectral peak occurs and A_n is the area underneath the peak. The second factor on the right-hand-side of Equation (22) is the square of the $T = 0$ wavefunction at $r = 0$ divided by the square of the mass of each state and is required to obtain the total number density from $R_{\ell\ell}^{\Psi_n}$, which only includes electromagnetic decays [17].

The final results from this entire procedure are plotted in Figure 4, together with the prediction by the statistical hadronisation model. Our analysis shows very good agreement with both the statistical model and the latest experimental results, strengthening the interpretation that charm quarks thermalise before reaching the freeze-out boundary.

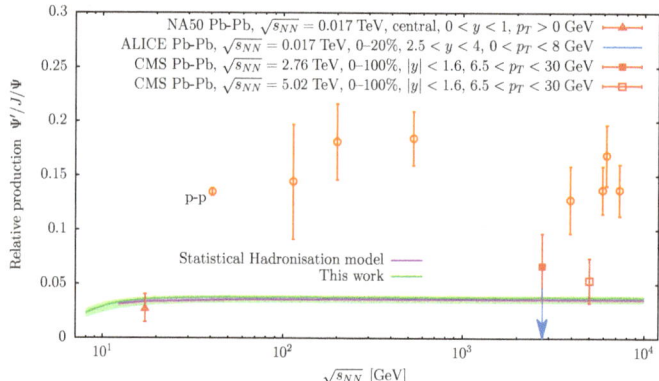

Figure 4. The prediction of this work (green) for the relative production yield of Ψ' to J/Ψ. We also include the statistical hadronisation model prediction [15] (purple) and experimental data measured by the NA50 [18], ALICE [19], and CMS [20,21] collaborations (red) for Pb–Pb collisions, as well as the pp baseline [15,22] (orange).

4. Conclusions

We have presented an improved parametrisation of the in-medium heavy quark potential by employing a generalised Gauss law ansatz in linear response theory. The resulting analytic expressions depended only on a single temperature dependent parameter and were able to quantitatively reproduce the lattice results for the real part of the potential. The resulting imaginary part showed an unphysical logarithmic divergence which we attributed to the equally unphysical unending linear rise of the vacuum Cornell potential. By regularising this artefact, we were able to give physically sound predictions for the imaginary part that in turn qualitatively matched the lattice data. Furthermore, our prescription can be easily extended to model a finite baryo-chemical potential, a region currently inaccessible to lattice QCD simulations. Using the values for μ_B obtained in the statistical model of hadronisation we computed Ψ' to J/Ψ production yield ratio for different beam energies. The extension of the Gauss-law parametrisation to finite velocity remains work in progress.

Author Contributions: conceptualization, A.R.; formal analysis, D.L.; writing–original draft preparation, D.L.; writing–review and editing, A.R.

Funding: This study is part of and supported by the DFG Collaborative Research Centre "SFB 1225 (ISOQUANT)"

Acknowledgments: We are grateful to Anton Andronic for providing the statistical hadronisation model results.

Conflicts of Interest: The authors declare no conflict of interest.

References

1. Andronic, A.; Arleo, F.; Arnaldi, R.; Beraudo, A.; Bruna, E.; Caffarri, D.; Conesa del Valle, Z.; Contreras, J.G.; Dahms, T.; Dainese, A.; et al. Heavy-flavour and quarkonium production in the LHC era: From proton–proton to heavy-ion collisions. *Eur. Phys. J.* **2016**, *76*, 107. [CrossRef] [PubMed]
2. Brambilla, N.; Pineda, A.; Soto, J.; Vairo, A. Effective-field theories for heavy quarkonium. *Rev. Mod. Phys.* **2005**, *77*, 1423. [CrossRef]
3. Burnier, Y.; Rothkopf, A. Disentangling the timescales behind the nonperturbative heavy quark potential. *Phys. Rev. D* **2012**, *86*, 051503. [CrossRef]
4. Laine, M.; Philipsen, O.; Tassler, M.; Romatschke, P. Real-time static potential in hot QCD. *J. High Energy Phys.* **2007**, *3*, 054. [CrossRef]
5. Brambilla, N.; Ghiglieri, J.; Vairo, A.; Petreczky, P. Static quark-antiquark pairs at finite temperature. *Phys. Rev. D* **2008**, *78*, 014017. [CrossRef]

6. Kajimotoa, S.; Akamatsua, Y.; Asakawaa, M.; Rothkopf, A. Quantum dynamical dissociation of quarkonia by wave function decoherence in quark-gluon plasma. *Nucl. Phys. A* **2019**, *982*, 711–714. [CrossRef]
7. Petreczky, P.; Rothkopf, A.; Weber, J. Realistic in-medium heavy-quark potential from high statistics lattice QCD simulations. *Nucl. Phys. A* **2019**, *982*, 735–738. [CrossRef]
8. Burnier, Y.; Rothkopf, A. A gauge invariant Debye mass and the complex heavy-quark potential. *Phys. Lett. B* **2016**, *753*, 232–236. [CrossRef]
9. Kapusta, J.I.; Gale, C. Finite-Temperature Field Theory. Cambridge University Press: Cambridge, UK, 2006.
10. Thakur, L.; Kakade, U.; Patra, B.K. Dissociation of quarkonium in a complex potential. *Phys. Rev. D* **2014**, *89*, 094020. [CrossRef]
11. Burnier, Y.; Kaczmarek, O.; Rothkopf, A. Static Quark-Antiquark Potential in the Quark-Gluon Plasma from Lattice QCD. *Phys. Rev. Lett.* **2015**, *114*, 082001. [CrossRef]
12. Bazavov, A.; Bhattacharya, T.; Cheng, M.; DeTar, C.; Ding, H.-T.; Gottlieb, S.; Gupta, R.; Hegde, P.; Heller, U.M.; Karsch, F.; et al. The chiral and deconfinement aspects of the QCD transition. *Phys. Rev. D* **2012**, *85*, 054503. [CrossRef]
13. Burnier, Y.; Kaczmarek, O.; Rothkopf, A. Quarkonium at finite temperature: towards realistic phenomenology from first principles. *J. High Energy Phys.* **2015**, *12*, 1–34. [CrossRef]
14. Burnier, Y.; Laine, M.; Vepsalainen, M. Heavy quarkonium in any channel in resummed hot QCD. *J. High Energy Phys.* **2008**, *1*, 43. [CrossRef]
15. Andronic, A.; Braun-Munzinger, P.; Redlich, K., Stachel, J. Decoding the phase structure of QCD via particle production at high energy. *Nature* **2018**, *561*, 321–330. [CrossRef]
16. Laine, M.; Vuorinen, A. Basics of Thermal Field Theory. *Lect. Notes Phys.* **2016**, *925*, 1–281.
17. Bodwin, G.T.; Braaten, E.; Lepage, G.P. Rigorous QCD analysis of inclusive annihilation and production of heavy quarkonium. *Phys. Rev. D* **1995**, *51*, 1125–1171. [CrossRef]
18. Alessandro, B. et al. [NA50 Collaboration] Ψ' production in Pb–Pb collisions at 158 GeV/nucleon. *Eur. Phys. J. C* **2007**, *49*, 559–567. [CrossRef]
19. Adam, J. et al. [ALICE Collaboration] Differential studies of inclusive J/Ψ and $\Psi(2S)$ production at forward rapidity in Pb-Pb collisions at $\sqrt{s_{NN}} = 2.76$ TeV. *J. High Energy Phys.* **2016**, *5*, 179. [CrossRef]
20. Khachatryan, V. et al. [CMS Collaboration] Measurement of Prompt $\psi(2S)$ to J/ψ Yield Ratios in Pb-Pb and p-p Collisions at $\sqrt{s_{NN}} = 2.76$ TeV. *Phys. Rev. Lett.* **2014**, *113*, 262301. [CrossRef]
21. Sirunyan, A.M. et. al. [CMS Collaboration] Relative Modification of Prompt $\psi(2S)$ and J/ψ Yields from pp to PbPb Collisions at $\sqrt{s_{NN}} = 5.02$ TeV. *Phys. Rev. Lett.* **2017**, *118*, 162301. [CrossRef]
22. Drees, A. et al. [PHENIX Collaboration] Relative Yields and Nuclear Modification of Ψ' to J/Ψ mesons in p+p, p(3He)+A Collisions at $\sqrt{s_{NN}} = 200$ GeV, measured in PHENIX. *Nucl. Particle Phys. Proc.* **2017**, *289–290*, 417–420. [CrossRef]

© 2019 by the authors. Licensee MDPI, Basel, Switzerland. This article is an open access article distributed under the terms and conditions of the Creative Commons Attribution (CC BY) license (http://creativecommons.org/licenses/by/4.0/).

Communication

Correlation of Heavy and Light Flavors in Simulations †

Eszter Frajna [1,2,*] and Róbert Vértesi [2]

[1] Department of Physics, Budapest University of Technology and Economics, Műegyetem rkp. 3, 1111 Budapest, Hungary
[2] Wigner Research Centre for Physics, Konkoly-Thege Miklós út 29-33, 1121 Budapest, Hungary; vertesi.robert@wigner.mta.hu
* Correspondence: frajna.eszter@wigner.mta.hu
† This paper is based on the talk at the 18th Zimányi School, Budapest, Hungary, 3–7 December 2018.

Received: 25 March 2019; Accepted: 16 May 2019; Published: 20 May 2019

Abstract: The ALICE experiment at the Large Hadron Collider (LHC) ring is designed to study the strongly interacting matter at extreme energy densities created in high-energy heavy-ion collisions. In this paper we investigate correlations of heavy and light flavors in simulations at LHC energies at mid-rapidity, with the primary purpose of proposing experimental applications of these methods. Our studies have shown that investigating the correlation images can aid the experimental separation of heavy quarks and help understanding the physics that create them. The shape of the correlation peaks can be used to separate the electrons stemming from b quarks. This could be a method of identification that, combined with identification in silicon vertex detectors, may provide much better sample purity for examining the secondary vertex shift. Based on a correlation picture it is also possible to distinguish between prompt and late contributions to D meson yields.

Keywords: angular correlations; jet structure; heavy flavor; high-energy collisions

1. Introduction

Quark-gluon plasma (QGP) is a state of matter which exists at extremely high temperatures and densities, where quarks are no longer confined to hadrons [1]. Initially, the Universe was filled with this hot and dense matter. Quark-gluon plasma can be recreated in high-energy heavy-ion collisions at large accelerator rings such as the LHC.

The angular correlation data of the STAR experiment [2] shows a strong suppression of the away-side correlation peak in central Au+Au collisions, while no such effect can be observed in p+p collisions. This indicates a strong quenching of jets that traverse the QGP. In the case of d+Au collision, we do not experience jet-suppression even though there is cold hadronic nuclear matter present. It was among the first convincing shreds of evidence of hot and dense strongly interacting nuclear matter in the final state. The interaction of partons with quark–gluon plasma is often studied by full jet reconstruction. However, in heavy-ion collisions, high background from the underlying event makes it difficult to reconstruct jets below a certain momentum. Measuring the angular correlation of particles is a technique that solves this problem. Comparison of angular correlations in small and large collision systems reveals information of jet modification by the strongly interacting medium.

Heavy-flavor (charm and beauty) quarks are an excellent tool to study heavy-ion collisions. Most of them are created in the initial stages of the reaction, and their long lifetime ensures that they interact both with the hot and dense medium as well as with the cold hadronic matter before they decay. Identifying characteristic correlation images of heavy and light quarks can help understand flavor-dependent fragmentation. Furthermore, finding these characteristic shapes can be used as an alternative method to pin down feed-down of beauty hadrons into charm without the need to find

Universe **2019**, *5*, 118; doi:10.3390/universe5050118   www.mdpi.com/journal/universe

the secondary vertex. We carried out studies with the PYTHIA 8.1 Monte Carlo event generator [3] to compare the near-side and away-side correlation peaks associated with hadrons from different heavy quarks from a given p_T. The following simulation results serve as a case study that the techniques we developed can be successfully applied in today's large experiments such as ALICE at the LHC to aid and interpret heavy-flavor correlation measurements.

2. Analysis Method

We used the PYTHIA 8.1 Monte Carlo event generator [3] to simulate hard QCD events using the default Monash 2013 [4] settings for LHC p + p data. Five million p − p collision events were simulated at a time, at the center-of-mass energy of $\sqrt{s} = 7$ TeV. The phase space has been reduced so that the leading hard process has at least 5 GeV/c momentum. Heavy flavor scale settings were used similarly to recent STAR analyses, e.g., that in [5]. For beauty and charm simulations, only the two-to-two processes were enabled ($gg \to b\bar{b}$ and $qq \to b\bar{b}$, as well as $gg \to c\bar{c}$ and $qq \to c\bar{c}$, respectively), while light hadron correlations were simulated by allowing all hard quantum color dynamics processes.

The variables we use to describe particle kinematics are the three-momentum (p_x, p_y, p_z), the azimuth angle ($\varphi = \arctan(\frac{p_y}{p_x})$) and the pseudorapidity ($\eta = \frac{1}{2}\ln\frac{p+p_z}{p-p_z}$, where $p_T = \sqrt{p_x^2 + p_y^2}$ is the transverse momentum).

To examine the correlation of particles, we followed a technique similar to [6]. We selected a trigger particle from a given momentum (p_T) range, and associated particles from a lower momentum (p_T) window to examine all other particles from the same event. Then we calculated the differences in the azimuth angles as well as the pseudorapidities of the trigger and associated particles within each event. The plane ($\Delta\eta, \Delta\varphi$) can be divided into two parts along $\Delta\varphi$: near-side and away-side peak region. We defined near-side as the range from $-\pi/2$ to $\pi/2$, and the away-side as the range from $\pi/2$ to $3\pi/2$. Near-side correlations give information about the structure of jets since, after background subtraction, most trigger and associated particles come from the same jet. Differences between correlations of beauty, charm and light flavor provide valuable information about flavor-dependent jet fragmentation. Away-side correlation is mostly from back-to-back jet pairs and it is sensitive to the underlying hard processes.

We examined the width of fitted functions in the $5 < p_T^{trigger} < 8$ GeV/c trigger particle transverse momentum range, and in different associated particle transverse momentum (p_T^{assoc}) ranges ($1 < p_T^{assoc} < 2$ GeV/c, $2 < p_T^{assoc} < 3$ GeV/c, $3 < p_T^{assoc} < 5$ GeV/c and $5 < p_T^{assoc} < 8$ GeV/c). In the latter case we ensured that $p_T^{trigger} > p_T^{assoc}$.

Correlation analyses usually use the mixed-event technique to correct for the finite size of the detector in η [6]. However, at mid-rapidity it is sufficient to assume a uniform track distribution and therefore we apply a reweighting of events with a "tent-shaped" function $\frac{dN}{d\eta} = \frac{1}{2A} - \frac{|\eta|}{4A^2}$ where $A = 2$ is the maximal acceptance in which tracks were recorded. In the analysis, we only used the $|\Delta\eta| < 1.6$ range.

Processes that fulfilled the conditions of the central limit theorem yield Gaussian-shaped correlation peaks. In case of weakly decaying resonances corresponding to secondary vertices that can be displaced with several millimeters, however, the Gaussian shape is not necessarily adequate. Therefore we also apply a Generalized Gaussian fit on the near- and away-side correlation peaks. Gaussian functions of the different projections are

$$f(\Delta\varphi) = N \cdot \frac{1}{\sqrt{2\pi}\sigma_{\Delta\varphi}} \cdot e^{-(\frac{\Delta\varphi^2}{2\sigma_{\Delta\varphi}^2})}, \quad f(\Delta\eta) = N \cdot \frac{1}{\sqrt{2\pi}\sigma_{\Delta\eta}} \cdot e^{-(\frac{\Delta\eta^2}{2\sigma_{\Delta\eta}^2})}, \quad (1)$$

and the generalized Gaussian functions of the different projections are

$$g(\Delta\varphi) = N \cdot \frac{\gamma_{\Delta\varphi}}{2\omega_{\Delta\varphi}\Gamma(\frac{1}{\gamma_{\Delta\varphi}})} \cdot e^{-(\frac{|\Delta\varphi|}{\omega_{\Delta\varphi}})^{\gamma_{\Delta\varphi}}}, \quad g(\Delta\eta) = N \cdot \frac{\gamma_{\Delta\eta}}{2\omega_{\Delta\eta}\Gamma(\frac{1}{\gamma_{\Delta\eta}})} \cdot e^{-(\frac{|\Delta\eta|}{\omega_{\Delta\eta}})^{\gamma_{\Delta\eta}}}, \quad (2)$$

where N is the normalization factor, $\sigma_{\Delta\varphi}$ is the width of the peak in the direction of $\Delta\varphi$, and $\sigma_{\Delta\eta}$ is the width of the peak in the direction of $\Delta\eta$.

The generalized Gaussian function has an extra parameter γ compared to the Gaussian. If $\gamma = 1$, then the generalized Gaussian function is an exponential function. If $\gamma = 2$, it was reduced to a regular Gaussian function. If γ was greater than two then the top of the function was flattened.

3. Results

3.1. Correlations of Light Charged Hadrons

As a first test we reproduced the near- and away-side correlation peaks of light charged hadrons ($\pi^{\pm}$, $K^{\pm}$, p and $\bar{p}$) both in the $\Delta\eta$ and the $\Delta\varphi$ directions. For all the particles we investigated in this study, we only examined the near-side peak in the direction of $\Delta\eta$, and in the direction of $\Delta\varphi$ we investigated both peaks. Below, the parameters were shown in different p_T^{assoc} ranges.

The left panel in Figure 1 represents the peak in $\Delta\eta$ with near-side cut ($|\Delta\Phi|<\pi/2$) with generalized Gaussian fit. The parameter γ was unity within uncertainties, indicating a distribution that was significantly sharper than Gaussian and consistent with an exponential function. The right panel in Figure 1 shows the near- and away-side peaks with Gaussian fits in $\Delta\varphi$ direction. The shape of the peaks were well described by a Gaussian.

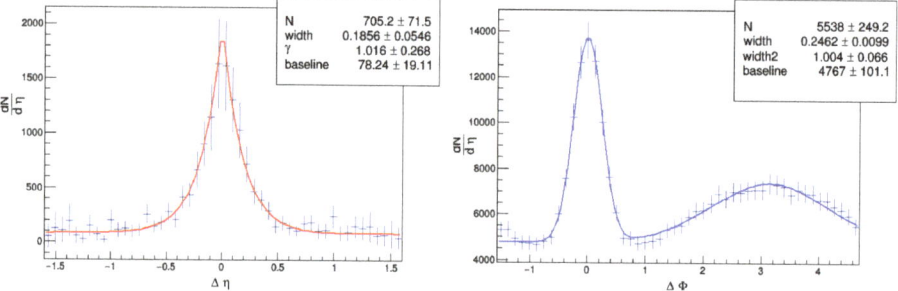

Figure 1. The peak in $\Delta\eta$ with near-side cut ($|\Delta\Phi|<\pi/2$) of light charged hadrons with generalized Gaussian fitting (on the **left** side) and the near-and away-side peaks of light charged hadrons in $\Delta\varphi$ plane with Gaussian fitting (on the **right** side).

To get a more comprehensive picture, we also showed the fit parameters in function of p_T. The left panel in Figure 2 shows the peak width for Gaussian fitting, while the right panel in Figure 2 shows the peak width for generalized Gaussian fitting and the γ parameter of the function. Error bars represent the uncertainties of the fit parameters. The correlation peaks of the light charged hadrons were getting narrower towards higher p_T and γ was constant within uncertainties.

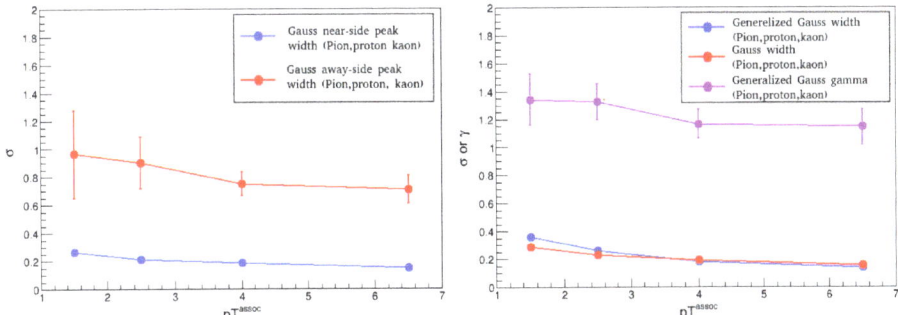

Figure 2. The peak width for Gaussian fitting (on the **left** side) and for generalized Gaussian fitting (on the **right** side) for $5 < p_T^{trigger} < 8$ GeV/c and different p_T^{assoc} values.

3.2. Prompt Production of Heavy Flavor Mesons

We examined the direct decay cases of D mesons from c quarks and B mesons from b quarks, without feed-down. In Figure 3 both peaks were consistent with a Gaussian. The only exception is B mesons with $5 < p_T^{assoc} < 8$ GeV, where the fit is significantly narrower than a Gaussian.

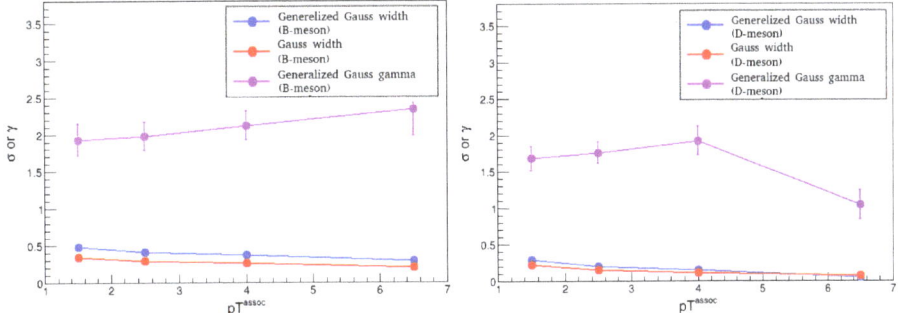

Figure 3. B mesons from b quarks (on the **left** side) and D mesons from c quarks (on the **right** side) and for $5 < p_T^{trigger} < 8$ GeV/c and different p_T^{assoc} values.

3.3. D Meson from the Decay of the B Meson

We investigated D mesons from the decay of B mesons. The left panel in Figure 4 shows a strong dependence of the away-side peak on p_T (not observed in light flavor or prompt heavy flavor production). The right panel in Figure 4 γ decreases with p_T, together with γ. (Peaks are getting both narrower and γ is less than two towards high p_T).

Based on Figures 3 and 4, it is possible to distinguish between prompt D mesons and non-prompt D mesons from decays of B mesons. The statistical separation of these two contributions allows for the understanding of the flavor-dependence of heavy-quark energy loss within the Quark-gluon plasma. The effect of Quark-gluon plasma on heavy quarks can be investigated by the ratio of D meson from b quarks and c quarks.

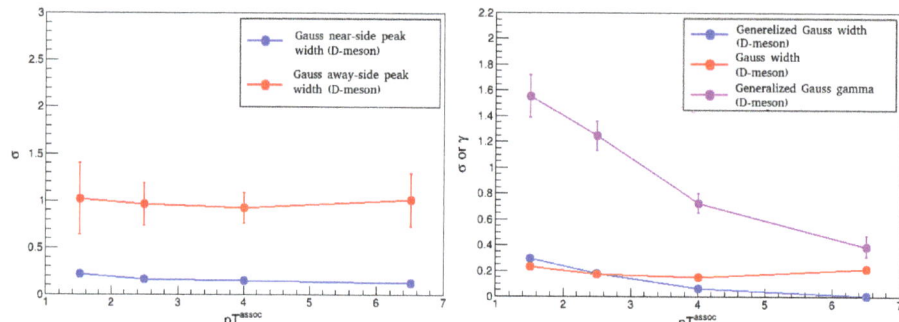

Figure 4. The peak width for Gaussian fitting (on the **left** side) and for generalized Gaussian fitting (on the **right** side) for D meson from the decay of the B meson for $5 < p_T^{trigger} < 8$ GeV/c and different p_T^{assoc} values.

3.4. Investigation of Electrons from B Mesons

We investigated the electrons from B mesons. Electrons can come directly from semileptonic B-decays, as well as semileptonic decays of charmed mesons that B feeds down into. Still, these branching ratios are in the order of a couple of percents [7], so the electron yield is relatively low. Therefore, the lack of statistics was a limit in the analysis. However, the results are suitable for drawing conclusions.

Figure 5 shows that correlations of B meson decay electron with hadron produce wider correlation peaks than in the c quark decay electron case. There's no significant dependence of γ on p_T^{assoc} and $\gamma \sim 2$.

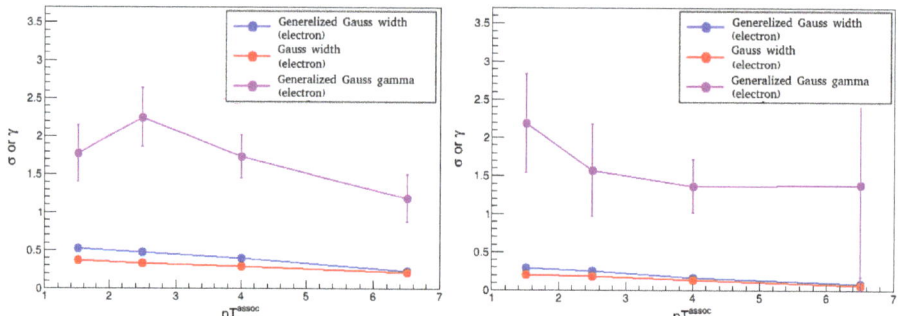

Figure 5. Generalized Gaussian fitting function for electrons from B mesons (**left**) and c quarks (**right**), for $5 < p_T^{trigger} < 8$ GeV/c and different p_T^{assoc} values.

It can be assumed that the significant deviation of the peaks from Gaussian and the strong momentum dependence of the parameters can be traced back to the decay kinematics, in which the momentum of the b quark and the location of the secondary vertex play a role. It is a well-known phenomenon that long-range correlation components from long-lived resonances lead to Lévy-like distributions rather than Gaussian [8]. It seems possible to separate the electrons coming from B mesons simply by the shape of the correlation peaks. This could be a method of identification that, combined with particle identification based on secondary vertex reconstruction in silicon tracking detectors such as the ALICE ITS [9], may provide a much better sample purity than what we can currently achieve.

3.5. Comparison of B-Meson and b-Quark Correlations

Finally, we compared correlations of B mesons with hadrons to b quarks with hadrons. The b quarks were taken directly from parton-level Monte Carlo truth information. In an experiment, correlations with b quarks can be constructed by taking the jet axis of b-tagged jets [10], a process that is very problematic at low momenta, especially in heavy ion collisions. We expect that there will be no significant difference between the two, as the only source of B meson is the b quark decay, besides the b quark direction is close to the axis of the b-jet, and due to its large mass the direction of the b quark momentum will determine the direction of the B meson momentum. However, higher precision is needed to verify whether the away-side peak of b quark to hadron correlations follows a similar trend to the away-side peak of B meson to hadron correlations.

Figure 6 shows that evolution of correlation pictures with momentum match within uncertainties. We have a characteristic b correlation image, which is present in both b quarks and B mesons, which is further supports that the B meson is a good proxy for the b quark.

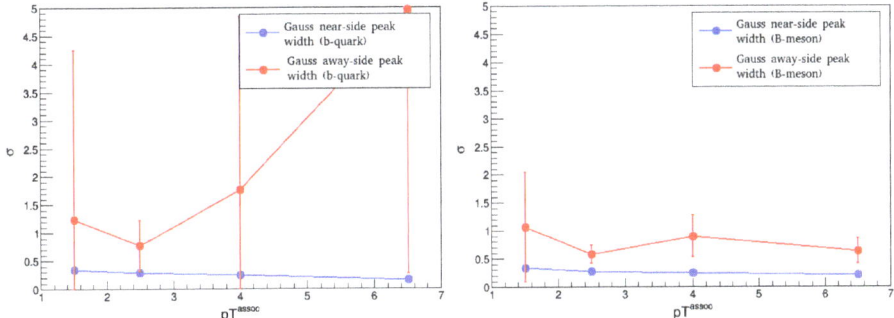

Figure 6. Comparison of b quark (on the **left**) and B meson (on the **right**) correlations to hadrons.

4. Conclusions

In summary, we have shown that a detailed analysis of heavy-flavor correlations can help in understanding flavor-dependent fragmentation as well as aid particle identification. The shape of the correlation peaks can be used to separate the electrons coming from b quark decays. This could be a method of identification that, combined with particle identification in secondary vertex detectors, may provide a much better sample purity than traditional methods. Correlation images are sensitive to the distribution of secondary vertices of heavy-quark decays, and the latter processes can be statistically separated from light quarks. Based on a correlation picture it is possible to distinguish between prompt D mesons and non-prompt D mesons from decays of B mesons. The statistical separation of these two contributions allows for the understanding of the flavor-dependence of heavy-quark energy loss within the quark-gluon plasma. We also see a characteristic b-correlation image, which is present in both b quarks and B mesons. B mesons can be used to study b quarks in the momentum regime where the reconstruction of b jets is not feasible.

Author Contributions: This research was carried out by E.F. under the supervision of R.V.

Funding: This work has been supported by the Hungarian National Research Fund (OTKA) grant K120660. Author RV thanks for the support of the János Bolyai fellowship of the Hungarian Academy of Sciences.

Conflicts of Interest: The author declares no conflict of interest.

References

1. Shuryak, E.V. Quantum Chromodynamics and the Theory of Superdense Matter. *Phys. Rep.* **1980**, *61*, 71–158. [CrossRef]
2. Adams, J.; et al. [STAR Collaboration] Evidence from d+Au measurements for final state suppression of high p_T hadrons in Au+Au collisions at RHIC. *Phys. Rev. Lett.* **2003**, *91*, 072304. [CrossRef] [PubMed]
3. Sjostrand, T.; Mrenna, S.; Skands, P.Z. A Brief Introduction to PYTHIA 8.1. *Comput. Phys. Commun.* **2008**, *178*, 852–867. [CrossRef]
4. Skands, P.; Carrazza, S.; Rojo, J. Tuning PYTHIA 8.1: The Monash 2013 Tune. *Eur. Phys. J. C* **2014**, *74*, 3024. [CrossRef]
5. Adamczyk, L.; et al. [STAR Collaboration] Υ production in U + U collisions at $\sqrt{s_{NN}}$ = 193 GeV measured with the STAR experiment. *Phys. Rev. C* **2016**, *6*, 064904. [CrossRef]
6. Adam, J.; et al. [ALICE Collaboration] Anomalous evolution of the near-side jet peak shape in Pb-Pb collisions at $\sqrt{s_{NN}}$ = 2.76 TeV. *Phys. Rev. Lett.* **2017**, *119*, 102301. [CrossRef] [PubMed]
7. Tanabashi, M.; et al. [Particle Data Group] Review of Particle Physics. *Phys. Rev. D* **2018**, *98*, 030001. [CrossRef]
8. Vértesi, R. [PHENIX Collaboration] THERMINATOR simulations and PHENIX images of a heavy tail of particle emission in 200-GeV Au+Au collisions. In Proceedings of the 23rd Winter Workshop on Nuclear Dynamics, Big Sky, MT, USA, 11–18 February 2007; Bauer, W., Bellwied, R., Harris, J.W., Eds.; EP Systema: Budapest, Hungary, 2007.
9. Yang, P.; Aglieri, G.; Cavicchioli, C.; Chalmet, P.L.; Chanlek, N.; Collu, A.; Gao, C.; Hillemanns, H.; Huang, G.; Junique, A.; et al. MAPS development for the ALICE ITS upgrade. *J. Instrum.* **2015**, *10*, C03030. [CrossRef]
10. Jung, K. [CMS Collaboration] Measurements of b-jet Nuclear Modification Factors in pPb and PbPb Collisions with CMS. *Nucl. Phys. A* **2014**, *931*, 470–474. [CrossRef]

© 2019 by the authors. Licensee MDPI, Basel, Switzerland. This article is an open access article distributed under the terms and conditions of the Creative Commons Attribution (CC BY) license (http://creativecommons.org/licenses/by/4.0/).

Communication
Quarkonium Production in the QGP

Alexander Rothkopf

Faculty of Science and Technology, University of Stavanger, 4021 Stavanger, Norway; alexander.rothkopf@uis.no

Received: 17 April 2019; Accepted: 14 May 2019; Published: 16 May 2019

Abstract: We report on recent theory progress in understanding the production of heavy quarkonium in heavy-ion collisions based on the in-medium heavy-quark potential extracted from lattice QCD simulations. On the one hand, the proper in-medium potential allows us to study the spectral properties of heavy quarkonium in thermal equilibrium, from which we estimate the ψ' to J/ψ ratio in heavy-ion collisions. On the other hand, the potential provides a central ingredient in the description of the real-time evolution of heavy-quarkonium formulated in the open-quantum-systems framework.

Keywords: quarkonium; quark-gluon plasma; heavy-ion collisions

1. Introduction

The bound states of heavy quarks and antiquarks, so-called heavy quarkonia, have matured into a high precision tool in heavy-ion collisions (HIC) at accelerator facilities, such as the Relativistic Heavy Ion Collider (RHIC) and the Large Hadron Collider (LHC). The availability of experimental data of unprecedented accuracy for both bottomonium ($b\bar{b}$) and charmonium ($c\bar{c}$), collected during the past five years, provides us access to different stages of the evolution of the quark–gluon plasma (QGP) created in the collision center.

The STAR collaboration at RHIC has observed overall suppression of bottomonium states in $\sqrt{s_{NN}} = 193$ MeV collisions [1]. At LHC, the most recent dimuon measurements of the CMS collaboration at $\sqrt{s_{NN}} = 5.02$ TeV furthermore resolve a clear sign of excited states suppression [2]. These are compatible with phenomenological models that describe the bottom–anti-bottom pair as non-equilibrium test-particle traversing the QGP while sampling its full time evolution (see, e.g., [3]). On the other hand, novel measurements by the ALICE collaboration have by now established an unambiguous signal for a finite elliptic flow, and even triangular flow of the charmonium vector channel ground state, the J/ψ particle [4]. This tells us that the charm quarks must at least be in partial kinetic equilibrium with the bulk matter to participate in its collective motion. In turn, equilibration entails a loss of memory of the initial conditions, positioning charmonium as probe of the late stages of the collision.

The goal for theory thus must be to provide a first principles description of this intricate phenomenology. As the temperatures encountered in current heavy-ion collisions are relatively close to the chiral crossover transition, genuinely non-perturbative methods are called for and, in this article, I discuss one possible route how first principles lattice QCD simulations can contribute to gain insight into the equilibrium and non-equilibrium properties of heavy quarkonium in HIC.

2. Quarkonium in Thermal Equilibrium

Let us start with the question of what are the properties of heavy quarkonium in thermal equilibrium? That is, we consider the idealized setting of immersing a heavy quark and antiquark pair in an infinitely extended QCD medium at a fixed temperature and wait until full kinetic equilibration is achieved. Then, we ask for the presence or absence of in-medium bound eigenstates and their properties, such as their in-medium mass and stability.

These questions may be answered in the modern language of quantum field theory by computing so-called in-medium meson spectral functions, which encode the particle properties as well defined peak structures. The position of the peaks along the frequency axis encodes the mass of the particle, while their width is directly related to the inverse lifetime of the state. At higher frequencies, the open heavy-flavor threshold manifests itself in the spectral function as broad continuous structures, often with a steep onset. In a thermal setting, the peak width not only encodes the decay of the bound state into gluons but also carries a contribution from processes that: (1) excite the color singlet bound state into another singlet state due to thermal fluctuations; and (2) transform the singlet state into a color octet state due to the absorption of a medium gluon. On the level of the spectral function, these three contributions cannot be disentangled. Once we have access to the in-medium meson spectral function, we argue that phenomenologically relevant processes, such as the production of J/ψ particles at hadronization, may be estimated from inspecting the in-medium spectral structures.

There are currently two viable options to determine the in-medium quarkonium spectra in QCD and both involve lattice QCD simulations. For the first and direct one, we can compute the current–current correlators of a heavy meson in the Euclidean time domain, in which the simulation is carried out. In particular, for bottomonium, it is customary to use a discretization of the heavy quarks, which is derived from a non-relativistic effective field theory (EFT) (see, e.g., [5,6]). A fully relativistic description of bottomonium still requires too fine of a lattice spacing, which in turn would make simulation in dynamical QCD prohibitively expensive. For charmonium, relativistic formulations have been considered in, e.g., [7,8]. From the Euclidean correlation function obtained in that way, the spectral function may be extracted using Bayesian inference. Due to the intricate structures encoded in the in-medium spectral function and the relatively small number of available simulated correlator points along the Euclidean time domain, this approach remains very challenging. Recent progress has been made in the robust determination of the ground state properties using the lattice NRQCD discretization at finite temperature [6]. It was shown that the ground state of both bottomonium and charmonium becomes lighter as temperature increases. An investigation of the excited state properties however is currently still out of reach.

The second possibility is to take a detour and instead of the spectral function compute first the potential acting in between a static quark and antiquark at finite temperature. Using this in general complex valued potential one can solve a Schrödinger equation for the unequal time correlation function of meson color singlet wavefunctions, i.e., for the meson forward current–current correlator, whose imaginary part then yields the in-medium spectral function. This approach on the one hand provides us with a very precise determination of the spectral function, however it does not yet include finite velocity or spin dependent corrections, since only the static potential is used in the computation. At $T = 0$, some of the correction terms to the heavy-quark potential have already been computed [9] and their determination at $T > 0$ is a work in progress. We show below that, to extract the in-medium potential from lattice QCD simulations, a spectral function also needs to be reconstructed. However, the benefit here lies in the fact that the structure of this Wilson correlator spectral function is much simpler than that of the full in-medium meson spectral function and thus its reconstruction can be achieved with much higher precision. In this article, I focus on the second strategy.

Today we are in the fortunate position of not having to rely anymore on model potential for the description of heavy quarkonium. Indeed, over the past decade, it has become possible to derive the inter quark potential directly from QCD using a chain of EFTs [10]. An EFT provides a systematic prescription of how to exploit the inherent separation of scales between the heavy quark rest mass and the temperature, as well as the characteristic scale of quantum fluctuations in QCD Λ_{QCD} to simplify the language needed to describe the relevant physics of the in-medium two-body system. Starting out from the relativistic field theory QCD where heavy quarks are described by four-component Dirac spinors, one may go over to Non-Relativistic QCD (NRQCD), a theory of two-component Pauli spinors. Subsequently, we can leave the language of fermion fields all together and go over to an EFT called potential NRQCD (pNRQCD). The latter describes the quark antiquark pair in terms of color singlet and color octet wavefunctions with in general coupled equations of motions, containing both potential and non-potential effects. The potential in pNRQCD is nothing but a matching (Wilson) coefficient in the Lagrangian of the EFT.

It is the process of matching that allows us to connect back to QCD. We need to select a correlation function in the EFT and find the corresponding correlation function in QCD with the same physics content. Once we set them equal at the scale at which the EFT is supposed to reproduce the microscopic physics, we can express the non-local Wilson coefficients of the former in terms of correlation functions of QCD. For static quarks, it can be shown that the unequal time singlet wave function correlation function is related to the rectangular Wilson loop

$$\langle \psi_s(t,r)\psi_s^*(0,r)\rangle_{pNRQCD} \stackrel{m\to\infty}{\equiv} W(t,r) = \left\langle \text{Tr}\left[\mathcal{P}\exp\left(-ig\int dx^\mu A_\mu(x)\right)\right]\right\rangle_{QCD}, \quad (1)$$

which obeys an equation of motion of the following type

$$i\partial_t W(t,r) = \Phi(t,r)W(t,r) \quad \Phi(t,r) \in \mathbb{C}; \quad V(r) = \lim_{t\to\infty}\Phi(t,r) = \lim_{t\to\infty}\frac{i\partial_t W(t,r)}{W(t,r)} \quad (2)$$

If the function Φ at late times converges to a constant, we may use its value to define what we mean by the interquark potential [11].

This genuinely real-time definition of the potential was first evaluated at high temperature in resummed perturbation theory by Laine et al. [12] who found it to be complex valued. The physics of the imaginary part has since been related to the phenomenon of Landau damping and gluo-dissociation [13,14]. Note that this complex potential does not evolve the wavefunction itself but instead a correlation function of wavefunctions. Thus, the presence of an imaginary part is by no means related to the disappearance of the heavy quarks (since they are static they cannot disappear from the system) but instead encodes the decoherence of the evolving in-medium system from its initial conditions [15].

We may now ask how to evaluate the real-time definition of the potential in non-perturbative lattice QCD, as these simulations are carried out in artificial Euclidean time. It is here that the technical concept of spectral function again finds application [16,17]. Indeed, we may express the real-time Wilson line correlator $W(t,r)$ as a Fourier transform over its real-valued and positive definite spectral function $\rho(\omega,r)$

$$W(t,r) = \int d\omega e^{i\omega t}\rho(\omega,r) \quad \Leftrightarrow \quad W(\tau,r) = \int d\omega e^{-\omega\tau}\rho(\omega,r) \quad (3)$$

The quantity accessible on the lattice is the imaginary time Wilson correlator, which is governed by the same spectral function, just with a different integral transform. Let me first note that using the spectral decomposition, inserted in the r.h.s. of Equation (2), we can relate $\rho(\omega,r)$ and $V(r)$. A careful inspection of the relation between the two reveals that a potential picture is applicable as long as we can identify a well defined lowest lying peak structure in ρ (for details see [11]). Its position is related to the real part of V, its width to the imaginary part. (In practice, we use the Wilson line correlators in Coulomb gauge instead of the Wilson loop in order to avoid the cusp divergences present in the latter.)

The central challenge lies in extracting the spectral function from lattice simulations, which amounts to solving an ill-posed inverse problem. In the past, this required the application of Bayesian inference [18], which uses additional prior information available on the spectral function to regularize the inversion task. The benefit of the Bayesian strategy is that it is applicable to simulation data with moderate statistical uncertainty ($\Delta W/W \approx 10^{-2}$). One challenging aspect on the other hand is that the influence of the prior information on the end result must be carefully investigated. In this article, we present the most recent results obtained for the potential using very high statistics simulations ($\Delta W/W < 10^{-2}$). In that case, another method for spectral reconstruction becomes feasible, the Pade approximation [19]. The simulation data are interpolated with an optimal Pade rational approximant, which in turn is analytically continued to give the retarded current–current propagator in real-time frequencies. Taking the imaginary part of this object yields the spectral function of interest, from which the values of Re[V] may be read off. Through mock datatests, we have found that, based on $N_\tau = 12, 16$ data points, the Pade is yet unable to faithfully reconstruct the width of the potential peak, which is why we have resorted to extracting tentative values of Im[V] using standard methods of Bayesian inference [20]. (For an alternative analysis based on the concept of effective potential, see [21].)

The lattice data on which the latest determination of the potential is based were obtained in a collaboration with the HotQCD and TUMQCD collaboration [22,23]. We compute the Wilson correlators on realistic 48^3 and $48^3 \times 16$ lattices, featuring $N_f = 2+1$ flavors of dynamical light quarks in the medium. These ensembles are deemed realistic, as the pion mass $m_\pi = 161$ MeV lies close to its physical value. Temperature is changed via the lattice spacing in a large range of $T \in [151, 1451]$ MeV.

In Figure 1 (left), we present the latest results for Re[V] from the aforementioned lattices [24]. The values shown are shifted manually in the y-direction for better readability. A qualitative inspection reveals that, while the potential in the hadronic phase at $T = 151$ MeV is well described by a Cornell type potential (Coulombic at small distances, linear rising at larger distances), it quickly becomes weakened as one passes into the QGP regime. Well above $T = 155$ MeV, Re[V] flattens off asymptotically and exhibits a form compatible with Debye screening. Figure 1 (right) contains a selection of results for the imaginary part. As the extraction of spectral widths is much more challenging than that of the peak positions, the values shown are only tentative. (Since the Pade method is known to underestimate Im[V] based on $N_\tau = 16$ data points, we show here results utilizing the Bayesian BR method instead.)

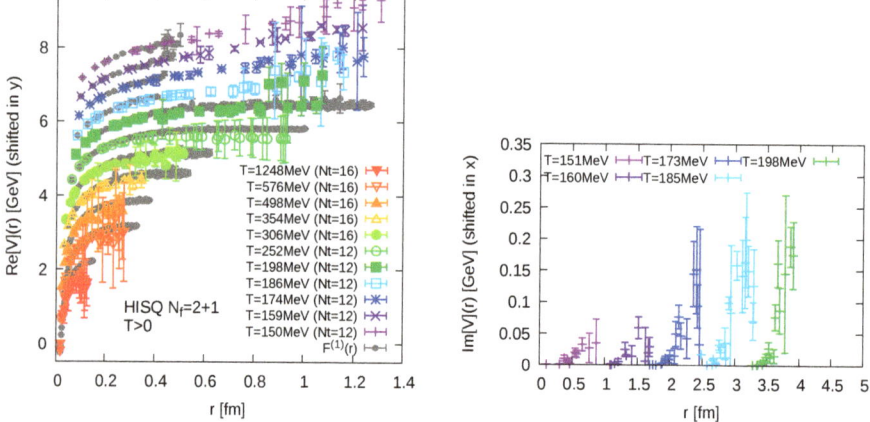

Figure 1. (**left**) Re[V] obtained from Pade reconstructed spectral functions of the Wilson line correlator in Coulomb gauge on $48^3 \times 12, 16$ lattices with $N_f = 2+1$ light quarks. The values are shifted by hand in y-direction for better readability from lowest temperature $T = 151$ MeV on top to highest $T = 1451$ MeV bottom. The gray data points denote the color singlet free energy in Coulomb gauge on the same lattices. (**right**) Tentative values of Im[V] at a selection of temperatures extracted via Bayesian inference from the same lattice data.

While it might be tempting to use the lattice values of Re[V] and Im[V] directly for a subsequent computation of the in-medium spectral function, this is not admissible. The lattice results obtained here are not yet extrapolated to the continuum limit and thus will not lead to consistent phenomenological results. Obtaining a genuine extrapolation is a work in progress but has thus far not yet been achieved.

Therefore, continuum corrections need to be used as laid out in detail, e.g., in Ref. [25]. To utilize the discrete values of the in-medium potential to solve a Schrödinger equation requires an analytic parametrization of $\text{Re}[V]$ and $\text{Im}[V]$ that can faithfully reproduce the lattice data. A novel derivation of such a parameterization, based on the generalized Gauss law, has been presented at the 2018 Zimanyi workshop (see Ref. [26]).

In Figure 2 (left), we show the in-medium spectral functions computed from the continuum corrected in-medium heavy quark potential obtained in [25]. One can clearly see the characteristic in-medium modification consisting of a shift of the peaks to lower frequencies and a concurrent broadening before they are dissolved into the continuum structure, whose onset moves to lower and lower frequencies. Consistent with intuition, the more weakly bound excited state is more strongly affected by the medium than the deeply bound ground state.

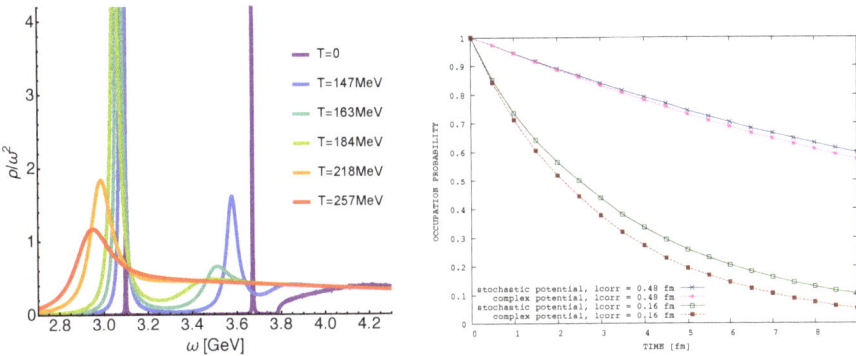

Figure 2. (left) Charmonium in-medium spectral functions from the continuum corrected in-medium heavy quark potential [25]. (right) Survival probabilities of the ground state in a one-dimensional model calculation of the real-time dynamics of bottomonium in the open-quantum systems approach [27]. The blue and green curve correspond to the stochastic potential computation with different correlation lengths. The pink and dark red curves arise from a naive Schrödinger equation with complex potential.

How can such spectral functions help us to learn about quarkonium production in HICs? Note that we are considering a fully thermalized scenario here, which applies, if at all, for charmonium. Note further that what is measured in experiment are not the decay dileptons from the in-medium states but the decays of vacuum states long after the QGP ceases to exist. Thus, any information of in-medium quarkonium needs to be translated into a modification of the yields of produced vacuum states at hadronization. The process of hadronization is among the least well known stages of a HIC and a first principles understanding of its dynamics has thus far not been achieved. Therefore, we continue with the phenomenological ansatz of instantaneous freezeout introduced in [25]. That is, we assume that at the phase boundary the in-medium states convert into vacuum states. The question we then wish to answer is: How many vacuum states does the in-medium spectral peak correspond to? The answer may be given in units of dilepton emission $R_{\ell\bar{\ell}} \propto \int dp_0 d^3\mathbf{p} \frac{\rho(P)}{P^2} n_B(p_0)$, which relates to the area under the spectral peaks.

That is, we compute the weighted area under the in-medium J/ψ peak and divide by the area of the vacuum spectral peak. This is our estimate for the number of J/ψ particles produced in this scenario. Carrying out the same computation for the ψ' peak, we may form the ratio of the two results, which constitutes our estimate for the in-medium ψ' to J/ψ ratio. The value obtained in Ref. [25] reads

$$R_{\ell\bar{\ell}}^{\psi'}/R_{\ell\bar{\ell}}^{J/\psi} = 0.023 \pm 0.004. \qquad (4)$$

and agrees within uncertainty with the value predicted by the statistical model of hadronization [28], as well as with the most recent determination of the ratio by the ALICE collaboration at the LHC (see, e.g., [29]).

3. In-Medium Quarkonium Real-Time Dynamics

Up to this point, we have only considered equilibrium aspects of quarkonium. In a HIC, this will always constitute only an approximation to the genuine non-equilibrium physics occurring. Therefore, we wish to learn more about the real-time dynamics of quarkonium states exploiting the fact that we already have access to the in-medium potential extracted on the lattice. A promising route towards a microscopic understanding of quarkonium real-time dynamics is offered by the open-quantum systems approach, a technique developed originally in the context of condensed matter theory.

The overall system consisting of the heavy quark and antiquark, as well as the medium degrees of freedom is of course closed and described by a hermitean Hamiltonian. The overall density matrix evolves according to the von Neumann equation

$$H = H_{Q\bar{Q}} \otimes I_{med} + I_{Q\bar{Q}} \otimes H_{med} + H_{int}, \quad \frac{d\rho}{dt} = -i[H,\rho]. \tag{5}$$

Our goal however is to investigate the properties and dynamics of the heavy quarkonium sub system coupled to the thermal bath. To this end, we may trace out all medium degrees of freedom from the density matrix of the full system, ending up with $\rho_{Q\bar{Q}} = \text{Tr}_{med}[\rho]$. The question then is: What kind of equation of motion does this reduced density matrix obey?

Over the past five years, it has become possible to derive the master equation for $\rho_{Q\bar{Q}}$ from QCD, based on a limited number of assumptions [30,31]. Starting from the path integral representation of the density matrix on the Schwinger–Keldysh contour, the integrating out of the medium degrees of freedom may be implemented in a functional sense. This leads to a path integral for the reduced density matrix, in which only the heavy quark degrees of freedom appear explicitly. In addition to the heavy quark action on the forward and backward contour, an additional effective action emerges, the so-called Feynman–Vernon influence functional S_{FV}. It encodes all interactions between the subsystem and the traced out medium. S_{FV} in general is a very complicated object but it may be simplified using the separation of scales in the system. As shown in Ref. [31], at high temperatures, where at intermediate steps of the derivation a weak coupling ansatz has been used, the Feynman–Vernon influence functional takes the explicit form

$$S_{FV} \approx S_{pot}[\text{Re}[V]] + S_{fluct}[\text{Im}[V]] + S_{diss}[\text{Im}[V]] + S_{LB}. \tag{6}$$

The first part is related to a real valued in-medium potential term, while the second and third implement the fluctuation–dissipation relation for the heavy quarkonium. They are intimately related to the imaginary part of the interquark potential. The last term assures that the master equation for $\rho_{Q\bar{Q}}$ preserves the positivity of its eigenvalues. (For other recent studies of the open-quantum systems approach for quarkonium, see [32–36].)

The above expression for S_{FV} leads to Markovian dynamics for $\rho_{Q\bar{Q}}$, described by a so called Lindblad equation.

$$\frac{d}{dt}\rho_{Q\bar{Q}}(t) = -i[H_{Q\bar{Q}},\rho_{Q\bar{Q}}] + \sum_{i=1}^{N_{LB}} \gamma_i \left(\hat{L}_i \rho_{Q\bar{Q}} \hat{L}_i^\dagger - \frac{1}{2}\hat{L}_i\hat{L}_i^\dagger \rho_{Q\bar{Q}} - \frac{1}{2}\rho_{Q\bar{Q}} \hat{L}_i \hat{L}_i^\dagger \right) \tag{7}$$

The operators L_i are called Lindblad operators and encode the interactions between the quarkonium subsystem and the surrounding environment. They may be expressed in terms $\text{Im}[V]$. It is important to note that the Lindblad equation cannot be implemented (unraveled) in terms of a deterministic evolution of a microscopic wave function. Instead, one is led to stochastic dynamics for an ensemble of wavefunctions.

Together with collaborators from Japan, we have investigated the effects of the Lindblad operators on the real-time dynamics of heavy quarkonium in a simple one-dimensional setting [27]. As a first step, we considered only the leading order gradient expansion of $S_{FV} \approx S_{pot}[\text{Re}[V]] + S_{fluct}[\text{Im}[V]]$, which leads to the notion of a stochastic potential. That is, it allows implementing unitary time evolution via $\text{Re}[V]$, which is stochastically disturbed with noise η, whose correlations are governed by $\text{Im}[V]$.

$$\psi_{Q\bar{Q}}(t) = \exp\left[-\frac{\nabla^2}{M} + \text{Re}[V] + \eta(t)\right]\psi_{Q\bar{Q}}(0), \quad i\partial_t \langle\psi_{Q\bar{Q}}(t)\rangle = \left(-\frac{\nabla^2}{M} + \text{Re}[V] - i|\text{Im}[V]|\right)\langle\psi_{Q\bar{Q}}(t)\rangle \quad (8)$$

While the evolution of each realization of the ensemble proceeds via a norm preserving evolution operator, the ensemble average of the wave function washes out according to a Schrödinger equation with a complex valued potential. This mechanism provides a unitary microscopic implementation of quarkonium real-time dynamics, which reproduces the imaginary part of the interquark potential for the unequal time correlation function of wavefunctions.

Note that there is a new physical scale present in this approach, which is the correlation length of the noise induced by the medium. Depending on the size of the quarkonium bound state compared to this correlation length, the noise may be able to efficiently destabilize the bound state or not. This phenomenon is known as decoherence. That is, the noise provides an additional mechanism to dissociate a heavy quarkonium particle over time, which acts in addition to the screening of the real-valued potential.

In Figure 2 (right), we show an example computation of the survival probabilities of the bottomonium ground state in a one dimensional setup based on perturbative values for the in-medium $\text{Re}[V]$ and $\text{Im}[V]$. We draw two conclusions. First, the survival crucially depends on the value of the medium correlation length. Secondly, using the more realistic description in terms of a stochastic potential instead of a naive Schrödinger equation with a complex potential leads to significantly different survival. The naive approach systematically underestimates the survival.

While the stochastic potential provides a conceptually attractive microscopic implementation of the complex inter-quark potential, it can only be the first step towards understanding heavy quarkonium in-medium dynamics. It does not account for dissipation effects and thus does not allow the quarkonium to thermalize with its surroundings. This means that the stochastic potential description is only applicable to early times in the evolution. Incorporation of the full Lindblad equation is work in progress and we have successfully tested it in the single heavy quark case [37]. The extension to quarkonium is under way.

4. Summary

In this article, I have showcased recent progress in our understanding of in-medium heavy quarkonium in the context of heavy-ion collisions. In thermal equilibrium, it has become possible to derive a complex valued real-time in-medium potential from QCD based on EFT methods. Its evaluation in lattice QCD simulations is challenging as it involves the reconstruction of spectral functions from Wilson correlators. The most recent determination has been performed on realistic $m_\pi = 161$ MeV ensembles by the HotQCD and TUMQCD collaboration. From the continuum corrected potential, one may compute in-medium quarkonium spectral functions, which have been used to estimate the ψ' to J/ψ ratio, showing good agreement with the statistical model of hadronization and the most recent measurements by the ALICE collaboration. To implement the microscopic dynamics of heavy quarkonium based on the complex in-medium potential, the open-quantum-systems approach is promising. Using a clear set of assumptions, one may derive a Lindblad master equation for the reduced density matrix, which to first order leads to unitary time evolution with a stochastic potential. The medium induced noise leads to decoherence of the in-medium quarkonium, which provides an additional mechanism to the dissolution of the in-medium state besides Debye screening. The implementation of the full Lindblad equation for quarkonium remains work in progress.

Funding: This work is part of and partially supported by the DFG Collaborative Research Centre "SFB 1225 (ISOQUANT)".

Conflicts of Interest: The author declares no conflict of interest.

References

1. Adamczyk, L. et al. [STAR Collaboration]. Υ production in U + U collisions at $\sqrt{s_{NN}}$ = 193 GeV measured with the STAR experiment. *Phys. Rev. C* **2016**, *94*, 064904. [CrossRef]
2. Sirunyan, A.M. et al. [CMS Collaboration]. Suppression of Excited Υ States Relative to the Ground State in Pb-Pb Collisions at $\sqrt{s_{NN}}$ = 5.02 TeV. *Phys. Rev. Lett.* **2018**, *120*, 142301. [CrossRef]
3. Krouppa, B.; Rothkopf, A.; Strickland, M. Bottomonium suppression using a lattice QCD vetted potential. *Phys. Rev. D* **2018**, *97*, 016017. [CrossRef]
4. Acharya, S. et al. [ALICE Collaboration]. Study of J/ψ azimuthal anisotropy at forward rapidity in Pb-Pb collisions at $\sqrt{s_{NN}}$ = 5.02 TeV. *J. High Energy Phys.* **2019**, *2019*, 12. [CrossRef]
5. Aarts, G.; Allton, C.; Harris, T.; Kim, S.; Lombardo, M.P.; Ryan, S.M.; Skullerud, J.I. The bottomonium spectrum at finite temperature from N_f = 2 + 1 lattice QCD. *J. High Energy Phys.* **2014**, *2014*, 97. [CrossRef]
6. Kim, S.; Petreczky, P.; Rothkopf, A. Quarkonium in-medium properties from realistic lattice NRQCD. *J. High Energy Phys.* **2018**, *2018*, 88. [CrossRef]
7. Ding, H.T.; Francis, A.; Kaczmarek, O.; Karsch, F.; Satz, H.; Soeldner, W. Charmonium properties in hot quenched lattice QCD. *Phys. Rev. D* **2012**, *86*, 014509. [CrossRef]
8. Kelly, A.; Rothkopf, A.; Skullerud, J.I. Bayesian study of relativistic open and hidden charm in anisotropic lattice QCD. *Phys. Rev. D* **2018**, *97*, 114509. [CrossRef]
9. Koma, Y.; Koma, M.; Wittig, H. Relativistic corrections to the static potential at O(1/m) and O(1/m**2). *PoS LATTICE 2007*, *2007*, 111.
10. Brambilla, N.; Pineda, A.; Soto, J.; Vairo, A. Effective field theories for heavy quarkonium. *Rev. Mod. Phys.* **2005**, *77*, 1423. [CrossRef]
11. Burnier, Y.; Rothkopf, A. Disentangling the timescales behind the non-perturbative heavy quark potential. *Phys. Rev. D* **2012**, *86*, 051503. [CrossRef]
12. Laine, M.; Philipsen, O.; Romatschke, P.; Tassler, M. Real-time static potential in hot QCD. *J. High Energy Phys.* **2007**, *2007*, 054. [CrossRef]
13. Beraudo, A.; Blaizot, J.-P.; Ratti, C. Real and imaginary-time Q anti-Q correlators in a thermal medium. *Nucl. Phys. A* **2008**, *806*, 312. [CrossRef]
14. Brambilla, N.; Ghiglieri, J.; Vairo, A.; Petreczky, P. Static quark-antiquark pairs at finite temperature. *Phys. Rev. D* **2008**, *78*, 014017. [CrossRef]
15. Akamatsu, Y.; Rothkopf, A. Stochastic potential and quantum decoherence of heavy quarkonium in the quark-gluon plasma. *Phys. Rev. D* **2012**, *85*, 105011. [CrossRef]
16. Rothkopf, A.; Hatsuda, T.; Sasaki, S. Proper heavy-quark potential from a spectral decomposition of the thermal Wilson loop. *Soryushiron Kenkyu Electron.* **2010**, *118*, A145–A147.
17. Rothkopf, A.; Hatsuda, T.; Sasaki, S. Complex Heavy-Quark Potential at Finite Temperature from Lattice QCD. *Phys. Rev. Lett.* **2012**, *108*, 162001. [CrossRef]
18. Rothkopf, A. Bayesian techniques and applications to QCD. *arXiv* **2019**, arXiv:1903.02293.
19. Tripolt, R.A.; Gubler, P.; Ulybyshev, M.; von Smekal, L. Numerical analytic continuation of Euclidean data. *Comput. Phys. Commun.* **2019**, *237*, 129. [CrossRef]
20. Burnier, Y.; Rothkopf, A. Bayesian Approach to Spectral Function Reconstruction for Euclidean Quantum Field Theories. *Phys. Rev. Lett.* **2013**, *111*, 182003. [CrossRef] [PubMed]
21. Petreczky, P. et al. [TUMQCD Collaboration]. Lattice Calculations of Heavy Quark Potential at Finite Temperature. *Nucl. Phys. A* **2017**, *967*, 592–595. [CrossRef]
22. Bazavov, A.; Petreczky, P.; Weber, J.H. Equation of State in 2 + 1 Flavor QCD at High Temperatures. *Phys. Rev. D* **2018**, *97*, 014510. [CrossRef]
23. Bazavov, A. et al. [HotQCD Collaboration]. Equation of state in (2 + 1)-flavor QCD. *Phys. Rev. D* **2014**, *90*, 094503. [CrossRef]
24. Petreczky, P.; Rothkopf, A.; Weber, J. Realistic in-medium heavy-quark potential from high statistics lattice QCD simulations. *Nucl. Phys. A* **2019**, *982*, 735–738. [CrossRef]

25. Burnier, Y.; Kaczmarek, O.; Rothkopf, A. Quarkonium at finite temperature: Towards realistic phenomenology from first principles. *J. High Energy Phys.* **2015**, *2015*, 101. [CrossRef]
26. Lafferty, D.; Rothkopf, A. Quarkonium phenomenology from a generalised Gauss-law. *Universe* **2019**, in press.
27. Kajimoto, S.; Akamatsu, Y.; Asakawa, M.; Rothkopf, A. Dynamical dissociation of quarkonia by wave function decoherence. *Phys. Rev. D* **2018**, *97*, 014003. [CrossRef]
28. Andronic, A.; Beutler, F.; Braun-Munzinger, P.; Redlich, K.; Stachel, J. Statistical hadronization of heavy flavor quarks in elementary collisions: Successes and failures. *Phys. Lett. B* **2009**, *678*, 350. [CrossRef]
29. Andronic, A.; Braun-Munzinger, P.; Köhler, M.K.; Stachel, J. Testing charm quark thermalisation within the Statistical Hadronisation Model. *Nucl. Phys. A* **2019**, *982*, 759–762. [CrossRef]
30. Akamatsu, Y. Real-time quantum dynamics of heavy quark systems at high temperature. *Phys. Rev. D* **2013**, *87*, 045016. [CrossRef]
31. Akamatsu, Y. Heavy quark master equations in the Lindblad form at high temperatures. *Phys. Rev. D* **2015**, *91*, 056002. [CrossRef]
32. De Boni, D. Fate of in-medium heavy quarks via a Lindblad equation. *J. High Energy Phys.* **2017**, *2017*, 64. [CrossRef]
33. Blaizot, J.P.; Escobedo, M.A. Quantum and classical dynamics of heavy quarks in a quark-gluon plasma. *J. High Energy Phys.* **2018**, *2018*, 34. [CrossRef]
34. Brambilla, N.; Escobedo, M.A.; Soto, J.; Vairo, A. Heavy quarkonium suppression in a fireball. *Phys. Rev. D* **2018**, *97*, 074009. [CrossRef]
35. Yao, X.; Mehen, T. Quarkonium in-Medium Transport Equation Derived from First Principles. *arXiv* **2019**, arXiv:1811.07027.
36. Brambilla, N.; Escobedo, M.A.; Vairo, A.; Griend, P.V. Transport coefficients from in medium quarkonium dynamics. *arXiv* **2019**, arXiv:1903.08063.
37. Akamatsu, Y.; Asakawa, M.; Kajimoto, S.; Rothkopf, A. Quantum dissipation of a heavy quark from a nonlinear stochastic Schrödinger equation. *J. High Energy Phys.* **2018**, *2018*, 29. [CrossRef]

© 2019 by the authors. Licensee MDPI, Basel, Switzerland. This article is an open access article distributed under the terms and conditions of the Creative Commons Attribution (CC BY) license (http://creativecommons.org/licenses/by/4.0/).

Communication

Study of Jet Shape Observables in Au+Au Collisions at $\sqrt{s_{NN}}$ = 200 GeV with JEWEL

Veronika Agafonova

Nuclear Physics Institute of the Czech Academy of Sciences, CZ-25068 Řež, Czech Republic; agafonova@ujf.cas.cz

Received: 31 March 2019; Accepted: 8 May 2019; Published: 11 May 2019

Abstract: Nuclear–nuclear collisions at energies attainable at the large accelerators RHIC and the LHC are an ideal environment to study nuclear matter under extreme conditions of high temperature and energy density. One of the most important probes of such nuclear matter is the study of production of jets. In this article, several jet shape observables in Au+Au collisions at the center of mass energy per nucleon–nucleon pair of $\sqrt{s_{NN}}$ = 200 GeV simulated in the Monte Carlo generator JEWEL are presented. Jets were reconstructed using the anti-k_T algorithm and their shapes were studied as a function of the jet-resolution parameter R, transverse momentum p_T and collision centrality.

Keywords: jet; jet algorithm; jet shapes; ALICE; LHC; RHIC; JEWEL

1. Introduction

The study of production of jets is one of the most important probes of nuclear matter under extreme conditions of high temperature and energy density. The jet is a collimated spray of hadrons originating from fragmentation of a hard parton created in the initial stage of the nucleus–nucleus collision and can be used for tomography of the nuclear matter (Figure 1). As jets mostly conserve the energy and the direction of the originating parton, they are measured in particle detectors and studied to determine the properties of the original quarks.

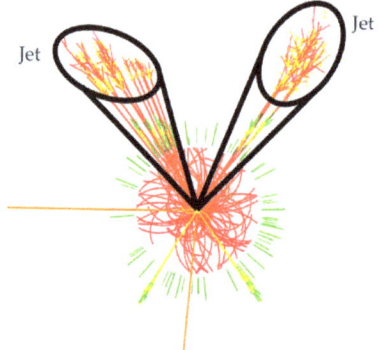

Figure 1. A schematic view of jet created in a heavy-ion collision [1].

To probe the complimentary aspects of the jet fragmentation and constrain theoretical description of jet–medium interactions, different observables related to shapes of jets are studied at the CERN Large Hadron Collider (LHC) [2–4]. It is important to perform similar measurements at lower collision energies at the Relativistic Heavy Ion Collider (RHIC) taking advantage of new high statistics data. The jet substructure observables are the perfect tool to understand what is happening when the particles interact with Quark–Gluon Plasma (QGP) medium. This article is focused on two jet shape observables: the girth and momentum dispersion that will be described in more detail below.

2. Jet Shape Observables

The *first radial moment* (alternatively *angularity* or *girth*), g, probes the radial distribution of radiation inside a jet. It is defined as

$$g = \sum_{i \in jet} \frac{p_T^i}{p_{T,jet}} |\Delta R_{i,jet}|. \tag{1}$$

Here p_T^i represents the momentum of the ith jet constituent and $\Delta R_{i,jet}$ is the distance in $\eta \times \phi$ plane between the constituent i and the jet axis [5], where η is the pseudorapidity and ϕ is the azimuthal angle. This type of shape is sensitive to the radial energy profile or broadening of the jet. In the collinear limit for the polar angle $\theta \to 0$ the radial moment becomes equivalent to jet broadening.

The next observable is *momentum dispersion*, $p_T D$. It measures the second moment of the constituent p_T distribution in the jet and is connected to hardness or softness of the jet fragmentation. It means that in the case of a large number of constituents and softer momentum the $p_T D$ tends to 0, while in the opposite situation the $p_T D$ will be close to 1. Its definition is given by the equation:

$$p_T D = \frac{\sqrt{\sum_{i \in jet} p_{T,i}^2}}{\sum_{i \in jet} p_{T,i}}. \tag{2}$$

These two jet shape observables are infrared and collinear (IRC) safe. It means that if one modifies an event by a collinear splitting or the addition of a soft emission, the set of hard jets that are found in the event should remain unchanged [6].

3. The Anti-k_T Algorithm

Jets are commonly reconstructed using the anti-k_T clustering algorithm [7]. The anti-k_T algorithm is a sequential-clustering algorithm. The algorithm is based on successive pair-wise recombination of particles and it works as follows. Firstly, the distance, d_{ij}, between particles i and j is found as

$$d_{ij} = \min(k_{ti}^{-2}, k_{tj}^{-2}) \frac{\Delta_{ij}^2}{R^2}, \tag{3}$$

where $\Delta_{ij}^2 = (y_i - y_j)^2 + (\phi_i - \phi_j)^2$ and k_{ti} or k_{tj}, y_i, ϕ_i and R stand for the transverse momenta, rapidity, azimuth, and radius parameter of particle i respectively. Secondly, the algorithm calculates the distance, d_{iB} between the entity i and the beam B as

$$d_{iB} = k_{ti}^{-2}. \tag{4}$$

The next step of the anti-k_T jet algorithm is to find the minimum distance, d_{min}, between the distances d_{ij} and d_{iB}. In case the smallest distance is d_{ij}, the algorithm performs a recombination of the entities. In other situation i is called to be a jet and is subsequently removed from the list. All these steps are repeated until no particles are left.

4. JEWEL

Jet Evolution With Energy Loss (JEWEL) is a Monte Carlo event generator that describes the Quantum Chromodynamics (QCD) evolution of jets in vacuum and in a medium in a perturbative approach [8–10]. In this section, the simulation in JEWEL will be described. For this research 50 million events were simulated for the interaction in vacuum and 20 million events for the interaction in medium. The simulation was made for 0–10% central and 60–80% peripheral Au+Au collisions with additional "recoils on/off" option for interaction with medium. "Recoils on" option in JEWEL keeps the thermal partons recoiling against interactions with the jet in the event and let them hadronize together with the jet, while the "recoils off" option ignore the medium response [11]. All events were required to have the center-of-mass (CMS) energy $\sqrt{s_{NN}}$ = 200 GeV.

Table 1 contains the parameters used for the vacuum model. Additional parameters for the simulation with the medium can be found in Table 2.

Table 1. Parameters of the JEWEL vacuum simulation for central and peripheral collisions [8].

Name of Parameter	Name in JEWEL	Value
Parton Distribution Function set	PDFSET	10,100
Number of events	NEVENT	100,000
Mass number of Au nucleus	MASS	197
The CMS energy of the colliding system	SQRTS, [GeV]	200
Minimum p_T in matrix element	PTMIN, [GeV]	3
Maximum p_T in matrix element	PTMAX, [GeV]	−1
The rapidity range	ETAMAX	2.5

Table 2. Parameters of the JEWEL simulation with medium for central and peripheral "recoils on/off" collisions [8].

Name of Parameter	Name in JEWEL	Value	
The initial (mean) temperature	TI, [GeV]	0.28	
The initial time τ_i	TAUI, [fm]	0.6	
An integer mass number of colliding nuclei	A	197	
The lower end of centrality range	CENTRMIN, [%]	0	60
The upper end of centrality range	CENTRMAX, [%]	10	80
The switch of keeping recoils	KEEPRECOILS	T	F
The nucleus–nucleus cross-section	SIGMANN, [fm^2]	4.2	

A resolution parameter, R, quantifies the size of the jet. For this study values of the resolution parameter were chosen to be $R = 0.2$ and $R = 0.4$, respectively. The charged particles were simulated in pseudorapidity $\eta_{cent} = 2.5$ and full azimuth. Jets were reconstructed with the anti-k_T algorithm included in FastJet software package [12].

5. Results

In this section, only the JEWEL results for central Au+Au collisions at $\sqrt{s_{NN}}$ = 200 GeV are presented as they are more appealing from the physical point of view. The jet shape observables are calculated for different values of the resolution parameter R and charged jet p_T separately for vacuum and medium with "recoils on/off" option. The distributions will be further compared to the results and the JEWEL simulation from the ALICE collaboration [13].

Figure 2 shows the measured jet shape distributions in 0–10% central Pb–Pb collisions at $\sqrt{s_{NN}}$ = 2.76 TeV for anti-k_T charged jets at ALICE compared to JEWEL simulation with and without recoils [13]. As the resolution parameter is small, $R = 0.2$, the effects of medium recoils are also small. It means that the measurement is constrained by purely radiative aspects of the JEWEL shower modification. A good agreement between the data and the model, especially in momentum dispersion, can be observed.

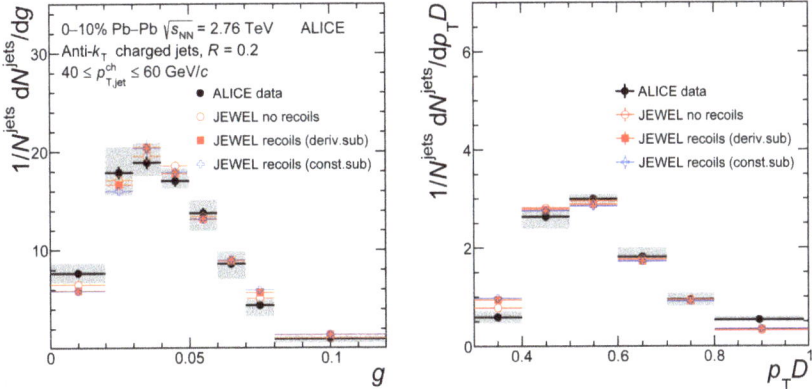

Figure 2. Jet shape distributions g (**left**) and $p_T D$ (**right**) in 0–10% central Pb–Pb collisions at $\sqrt{s_{NN}}$ = 2.76 TeV for R = 0.2 in range of jet $p_{T,jet}^{ch}$ of 40–60 GeV/c compared to JEWEL with and without recoils with different subtraction methods. The colored boxes represent the experimental uncertainty on the jet shapes [13].

Figures 3 and 4 compare the distributions of angularity for vacuum and medium "recoils on/off" central Au+Au collisions in two different p_T ranges $10 < p_T < 20$ GeV/c and $20 < p_T < 30$ GeV/c, respectively. As it can be seen, the first radial moment has the same behavior for R = 0.2 as the results from the ALICE experiment (Figure 2). Nevertheless, peaks for the medium "recoils on" and medium "recoils off" simulation of angularity with R = 0.4 are shifted to the right and left, respectively. Distributions for medium "recoils on" collisions with R = 0.4 have a longer tail than others. Also, the spike for g = 0.01 in the case of jets with $10 < p_T < 20$ GeV/c can be observed for both resolution parameters. That signals the presence of jets with only one constituent. To probe this, the dependence of the number of constituents on the angularity is shown in Figure 5. It can be clearly seen that there is a larger number of particles for R = 0.4 than for R = 0.2 jets.

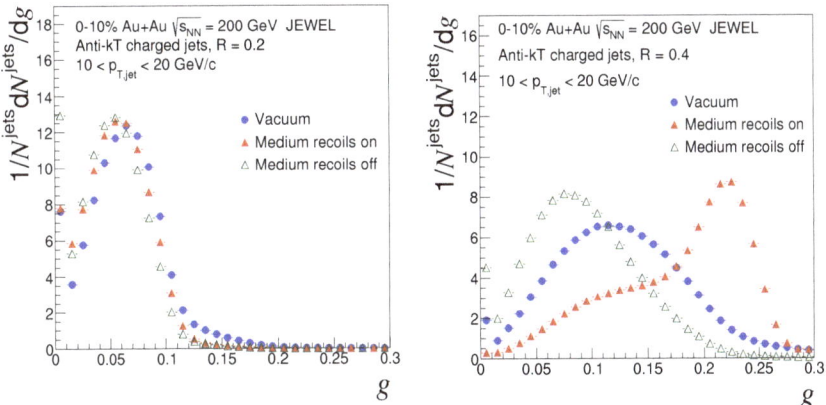

Figure 3. Girth for jets with p_T of 10–20 GeV/c and R = 0.2 (**left**) and R = 0.4 (**right**) in central Au+Au collisions at $\sqrt{s_{NN}}$ = 200 GeV.

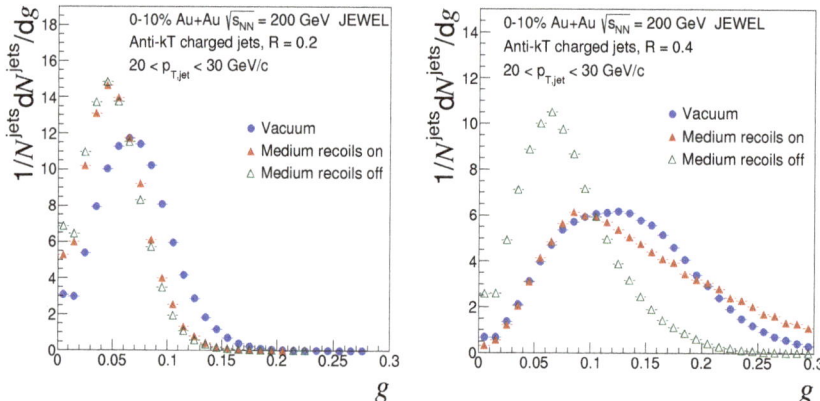

Figure 4. Girth for jets with p_T of 20–30 GeV/c and $R = 0.2$ (**left**) and $R = 0.4$ (**right**) in central Au+Au collisions at $\sqrt{s_{NN}} = 200$ GeV.

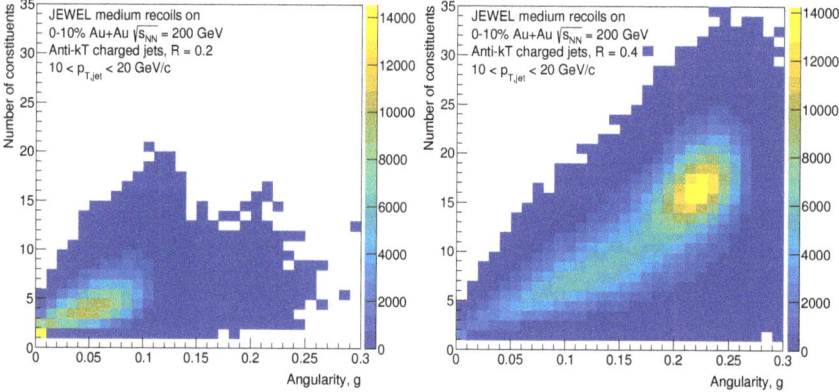

Figure 5. 2D statistics for jets with p_T of 10–20 GeV/c and $R = 0.2$ (**left**) and $R = 0.4$ (**right**) in central "recoils on" Au+Au collisions at $\sqrt{s_{NN}} = 200$ GeV simulated with medium.

Figures 6 and 7 compare the results for the momentum dispersion for jets with $10 < p_T < 20$ GeV/c and $20 < p_T < 30$ GeV/c, respectively. As for previous observable, there is a better agreement between the models in $20 < p_T < 30$ GeV/c p_T range. However, in contradiction to the ALICE results, the obtained distributions for the momentum dispersion start form $p_T D = 0$ (for $R = 0.4$ in central and peripheral collisions) and $p_T D = 0.1$ (for $R = 0.2$ in central collisions) instead of $p_T D = 0.3$. That can be a consequence of the use of different centrality ranges. Also, a shift of the distribution to lower values for the central medium "recoils on" setting for $R = 0.4$ and $10 < p_T < 20$ GeV/c can be observed.

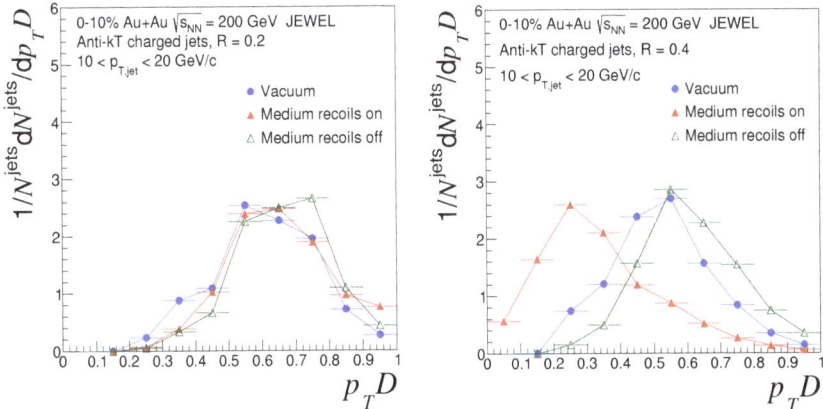

Figure 6. Momentum dispersion for jets with p_T of 10–20 GeV/c and $R = 0.2$ (**left**) and $R = 0.4$ (**right**) in central Au+Au collisions at $\sqrt{s_{NN}} = 200$ GeV.

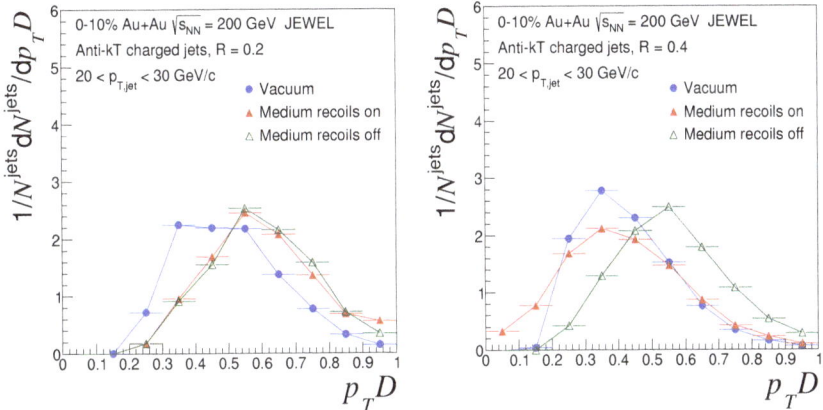

Figure 7. Momentum dispersion for jets with p_T of 20–30 GeV/c and $R = 0.2$ (**left**) and $R = 0.4$ (**right**) in central Au+Au collisions at $\sqrt{s_{NN}} = 200$ GeV.

6. Conclusions

In this article, the results of study of two jet shape observables, girth and momentum dispersion, in central Au+Au collisions at CMS energy of 200 GeV per nucleon–nucleon pair with the JEWEL Monte Carlo generator were presented. The jet shapes were calculated using the anti-k_T jet finding algorithm implemented in the FastJet software package. The chosen observables were studied as a function of the transverse momentum, jet-resolution parameter, and collision centrality. All the obtained results have the same behavior as the results from the ALICE collaboration [13]. In this study, it was shown that the spike in the girth results for $g = 0.01$ for jets with p_T 10–20 GeV/c for both values of resolution parameter, $R = 0.2$ and $R = 0.4$, (Figure 3) is due to presence of jets with only one constituent (Figure 5). As for the momentum dispersion, the obtained distributions (Figures 6 and 7) are wider in comparison to the ALICE results.

One of the goals of future work is to perform the background subtraction similarly to the ALICE experiment. It is expected that after the background subtraction the points for medium "recoils on/off" and vacuum models will be closer to each other.

Funding: This research was funded by INTER-EXCELLENCE INTER-TRANSFER of Ministry of Education, Youth and Sports of the Czech Republic grant number LTT18002.

Conflicts of Interest: The authors declare no conflict of interest.

Abbreviations

The following abbreviations are used in this manuscript:

LHC	Large Hadron Collider
RHIC	Relativistic Heavy-Ion Collider
CERN	Conseil européen pour la recherche nucléaire
QGP	Quark Gluon Plasma
QCD	Quantum Chromodynamics
CMS	center-of-mass
JEWEL	Jet Evolution With Energy Loss
IRC safe	Infrared and Collinear safe
ALICE	A Large Ion Collider Experiment

References

1. Agafonova, V. Study of Jet Shape Observables at RHIC. Available online: https://physics.fjfi.cvut.cz/publications/ejcf/bp_ejcf_17_agafonova.pdf (accessed on 11 May 2019).
2. Aad, G.; Abbott, B.; Abdallah, J.; Abdelalim, A.A.; Abdesselam, A.; Abdinov, O.; Abi, B.; Abolins, M.; Abramowicz, H.; Abreu, H.; et al. Observation of a Centrality-Dependent Dijet Asymmetry in Lead-Lead Collisions at $\sqrt{s_{NN}}$ = 2.77 TeV with the ATLAS Detector at the LHC. *Phys. Rev. Lett.* **2010**, *105*, 252303. [CrossRef] [PubMed]
3. Chatrchyan, S. et al. [CMS Collaboration] Observation and studies of jet quenching in PbPb collisions at nucleon-nucleon center-of-mass energy = 2.76 TeV. *Phys. Rev. C* **2011**, *84*, 024906. [CrossRef]
4. Vitev, I.; Wicks, S.; Zhang, B.W. A Theory of jet shapes and cross sections: From hadrons to nuclei. *J. High Energy Phys.* **2008**, *11*, 93. [CrossRef]
5. Cunqueiro, L. et al. [ALICE Collaboration] Jet shapes in pp and Pb–Pb collisions at ALICE. *Nucl. Phys. A* **2016**, *956*, 593–596. [CrossRef]
6. Salam, G.P. Towards Jetography. *Eur. Phys. J. C* **2010**, *167*, 637–686. [CrossRef]
7. Cacciari, M.; Salam, G.P.; Soyez, G. The Anti-k(t) jet clustering algorithm. *J. High Energy Phys.* **2008**, *4*, 63. [CrossRef]
8. Zapp, K.C. JEWEL 2.0.0: Directions for use. *Eur. Phys. J.* **2014**, *74*, 2762. [CrossRef]
9. Zapp, K.C. Geometrical aspects of jet quenching in JEWEL. *Phys. Lett. B* **2014**, *735*, 157–163. [CrossRef]
10. Zapp, K.C.; Ingelman, G.; Rathsman, J.; Stachel, J.; Wiedemann, U.A. A Monte Carlo Model for 'Jet Quenching'. *Eur. Phys. J. C* **2009**, *60*, 617–632. [CrossRef]
11. Zapp, K.C.; Kunnawalkam Elayavalli, R. Medium response in JEWEL and its impact on jet shape observables in heavy ion collisions. *J. High Energy Phys.* **2017**, *7*, 141. [CrossRef]
12. Cacciari, M.; Salam, G.P.; Soyez, G. FastJet user manual. *Eur. Phys. J. C* **2012**, *72*, 1896. [CrossRef]
13. Acharya, S.; Adamová, D.; Adler, A.; Adolfsson, J.; Aggarwal, M.M.; Rinella, G.A.; Agnello, M.; Agrawal, N.; Ahammed, Z.; Ahn, S.U.; et al. Medium modification of the shape of small-radius jets in central Pb-Pb collisions at $\sqrt{s_{NN}}$ = 2.76 TeV. *J. High Energy Phys.* **2018**, *10*, 139. [CrossRef]

© 2019 by the authors. Licensee MDPI, Basel, Switzerland. This article is an open access article distributed under the terms and conditions of the Creative Commons Attribution (CC BY) license (http://creativecommons.org/licenses/by/4.0/).

Article

Viscous Hydrodynamic Description of the Pseudorapidity Density and Energy Density Estimation for Pb+Pb and Xe+Xe Collisions at the LHC

Xiong-Tao Gong [1,2], Ze-Fang Jiang [1,3,4,*], Duan She [1,4] and C. B. Yang [1,4,*]

1. College of Physical Science and Technology, Central China Normal University, Wuhan 430079, China; xggxt@mails.ccnu.edu.cn (X.-T.G.); sheduan@mails.ccnu.edu.cn (D.S.)
2. Hubei Polytechnic Institute, Xiaogan 432000, China
3. Department of Physics and Electronic-Information Engineering, Hubei Engineering University, Xiaogan 432000, China
4. Key Laboratory of Quark and Lepton Physics (MOE) & Institute of Particle Physics, Wuhan 430079, China
* Correspondence: jiangzf@mails.ccnu.edu.cn (Z.-F.J.); cbyang@mail.ccnu.edu.cn (C.-B.Y.)

Received: 23 March 2019; Accepted: 9 May 2019; Published: 10 May 2019

Abstract: Based on the analytical solution of accelerating relativistic viscous fluid hydrodynamics and Buda–Lund model, the pseudorapidity distributions of the most central Pb+Pb and Xe+Xe collisions are presented. Inspired by the CNC model, a modified energy density estimation formula is presented to investigate the dependence of the initial energy density estimation on the viscous effect. This new energy density estimation formula shows that the bulk energy is deposited to the neighboring fluid cells in the presence of the shear viscosity and bulk viscosity. In contrast to the well-known CNC energy density estimation formula, a 4.9% enhancement of the estimated energy density at the LHC kinematics is shown.

Keywords: viscous hydrodynamics; pseudorapidity distribution; energy density estimation

1. Introduction

Relativistic hydrodynamics is one of the most useful tools to investigate the space-time evolution and transport properties of the quark-gluon plasma (QGP) produced in high-energy heavy-ion collisions [1,2]. Besides numerical simulations, analytical solutions with simplified initial conditions are also useful in understanding the properties of this strongly coupled quantum chromodynamics (QCD) matter, such as the famous Hwa–Bjorken solution [3,4], Gubser solution [5], CGHK solution [6], CCHK solution [7], CNC solution [8,9], CKCJ solutions [10], and other interesting solutions [11–13]. In this paper, based on the well-known Buda–Lund model [14], an analytical solution of accelerating viscous relativistic hydrodynamics [15] is applied to investigate the final hadron pseudorapidity distribution and the energy density estimation. The charged particle pseudorapidity distributions ($dN/d\eta_p$) for the most central $\sqrt{s_{NN}} = 2.76$ TeV Pb+Pb collisions [16], $\sqrt{s_{NN}} = 5.02$ TeV Pb+Pb collisions [17] and $\sqrt{s_{NN}} = 5.44$ TeV Xe+Xe collisions [18] are presented. Based on this hydrodynamic model with longitudinal accelerating flow effect, the longitudinal acceleration parameters (λ) are extracted from those experimental systems. Based on the CNC (Csörgő, Nagy, Csanád.) energy density estimation model [8,9] and its new results [10,19], a possible relationship between the energy density estimation and viscosity effect is also investigated.

This paper is organized as follows: in Section 2, we describe the hydrodynamic solutions and calculate the pseudorapidity densities. In Section 3, the energy density estimation and its viscosity dependence are investigated. A summary and discussion are given in Section 4.

2. Pseudorapidity Distribution from Hydrodynamics

The basic formulation of relativistic hydrodynamics can be found in the literature [20,21]. In this paper, we consider a system with net conservative charge ($\mu_i = 0$). The flow velocity field is normalized to unity, $u^\mu u_\mu = 1$ and the metric tensor is chosen as $g_{\mu\nu} = \text{diag}(1, -1, -1, -1)$.

Equations of hydrodynamics can be described by the following conservation laws

$$\partial_\mu(nu^\mu) = 0, \quad \partial_\mu T^{\mu\nu} = 0, \tag{1}$$

where the first one is the continuity equation of conserved charges and the second one is the energy-momentum conservation equations. n is a conserved charge and $T^{\mu\nu}$ is the energy-momentum tensor. In the Landau frame, the energy-momentum tensor $T^{\mu\nu}$ of the fluid, in the presence of viscosity, can be expressed as

$$T^{\mu\nu} = \varepsilon u^\mu u^\nu - P\Delta^{\mu\nu} + \Pi^{\mu\nu}. \tag{2}$$

In this expression, u^μ is the velocity field, ε is the energy density, P is the pressure, $\Pi^{\mu\nu} = \pi^{\mu\nu} - \Delta^{\mu\nu}\Pi$ is the viscous stress tensor with Π the bulk pressure and $\pi^{\mu\nu}$ the stress tensor [20]. The projector $\Delta^{\mu\nu} = g^{\mu\nu} - u^\mu u^\nu$ satisfies $\Delta^{\mu\nu} u_\nu = 0$. Please note that an Equation of State (EoS) is needed for the above conservation equations. For that, $\varepsilon = \kappa P$ is frequently used with a constant κ value.

The simplest way to satisfy the second law of thermodynamics (entropy must always increase locally) is to impose the linear relationships between the thermodynamic forces and fluxes (in the Navier–Stokes limit [20,21]),

$$\Pi = -\zeta\theta, \quad \pi^{\mu\nu} = 2\eta\sigma^{\mu\nu}, \tag{3}$$

where the bulk viscosity ζ and the shear viscosity η are two positive coefficients. Please note that throughout this work we denote the shear viscosity as η, the space-time rapidity as η_s and the particle pseudorapidity as η_p.

We solved the conservation equations $\partial_\mu T^{\mu\nu} = 0$ in the Rindler coordinates and obtained a perturbative analytical solution of the relativistic viscous hydrodynamics with a longitudinally accelerating flow with constant shear viscosity to entropy density ratio and constant bulk viscosity to entropy density ratio (see detailed derivations in Ref. [15]). This analytical solution describes a finite size plasma produced in heavy-ion collision and is obtained from viscous hydrodynamics in the so-called Rindler coordinates by demanding rotational invariance around z and existing longitudinal pressure gradient along the beam direction.

The perturbative solution expression from the Ref. [15] is

$$u^\mu = (\cosh\lambda\eta_s, 0, 0, \sinh\lambda\eta_s), \tag{4}$$

$$T(\tau,\eta_s) = T_0 \left(\frac{\tau_0}{\tau}\right)^{\frac{1+\lambda^*}{\kappa}} \left[\exp\left(-\frac{1}{2}\lambda^*\left(1-\frac{1}{\kappa}\right)\eta_s^2\right) + \frac{R_0^{-1}}{\kappa-1}\left(2\lambda^* + \exp\left[-\frac{1}{2}\lambda^*\left(1-\frac{1}{\kappa}\right)\eta_s^2\right]\right.\right.$$
$$\left.\left. - (2\lambda^*+1)\left(\frac{\tau_0}{\tau}\right)^{\frac{\kappa-\lambda^*-1}{\kappa}}\right)\right], \tag{5}$$

here T_0 is the temperature at the proper time τ_0, τ is a coordinate at proper time, η_s is the space-time rapidity, $\lambda = (1 + \lambda^*)$ controls the longitudinal acceleration, R_0 is the Reynolds number and $R_0^{-1} = \frac{\Pi_d}{T_0 \tau_0}$, and $\Pi_d \equiv \left(\frac{\zeta}{s} + \frac{4\eta}{3s}\right)$ [20]. The profile of $T(\tau,\eta_s)$ is a (1+1) dimensional scaling solution in (1+3) dimensions and it contains not only acceleration but also the viscosity dependent terms now, and the η_s dependence is of the Gaussian form. Please note that when $\lambda^* = 0$ and $R_0^{-1} = 0$, one obtains the same solutions as the ideal hydrodynamics [4]. When $\lambda^* = 0$ and $R_0^{-1} \neq 0$, one obtains the first order Bjorken solutions [20]. If $\lambda^* \neq 0$ and $R_0^{-1} = 0$, one obtains a special solution which is consistent with case (c) in the CNC solutions in [8,9]. Furthermore, the temperature profile (5) implies that for a non-vanishing acceleration λ^*, the cooling rate is larger compared to the ideal case. Meanwhile, a non-zero viscosity makes the cooling rate smaller than that of the ideal case.

Based on the Buda–Lund model and Cooper-Frye formula, the pseudorapidity distribution is calculated as follow [15]

$$\begin{aligned}\frac{dN}{d\eta_p} &= N_0 \int_{-\infty}^{+\infty} d\eta_s \int_0^{+\infty} dp_T \sqrt{1 - \frac{m^2}{m_T^2 \cosh^2 y}} m_T p_T \cosh((\lambda^* + 1)\eta_s - y) \\ &\times \exp\left[-\frac{m_T}{T(\tau, \eta_s)} \cosh((\lambda^* + 1)\eta_s - y)\right] \left(\tau_f \cosh^{\frac{1-\lambda^*}{\lambda^*}}(\lambda^*\eta_s) + \frac{1+\lambda^*}{T^3(\tau, \eta_s)}\right. \\ &\times \left.\left[\frac{1}{3}\frac{\eta}{s}(p_T^2 - 2m_T^2 \sinh^2((\lambda^*+1)\eta_s - y)) + \frac{1}{5}\frac{\zeta}{s}(p_T^2 + m_T^2 \sinh^2((\lambda^*+1)\eta_s - y))\right]\right),\end{aligned} \quad (6)$$

where N_0 is the normalization parameter, $m_T = \sqrt{p_T^2 + m^2}$ is the transverse mass, p_T is the transverse momentum, m is the particle mass, η_p is the pseudorapidity of the final hadron, y is the rapidity of the final particle, and we have the relationship: $y = \frac{1}{2}\ln\frac{\sqrt{m^2+p_T^2\cosh^2\eta_p}+p_T\sinh\eta_p}{\sqrt{m^2+p_T^2\cosh^2\eta_p}-p_T\sinh\eta_p}$.

3. Relationship between the Energy Density Estimation and Viscous Effect

As given in Bjorken's paper [4], the phenomenological formula of the initial energy density estimation ϵ_{Bj} is

$$\epsilon_{Bj} = \frac{1}{S_\perp \tau_0}\frac{d\langle E\rangle}{d\eta_p} = \frac{\langle E\rangle}{S_\perp \tau_0}\frac{dN}{dy}\bigg|_{y=y_0}, \quad (7)$$

where $S_\perp$ is area of the thin transverse slab at midrapidity. For the most central collisions of identical nucleii, the transverse area can be approximated as $S_\perp = \pi R^2$, with R being the nuclear radius, $R = 1.18A^{1/3}$ fm. $\langle E\rangle$ is the average energy of final particle, y_0 is the middle rapidity τ_0 is the proper time at thermalization. This energy density was traditionally estimated by Bjorken as $\tau_0 = 1\text{fm}/c$, though the exact value of τ_0 is still a matter of debate. The volume element is $dV = (R^2\pi)\tau d\eta_s$, where $d\eta_s$ is the space-time rapidity element corresponding to the slab $S_\perp$. The energy content in this slab is $dE = \langle m_t\rangle dN$, with $\langle m_t\rangle = \sqrt{\langle p_T\rangle^2 + m^2}$ from the $\pi^\pm$, $K^\pm$, p and $\bar{p}$ average transverse momenta at midrapidity.

Based on the CNC energy density formula [8,9], for accelerationless, boost-invariant Hwa–Bjorken flow [3,4], the initial and final state space-time rapidities η_s are on the average equal to the hadron rapidity y. Thus, for a longitudinal accelerating flow, one must take:

$$\epsilon_{\text{corr}} = \epsilon_{Bj}\frac{dy}{d\eta_s^f}\frac{d\eta_s^f}{d\eta_s^i} = \epsilon_{Bj}(2\lambda - 1)\left(\frac{\tau_f}{\tau_0}\right)^{\lambda-1}, \quad (8)$$

where the upscript i and f indicate the initial state and final state, $\lambda = \lambda^* + 1$.

In addition, inspired by the recent CNC results [8,9,19], for the ideal flow, the formula of the initial energy density with the pressure evolution taken into account is:

$$\epsilon_{\text{corr}}^{\text{CNC}} = \epsilon_{Bj}(2\lambda - 1)\left(\frac{\tau_f}{\tau_0}\right)^{\lambda-1}\left(\frac{\tau_f}{\tau_0}\right)^{(\lambda-1)(1-\frac{1}{\kappa})}, \quad (9)$$

here λ is the longitudinal acceleration parameter, κ is a constant from the EoS, τ_f is the freeze-out proper time.

For a viscous fluid, because the shear viscous tensor and bulk viscous pressure affect the pressure gradient (Equation (5)), the bulk energy is deposited to the neighboring fluid cells, which results in a system energy loss (or the so-called dissipative part in the midrapidity final yield). As one can see in

Equation (7), the initial energy density is calculated from the final state charged particle multiplicity from experiments at the midrapidity. Based on the final state spectrum (see Equation (6)), one finds that the viscosity effect reduces the particle multiplicity in the midrapidity ($\frac{dN}{dy}\big|_{y=y_0}$) because of the viscosity effect. In other words, if we take into account the viscosity effect for the energy density estimation from the midrapidity experimental data, the total energy at final state is lower than that in the initial state. Such dissipative effect can be calculated from Equation (5) and the EoS. Because of such difference, the energy density estimation based on the ideal fluid method would be lower than the viscous fluid method. Based on the above analysis, a possible energy density estimation, which considers the presence of accelerating flow effect and the viscous effect, can be presented as follows:

$$\epsilon_{\text{corr}}^{\text{viscous}} = \epsilon_{\text{Bj}}(2\lambda - 1)\left(\frac{\tau_f}{\tau_0}\right)^{\lambda-1}\left(\frac{\tau_f}{\tau_0}\right)^{(\lambda-1)(1-\frac{1}{\kappa})}\left[1 + \frac{(2\lambda-1)R_0^{-1}}{\kappa-1}\left(1-\left(\frac{\tau_0}{\tau_f}\right)^{\frac{\kappa-\lambda}{\kappa}}\right)\right]^{\kappa+1}, \quad (10)$$

where the square brackets term represents the enhancement from the viscosity (based on the thermodynamical evolution). From above expression, one can find: (a) if $\lambda > 1$ and viscosity ratio $\zeta/s = \eta/s = 0.0$, it returns to the CNC energy density estimation Equation (9), (b) if $\lambda \to 1$ and viscosity ratio $\zeta/s = \eta/s = 0.0$ (or $R_0^{-1} = 0.0$), it returns to the Bjorken energy density estimation Equation (7).

We present the numerical results for the pseudorapidity density and the energy density estimation in the Figure 1. In the left panel of Figure 1, the solid curves show the calculated pseudorapidity distribution. The normalization factor is determined from the most central multiplicity $dN/d\eta_p(\eta_p = \eta_0)$ with the parameters η/s=0.16 [22], ζ/s=0.015 [23]. The freeze-out temperature is $T_f = 140$ MeV. For simplicity, $\kappa \approx 7$ is assumed to be a constant in this study [24], m=220 ± 20 MeV is an approximate average mass of the final charged particle ($\pi^\pm$, $K^\pm$, $p^\pm$) and it is calculated by a weighted average from the published experimental data [16]. The freeze-out proper time is chosen as $\tau_f = 8$ fm. The rescatterings in the hadronic phase and the decays of hadronic resonance into stable hadrons are not included here. The acceptable integral region for each space-time rapidity in the model is $-5.0 \leq \eta_s \leq 5.0$ (where the perturbative condition $\lambda^* \eta_s \ll 1$ is satisfied). We then extracted the longitudinal acceleration parameters λ for 2.76 TeV Pb+Pb, 5.02 TeV Pb+Pb and 5.44 TeV Xe+Xe, the most central colliding systems without modifying any extra independent parameters. The values of λ for different colliding systems are listed in Table 1.

Having achieved a good description of the pseudorapidity distribution for Pb+Pb and Xe+Xe collisions, we move on to the calculation for the initial energy density. In Figure 1 right panel, the color bar shows the correction factor $\epsilon_{\text{corr}}/\epsilon_{\text{Bj}}$ as a function of the ratio of freeze-out time and thermalization time (τ_f/τ_0) for λ and viscosity ratio ζ/s and η/s. The value of initial energy density is not calculated here, and we will present detailed discussion of this interesting problem in the near future as we did in Refs. [19,25]. Based on Equation (9), we found that the viscous effect results in an almost 4.9% enhancement for the energy density estimation when $\tau_f/\tau_0 = 8$. The correction factors $\epsilon_{\text{corr}}/\epsilon_{\text{Bj}}$ are presented in Table 1 for different colliding systems.

Table 1. Parameters from hydrodynamic results in the text.

$\sqrt{s_{NN}}$	System	$\frac{dn}{d\eta_p}\big|_{\eta_p=\eta_{p0}}$	λ	$\epsilon_{\text{corr}}^{\text{CNC}}/\epsilon_{\text{Bj}}$	$\epsilon_{\text{corr}}^{\text{viscous}}/\epsilon_{\text{Bj}}$
2.76 TeV	Pb+Pb	1615 ± 39.0	1.035 ± 0.003	1.225 ± 0.022	1.285 ± 0.022
5.02 TeV	Pb+Pb	1929 ± 47.0	1.032 ± 0.002	1.204 ± 0.012	1.263 ± 0.015
5.44 TeV	Xe+Xe	1167 ± 26.0	1.030 ± 0.003	1.190 ± 0.021	1.248 ± 0.022

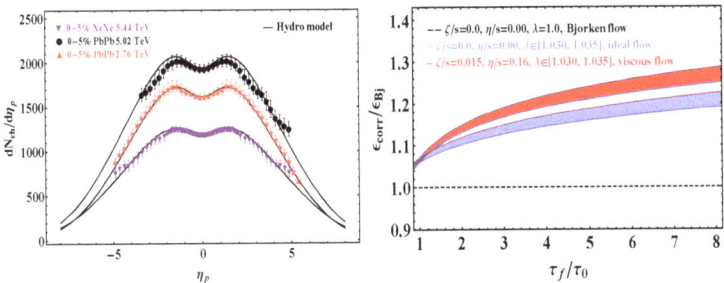

Figure 1. (**Left**): Pseudorapidity distribution from our model calculation (solid curves) compared to the LHC experimental data [16–18]. The black curves represent the pseudorapidity distribution for $\sqrt{s_{NN}} = 2.76$ TeV Pb+Pb collisions, $\sqrt{s_{NN}} = 5.02$ TeV Pb+Pb collisions and $\sqrt{s_{NN}} = 5.44$ TeV Xe+Xe collisions. (**Right**): the correction factor $\epsilon_{corr}/\epsilon_{Bj}$ as a function of the ratio of freeze-out time and thermalization time (τ_f/τ_0) for different λ and shear viscosity ratio η/s, bulk viscosity ratio ζ/s. The black dashed line is the result of Bjorken model, while the red band is the result that include the viscosity effect enhancement (Equation(10)), and the blue band is the result from Equation (9). The band width comes from the uncertainty of λ ($-1.030 \leq \lambda \leq 1.035$). For the viscous fluid, the viscosity ratio is assumed to be constant [22,23] here and the statistical analysis of viscosity ratio is not discussed here.

4. Conclusions and Discussion

In conclusion, the pseudorapidity densities and the viscosity dependence of the energy density estimation are presented in this paper based on an accelerating viscous hydro model and the experimental data for the Pb+Pb collisions and Xe+Xe collisions at the LHC energy region.

From the perturbative solution, one finds that the flow is generally decelerated due to the viscosity from the hydro solution, meanwhile, the longitudinal accelerating effect of the flow-element compensates for the decrease of pressure gradient. Those two opposite behaviors affect the thermodynamic evolution of the strong coupling QCD matter. Furthermore, from the final state expression and a good description of the experimental data at the LHC, one sees that the final state hadron spectrum is sensitive to the longitudinal flow effect. Simple modifications to the energy density estimation are proposed based on such two opposite behaviors, too. In contrast to the Bjorken model and CNC model, the viscosity effect results in a tiny enhancement for the energy density estimation. Detailed calculation of the energy density for different systems will be studied in next step.

In addition, it is also worth noting that we have made many simplifying assumptions in our hydrodynamic model, since our goal is only to show which longitudinal flow effect can be used to describe the pesudorapidity densities and to give a reasonable description for viscous effects dependence of the initial energy density estimation. For a more realistic study based on or beyond this study, the following physical effects are important and should be taken into account: the EoS, viscosity dependence (especially the bulk viscosity ratio taken from SU(3) pure-glue lattice which is of large uncertainty about which there are recently suggestions that the full QGP value may actually be significantly larger than the lattice QCD results), freeze-out hypersurface calculation, resonance decay, and rescatterings in the hadronic phase and so on. Those important effects and conditions should be studied in our future research.

Author Contributions: Z.-F.J., X.-T.G. and D.S. performed the conceptualization, methodology, and validation of the energy density estimation viscosity dependence. Hydro model formal analysis is performed by Z.-F.J. and X.-T.G.; Z.-F.J. writted the draft; C.B.Y., X.-T.G. and D.S. performed review and editing; All authors worked on finalization of the text of the manuscript.

Funding: This work was supported by the Sino-Hungarian bilateral cooperation program, under the Grand No.Te'T 12CN-1-2012-0016, by the financial supported from NNSF of China under grant No.11435004.

Acknowledgments: We especially thank T. Csörgő, M. Csanád and the Organizers of the ZIMÁNYI SCHOOL'2018 for their kind hospitality and for an inspiring and useful meeting at Budapest, Hungary. We thank Xin-Nian Wang for a useful suggestion about xeon+xeon collisions. X.-T. Gong would like to thank the staff of College of Physical Science and Technology for kind hospitality during his stay at Institute of Particle Physics, Wuhan, China. Z.-F. Jiang would like to thank Lévai Péter and Gergely Gábor Barnafoldi for kind hospitality during his stay at Winger RCP, Budapest, Hungary.

Conflicts of Interest: The authors declare no conflict of interest.

References

1. Bass, S.A.; Gyulassy, M.; Stöecker, H.; Greiner, W. Signatures of quark gluon plasma formation in high-energy heavy-ion collisions: A Critical review. *J. Phys. G Nucl. Part. Phys.* **1999**, *25* , R1–R57. [CrossRef]
2. Gyulassy, M.; McLerran, L. New forms of QCD matter discovered at RHIC. *Nucl. Phys. A* **2005**, *750*, 30–63. [CrossRef]
3. Hwa, R.C. Statistical Description of Hadron Constituents as a Basis for the Fluid Model of High-Energy Collisions. *Phys. Rev. D* **1974**, *10*, 2260. [CrossRef]
4. Bjorken, J.D. Highly Relativistic Nucleus-Nucleus Collisions: The Central Rapidity Region. *Phys. Rev. D* **1983**, *27*, 140. [CrossRef]
5. Gubser, S.S. Symmetry constraints on generalizations of Bjorken flow. *Phys. Rev. D* **2010**, *82*, 085027. [CrossRef]
6. Csörgő, T.; Grassi, F.; Hama, Y.; Kodama, T. Simple solutions of relativistic hydrodynamics for longitudinally expanding systems. *Acta Phys. Hung. A* **2003**, *21*, 53–62. [CrossRef]
7. Csörgő, T.; Csernai, L.P.; Hama, Y.; Kodama, T. Simple solutions of relativistic hydrodynamics for systems with ellipsoidal symmetry. *Acta Phys. Hung. A* **2004**, *21*, 73–84. [CrossRef]
8. Csörgő, T.; Nagy, M.I.; Csanád, M. A New family of simple solutions of perfect fluid hydrodynamics. *Phys. Lett. B* **2008**, *663*, 306. [CrossRef]
9. Nagy, M.I.; Csörgő, T.;Csanád, M. Detailed description of accelerating, simple solutions of relativistic perfect fluid hydrodynamics. *Phys. Rev. C* **2008**, *77*, 024908. [CrossRef]
10. Csörgő, T.; Kasza, G.; Csanád, M.; Jiang, Z.F. New exact solutions of relativistic hydrodynamics for longitudinally expanding fireballs. *Universe* **2018**, *4*, 69. [CrossRef]
11. Marrochio, H.; Noronha, J.; Denicol, G.S.; Luzum, M.; Jeon, S.; Gale, C. Solutions of Conformal Israel-Stewart Relativistic Viscous Fluid Dynamics. *Phys. Rev. C* **2015**, *91*, 051702. [CrossRef]
12. Hatta, Y.; Noronha, J.; Xiao, B.-W. Exact analytical solutions of second-order conformal hydrodynamics. *Phys. Rev. D* **2014**, *89*, 051702. [CrossRef]
13. Nopoush, M.; Ryblewski, R.; Strickland, M. Anisotropic hydrodynamics for conformal Gubser flow. *Phys. Rev. D* **2015**, *91*, 045007. [CrossRef]
14. Csörgő, T.; Lorstad, B. Bose-Einstein correlations for three-dimensionally expanding, cylindrically symmetric, finite systems. *Phys. Rev. C* **1996**, *54*, 1390. [CrossRef]
15. Jiang, Z.F.; Yang, C.B.; Ding, C.; Wu, X.-Y. Pseudo-rapidity distribution from a perturbative solution of viscous hydrodynamics for heavy ion collisions at RHIC and LHC. *Chin. Phys. C* **2018**, *42*, 123103. [CrossRef]
16. Adam, J.; et al. [ALICE Collaboration]. Centrality evolution of the charged-particlepseudorapidity density over a broad pseudorapidity range in Pb-Pb collisions at $\sqrt{s_{NN}}$ = 2.76 TeV. *Phys. Lett. B* **2016**, *754*, 373–385. [CrossRef]
17. Adam, J.; et al. [ALICE Collaboration]. Centrality dependence of the pseudorapidity density distribution for charged particles in Pb-Pb collisions at $\sqrt{s_{NN}}$ = 5.02 TeV. *Phys. Lett. B* **2016**, *722*, 567–577.
18. Acharya, S.; et al. [ALICE collaboration]. Centrality and pseudorapidity dependence of the charged-particle multiplicity density in Xe-Xe collisions at $\sqrt{s_{NN}}$ = 5.44 TeV. *arXiv* **2018**, arXiv:1805.04432.
19. Jiang, Z.-F.; Yang, C.B.; Csanád, M.; Csörgő, T. Accelerating hydrodynamic description of pseudorapidity density and the initial energy density in p+p , Cu + Cu, Au + Au, and Pb + Pb collisions at energies available at the BNL Relativistic Heavy Ion Collider and the CERN Large Hadron Collider. *Phys. Rev. C* **2018**, *97*, 064906.
20. Muronga, A. Causal theories of dissipative relativistic fluid dynamics for nuclear collisions. *Phys. Rev. C* **2004**, *69*, 034904. [CrossRef]

21. Romatschke, P. New Developments in Relativistic Viscous Hydrodynamics. *Int. J. Mod. Phys. E* **2010**, *19*, 1–53. [CrossRef]
22. Song, H.; Bass, S.A.; Heinz, U.; Hirano, T.; Shen, C. 200 A GeV Au+Au collisions serve a nearly perfect quark-gluon liquid. *Phys. Rev. Lett.* **2012**, *10*, 192301.
23. Meyer, H.B. Calculation of the bulk viscosity in SU(3) gluodynamics. *Phys. Rev. Lett.* **2008**, *100*, 162001. [CrossRef]
24. Pang, L.-G.; Petersen, H.; Wang, X.-N. Pseudorapidity distribution and decorrelation of anisotropic flow within CLVisc hydrodynamics. *Phys. Rev. C* **2018**, *97*, 064918. [CrossRef]
25. Jiang, Z.-F.; Csanád, M.; Kasza, G.; Yang, C.B.; Csörgő, T. Pseudorapidity and initial energy densities in p+p and heavy ion collisions at RHIC and LHC. *arXiv* **2018**, arXiv:1806.05750.

© 2019 by the authors. Licensee MDPI, Basel, Switzerland. This article is an open access article distributed under the terms and conditions of the Creative Commons Attribution (CC BY) license (http://creativecommons.org/licenses/by/4.0/).

Communication

Another Approach to Track Reconstruction: Cluster Analysis

Ferenc Siklér

Wigner Research Centre for Physics, 1525 Budapest, Hungary; sikler.ferenc@wigner.mta.hu

Received: 19 March 2019; Accepted: 3 May 2019; Published: 6 May 2019

Abstract: A novel combination of data analysis techniques is introduced for the reconstruction of primary charged particles and of daughters of photon conversions, created in high energy collisions. Instead of performing a classical trajectory building or an image transformation, efficient use of both local and global information is undertaken while keeping competing choices open. The measured hits in silicon-based tracking detectors are clustered with the help of a k-medians clustering. It proceeds by alternating between the hit-to-track assignment and the track-fit update steps, until convergence. The clustering is complemented with the possibility of adding new track hypotheses or removing unnecessary ones. A simplified model of a silicon tracker is employed to test the performance of the proposed method, showing good efficiency and purity characteristics.

Keywords: charged particle tracking; silicon trackers; cluster analysis

1. Introduction

The reconstruction of charged particles, of their trajectories, is an active area of research in high energy particle and nuclear physics. The task is usually computationally difficult (NP-hard). Detectors at today's particle colliders mostly employ large surface silicon-based tracking devices, which sample the trajectory of the emitted charged particles at several locations. When a charged particle crosses the semiconducting material, it deposits energy and creates a hit by exciting electrons to the valence band, producing electron-hole pairs. The electrons or holes, or both, are transported with an applied electric field, and their charge is read out, amplified, and digitized.

The silicon-based trackers are highly segmented; they consist of several millions of tiny pixels (dimensions of ~ 100 μm) and of narrow, but long strips (~ 10 cm in length). In a high energy collision event, several thousands of pixel and strip hits are created. Our task is to solve a mathematical puzzle: the goal is to identify particle trajectories by associating most of these hits with a limited number of true trajectories. The default solution for this problem is the combinatorial track finding and fitting [1] via the Kalman filter [2]. On the one hand, classical trajectory building utilizes mostly local information by extending the trajectory and picking up compatible hits. On the other hand, image transformation methods (e.g., variants of the Hough transform [3]) collect global information on the parameters of potential track candidates [4]. In the following, elements of an alternative track reconstruction method are outlined, with the aim of efficiently using both local and global information at the same time.

One of the goals of this study is to develop a reasonably efficient reconstruction method for converted photons, this way paving the way for a potential two-photon Bose–Einstein correlation measurement at LHC energies.

2. Methods

The k-medians clustering is a robust classification method [5,6]. It aims to partition the observations into k clusters where each observation belongs to the cluster with the nearest center. In our case, the observations are the pixel or strip hits, and the centers are the track candidates with parameters $(\eta, q/p_T, \phi_0, z_0, r_c)$, where η is the pseudorapidity, q is the electric charge, p_T is the momentum in the transverse plane, ϕ_0 is the initial azimuth angle, and z_0 is the longitudinal, while r_c is the radial coordinate of the emission point. The method consists of two alternating steps. First, each hit is assigned to the closest track candidate, and then, the parameters of the track candidates are updated by refitting their associated hits to an analytic model. The process is stopped if there are no hits changing their association (convergence) or if the number of steps exceeds a given limit.

It is important to choose a suitable measure of proximity. Because of outlier hits, the use of the sum of normalized hit-to-track distances (instead of the ordinary χ^2) provides a more robust method. In our implementation, the normalized distances are calculated through the global covariance of the measured hits; this way, no classical trajectory building through the Kalman filter is needed. This approach requires an analytic, but precise description of the main physical processes, such as multiple scattering, continuous energy loss, and bremsstrahlung, with conversion to electron-positron pairs for photons [7]. Details of such a setup are sketched in Figure 1. A photon created at point V is converted to electron-positron pair at point C, and one of the daughters is detected at point H on a cylindrical layer of the silicon detector.

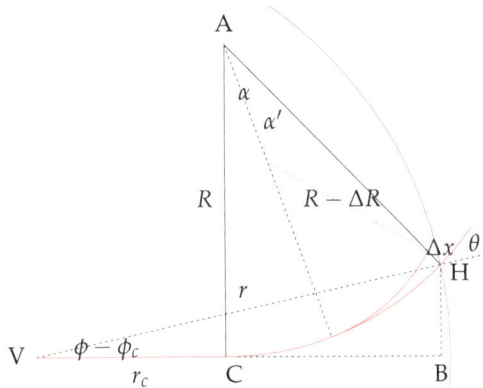

Figure 1. Geometry of a photon conversion in the plane transverse to the beam line.

The locations of trajectory hits are obviously highly correlated. The covariance between hits in layers i and j decays roughly proportionally to $\rho^{-|i-j|}$, where $\rho \approx 0.8$–0.9. With that approximation, the inverse of the covariance matrix (in the example below with four hits) is:

$$V^{-1} = \begin{pmatrix} \sigma_1^2 & \rho\sigma_1\sigma_2 & \rho^2\sigma_1\sigma_3 & \rho^3\sigma_1\sigma_4 \\ \rho\sigma_1\sigma_2 & \sigma_2^2 & \rho\sigma_2\sigma_3 & \rho^2\sigma_2\sigma_4 \\ \rho^2\sigma_1\sigma_3 & \rho\sigma_2\sigma_3 & \sigma_3^2 & \rho\sigma_3\sigma_4 \\ \rho^3\sigma_1\sigma_4 & \rho^2\sigma_2\sigma_4 & \rho\sigma_3\sigma_4 & \sigma_4^2 \end{pmatrix}^{-1} =$$

$$= \frac{1}{1-\rho^2} \begin{pmatrix} 1/\sigma_1^2 & -\rho/(\sigma_1\sigma_2) & 0 & 0 \\ -\rho/(\sigma_1\sigma_2) & (1+\rho^2)/\sigma_2^2 & -\rho/(\sigma_2\sigma_3) & 0 \\ 0 & -\rho/(\sigma_2\sigma_3) & (1+\rho^2)/\sigma_3^2 & -\rho/(\sigma_3\sigma_4) \\ 0 & 0 & -\rho/(\sigma_3\sigma_4) & 1/\sigma_4^2 \end{pmatrix}.$$

As can be seen, the inverse is tridiagonal, and in the calculation of the goodness-of-fit measure ($\sum x^T V^{-1} x$), only the differences between hits on neighboring layers have to be taken into account. Track fit to the associated hits is best accomplished by the downhill simplex method of Nelder and Mead [8]. It employs no function derivatives, but only function evaluations at the vertices of a simplex, in our case a five-simplex.

The choice for initial clusters (tracks) is an important one. The initial tracks could be chosen randomly, but much better performance can be achieved. We first find all mutual nearest hit neighbors in the angular distance, with respect to the nominal interaction point (center of the detector). Then, we take the chains of connected hits as initial clusters (Figure 2).

Figure 2. Chains of connected hits, taken as initial clusters in the k-medians clustering method.

3. Simulation Results

The above ideas are demonstrated on a very simple detector model, with cylindrical and disk-type layers of pixel and strip silicon sensors, in a barrel-and-end-cap layout (Table 1). The tracker detector was immersed in a homogeneous magnetic field of $B_z = 3.8$ T, where z was in the beam direction. Altogether, a thousand collision events with 24, 48, or 96 primary charged particles, and half as many converted photons, were generated. The primary interaction points were chosen on the z-axis, according to a normal distribution with a standard deviation of $\sigma_z = 5$ cm.

Table 1. Main characteristics of tracking detector (silicon layers) used in the simulation. For the barrel layers, the layer type is shown along with the radii (r) of the concentric cylinders and their longitudinal extent ($-z_{max}$–z_{max}) in the beam direction. For the end-cap layers, the layer type is shown along with their $|z|$ positions and with the inner (r_{min}) and outer radii (r_{max}) of their disks.

Barrel	r (cm)	z_{max} (cm)
pixels	4, 7, 10	25
strips	20, 30, 40, 50	55
strips	60, 70, 80, 90, 100, 110	55
End-Cap	$|z|$ (cm)	r_{min}–r_{max} (cm)
pixels	35, 45	5–15
strips	75, 90, 105	20–50
strips	125, 140, 155	20–110
strips	170, 185, 200	30–110
strips	220, 245	40–110
strips	270	50–110

The generated charged particles had a uniform distribution in pseudorapidity in the range $-2.5 < \eta < 2.5$ and in azimuthal angle ϕ. Their p_T distribution was proportional to $p_T^2 \exp(-p_T/p_0)$, where p_0 was chosen to be 0.2 GeV/c. Photons were generated with similar η, ϕ, and p_T distributions, but with $p_0 = 0.1$ GeV/c. Their conversion points were picked randomly in the active volume of the detector, while the momentum distribution of their conversion products (electrons are positrons) was chosen according to the simplified Tsai's formula [7].

The layer-to-layer tracking of charged particles in the homogeneous magnetic field was performed by piecewise helices. This time, instead of dealing with the details of the physical processes, the uncertainties were limited to the resolution of the local position measurement, which was modeled according to a normal distribution with a standard deviation of 1 mm. The efficiency of hit finding was taken to be 98%.

As the results of the track finding steps outlined in Section 2, hits and track candidates and their trajectories, after the first and the 30th (final) k-medians iterations, are shown in Figure 3. For primary particles, the tracking efficiency in the range $p_T > 0.5$ GeV/c was observed to be around of 90–95%. It decreased towards very low transverse momenta and reached 50% near 0.2 GeV/c. The purity was around 90%, independent of p_T. Photon conversions were found by searching for close positively- and negatively-charged track candidates in the (η, ϕ_0, rc) space (Figure 4), and the corresponding electron and positron tracks are plotted in Figure 5. For conversion electrons (and positrons), the tracking efficiency in the range $p_T > 0.6$ GeV/c was around 70%, with a slight decrease towards lower transverse momenta, and it reached 30% near 0.2 GeV/c.

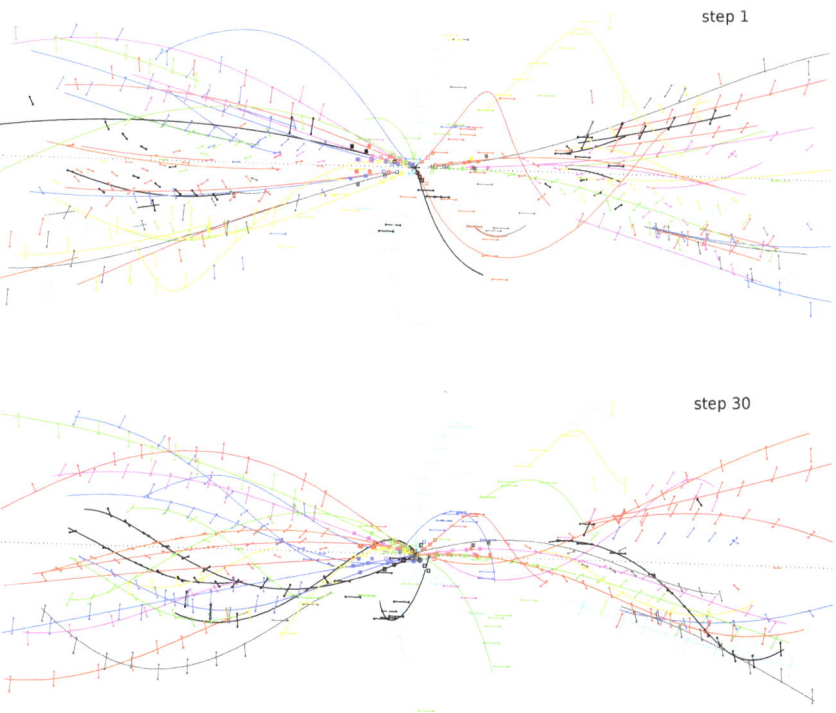

Figure 3. Hits and track candidates and their trajectories (colored curves), after the first iteration (**top**), and after the 30th iteration (**bottom**). The event is identical to the one displayed in Figure 2.

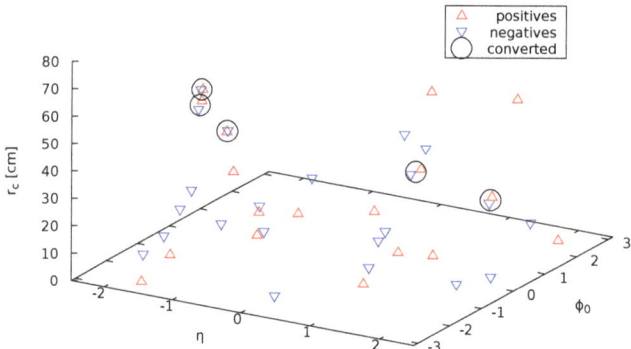

Figure 4. Identification of photon conversions in the (η, ϕ_0, rc) space of track candidates.

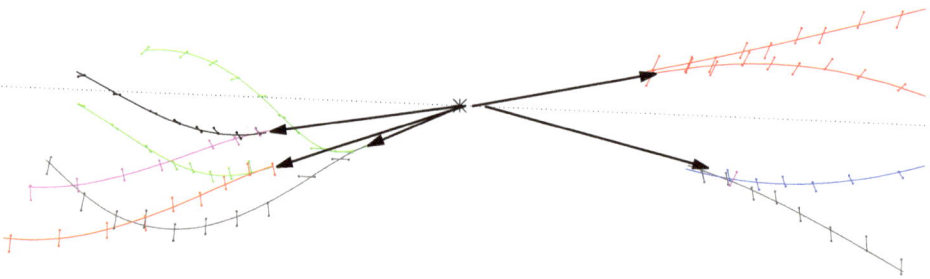

Figure 5. Hits and track candidates and their trajectories, corresponding to electron or positron tracks (colored curves) coming from photon conversions (thick black arrows). The event is identical to the one displayed in Figure 3.

According to these simple tests, the measures mentioned above were independent of the number of primary charged particles in the studied multiplicity range. The performance can be further increased by using elements from a more sophisticated Metropolis–Hastings MCMC algorithm [9], namely by sometimes adding new track hypotheses and removing unnecessary ones during the iteration process.

4. Conclusions

A novel combination of data analysis techniques was proposed for the reconstruction of all tracks of primary charged particles, as well as of daughters of displaced vertices, created in high energy collisions. Instead of performing a classical trajectory building or an image transformation, an efficient use of both local and global information was undertaken while keeping competing choices open.

The measured hits of adjacent tracking layers were clustered first with the help of a mutual nearest neighbor search in the angular distance. The resulting chains of connected hits were used as initial clusters and as input for a cluster analysis algorithm, the robust k-medians clustering. This latter proceeded by alternating between the hit-to-track assignment and the track-fit update steps, until convergence. The calculation of the hit-to-track distance and that of the track-fit χ^2 was performed through the global covariance of the measured hits. The clustering was complemented with elements from a more sophisticated Metropolis–Hastings MCMC algorithm, with the possibility of adding new track hypotheses or removing unnecessary ones.

Preliminary studies show that the proposed method provided reasonable efficiency and purity for the reconstruction of converted photons; this way, it opens the way towards an efficient identification of low momentum converted photons.

Funding: This work was supported by the National Research, Development and Innovation Office of Hungary (K 128786).

Conflicts of Interest: The author declares no conflict of interest. The funders had no role in the design of the study; in the collection, analyses, or interpretation of data; in the writing of the manuscript; nor in the decision to publish the results.

References

1. Frühwirth, R. Application of Kalman filtering to track and vertex fitting. *Nucl. Instrum. Meth. A* **1987**, *262*, 444–450. [CrossRef]
2. Kalman, R.E. A new approach to linear filtering and prediction problems. *J. Basic Eng.* **1960**, *82*, 35–45. [CrossRef]
3. Hough, P.V.C. Method and Means for Recognizing Complex Patterns. U.S. Patent 3069654, 18 December 1962.
4. Siklér, F. A combination of analysis techniques for efficient track reconstruction of high multiplicity events in silicon detectors. *Eur. Phys. J. A* **2018**, *54*, 113. [CrossRef]
5. Steinhaus, H. Sur la division des corps matériels en parties. *Bull. Acad. Pol. Sci. CL. III* **1957**, *4*, 801–804.
6. MacQueen, J. Some methods for classification and analysis of multivariate observations. In *Proceedings of the Fifth Berkeley Symposium on Mathematical Statistics and Probability, Volume 1: Statistics*; University of California Press: Berkeley, CA, USA, 1967; pp. 281–297.
7. Tanabashi, M.; Hagiwara, K.; Hikasa, K.; Nakamura, K.; Sumino, Y.; Takahashi, F.; Tanaka, J.; Agashe, K.; Aielli, G.; Amsler, C.; et al. Review of Particle Physics. *Phys. Rev. D* **2018**, *98*, 030001. [CrossRef]
8. Nelder, J.A.; Mead, R. A simplex method for function minimization. *Comput. J.* **1965**, *7*, 308–313. [CrossRef]
9. Hastings, W.K. Monte Carlo Sampling Methods Using Markov Chains and Their Applications. *Biometrika* **1970**, *57*, 97–109. [CrossRef]

© 2019 by the authors. Licensee MDPI, Basel, Switzerland. This article is an open access article distributed under the terms and conditions of the Creative Commons Attribution (CC BY) license (http://creativecommons.org/licenses/by/4.0/).

Communication

Highlights from NA61/SHINE: Proton Intermittency Analysis

Daria Prokhorova [1],*, Nikolaos Davis [2],* and on behalf of NA61/SHINE Collaboration

[1] Saint-Petersburg State University, 199034 St. Petersburg, Russia
[2] H. Niewodniczanski Institute of Nuclear Physics, Polish Academy of Sciences, 31-342 Warsaw, Poland
* Correspondence: daria.prokhorova@cern.ch (D.P.); nikolaos.davis@cern.ch (N.D.)

Received: 14 April 2019; Accepted: 29 April 2019; Published: 3 May 2019

Abstract: The NA61/SHINE experiment at CERN SPS searches for the critical point of strongly interacting matter via scanning the phase diagram by changing beam momenta ($13A$–$150A$ GeV/c) and system size (p + p, p + Pb, Be + Be, Ar + Sc, Xe + La). An observation of local proton-density fluctuations that scale as a power law of the appropriate universality class as a function of phase space bin size would signal the approach of the system to the vicinity of the possible critical point. An investigation of this phenomenon was performed in terms of the second-scaled factorial moments (SSFMs) of proton density in transverse momentum space with subtraction of a noncritical background. New NA61/SHINE preliminary analysis of Ar + Sc data at $150A$ GeV/c revealed a nontrivial intermittent behavior of proton moments. A similar effect was observed by NA49 in "Si" + Si data at $158A$ GeV/c. At the same time, no intermittency signal was detected in "C" + C and Pb + Pb events by NA49, as well as in Be + Be collisions by NA61/SHINE. EPOS1.99 also fails to describe the power-law scaling of SSFMs in Ar + Sc. Qualitatively, the effect is more pronounced with the increase of collision-peripherality and proton-purity thresholds, but a quantitative estimate is to be properly done via power-law exponent fit using the bootstrap method and compared to intermittency critical index ϕ_2, derived from 3D-Ising effective action.

Keywords: proton intermittency; power law; QCD critical point; NA61/SHINE experiment

1. Introduction

One of the most challenging tasks today is to determine the structure of the phase diagram of strongly interacting matter. State-of-the-art lattice quantum chromodynamics (QCD) calculations predict a crossover between confined and deconfined states at low baryochemical potential and high temperature at the freeze-out stage. On the other hand, at low temperatures and high baryochemical potentials, a phase transition occurs between nuclear liquid and gas. Beyond these established facts, experimental evidence [1,2] and theoretical predictions [3–5] give us a hint that nature may possess some distinct transition between hadron gas and quark–gluon plasma (QGP). The most common scenario [6,7] suggests these two regions to be separated at high baryochemical potentials and moderate temperatures by a first-order phase-transition line, which then ends at a critical point. However, the exact location of the critical end-point in the phase diagram is unknown. Moreover, some lattice QCD calculations suggest that there might be no critical point at all, with only a crossover separating the two phases.

The aim of the strong interactions program of NA61/SHINE [8], a fixed-target experiment at CERN SPS, is to study the properties of the onset of deconfinement and search for the critical point of strongly interacting matter. Since direct control of the freeze-out temperature and baryochemical potential is impossible, one can only vary the initial conditions. Therefore, the main strategy of the NA61/SHINE collaboration in this study [9] is to perform a comprehensive two-dimensional scan of the phase diagram of strongly interacting matter by changing the energy (beam momentum 13A–150A GeV/c) and the size of colliding systems (p + p, p + Pb, Be + Be, Ar + Sc, Xe + La). The characteristic signatures of the critical point could be observed, provided the system freezes out close enough to it, in the parameter space of temperature and baryochemical potential. This brings hope that critical fluctuations would not be washed out during the evolution of the system. Thus, if the critical point exists and can be reached within the NA61/SHINE phase-diagram-scan program, then, at some values of collision energy and system size, an enhancement of fluctuation signals is believed to be observed [10].

The present analysis [11] was inspired by the possibility to detect the QCD critical point not only via study of event-by-event global fluctuations of integrated quantities [12–16], but also by investigating the local power-law fluctuations [17] of the order parameters of QCD, the chiral condensate $\langle \bar{q}q \rangle$, and net-baryon density. At finite baryochemical potentials, critical fluctuations are also transferred to the net-proton density and can additionally be detected in the intermittent behavior of antiproton or proton density [18].

In experimental data, one may expect to observe proton-density fluctuations with a power-law dependence on phase-space resolution if the system freezes out right in the vicinity of the critical point [19]. The behavior of second-scaled factorial moments (SSFMs) in transverse-momentum space [20,21] as a function of the number of (equal-size) cells in which it is partitioned, was chosen to be the measure of proton-density fluctuations. Therefore, this analysis approach allows us to search for detectable intermittent behavior originating from the critical behavior of the order parameter in NA61/SHINE experimental data. The expected power-law behavior of factorial moments is quantitatively described by means of intermittency critical index ϕ_2 [19]. Theoretical prediction of its value is provided by the 3D-Ising effective action [19] since fluctuations of the order parameter at the critical point are self-similar [22], belonging to the 3D-Ising universality class. Other effects, such as resonance decays, HBT (Hanbury Brown and Twiss effect or Bose-Einstein momentum correlations), and fragmentation of jets and minijets induced by conventional strong interactions, are not expected to lead to scaling behavior of factorial moments as evidenced by experimental studies in various A + A collision systems, for example, Pb + Pb at 158A GeV/c [23], and further supported by the lack of clear intermittent behavior in EPOS-simulated [24] Ar + Sc collisions at 150A GeV/c in the present work.

New experimental evidence comes from analysis [11] of Ar + Sc collisions at 150A GeV/c, which was performed at midrapidity for three different centralities and three thresholds of proton-purity selection. The results are compared with previous intermittency analyses performed on data collected by NA49 on "C" + C, "Si" + Si and Pb + Pb most central collisions (12%, 12%, 10%, respectively) [23], and by NA61/SHINE for 10% most central Be + Be [25]. The "C" beam as defined by the online trigger and offline selection was a mixture of ions with charge Z = 6 and 7 (intensity ratio 69:31); the "Si" beam of ions with Z = 13, 14, and 15 (intensity ratio 35:41:24) [26]. Only Ar + Sc and "Si" + Si data show nontrivial intermittent behavior, as determined by the power-law scaling of their corresponding SSFMs.

2. Method of Analysis

In quantum electrodynamics (QED) interactions of ordinary matter, the behavior of the system at the critical point may be described through the phenomenon of critical opalescence [27] by the examination of scattered-photon spectra. In QCD, our tool for probing the state of a system is to measure the momenta of particles produced by a chemically and thermally excited QCD vacuum [28]. In the vicinity of a critical point, the correlation length of the system diverges [29,30], and long-range correlations appear. Moreover, the power-law decay of correlations with distance in r-space leads, through Fourier transformation, to power-law singularity of the density–density correlation function in p-space in the limit of small-momentum transfer. The latter can be detected within the framework of intermittency analysis of proton-density fluctuations in transverse-momentum space [20,21] by use of scaled factorial moments at midrapidity.

For this purpose, the available region of transverse-momentum space is partitioned into a Lattice of M^2 equal-size cells (1):

$$F_2(M) = \frac{\left\langle \frac{1}{M^2} \sum_{i=1}^{M^2} n_i(n_i - 1) \right\rangle}{\left\langle \frac{1}{M^2} \sum_{i=1}^{M^2} n_i \right\rangle^2} \quad (1)$$

If the system exhibits critical fluctuations, second-scaled factorial moments $F_2(M)$ as a function of the cell size (or the number of cells) are expected to scale (2) with M for large values of M, as a power-law with ϕ_2 being the intermittency index, which happens if system freeze-out occurs exactly at the critical point [19]:

$$F_2(M) \approx M^{2\phi_2}, \phi_2 = \phi_{2,cr}^B = \frac{5}{6} \quad (2)$$

Note that the background of noncritical proton pairs must be subtracted at the level of factorial moments in order to eliminate trivial (baseline) and noncritical correlations (with a characteristic length scale that do not scale with bin size). Thus, we can define Formula (3) a correlator $\Delta F_2(M)$ in terms of moments of original and mixed events, as well as a cross term:

$$\Delta F_2(M) = F_2^{(d)}(M) - \lambda^2(M) F_2^{(m)}(M) - \lambda(M)(1 - \lambda(M)) f_{abc}, \quad \lambda(M) \equiv \frac{\langle n_b \rangle}{\langle n \rangle} \quad (3)$$

In the limit of tiny number of critical protons (when background is dominant), which corresponds to the case of $\lambda \lesssim 1$, the simplification of Formula (3) by omitting the cross-term [23] and an approximation of noncritical background by correlation-free generated mixed events gives us Equation (4):

$$\Delta F_2^{(e)}(M) = F_2^{(d)}(M) - F_2^{(m)}(M) \quad (4)$$

Now, experimentally measured intermittency index ϕ_2 can be compared with the theoretically predicted value derived from 3D-Ising effective action. The analysis of Ar + Sc collisions at $150A$ GeV/c was performed for different centralities and purity of proton selection.

The calculation of SSFMs is smoothed by averaging over many lattice positions (lattice averaged SSFMs, see Reference [23]) and an improved estimation of statistical errors of SSFMs is achieved by use of the bootstrap method [31–33], whereby the original set of events is re-sampled with replacement [23].

It is to be noted that, while individual $F_2(M)$ errors and confidence intervals can be estimated fairly well through the bootstrap, $F_2(M)$ errors for a different M are correlated, since the same dataset is used in the calculation of all $F_2(M)$. Additional information about error correlations is contained in the full $F_2(M)$ correlation matrix, which can also be estimated through the bootstrap (see, for example, Reference [31]). Furthermore, ϕ_2 and its accompanying uncertainties should be properly determined, not through a simple χ_2-fit, but through a correlated fit. Unfortunately, such fits are plagued by instabilities [34]. We therefore resort to other methods in order to estimate ϕ_2 uncertainties, such as individually fitting bootstrap samples to obtain a distribution of ϕ_2 values and corresponding confidence intervals; however, present quoted ϕ_2 uncertainties should be considered tentative. A proton-generating modification of the Critical Monte Carlo (CMC) code [17,19] is used to simulate a system of critically correlated protons, which are mixed with a noncritical background to study the effects on the quality of intermittency analysis.

There is a necessity to apply additional quality cuts to selected protons before intermittency analysis. In particular, one must guard against the possibility of split tracks, i.e., sections of a track that are erroneously identified as a pair of distinct tracks and could therefore compromise an analysis on correlations. To this end, we impose a minimum separation distance of accepted tracks in the detector. Additionally, we calculate, for original and mixed events, the distributions of invariant four-momentum difference, q_{inv}, of proton pairs (5):

$$q_{inv}(p_i, p_j) = \frac{1}{2}\sqrt{-(p_i - p_j)^2} \qquad (5)$$

Ratio of distributions $P(q_{data})/P(q_{mixed})$ is predicted [35] to have a peak around 20 MeV/c due to strong interactions and to be suppressed for lower q_{inv} due to Fermi–Dirac effects and Coulomb repulsion. Thus, any additional peaks at low q_{inv} indicate possible split-track contamination and must be removed. This procedure led us to impose a universal cutoff of $q_{inv} > 7$ MeV/c to all sets before analysis.

3. Results

Already published results on proton intermittency in "C" + C, "Si" + Si and Pb + Pb data (Figure 1) at the same beam momenta of 158A GeV/c show the distinct signal of intermittent behavior only in "Si"+Si, although with significant statistical errors. The intermittency index value for this collision system was estimated [23] through the bootstrap as $\phi_{2,B} = 0.96^{+0.38}_{-0.25}$.

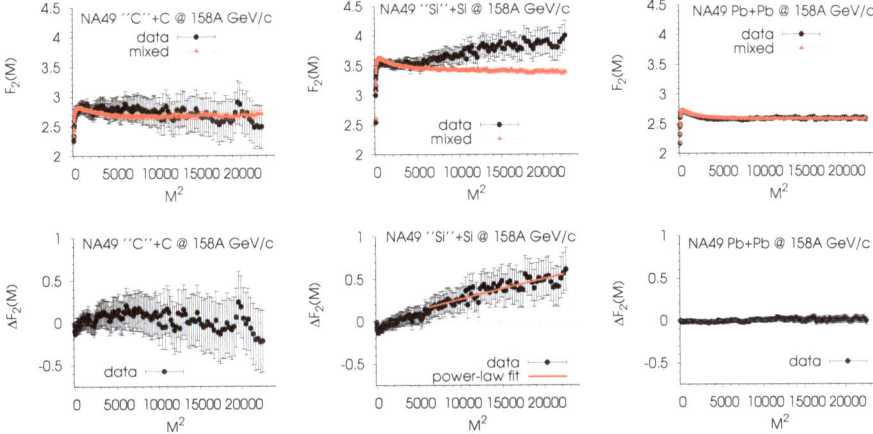

Figure 1. (**Top row**) $F_2(M)$ of original (filled circles) and mixed events (filled triangles) for NA49 "C" + C (left), "Si" + Si (middle), and Pb + Pb (right) most central collisions (12%, 12%, 10%, respectively) at 158A GeV/c ($\sqrt{s_{NN}} = 17.3$ GeV). (**Bottom row**) $\Delta F_2^{(e)}(M)$ for the corresponding systems. "Si" + Si system (middle) is fitted with a power law, $\Delta F_2^{(e)}(M;\mathcal{C},\phi_2) = e^{\mathcal{C}}(M^2)^{\phi_2}$, for $M^2 > 6000$.

After this success, analysis was extended to other intermediate-size systems for which collisions were performed by the NA61/SHINE experiment. In order to satisfy the requirements of high statistics, reliable proton identification and sufficient mean proton-multiplicity density in midrapidity (to study two-particle correlations), Be + Be and Ar + Sc at 150A GeV/c were chosen for analysis. Preliminary analysis of NA61/SHINE Be + Be events (Figure 2) at 150A GeV/c was presented in Reference [25]. $F_2(M)$ for data and mixed events overlap; thus, $\Delta F_2(M)$ fluctuates around zero, and no intermittency effect is observed.

Universe **2019**, *5*, 103

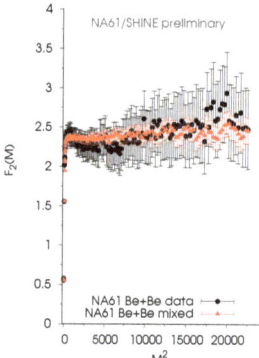

Figure 2. $F_2(M)$ of protons in NA61/SHINE 10% most central Be + Be collisions at $\sqrt{s_{NN}} = 16.8$ GeV, for data (black circles) and mixed events (red triangles).

Ar + Sc events at 150*A* GeV/*c* were analyzed [11] for three centrality bins: 0–5%, 5%–10%, and 10%–15% of most central collisions. Determination of centrality was performed by projectile spectator energy, which was deposited in PSD [36], the forward hadron calorimeter, located right at the beam line in the end of NA61/SHINE experimental facility. A scan on proton purity thresholds of 80%, 85%, and 90% was also carried out.

Figure 3 shows the results for 90% proton-purity selection for NA61/SHINE Ar + Sc datasets. One may clearly observe a significant separation of $F_2(M)$ of data from those of mixed events for the 10%–15% centrality case. For 5%–10% most central collisions, the effect is weaker, while central (0–5%) collisions show a total overlap of moments (no intermittency effect). At present, the uncertainties of $\Delta F_2(M)$, as well as the fact that they are correlated, do not permit a safe estimation of power-law quality or the calculation of confidence intervals for ϕ_2.

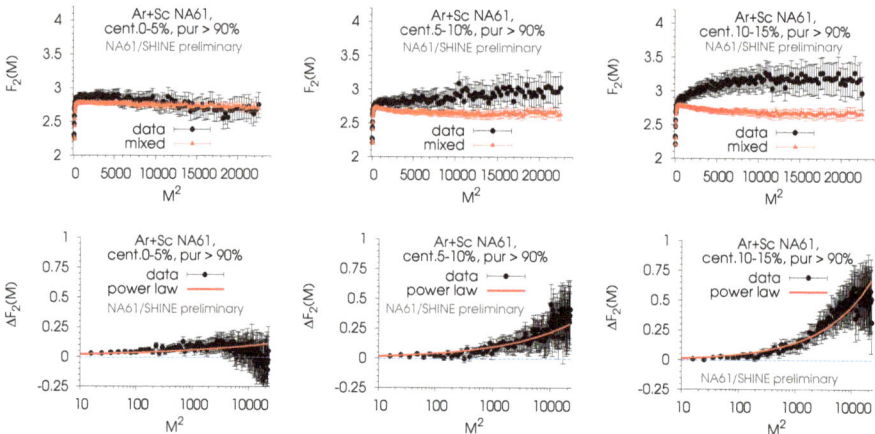

Figure 3. (**Top row**) $F_2(M)$ of original (filled circles) and mixed events (filled triangles) for NA61 Ar + Sc collisions at 0–5% (left), 5%–10% (middle), and 10%–15% (right) centrality at 150*A* GeV/*c* ($\sqrt{s_{NN}} = 16.8$ GeV). (**Bottom row**) $\Delta F_2^{(e)}(M)$ for corresponding systems. Solid curves are drawn to guide the eye and correspond to power-law scaling functions, $\Delta F_2^{(e)}(M; \mathcal{C}, \phi_2) = e^{\mathcal{C}}(M^2)^{\phi_2}$ with parameters: (left) $\phi_2 = 0.21$, $\mathcal{C} = -4.27$; (middle) $\phi_2 = 0.36$, $\mathcal{C} = -4.84$; (right) $\phi_2 = 0.49$, $\mathcal{C} = -5.4$.

In contrast, Figure 4 shows the corresponding $F_2(M)$ and $\Delta F_2(M)$ calculated within EPOS-simulated collisions. There is no prominent scaling of $\Delta F_2(M)$ for midcentral collisions: a significant overlap of data and mixed event moments merely allows to perform power-law fits (red solid lines) just to guide the eye, as the fits fail due to the prevalence of negative $\Delta F_2(M)$ values.

These results, however, are consistent with the simple but intuitive check that was performed for the ratio of $P(\Delta p_T^{data})/P(\Delta p_T^{mixed})$ distributions. The comparison of data and an EPOS event-generator simulation of the Ar + Sc system revealed a power-law-like structure for $\Delta p_T \to 0$ for middle-central (5%–10% and 10%–15%) NA61/SHINE Ar + Sc collisions, in contrast to the absence of any clear power-law structure in the corresponding EPOS spectra.

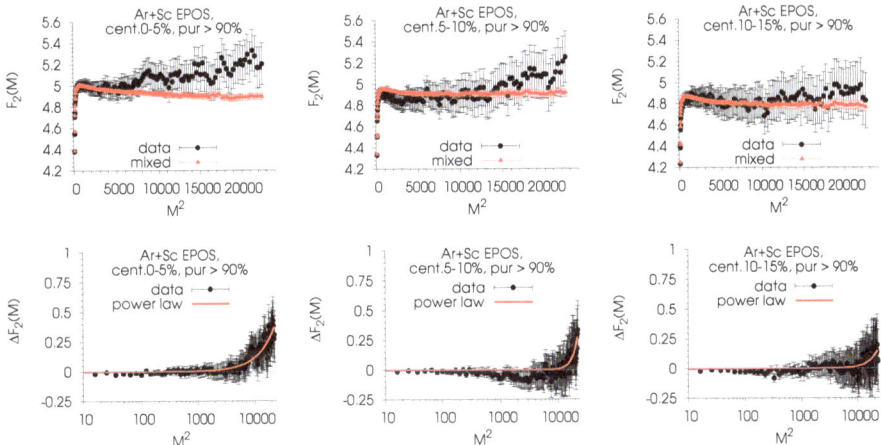

Figure 4. (**Top row**) $F_2(M)$ of original (filled circles) and mixed events (filled triangles) for EPOS Ar + Sc collisions at 0–5% (left), 5%–10% (middle) and 10%–15% (right) centrality at 150A GeV/c ($\sqrt{s_{NN}}$ = 16.8 GeV). (**Bottom row**) $\Delta F_2^e(M)$ for the corresponding systems. Solid curves are drawn to guide the eye and correspond to power-law scaling functions, $\Delta F_2^e(M; \mathcal{C}, \phi_2) = e^{\mathcal{C}}(M^2)^{\phi_2}$.

4. Discussion

The study of self-similar (power-law) fluctuations of proton density in transverse-momentum space through intermittency analysis provides us with a powerful tool for the detection of the QCD critical point. The strategy of NA61/SHINE, the successor of NA49, in this search is to perform a comprehensive two-dimensional scan of the phase diagram by changing the energy and size of colliding systems.

Up to now, no observed power-law behavior was detected for "C" + C, Pb + Pb (NA49), and Be + Be (NA61/SHINE) systems. However, a first indication of a nontrivial intermittency effect was detected in middle-central NA61/SHINE Ar + Sc collisions at 150A GeV/c, consistent with the one observed for 12% most-central "Si" + Si collisions at 158A GeV/c, but with large statistical uncertainties. For "Si" + Si, the estimated value of intermittency index $0.96^{+0.38}_{-0.25}$ overlaps with the critical QCD prediction.

Preliminary NA61/SHINE results exhibit power-law scaling of the second-scaled factorial moments $\Delta F_2(M)$ of proton density as a function of transverse-momentum bin size for Ar + Sc collisions at 150A GeV/c. Critical intermittency index ϕ_2 values are still to be properly evaluated, taking into account the magnitude of SSFM uncertainties, and the fact that $F_2(M)$ values for distinct M are correlated; the quality of $\Delta F_2(M)$ power-law scaling remains to be established, and an estimation of ϕ_2 confidence intervals is still pending. However, one may qualitatively observe that intermittent behavior in Ar + Sc shows centrality dependence possibly due to the change of baryochemical potential and the small extent of the critical region in the phase diagram [37]. The observed effect is also sensitive to proton-purity selection and increases with the increase of the purity threshold. We note that EPOS1.99 does not reproduce the observed phenomenon. NA61/SHINE continues the analysis of other systems (Xe + La and Pb + Pb) and SPS energies (Ar + Sc) in order to obtain a reliable interpretation of the observed intermittency signal.

Author Contributions: Formal analysis, N.D.; talk on behalf of NA61/SHINE Collaboration at the 18. Zimanyi School Winter Workshop on Heavy Ion Physics, 3–7 December 2018, Budapest, Hungary, D.P.; writing—original draft preparation, D.P.

Funding: This work was funded by the National Science Center, Poland (grant No. 2014/14/E/ST2/00018).

Conflicts of Interest: The authors declare no conflict of interest.

References

1. Grebieszkow, K. [NA49 Collaboration]. Search for the critical point of strongly interacting matter in NA49. *Nucl. Phys. A* **2009**, *830*, 547c–550c. [CrossRef]
2. Anticic, T.; Baatar, B.; Bartke, J.; Beck, H.; Betev, L.; Białkowska, H.; Blume, C.; Boimska, B.; Book, J.; Botje, M.; et al. [NA49 Collaboration]. Measurement of event-by-event transverse momentum and multiplicity fluctuations using strongly intensive measures $\Delta[P_T, N]$ and $\Sigma[P_T, N]$ in nucleus-nucleus collisions at the CERN Super Proton Synchrotron. *Phys. Rev. C* **2015**, *92*, 044905. [CrossRef]
3. Asakawa, M.; Yazaki, K. Chiral restoration at finite density and temperature. *Nucl. Phys. A* **1989**, *504*, 668–684.
4. Barducci, A.; Casalbuoni, R.; De Curtis, S.; Gatto, R.; Pettini, G. Chiral symmetry breaking in QCD at finite temperature and density. *Phys. Lett. B* **1989**, *231*, 463–470. [CrossRef]
5. Bialas, A.; Peschanski, R.B. Moments of rapidity distributions as a measure of short-range fluctuations in high-energy collisions. *Nucl. Phys. B* **1986**, *273*, 703–718. [CrossRef]
6. Bowman, E.S.; Kapusta, J.I. Critical points in the linear σ model with quarks. *Phys. Rev. C* **2009**, *79*, 015202. [CrossRef]
7. Stephanov, M.A. QCD Phase Diagram and the Critical Point. *Prog. Theor. Phys. Suppl.* **2004**, *153*, 139–156. [CrossRef]
8. Abgrall, N.; Andreeva, O.; Aduszkiewicz, A.; Ali, Y.; Anticic, T.; Antoniou, N.; Baatar, B.; Bay, F.; Blondel, A.; Blumer, J.; et al. [NA61 Collaboration]. NA61/SHINE facility at the CERN SPS: Beams and detector system. *J. Instrum.* **2014**, *9*, P06005. [CrossRef]
9. Gazdzicki, M. for the [NA49-future Collaboration]. A new SPS programme. *arXiv* **2006**, arXiv:nucl-ex/0612007.
10. Gazdzicki, M.; Seyboth, P. Search for critical behavior of strongly interacting matter at the CERN Super Proton Synchrotron. *Acta Phys. Pol. B* **2016**, *47*, 1201. [CrossRef]
11. Davis, N. Recent results from proton intermittency analysis in nucleus-nucleus collisions from NA61/SHINE at CERN SPS. In Proceedings of the CPOD 2018, Corfu Island, Greece, 24–28 September 2018.
12. Alt, C.; Anticic, T.; Baatar, B.; Barna, D.; Bartke, J.; Betev, L.; Bialkowska, H.; Blume, C.; Boimska, B.; Botje, M.; et al. Centrality and system size dependence of multiplicity fluctuations in nuclear collisions at 158A GeV. *Phys. Rev. C* **2007**, *75*, 064904. [CrossRef]
13. Alt, C.; Anticic, T.; Baatar, B.; Barna, D.; Bartke, J.; Betev, L.; Białkowska, H.; Blume, C.; Boimska, B.; Botje, M.; et al. Energy dependence of multiplicity fluctuations in heavy ion collisions at 20A to 158A GeV. *Phys. Rev. C* **2008**, *78*, 034914. [CrossRef]
14. Adamczyk, L.; Adkins, J.K.; Agakishiev, G.; Aggarwal, M.M.; Ahammed, Z.; Alekseev, I.; Alford, J.; Anson, C.D.; Aparin, A.; Arkhipkin, D.; et al. [STAR Collaboration]. Energy Dependence of Moments of Net-Proton Multiplicity Distributions at RHIC. *Phys. Rev. Lett.* **2014**, *112*, 032302. [CrossRef]
15. Anticic, T.; Baatar, B.; Barna, D.; Bartke, J.; Behler, M.; Betev, L.; Bialkowska, H.; Billmeier, A.; Blume, C.; Boimska, B.; et al. Transverse momentum fluctuations in nuclear collisions at 158A GeV. *Phys. Rev. C* **2004**, *70*, 034902. [CrossRef]
16. Anticic, T.; Baatar, B.; Barna, D.; Bartke, J.; Betev, L.; Białkowska, H.; Blume, C.; Boimska, B.; Botje, M.; Bunčić, P.; et al. [NA49 Collaboration]. Energy dependence of transverse momentum fluctuations in Pb+Pb collisions at the CERN Super Proton Synchrotron (SPS) at 20A to 158A GeV. *Phys. Rev. C* **2009**, *79*, 044904. [CrossRef]

17. Antoniou, N.G.; Contoyiannis, Y.F.; Diakonos, F.K.; Karanikas, A.I.; Ktorides, C.N. Pion production from a critical QCD phase. *Nucl. Phys. A* **2001**, *693*, 799–824. [CrossRef]
18. Hatta, Y.; Stephanov, M.A. Proton-Number Fluctuation as a Signal of the QCD Critical End Point. *Phys. Rev. Lett.* **2003**, *91*, 102003. [CrossRef]
19. Antoniou, N.G.; Diakonos, F.K.; Kapoyannis, A.S.; Kousouris, K.S. Critical Opalescence in Baryonic QCD Matter. *Phys. Rev. Lett.* **2006**, *97*, 032002. [CrossRef]
20. Antoniou, N.G.; Davis, N.; Diakonos, F.K. Fractality in momentum space: A signal of criticality in nuclear collisions. *Phys. Rev. C* **2016**, *93*, 014908. [CrossRef]
21. De Wolf, E.A.; Dremin, I.M.; Kittel, W. Scaling laws for density correlations and fluctuations in multiparticle dynamics. *Phys. Rep.* **1996**, *270*, 1–141. [CrossRef]
22. Vicsek, T. *Fractal Growth Phenomena*; World Scientific: Singapore, 1989; ISBN 9971-50-830-3.
23. Anticic, T.; Baatar, B.; Bartke, J.; Beck, H.; Betev, L.; Białkowska, H.; Blume, C.; Bogusz, M.; Boimska, B.; Book, J.; et al. Critical fluctuations of the proton density in A+A collisions at 158*A* GeV. *Eur. Phys. J. C* **2015**, *75*, 587. [CrossRef]
24. Werner, K.; Pierog, T.; Liu, F.M. Parton ladder splitting and the rapidity dependence of transverse momentum spectra in deuteron-gold collisions at the BNL Relativistic Heavy Ion Collider. *Phys. Rev. C* **2006**, *74*, 044902. [CrossRef]
25. Davis, N.; Antoniou, N.; Diakonos, F.K. Search for the critical point of strongly interacting matter through power-law fluctuations of the proton density in NA61/SHINE. In Proceedings of the CPOD 2017, Stony Brook, NY, USA, 7–11 August 2017.
26. Anticic, T.; Baatar, B.; Barna, D.; Bartke, J.; Beck, H.; Betev, L.; Białkowska, H.; Blume, C.; Bogusz, M.; Boimska, B.; et al. $K^*(892)^0$ and $\overline{K}^*(892)^0$ production in central Pb+Pb, Si+Si, C+C, and inelastic p+p collisions at 158*A* GeV. *Phys. Rev. C* **2011**, *84*, 064909. [CrossRef]
27. Csorgo, T. Correlation Probes of a QCD Critical Point. *arXiv* **2008**, arXiv:0903.0669.
28. Diakonos, F.K.; Antoniou, N.G.; Mavromanolakis, G. Searching for the QCD critical point in nuclear collisions. *arXiv* **2006**, arXiv:1701.02105.
29. Wosiek, J. Intermittency in the Ising Systems. *Acta Phys. Pol. B* **1988**, *19*, 863. Available online: http://www.actaphys.uj.edu.pl/fulltextseries=Regvol=19page=863 (accessed on 30 April 2019).
30. Bialas, A.; Hwa, R.C. Intermittency parameters as a possible signal for quark-gluon plasma formation. *Phys. Lett. B* **1991**, *253*, 436–438. [CrossRef]
31. Metzger, W.J. Estimating the Uncertainties of Factorial Moments. HEN-455. 2004. Available online: https://repository.ubn.ru.nl/bitstream/handle/2066/60397/60397.pdf (accessed on 30 April 2019).
32. Efron, B. Bootstrap Methods: Another Look at the Jackknife. *Ann. Stat.* **1979**, *7*, 1–26. [CrossRef]
33. Hesterberg, T.; Moore, D.S.; Monaghan, S.; Clipson, A.; Epstein, R. *Bootstrap Method and Permutation Tests*; W. H. Freeman & Co.: New York, NY, USA, 2003; ISBN 0716757265.
34. Michael, C. Fitting correlated data. *Phys. Rev. D* **1994**, *49*, 2616–2619. [CrossRef]
35. Koonin, S.E. Proton pictures of high-energy nuclear collisions. *Phys. Let. B* **1977**, *70*, 43–47. [CrossRef]
36. NA61/SHINE PSD Acceptance Map. Available online: https://edms.cern.ch/document/1867336/1 (accessed on 30 April 2019).
37. Antoniou, N.G.; Diakonos, F.K.; Maintas, X.N.; Tsagkarakis, C.E. Locating the QCD critical endpoint through finite-size scaling. *Phys. Rev. D* **2018**, *97*, 034015. [CrossRef]

© 2019 by the authors. Licensee MDPI, Basel, Switzerland. This article is an open access article distributed under the terms and conditions of the Creative Commons Attribution (CC BY) license (http://creativecommons.org/licenses/by/4.0/).

Article

Polarized Baryon Production in Heavy Ion Collisions: An Analytic Hydrodynamical Study

Bálint Boldizsár, Márton I. Nagy and Máté Csanád *

Department of Atomic Physics, Eötvös Loránd University, Pázmány P. s. 1/A, H-1117 Budapest, Hungary; boldizsar.balint@hotmail.com (B.B.); nmarci@elte.hu (M.I.N.)
* Correspondence: csanad@elte.hu

Received: 19 March 2019; Accepted: 28 April 2019; Published: 1 May 2019

Abstract: In this paper, we utilize known exact analytic solutions of perfect fluid hydrodynamics to analytically calculate the polarization of baryons produced in heavy-ion collisions. Assuming local thermodynamical equilibrium also for spin degrees of freedom, baryons get a net polarization at their formation (freeze-out). This polarization depends on the time evolution of the Quark-Gluon Plasma (QGP), which can be described as an almost perfect fluid. By using exact analytic solutions, we can thus analyze the necessity of rotation (and vorticity) for non-zero net polarization. In this paper, we give the first analytical calculations for the polarization four-vector. We use two hydrodynamical solutions; one is the spherically symmetric Hubble flow (a somewhat oversimplified model, to demonstrate the methodology); and the other solution is a somewhat more involved one that corresponds to a rotating and accelerating expansion, and is thus well-suited for the investigation of some of the main features of the time evolution of the QGP created in peripheral heavy-ion collisions (although there are still numerous features of real collision geometry that are beyond the scope of this simple model). Finally, we illustrate and discuss our results on the polarization.

Keywords: hydrodynamics; heavy ion collisions; polarization

1. Introduction

Our aim is to give analytical results for the polarization four-vector of massive spin 1/2 particles produced in heavy-ion collisions from hydrodynamical models. The motivation for this work was the recently observed non-vanishing polarization of Λ baryons at the STAR (Solenoidal Tracker at the Relativistic Heavy Ion Collider) experiment [1,2] that hints at local thermal equilibrium also for spin degrees of freedom in the Quark Gluon Plasma (QGP) produced in heavy-ion collisions. The assumption of thermal equilibration for spin is at the core of the current understanding of polarization of particles produced from a thermal ensemble (such as the QGP), and almost all studies aimed at describing it in terms of collective models utilize the formula derived from this assumption by Becattini et al. [3].

Although many numerical hydrodynamical models do indeed predict non-zero polarization of produced spin 1/2 particles [4–7], a clear connection between the initial state, the final state, and the observable polarization is to be expected from analytical studies, on which topic we do the first calculations here (to our best knowledge).

The observable quantities at the final state of the hydrodynamical evolution can be described by utilizing the kinetic theory. At local thermodynamical equilibrium, for spin 1/2 particles, such a description can be based on the the Fermi–Dirac distribution:

$$f(x,p) \propto \frac{1}{\exp\left(\frac{p_\mu u^\mu(x)}{T(x)} - \frac{\mu(x)}{T(x)}\right) + 1}, \qquad (1)$$

where p_μ is the four-momentum of the produced particle, and $u^\mu(x)$, $\mu(x)$, and $T(x)$ are the four-velocity, the chemical potential, and the temperature field of the fluid, respectively.

Assuming local thermal equilibrium for the spin degrees of freedom, for the space-time- and momentum-dependent polarization four-vector $\langle S(x,p) \rangle^\mu$ of the produced particles, the following formula is given in Ref. [3]:

$$\langle S(x,p)\rangle^\mu = \frac{1}{8m}(1-f(x,p))\varepsilon^{\mu\nu\rho\sigma}p_\sigma\partial_\nu\beta_\rho, \tag{2}$$

where m is the mass of the investigated particle, and the inverse temperature field $\beta^\mu = u^\mu/T(x)$ is introduced. Here, $\varepsilon^{\mu\nu\rho\sigma}$ is the totally antisymmetric Levi-Civita-symbol, where the $\varepsilon^{0123} = 1$ convention is used. In this paper, we use this formula to calculate the polarization four-vector at the freeze-out from analytical, relativistic, hydrodynamical solutions.

The general consensus is that the appearance of polarization strongly depends on the rotation of the expanding QGP fireball. However, the Equation of State (EoS) of the QGP influences the rotation; thus, by measuring the polarization, we can get information about the EoS of the QGP. Analytic hydrodynamic calculations may provide special insight by yielding analytic formulas for the connections of the aforementioned physical quantities.

We investigate two hydrodynamical solutions: the spherically symmetric Hubble flow [8,9] and a rotating and accelerating solution (first reported in Ref. [10], then in a different context in [11]). We expect to obtain zero polarization in the case of the spherical symmetric Hubble-flow, as it has no rotation, so the study of this solution can be regarded as a simple cross-check of our methodology. The second one, however, being a rotating and expanding solution, could be a well-usable model of peripheral heavy-ion collisions, and it is expected that one gets non-zero polarization out of it. Thus, this rotating expanding solution constitutes the core point of the reported work.

2. Basic Equations and Assumptions

We use the $c=1$ notation. Let us denote the space–time coordinate by $x^\mu \equiv (t, \mathbf{r})$, and the Minkowskian metric tensor by $g^{\mu\nu} = \mathrm{diag}(1,-1,-1,-1)$. The convention for the Levi-Civita symbol is $\varepsilon^{0123} = 1$. Greek letters denote Lorentz indices, and Latin letters denote three-vector indices. For repeated Greek indices, we use the Einstein summation convention. We denote the space dimension by d; this implies $g^\mu{}_\mu = d+1$. In reality, $d = 3$, but it is useful to retain the d notation wherever possible, in order to see whether the reason for a specific numeric constant in the formulas is the dimensionality of space. The four-velocity of the fluid is $u^\mu = \gamma(1,\mathbf{v})$, where $\gamma = \sqrt{1-v^2}$ is the Lorentz factor. The velocity three-vector is then $\mathbf{v} = u^k/u^0$. With p^μ, we denote the four-momentum of a produced particle; we also use the three-momentum $\mathbf{p}$, whose magnitude we simply denote by p (whenever there is no risk of confusion). The energy of the particle is denoted by E; the mass shell condition then reads as $E = \sqrt{p^2+m^2}$, with m being the particle mass.

The usability of hydrodynamics in heavy ion physics phenomenology relies on the assumption of local thermodynamical equilibrium of the matter. For describing particles with spin $1/2$, we use the source function as written up in Equation (1). Hadronic final state observables can then be calculated by integrating over the freeze-out hypersurface; for example, in the case of the invariant momentum distribution, the driving formula is

$$E\frac{\mathrm{d}N}{\mathrm{d}^3\mathbf{p}} = \int \mathrm{d}^3\Sigma_\mu(x) p^\mu f(x,p). \tag{3}$$

Here, $\mathrm{d}^3\Sigma_\nu$ is the three-dimensional vectorial integration measure of the freeze-out hypersurface, the appearance of which is the so-called Cooper-Frye prescription [12] for calculating the invariant momentum distribution. Of the two solutions (mentioned above) which we investigate in this work, in the case of the rotating and expanding accelerating solution, we also calculate the invariant momentum distribution, as this has not been done before.

The formula given in Ref. [3] for the polarization of spin 1/2 particles, as written up in Equation (2), may be utilized for any given $\beta^\mu = u^\mu/T$ field that one gets from a given solution of the hydrodynamical equations. We are interested in calculating the polarization at the final state of the hydrodynamical evolution, so we must integrate the $\langle S(x,p)\rangle^\mu$ field over the freeze-out hypersurface. The formula to be analyzed further, that is, that for the observed polarization $\langle S(p)\rangle^\mu$ of particles with momentum p, thus becomes

$$\langle S(p)\rangle^\mu = \frac{\int d^3\Sigma_\nu p^\nu f(x,p) \langle S(x,p)\rangle^\mu}{\int d^3\Sigma_\nu p^\nu f(x,p)}, \qquad (4)$$

as written up in, for example, [7]. For being able to perform analytical calculations, we had to make some assumptions. We used saddle-point integration, in which one assumes that the integrand is of the form $f(\mathbf{r})g(\mathbf{r})$, where $f(\mathbf{r})$ is a slowly changing function, while $g(\mathbf{r})$ has a unique and sharp maximum; then the integral can be calculated with a Gaussian approximation, as

$$\int d^d\mathbf{r}\, f(\mathbf{r})g(\mathbf{r}) \approx f(\mathbf{R}_0)g(\mathbf{R}_0)\sqrt{\frac{(2\pi)^d}{\det \mathbf{M}}}, \quad \text{where} \quad \begin{aligned} \mathbf{M}_{ij} &= \partial_i\partial_j g(\mathbf{r})|_{\mathbf{r}=\mathbf{R}_0}, \\ \partial_k g(\mathbf{R}_0) &= 0; \end{aligned} \qquad (5)$$

that is, $\mathbf{R}_0$ is the location of the unique maximum of $g(\mathbf{r})$, and $\mathbf{M}$ is the second derivative matrix.

Another assumption concerns the expression of $\langle S(x,p)\rangle^\mu$, Equation (2): if the exponent in the Fermi–Dirac distribution is large (i.e., phase space occupancy is small), we can use the Maxwell–Boltzmann distribution instead:

$$f(x,p) \ll 1 \quad \Rightarrow \quad f(x^\mu, p^\mu) = \frac{g}{(2\pi\hbar)^d}\exp\Big(\frac{\mu(x)}{T(x)} - \frac{p_\mu u^\mu}{T(x)}\Big). \qquad (6)$$

Here, g is the spin-degeneracy factor; for spin 1/2 baryons, $g = 2$.

In high-energy heavy ion phenomenology (when the collision energy is high enough, say for collisions at the Relativistic Heavy Ion Collider or the Large Hadron Collider), the μ/T factor can be (and usually is) neglected; we use this approximation here[1]. With this, we have

$$f(x^\mu, p^\mu) = C_0 \exp\big(-p_\mu \beta^\mu(x)\big), \quad \text{where} \quad \beta^\mu(x) = \frac{u^\mu(x)}{T(x)}, \quad \text{and} \quad C_0 = \frac{g}{(2\pi\hbar)^d}. \qquad (7)$$

If the Maxwell-Boltzmann approximation is justified, it means that $f(x,p) \ll 1$ indeed, and Equations (2) and (4) then also become simpler:

$$\langle S(x,p)\rangle^\mu = \frac{1}{8m}\varepsilon^{\mu\nu\rho\sigma} p_\sigma \partial_\nu \beta_\rho, \qquad (8)$$

and in the saddle-point approximation, the polarization of particles with momentum p simply becomes

$$\langle S(p)\rangle^\mu \approx \frac{1}{8m}\varepsilon^{\mu\nu\rho\sigma} p_\sigma \partial_\nu \beta_\rho\Big|_{\mathbf{r}=\mathbf{R}_0}, \qquad (9)$$

---

[1] The vanishing of μ can also be interpreted as an absence of a conserved particle number density n. All our conclusions would change only by a proportionality factor if we said μ/T = const instead of $\mu/T = 0$; if $\mu \neq 0$, we would have had to introduce n. Depending on the EoS (Equation of State) of the matter (one that also contains the conserved particle density n), one could write the $f(x,p)$ function in another form, where the normalization $\int dp\, f(x,p) = n(x)$ is evident. For example, if one chooses an ultra-relativistic ideal gas, with $p = nT$, $\varepsilon = \kappa p$, with $\kappa = d$ as EoS, one has $\frac{g}{(2\pi\hbar)^d}e^{\mu/T} = \frac{n}{4\pi T^3}$. Indeed, in the solutions discussed below, μ/T =const is satisfied, which means $n \propto T^d$, which is the well-known condition for an adiabatic expansion.

since in the saddle-point approximation, in the numerator of Equation (4), $\langle S(x,p) \rangle^\mu$ can be considered the "smooth"' function, and the determinant factors cancel out.

3. Some Exact Hydrodynamical Solutions and Polarization

In this section, we first specify and recapitulate the investigated hydrodynamical solutions, then give the analytical formulas for the polarization four-vector calculated from them. The equations of perfect fluid relativistic hydrodynamics utilized here are

$$(\varepsilon+p)u^\nu \partial_\nu u^\mu = (g^{\mu\nu} - u^\mu u^\nu)\partial_\nu p \quad \text{(Euler equation)},$$
$$(\varepsilon+p)\partial_\mu u^\mu = -u^\mu \partial_\mu \varepsilon \quad \text{(energy conservation equation)},$$
$$n\partial_\mu u^\mu = -u^\mu \partial_\mu n \quad \text{(particle number/charge conservation)},$$

and we specify the simple EoS:

$$\varepsilon = \kappa p, \tag{10}$$

where the notations are: ε, p, and n are the energy density, pressure, and particle number density, respectively. Concerning the n density: if it was assumed to be non-vanishing, we set the EoS as $p = nT$. However, the solutions presented below are valid also if $n = 0$ (i.e., if $\mu = 0$). Thus, the expressions for n that we wrote up for the solutions can be regarded as supplemental to the solutions that work for $\mu = 0$.

We note that the simple analytic solutions of perfect fluid hydrodynamics that we utilize in this manuscript all assume this simple form of EoS, $\varepsilon = \kappa p$. Finding exact analytic relativistic solutions for a more complex equation of state is a daunting task (however, some simple developments have gradually been made in this direction, see e.g., Ref. [13]), but would be nevertheless required if one wants to use the methodology presented here to give constraints on the equation of state from the measured polarization effect of baryons. Such more general studies are beyond the scope of the present work, wherein we lay the groundwork for the analytic calculation of polarization. So we stick to the simple solutions (and their simple equation of state, $\varepsilon = \kappa p$) as discussed below.

We also note that there is recent development on taking the effect that polarization of the constituents of the fluid has on the fluid dynamics itself [14], along with some numerical calculations of how this modified hydrodynamical picture affects final state polarization [15]. We do not investigate this possibility here; we restrict ourselves to the simple and well-known basic equations written up above.

3.1. Hubble Flow

We do not go into the details about the method for finding or verifying that the solutions presented below are indeed solutions of the perfect fluid hydrodynamical equations; we refer back to the original publications of the solutions.

We investigate the Hubble-like relativistic hydrodynamical solution, firstly fully described in Ref. [8]. This solution has the following velocity, particle density, and temperature fields:

$$u^\mu = \frac{x^\mu}{\tau}, \qquad n = n_0\left(\frac{\tau_0}{\tau}\right)^d, \qquad T = T_0\left(\frac{\tau_0}{\tau}\right)^{d/\kappa}, \tag{11}$$

where $\tau = \sqrt{t^2 - r^2}$, and κ is the inverse square speed of sound (constant in the case of this exact solution). The $\kappa = 3$ case corresponds to ultra-relativistic ideal gas, and $\kappa = 3/2$ corresponds to a non-relativistic gas; however, this solution is valid for any arbitrary constant κ value[2]

To calculate the polarization four-vector, as of now we investigate the simplest case, the spherical symmetric expansion. For the freeze-out hypersurface, the $\tau = \tau_0 =$ const. hypersurface was chosen (which, in the case of the investigated solution, equals the constant temperature freeze-out hypersurface), and a given point of this hypersurface can simply be parametrized by the $\mathbf{r}$ coordinate three-vector, and the time coordinate on the hypersurface is $t(\mathbf{r}) \equiv \sqrt{\tau_0^2 + r^2}$. The integration measure and the resulting expression for the Cooper–Frye formula can then be written as

$$d^3\Sigma_\mu = \frac{1}{t(\mathbf{r})}\begin{pmatrix} t(\mathbf{r}) \\ \mathbf{r} \end{pmatrix} d^3\mathbf{r} \quad \Rightarrow \quad E\frac{dN}{d^3\mathbf{p}} = C_0 \int d^3\mathbf{r}\, \frac{Et(\mathbf{r}) - \mathbf{p}\mathbf{r}}{t(\mathbf{r})} \exp\left(-\frac{Et(\mathbf{r}) - \mathbf{p}\mathbf{r}}{T_0}\right). \quad (12)$$

As we are discussing massive particles, this integral always exists. The T_0 constant (an arbitrary parameter of the solution) can simply be taken as the temperature at freeze-out, and we did so.

The position of the saddle-point ($\mathbf{R}_0$), as well as the second derivative matrix M_{kl} is calculated as:

$$\partial_k \frac{Et - \mathbf{p}\mathbf{r}}{T_0}\bigg|_{\mathbf{r}=\mathbf{R}_0} \stackrel{!}{=} 0 \quad \Rightarrow \quad \mathbf{R}_0 = \frac{\tau_0}{m}\mathbf{p}. \qquad M_{kl} \equiv -\partial_k \partial_l \frac{Et - \mathbf{p}\mathbf{r}}{T_0}\bigg|_{\mathbf{r}=\mathbf{R}_0} = \frac{m}{T_0 \tau_0}\left(\delta_{kl} - \frac{p_k p_l}{E^2}\right). \quad (13)$$

With this, we can get an approximation for the invariant single-particle momentum distribution:

$$\det \mathbf{M} = \frac{m^2}{E^2}\left(\frac{m}{T_0 \tau_0}\right)^3 \quad \Rightarrow \quad E\frac{dN}{d^3\mathbf{p}} = \frac{n_0}{4}\sqrt{\frac{\pi \tau_0^3}{m T_0^3}} \exp\left(-\frac{\tau_0 m}{T_0}\right). \quad (14)$$

The formula is independent of momentum. This was expected because this hydrodynamical solution (in the $\mathcal{V}(S) = 1$ case) is boost invariant.

To use (9) to determine the polarization four-vector in the hydrodynamical solution of the Hubble-flow, first we give the expression for the $\partial_\nu \beta_\rho$ derivative:

$$\partial_\nu \beta_\rho = \partial_\nu \left(\frac{r_\rho}{\sqrt{\tau_0^2 + r^2}\, T_0}\right) = \frac{g_{\nu\rho}}{\sqrt{\tau_0^2 + r^2}\, T_0} + \frac{r_\nu r_\rho}{(\tau_0^2 + r^2)^{3/2} T_0}. \quad (15)$$

Then, for the time component, we get:

$$\langle S(p) \rangle^0 = \frac{1}{8mT_0}\varepsilon^{0ikl} p_l \partial_i \beta_k \bigg|_{\mathbf{r}=\mathbf{R}_0} = \frac{1}{8mT_0}\varepsilon_{ikl} p_l \left(\frac{g_{ik}}{\sqrt{\tau_0^2 + r^2}\, T_0} + \frac{r_i r_k}{(\tau_0^2 + r^2)^{3/2} T_0}\right)\bigg|_{\mathbf{r}=\mathbf{R}_0} = 0, \quad (16)$$

as ε^{0ikl} is antisymmetric, whereas g_{ik} and $r_i r_k$ are symmetric to the change in the $i \leftrightarrow k$ indices.

---

[2] We note that a more general class of solutions is possible [8,9,16] in which the temperature and density fields are supplemented with an arbitrary $\mathcal{V}$ function of a "scaling variable" S:

$$n = n_0 \left(\frac{\tau_0}{\tau}\right)^d \mathcal{V}(S), \qquad T = T_0 \left(\frac{\tau_0}{\tau}\right)^{d/\kappa} \frac{1}{\mathcal{V}(S)},$$

and the S variable is any function of S_x, S_y, and S_z:

$$S \equiv S(S_x, S_y, S_z), \quad \text{where} \quad S_x \equiv \frac{r_x^2}{\dot{X}_0^2 t^2}, \quad S_y \equiv \frac{r_y^2}{\dot{X}_0^2 t^2}, \quad S_z \equiv \frac{r_z^2}{\dot{X}_0^2 t^2}, \quad \text{for example:} \quad S = \frac{r_x^2}{\dot{X}_0^2 t^2} + \frac{r_y^2}{\dot{Y}_0^2 t^2} + \frac{r_z^2}{\dot{Z}_0^2 t^2}.$$

Here $\dot{X}_0$, $\dot{Y}_0$ and $\dot{Z}_0$ are arbitrary constants. In the given example, the $S =$ const surfaces are ellipsoids, and $\dot{X}_0$, $\dot{Y}_0$, $\dot{Z}_0$ are time derivatives of the principal axes of them.

Similarly for the spatial coordinates:

$$\langle S(p)\rangle^i = \frac{1}{8mT_0}\left(-\varepsilon_{ikl}p_l\partial_0\beta_k + \varepsilon_{ikl}p_l\partial_k\beta_0 - \varepsilon_{ikl}p_0\partial_k\beta_l\right)\bigg|_{\mathbf{r}=\mathbf{R}_0} = 0. \tag{17}$$

In conclusion, the polarization four-vector in the spherical symmetric Hubble-flow is

$$\langle S(p)\rangle^\mu = \begin{pmatrix} 0 \\ 0 \end{pmatrix}, \tag{18}$$

which is consistent with our expectations.

3.2. Rotating and Accelerating Expanding Solution

Another hydrodynamical solution of particular interest to us is a rotating and accelerating expanding solution, first written up in Ref. [10]. This solution has the following velocity, temperature, and particle density profiles:

$$\mathbf{v} = \frac{2t\mathbf{r} + \tau_0^2 \mathbf{\Omega} \times \mathbf{r}}{t^2 + r^2 + \rho_0^2}, \qquad T = \frac{T_0\tau_0^2}{\sqrt{(t^2 - r^2 + \rho_0^2)^2 + 4\rho_0^2 r^2 - \tau_0^4(\mathbf{\Omega} \times \mathbf{r})^2}}, \qquad n = n_0\left(\frac{T}{T_0}\right)^3, \tag{19}$$

where ρ_0 and τ_0 are arbitrary parameters, and $\mathbf{\Omega}$ is an arbitrary angular velocity three-vector that indicates the axis and magnitude of rotation. The ρ_0 parameter tells about the initial spatial extent of the expanding matter; however, the τ_0 parameter is just there for the sake of consistency of physical units; in this way, the unit of $\mathbf{\Omega}$ is c/fm, as it should be for an angular velocity-like quantity[3], and T_0 is a temperature constant. In the case of $\mathbf{\Omega} = 0$, we recovered an acceleratingly expanding but non-rotating spherically symmetric solution.

We note that in general $\mathbf{\Omega} \neq 0$, hence this solution has non-vanishing acceleration and rotation, as well as spatially non-trivial (i.e., not spherically symmetric) temperature distribution (and temperature gradient). In the $\mathbf{\Omega} \to 0$-limiting case (as noted above), the accelerating nature persists. However, in this case, the temperature distribution becomes spherically symmetric, and *at the same time*, the vorticity of the flow vanishes. In the case of this simple solution, we thus cannot choose the free parameters in a way to separately turn on and off these features, and thus cannot analytically disentangle the effects that these features have on the final state polarization. (Some numerical calculations of polarization, e.g., the one found in Ref. [17] state that different components of the polarization vector are influenced differently by these features of a realistic hydrodynamical expansion).

Turning to the calculation of polarization, it is convenient to write up the investigated solution with the following notation:

$$\frac{u^\mu}{T} \equiv \beta^\mu = a^\mu + F^{\mu\nu}x_\nu + (x^\nu b_\nu)x^\mu - \frac{x^\nu x_\nu}{2}b^\mu, \tag{20}$$

with $\quad a^\mu = \dfrac{\rho_0^2}{2T_0\tau_0^2}\begin{pmatrix}1\\0\end{pmatrix}, \qquad b^\mu = \dfrac{1}{T_0\tau_0^2}\begin{pmatrix}1\\0\end{pmatrix}, \qquad F_{0k} = F_{k0} = F_{00} = 0, \qquad F_{kl} = \varepsilon_{klm}\dfrac{\Omega_m}{2T_0}. \tag{21}$

To calculate final state observables, we choose the constant proper time ($\tau_0 = \text{const}$) hypersurface here as well. The solution itself allows for a re-scaling of the arbitrary constants in the formulas; just as in the previous case, here too we can treat the T_0 quantity as the temperature at freeze-out (at the $\mathbf{r} = 0$ center of the expanding matter). We use the notation introduced in Equation (12) for the

---

[3] Here we changed the notation of Ref. [10]. The rather unfortunate **B** notation used there is now written as $\tau_0^2\mathbf{\Omega}$.

Maxwell–Boltzmann distribution. To derive the saddle-point for the calculation of the polarization four-vector, we shall use the expression of the invariant momentum spectrum:

$$E\frac{dN}{d^3\mathbf{p}} = C_0 \int d^3\mathbf{r} \left(E - \frac{\mathbf{p}\mathbf{r}}{\sqrt{\tau_0^2+r^2}}\right) \exp\left\{-\frac{E(2r^2+\tau_0^2+\rho_0^2) - 2\sqrt{\tau_0^2+r^2}\mathbf{p}\mathbf{r} - \tau_0^2 \mathbf{r}(\mathbf{p}\times\mathbf{\Omega})}{T_0\tau_0^2}\right\}. \quad (22)$$

This integral always exists (in the case of massive particles). In order to utilize the saddle-point integration method, we determine the position of the saddle-point ($\mathbf{R}_0$) and the second derivative matrix at the saddle-point:

$$\text{for } \mathbf{R}_0: \quad \nabla\left\{-\frac{1}{T_0\tau_0^2}\left(E(2r^2+\tau_0^2+\rho_0^2) - 2\sqrt{\tau_0^2+r^2}\mathbf{r}\mathbf{p} - \tau_0^2 \mathbf{r}(\mathbf{p}\times\mathbf{\Omega})\right)\right\}\bigg|_{\mathbf{r}=\mathbf{R}_0} \stackrel{!}{=} 0, \quad (23)$$

$$M_{kl} = \partial_k \partial_l \left\{\frac{1}{T_0\tau_0^2}\left(E(2r^2+\tau_0^2+\rho_0^2) - 2\sqrt{\tau_0^2+r^2}\mathbf{r}\mathbf{p} - \tau_0^2 \mathbf{r}(\mathbf{p}\times\mathbf{\Omega})\right)\right\}\bigg|_{\mathbf{r}=\mathbf{R}_0}. \quad (24)$$

We leave the detailed calculations to Appendix A; the results are the following. The $\mathbf{R}_0$ saddle-point (for a given $\mathbf{p}$ momentum) is in the plane spanned by the $\mathbf{p}$ and $\mathbf{p}\times\mathbf{\Omega}$ vectors. In the following, we use the $\hat{\mathbf{p}} \equiv \mathbf{p}/p$ notation for the unit vector pointing in the direction of $\mathbf{p}$. For the saddle-point, we get

$$\mathbf{R}_0 = \frac{\tau_0}{2p}\sqrt{\frac{E-m}{2m}}\sqrt{\tau_0^2(\hat{\mathbf{p}}\times\mathbf{\Omega})^2(E-m)^2+4p^2} \cdot \hat{\mathbf{p}} + \tau_0^2 \frac{E-m}{2p} \cdot \hat{\mathbf{p}}\times\mathbf{\Omega}. \quad (25)$$

Concerning the second derivative matrix, we need it only for the calculation of the invariant momentum distribution, where its determinant is invoked. It turns out that this quantity is

$$\det M_{kl} = \frac{32m^2}{T_0^3 \tau_0^6}(E+m)p. \quad (26)$$

Using this result, we get the invariant single-particle momentum distribution[4] as

$$E\frac{dN}{d^3\mathbf{p}} \propto \sqrt{\frac{\pi^3 T_0^3 \tau_0^3}{32p(m+E)}} \exp\left(-\frac{E_{\text{eff}}}{T_0}\right), \quad \text{with} \quad E_{\text{eff}} = m + \frac{\rho_0^2 E}{\tau_0^2} + \frac{\tau_0^2}{4}(\Omega^2 - (\hat{\mathbf{p}}\mathbf{\Omega})^2)(E-m). \quad (27)$$

Equivalently, by defining a "local slope" T_{eff}, the result can be expressed as

$$E\frac{dN}{d^3\mathbf{p}} \propto \sqrt{\frac{\pi^3 T_0^3 \tau_0^3}{32p(m+E)}} \exp\left(-\frac{E}{T_{\text{eff}}}\right), \quad \text{with} \quad T_{\text{eff}} = \frac{T_0}{\frac{m}{E} + \frac{\rho_0^2}{\tau_0^2} + \frac{\tau_0^2}{4}(\Omega^2-(\hat{\mathbf{p}}\mathbf{\Omega})^2)\left(1-\frac{m}{E}\right)}. \quad (28)$$

Proceeding to the polarization of the produced baryons, we calculated the derivative of the inverse temperature field for this solution from the form given in Equation (20), and then substituted it into the expression of the polarization, Equation (9). The result is

$$\partial_\nu \beta_\rho = F_{\rho\nu} + x^\alpha b_\alpha g_{\nu\rho} + x_\rho b_\nu - x_\nu b_\rho \quad \Rightarrow \quad \langle S(p)\rangle^\mu = \frac{1}{8m}\varepsilon^{\mu\nu\rho\sigma}p_\sigma\left(F_{\rho\nu} + x_\rho b_\nu - x_\nu b_\rho\right)\bigg|_{\mathbf{r}=\mathbf{R}_0}. \quad (29)$$

(The second term was cancelled owing to the symmetry of $g_{\nu\rho}$ and the antisymmetry of $\varepsilon^{\mu\nu\rho\sigma}$, and x^μ is understood as the four-coordinate of the freeze-out hypersurface whose three-coordinate is

---

[4] This has not yet been calculated for this hydrodynamical solution.

the $\mathbf{r} = \mathbf{R}_0$ three-vector). Remembering the expression of the introduced $F^{\mu\nu}$ tensor and b^μ vector from Equation (20), in particular that of $F^{0k} = 0$, and $b^k = 0$, we got the following expressions for the the time-like and space-like components:

$$\langle S(p)\rangle^0 = -\frac{1}{8m}\varepsilon^{0klm}p_m(F_{kl}+x_lb_k-x_kb_l)\Big|_{\mathbf{r}=\mathbf{R}_0} = -\frac{1}{16m}\varepsilon_{klm}\varepsilon_{klq}p_m\frac{\Omega_q}{T_0} = \frac{1}{8m}\frac{\mathbf{p}\Omega}{T_0},$$

$$\langle S(p)\rangle^k = \frac{1}{8m}\left(\varepsilon^{k0lr}p_r(F_{l0}+x_lb_0-x_0b_l)+\varepsilon^{kl0r}p_r(F_{0l}+x_0b_l-x_lb_0)+\varepsilon^{klr0}p_0(F_{rl}+x_rb_l-x_lb_r)\right)\Big|_{\mathbf{r}=\mathbf{R}_0}$$

$$= -\frac{1}{8m}\left(2b_0\varepsilon_{klm}x_lp_m+E\varepsilon_{klm}\varepsilon_{mlq}\frac{\Omega_q}{2T_0}\right)\Big|_{\mathbf{r}=\mathbf{R}_0} = \frac{1}{8mT_0}\left(E\Omega - \frac{2}{\tau_0^2}\mathbf{R}_0\times\mathbf{p}\right)_k$$

$$= \frac{m\Omega_k+(E-m)\hat{p}_l\Omega_l\hat{p}_k}{8mT_0}.$$

Summarizing this result, the polarization four-vector for the investigated rotating and accelerating expanding solution is the following:

$$\langle S(p)\rangle^\mu = \frac{1}{8mT_0}\left(m\Omega + \frac{E-m}{p^2}(\Omega\mathbf{p})\mathbf{p}\,\Big|\,\frac{\mathbf{p}\Omega}{}\right). \tag{30}$$

In the case of $\Omega = 0$, there is no rotation, and we got $\langle S(p)\rangle^\mu = 0$. Thus, in this model, polarization is very transparently connected to the presence of rotation.

It is useful to transform the polarization four-vector into the rest frame of the particle. The result is[5], with (r.f. standing for "rest frame"):

$$\langle S(p)\rangle^\mu_{\text{r.f.}} = \begin{pmatrix} 0 \\ \mathbf{S}_{\text{r.f.}}\end{pmatrix}, \quad \text{where} \quad \mathbf{S}_{\text{r.f.}} = \frac{1}{8T_0}\Omega. \tag{31}$$

We can also compute the helicity of the produced spin 1/2 particles in this solution from this formula (the $\mathbf{S}$ polarization vector is taken in the laboratory frame):

$$H := \hat{\mathbf{p}}\mathbf{S} = \frac{E}{8mT_0}\hat{\mathbf{p}}\Omega. \tag{32}$$

4. Illustration and Discussion

In this section, we would like to illustrate our simple analytical results for the polarization vector. We use the same type of plots that were used to visualize some existing numerical simulations (e.g., those presented in Ref. [7]). We plot the components of the polarization vector with respect to the momentum components in the transverse plane (that is, w.r.t. momentum components p_x and p_y). On Figure 1, we plot the polarization vector in the laboratory frame. For the sake of plotting, the mass of the Λ baryon ($m_\Lambda c^2 = 1115$ MeV) was chosen. For the sake of this illustration, we chose a moderate value for the magnitude of the Ω vector as $|\Omega| = 0.1\,c/\text{fm}$.

---

[5] The Lorentz matrix performing this boost transformation is the following (in usual 1+3 dimensional block matrix notation):

$$\Lambda^\mu{}_\nu = \begin{pmatrix} \cosh\chi & -\hat{p}_l\sinh\chi \\ -\hat{p}_k\sinh\chi & \delta_{kl}+(\cosh\chi-1)\hat{p}_k\hat{p}_l \end{pmatrix} = \frac{1}{m}\begin{pmatrix} E & -p_l \\ -p_k & m\delta_{kl}+\frac{E-m}{p^2}p_kp_l \end{pmatrix},$$

where E and p could be parametrized with the velocity parameter χ as $E = m\cosh\chi$ and $p = m\sinh\chi$, respectively. It indeed can be checked that this matrix takes the $(E,\mathbf{p})$ four-momentum vector into $(m,\mathbf{0})$, as it should.

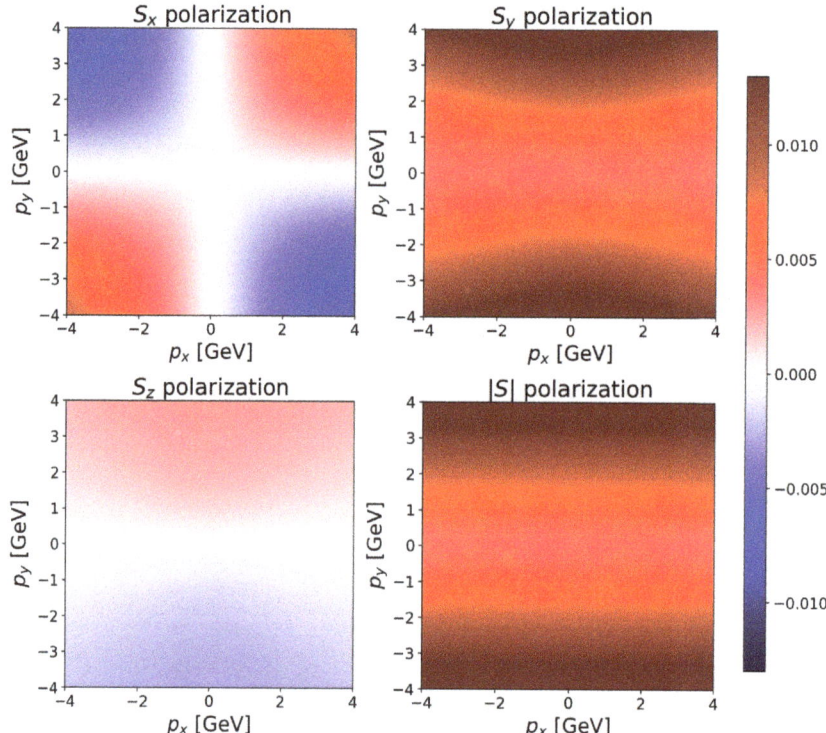

Figure 1. The components of the polarization four-vector in the rotating and accelerating expanding solution with respect to the momentum. Plots were made with the mass of the Λ baryon ($m_\Lambda = 1115 \, \text{MeV}/c^2$), and with $|\mathbf{\Omega}| = 0.1 \, c/\text{fm}$.

In our case, as a special coincidence owing purely to the specific algebraic form of the presented analytic solution, it turned out that the polarization in the rest frame of the produced baryons was independent of momentum **p**; see Equation (31). This coincidence is expected to be relieved in the case of more involved (complicated) solutions (that are left for future investigations). Figure 2, nevertheless shows the value of the S_y component in the baryon rest frame.

The helicity of the produced baryons (being proportional to the **pS** scalar product), however, *does* depend on the momentum, even in the case of our very simple solution. We plot it on Figure 3, with the same parameter values as in the foregoing two plots.

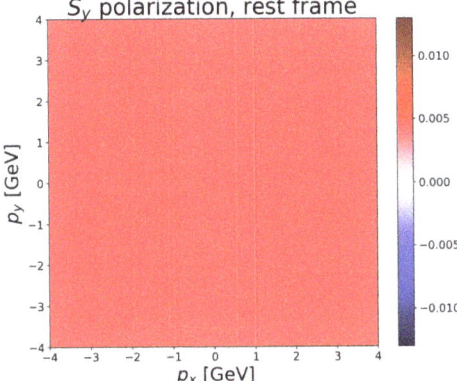

Figure 2. The only non-vanishing component of the polarization vector in the rest frame of the baryon is S_y in the investigated simple solution; in this case, its value is uniquely determined by the magnitude of the Ω vector. More involved types of analytic solutions would yield some dependence on the momentum components, p_x and p_y. For the plotted value of S_y (a constant, as seen in the plot), the same input parameters were used as above: $m_\Lambda = 1115 \,\mathrm{MeV}/c^2$, and $|\Omega| = 0.1 \, c/\mathrm{fm}$.

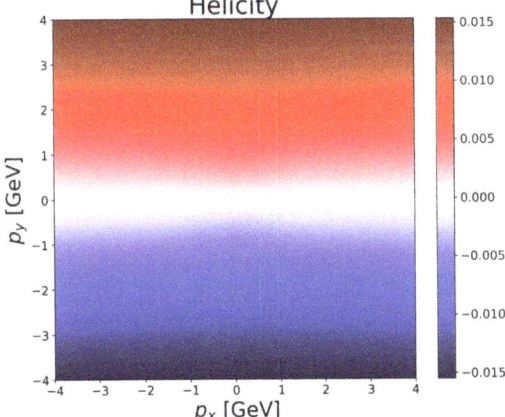

Figure 3. Helicity of the produced baryons calculated in the rotating and accelerating expanding solution. Parameter values as above: $m_\Lambda = 1115 \,\mathrm{MeV}/c^2$, $|\Omega| = 0.1 \, c/\mathrm{fm}$.

5. Summary and Outlook

In this paper, we gave the first analytical formulas for the polarization of baryons produced from a thermal ensemble corresponding to rotating and expanding exact hydrodynamical solutions. These arise as descriptions of the final state of non-central high energy heavy-ion collisions. We investigated two exact relativistic hydrodynamical solutions. One was the spherically symmetric Hubble flow (an overly simplistic one, the study of which can be regarded as a check of the methodology), in which the polarization turns out to be exactly zero (as is naturally expected from symmetry considerations). The other solution we investigated was one describing rotating and accelerating expansion. In this case, we obtained the first-ever analytical formulas that connected dynamical quantities of the expansion (i.e., magnitude of rotation, acceleration) with the observable final state polarization of spin 1/2 particles (baryons), which turned out to be non-zero in this case.

Our results are simple and straightforward. The calculations presented here yielded the first results in terms of exact formulas for the polarization. Nevertheless, many more solutions (more involved ones), as well as more complicated final state parametrizations can be investigated in the future. The motivation is that the simple solution that we used here that yields non-zero polarization is one that features acceleration, temperature gradient, as well as vorticity. However, these cannot be tuned (or turned on and off) separately by a continuous change of the parameters of the solution. Solutions that allow this to be done are to be investigated in a later stage of this research effort. Such future studies are needed to disentangle the effects that rotation, acceleration, and temperature gradient have on the observable final state polarization of baryons produced in heavy-ion collisions. Such studies have the potential of a better understanding of what phenomenological implications polarization measurements (such as was recently done by the STAR experiment [1]) can have on the properties (such as the Equation of State) of the strongly coupled Quark Gluon Plasma produced in heavy-ion collisions.

Author Contributions: Conceptualization, M.C.; Investigation, B.B. and M.I.N.; Supervision, M.I.N. and M.C.; Visualization, B.B.; Writing-Original Draft, B.B. and M.I.N.; Writing-Review & Editing, M.I.N. and M.C.

Funding: Our research has been partially supported by the Hungarian NKIFH grants No. FK-123842 and FK-123959, the Hungarian EFOP 3.6.1-16-2016-00001 project. M. Csanád and M. Nagy was supported by the János Bolyai Research Scholarship of the Hungarian Academy of Sciences and the ÚNKP New National Excellence Program of the Hungarian Ministry of Human Capacities.

Conflicts of Interest: The authors declare no conflict of interest.

Appendix A. Additional Calculations

Here we discuss some additional calculations used in Section 3.2 pertaining to the case of rotating and accelerating solution.

For a given momentum $\mathbf{p}$, the position of the saddle-point $\mathbf{R}_0$ (to be applied in the approximate calculation of the momentum spectrum and the polarization) was written up in Equation (25); we provide some additional details of the derivation of that formula here. The defining equation was Equation (23), of which the following equation for $\mathbf{R}_0$ is obtained:

$$4E\mathbf{R}_0 - 2\sqrt{\tau_0^2+R_0^2}\,\mathbf{p} - \frac{2(\mathbf{p}\mathbf{R}_0)}{\sqrt{\tau_0^2+R_0^2}}\mathbf{R}_0 - \tau_0^2(\mathbf{p}\times\mathbf{\Omega}) = 0, \tag{A1}$$

where $R_0^2 \equiv \mathbf{R}_0\mathbf{R}_0$. From this equation one readily sees that $\mathbf{R}_0$ must be a linear combination of $\mathbf{p}$ and the $\mathbf{p}\times\mathbf{\Omega}$ vector. We substitute this assumption into the equation above. We note that $\mathbf{p}$ and $\mathbf{p}\times\mathbf{\Omega}$ are orthogonal to each other, which leads to some intermediate simplifications, as well as enables us to rearrange the obtained condition into the following form:

$$\mathbf{R}_0 := \alpha\mathbf{p} + \beta\tau_0^2\mathbf{p}\times\mathbf{\Omega} \quad \Rightarrow \quad 2\left\{\left(2E-\frac{\alpha p^2}{A}\right)\alpha - A\right\}\mathbf{p} = \tau_0^2\left\{1-2\beta\left(2E-\frac{\alpha p^2}{A}\right)\right\}(\mathbf{p}\times\mathbf{\Omega}).$$

where we temporarily introduced the $A \equiv \sqrt{\tau_0^2 + \alpha^2 p^2 + \beta^2\tau_0^4(p^2\Omega^2-(\mathbf{p}\mathbf{\Omega})^2)}$ notation. Because of the orthogonality of $\mathbf{p}$ and $\mathbf{p}\times\mathbf{\Omega}$, both sides here have to vanish identically, from which we get

$$A = \alpha\left(2E-\frac{\alpha p^2}{A}\right), \qquad 4E-\frac{2\alpha p^2}{A} = \frac{1}{\beta}. \tag{A2}$$

One divides these equations to obtain a simple relation, the substituting back one gets a quadratic equation for β, the solution of which is

$$\frac{\alpha}{\beta} = 2A \quad \Rightarrow \quad 4E-4\beta p^2 = \frac{1}{\beta} \quad \Rightarrow \quad \beta = \frac{E}{2p^2} \pm \sqrt{\frac{E^2}{4p^2} - \frac{p^2}{4p^2}} = \frac{E\pm m}{2p^2}, \tag{A3}$$

where we used the $E^2 = p^2 + m^2$ relation. To find α we substitute this back into the expression of A:

$$\alpha = 2\beta A \quad \Rightarrow \quad \alpha^2 = 4\beta^2 \left\{ \tau_0^2 + \alpha^2 p^2 + \beta^2 \tau_0^4 (p^2 \Omega^2 - (\mathbf{p}\boldsymbol{\Omega})^2) \right\} \quad \Rightarrow \quad \alpha = 2\beta \tau_0 \sqrt{\frac{1 + \beta^2 \tau_0^2 (p^2 \Omega^2 - (\mathbf{p}\boldsymbol{\Omega})^2)}{1 - 4 p^2 \beta^2}}.$$

Using the above expression of β (with the yet undetermined sign) we get $1 - 4p^2\beta^2 = -\frac{2m}{p^2}(m \pm E)$, and see that the expression for α will be valid only in the case when $1 - 4\beta^2 p^2 > 0$, thus conclude that the bottom sign is the proper choice. We thus arrive at the following expressions:

$$\beta = \frac{E-m}{2p^2}, \quad \alpha = 2\beta \tau_0 \sqrt{\frac{1 + \beta^2 \tau_0^2 (p^2 \Omega^2 - (\mathbf{p}\boldsymbol{\Omega})^2)}{1 - 4p^2\beta^2}} = \frac{\tau_0}{2} \sqrt{\frac{E-m}{2m}} \sqrt{\tau_0^2 (\hat{\mathbf{p}} \times \boldsymbol{\Omega})^2 (E-m)^2 + 4p^2}. \quad (A4)$$

From these formulas the expression of $\mathbf{R}_0$ shown in Equation (25) readily follows. The other ingredient in the saddle-point integration necessary for getting the momentum spectrum is the determinant of the second derivative matrix of the source function. Here we outline the main steps of the derivation of Equation (26). From Equation (24) the second derivative matrix itself turns out to be

$$M_{kl} = \frac{1}{T_0 \tau_0^2} \left\{ \left(4E - \frac{2(\mathbf{p}\mathbf{r})}{A} \right) \delta_{kl} - \frac{2}{A} (p_k r_l + r_k p_l) + 2(\mathbf{p}\mathbf{r}) \frac{r_k r_l}{A^3} \right\} \Big|_{\mathbf{r} = \mathbf{R}_0}, \quad (A5)$$

where we use the notation A as above. We should use the expression of $\mathbf{R}_0$ as calculated above.

The determinant of this $\mathbf{M}$ matrix is the product of its eigenvalues. In our case the particular spatial directions are: $\mathbf{p}$, $\mathbf{p} \times \boldsymbol{\Omega}$, and the vector orthogonal to both these, that is, $\mathbf{p} \times (\mathbf{p} \times \boldsymbol{\Omega})$. One recognizes that the vector $\mathbf{p} \times (\mathbf{p} \times \boldsymbol{\Omega})$ is an eigenvector of the $\mathbf{M}$ second derivative matrix:

$$\mathbf{M}(\mathbf{p} \times (\mathbf{p} \times \boldsymbol{\Omega})) = \cdots = \frac{1}{\beta} \mathbf{p} \times (\mathbf{p} \times \boldsymbol{\Omega}). \quad (A6)$$

The corresponding eigenvalue is thus $1/\beta$. Owing to the symmetric nature of $\mathbf{M}$, the other two eigenvectors must be in the orthogonal complementer subspace of this vector, so they are linear combinations of $\mathbf{p}$ and $\mathbf{p} \times \boldsymbol{\Omega}$. Let us thus look for these eigenvectors in the form $\mathbf{a} = \mu \mathbf{p} + \nu \mathbf{r}$, with yet to be determined μ and ν coefficients. Substituting this expression, we get

$$\mathbf{M}\mathbf{a} = \lambda \mathbf{a} \quad \Rightarrow \quad \left(4E - \frac{2(\mathbf{p}\mathbf{R}_0)}{A} \right) \mathbf{a} - \frac{2}{A} \left(\mathbf{R}_0(\mathbf{a}\mathbf{p}) + \mathbf{p}(\mathbf{a}\mathbf{R}_0) \right) + 2(\mathbf{p}\mathbf{R}_0) \frac{\mathbf{R}_0(\mathbf{a}\mathbf{R}_0)}{A^3} = \lambda \mathbf{a}, \quad (A7)$$

where λ is the eigenvalue (the values of which we are looking for). By substituting the assumed form of $\mathbf{a}$ and inferring the components of this equation in the $\mathbf{p}$ and $\mathbf{p} \times \boldsymbol{\Omega}$ directions, we get the following equation for the μ and ν coefficients:

$$\frac{2}{A^3} \begin{pmatrix} 2EA^3 - 2A^2 \mathbf{p}\mathbf{R}_0 & -A^2 R_0^2 \\ -A^2 p^2 + (\mathbf{p}\mathbf{R}_0)^2 & 2EA^3 - 2A^2 \mathbf{p}\mathbf{R}_0 + R_0^2 \mathbf{p}\mathbf{R}_0 \end{pmatrix} \begin{pmatrix} \mu \\ \nu \end{pmatrix} = \lambda \begin{pmatrix} \mu \\ \nu \end{pmatrix}. \quad (A8)$$

We immediately infer the product of the two $\lambda_{1,2}$ eigenvalues as the determinant of this 2×2 matrix. Taking the third eigenvalue (calculated above) into account, after some simplifications, we indeed get the following expression for the determinant of the $\mathbf{M}$ matrix (the expression we used in Equation (26)):

$$\det \mathbf{M} = \left(\frac{1}{T_0 \tau_0^2} \right)^3 32 p m^2 (m + \sqrt{p^2 + m^2}). \quad (A9)$$

References

1. Adam, J.; Adamczyk, L.; Adams, J.R.; Adkins, J.K.; Agakishiev, G.; Aggarwal, M.M.; Ahammed, Z.; Alekseev, I.; Anderson, D.M.; Aoyama, R.; et al. STAR Collaboration. Global Λ hyperon polarization in nuclear collisions: Evidence for the most vortical fluid. *Nature* **2017**, *548*, 62.
2. Adam, J.; Adamczyk, L.; Adams, J.R.; Adkins, J.K.; Agakishiev, G.; Aggarwal, M.M.; Ahammed, Z.; Alekseev, I.; Anderson, D.M.; Aoyama, R.; et al. STAR Collaboration. Global polarization of Λ hyperons in Au+Au collisions at $\sqrt{s_{NN}}$ = 200 GeV. *Phys. Rev. C* **2018**, *98*, 014910. [CrossRef]
3. Becattini, F.; Chandra, V.; Del Zanna, L.; Grossi, E. Relativistic distribution function for particles with spin at local thermodynamical equilibrium. *Ann. Phys.* **2013**, *338*, 32. [CrossRef]
4. Csernai, L.P.; Becattini, F.; Wang, D.J. Turbulence, Vorticity and Lambda Polarization. *J. Phys. Conf. Ser.* **2014**, *509*, 012054. [CrossRef]
5. Xie, Y.L.; Bleicher, M.; Stöcker, H.; Wang, D.J.; Csernai, L.P. Λ polarization in peripheral collisions at moderate relativistic energies. *Phys. Rev. C* **2016**, *94*, 054907. [CrossRef]
6. Xie, Y.; Wang, D.; Csernai, L.P. Global Λ polarization in high energy collisions. *Phys. Rev. C* **2017**, *95*, 031901. [CrossRef]
7. Karpenko, I.; Becattini, F. Study of Λ polarization in relativistic nuclear collisions at $\sqrt{s_{NN}}$ = 7.7 –200 GeV. *Eur. Phys. J. C* **2017**, *77*, 213. [CrossRef]
8. Csörgő, T.; Grassi, F.; Hama, Y.; Kodama, T. Simple solutions of relativistic hydrodynamics for longitudinally and cylindrically expanding systems. *Phys. Lett. B* **2003**, *565*, 107–115. [CrossRef]
9. Csörgő, T.; Csernai, L.P.; Hama, Y.; Kodama, T. Simple solutions of relativistic hydrodynamics for systems with ellipsoidal symmetry. *Acta Phys. Hung. A* **2004**, *21*, 73. [CrossRef]
10. Nagy, M.I. New simple explicit solutions of perfect fluid hydrodynamics and phase-space evolution. *Phys. Rev. C* **2011**, *83*, 054901. [CrossRef]
11. Hatta, Y.; Noronha, J.; Xiao, B.W. Exact analytical solutions of second-order conformal hydrodynamics. *Phys. Rev. D* **2014**, *89*, 051702. [CrossRef]
12. Cooper, F.; Frye, G. Comment on the Single Particle Distribution in the Hydrodynamic and Statistical Thermodynamic Models of Multiparticle Production. *Phys. Rev. D* **1974**, *10*, 186. [CrossRef]
13. Csanád, M.; Nagy, M.I.; Lökös, S. Exact solutions of relativistic perfect fluid hydrodynamics for a QCD equation of state. *Eur. Phys. J. A* **2012**, *48*, 173. [CrossRef]
14. Florkowski, W.; Ryblewski, R. Hydrodynamics with spin—Pseudo-gauge transformations, semi-classical expansion, and Pauli-Lubanski vector. *arXiv* **2018**, arXiv:1811.04409.
15. Florkowski, W.; Kumar, A.; Ryblewski, R.; Singh, R. Spin polarization evolution in a boost invariant hydrodynamical background. *arXiv* **2019**, arXiv:1901.09655.
16. Csanád, M.; Szabó, A. Multipole solution of hydrodynamics and higher order harmonics. *Phys. Rev. C* **2014**, *90*, 054911. [CrossRef]
17. Karpenko, I.; Becattini, F. Lambda polarization in heavy ion collisions: From RHIC BES to LHC energies. *Nucl. Phys. A* **2019**, *982*, 519. [CrossRef]

© 2019 by the authors. Licensee MDPI, Basel, Switzerland. This article is an open access article distributed under the terms and conditions of the Creative Commons Attribution (CC BY) license (http://creativecommons.org/licenses/by/4.0/).

Article

Nuclear Physics at the Energy Frontier: Recent Heavy Ion Results from the Perspective of the Electron Ion Collider

Astrid Morreale [1,2]

1. Institute for Globally Distributed Open Research, 75014 Paris, France; astrid.morreale@igdore.org or astrid.morreale@gmail.com
2. Center for Frontiers in Nuclear Science, Stony Brook, NY 11790, USA

Received: 9 April 2019; Accepted: 24 April 2019; Published: 28 April 2019

Abstract: Quarks and gluons are the fundamental constituents of nucleons. Their interactions rather than their mass are responsible for 99% of the mass of all visible matter in the universe. Measuring the fundamental properties of matter has had a large impact on our understanding of the nucleon structure and it has given us decades of research and technological innovation. Despite the large number of discoveries made, many fundamental questions remain open and in need of a new and more precise generation of measurements. The future Electron Ion Collider (EIC) will be a machine dedicated to hadron structure research. It will study the content of protons and neutrons in a largely unexplored regime in which gluons are expected to dominate and eventually saturate. While the EIC will be the machine of choice to quantify this regime, recent surprising results from the heavy ion community have begun to exhibit similar signatures as those expected from a regime dominated by gluons. Many of the heavy ion results that will be discussed in this document highlight the kinematic limitations of hadron–hadron and hadron–nucleus collisions. The reliability of using as a reference proton–proton (pp) and proton–ion (pA) collisions to quantify and disentangle vacuum and Cold Nuclear Matter (CNM) effects from those proceeding from a Quark Gluon Plasma (QGP) may be under question. A selection of relevant pp and pA results which highlight the need of an EIC will be presented.

Keywords: QGP; EIC; gluon saturation; nPDF; DIS; CNM

---

1. Introduction

Quarks and gluons, collectively called partons, are the fundamental constituents of protons, neutrons, the atomic nucleus as well as other hadrons. Their interaction is governed by Quantum Chromodynamics (QCD). Understanding QCD, and in particular the confinement of quarks and gluons inside hadrons, is one of today's greatest physics challenges. QCD is the theory of strong interactions and it is expected to describe building blocks of visible matter and their binding in nuclei. While QCD is a well established theory, it contains elements that cannot be calculated and rely mostly on experimental input.[1] As of today, many fundamental aspects of the theory have not yet been quantified. These aspects include the quantified contribution of partons (and their interactions) to the proton spin, or the mechanisms that permits us to transition from point-like to non-point-like physics.

Since the discovery of quarks and gluons and the confirmation that they carried color and spin, QCD and related sub-fields have continuously given us discoveries. One of these discoveries is the

---

1. While lattice calculations address these problems directly, results emerging from the lattice typically require large time scales. The accuracy of the obtained results is largely correlated with the amount of computing power allocated to pursuing these calculations.

Quark Gluon Plasma (QGP) formation. This discovery uncovered a new state of matter in which partons were no longer confined to the boundaries of a hadron, but rather acted as free particles.

Evidence of this new state of matter was observed in heavy-ion collisions at the Relativistic Heavy Ion Collider (RHIC) with the discovery of a suppression of high transverse momentum hadrons, also called "jet quenching" [1]. Jet quenching is attributed to a decrease of the energy of the hard partons created during the first stages of a high-energy heavy-ion collision. The formation of the QGP is now understood as being responsible for this loss of energy via interactions with its constituting hot and dense medium. Since its discovery, we have learned many of the interesting properties governing the QGP:

- The QGP behaves as a near-ideal Fermi liquid (almost no frictional resistance or viscosity) [2].
- The mean free path of partons in the QGP is comparable to inter-particle spacing [3].
- Experimental evidence points towards collective motion of particles during the QGP expansion [4].

While more precision measurements are needed, some revealing information has been obtained regarding the QGP onset [5] as it is illustrated in Figure 1 from the The Solenoidal Tracker At RHIC (STAR) experiment. This figure illustrates a classic QGP measurement: particle suppression in heavy ion collisions observed via the central-to-peripheral nuclear modification factor ratio R_{cp}, as a function of transverse momentum and center of mass collision energy per nucleon-nucleon collision ($\sqrt{s_{NN}}$). A smooth transition is seen as a function of $\sqrt{s_{NN}}$ between enhancement and particle suppression, the latter a signature of the presence of a QGP.

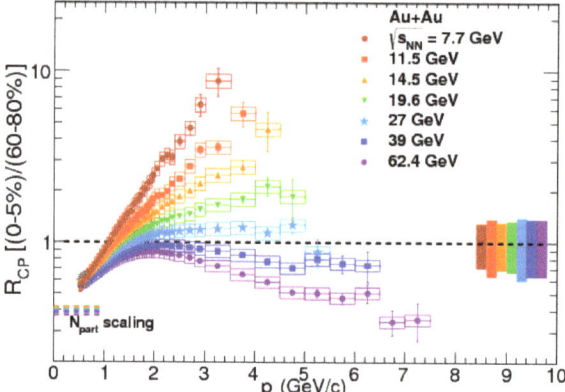

Figure 1. Nuclear modification factor (R_{CP}) of high-p_T hadrons produced in central collisions relative to those produced in peripheral collisions. A QGP onset is observed at collision energies $\sqrt{s_{NN}} > 30$ GeV while an enhancement is observed at lower energies [5].

Despite the plethora of information we have obtained regarding the QGP, many questions remain open—as an example: (1) How precisely does the plasma acquire its Fermi like fluid (i.e., almost no frictional resistance or viscosity) characteristics? (2) What are the processes in which color-charged quarks and gluons and colorless jets interact with a nuclear medium? (3) Is there a smooth transition for the physics involved in small systems to that of large systems? (4) Finally, when does one transition from a regime of partons to a regime in which gluons dominate?

Indeed, recent puzzling results from proton–proton (pp) and proton–ion (pA) collisions seem to insist we address the above.

2. A New Physics Regime

The interaction between partons is usually described as a function of at least two quantities: the momentum fraction x of the parent nucleon carried by the partons under consideration and the energy/length scale Q^2 at which the interaction between partons is probed. These two quantities allow one to identify several regimes for QCD, constituting what one calls the QCD landscape and illustrated on Figure 2 (left). For a given Q^2, as we decrease towards smaller values of x, the number of partons is increased—while, for a given x and as we decrease towards smaller values of Q^2 (reduce the resolution), the size of partons increase. Now, if we vary our kinematics towards small x and small Q, one enters a regime characterized by a large number of partons (gluons rather), with overlapping wave functions. This is the phenomena that is known as gluon saturation [6].

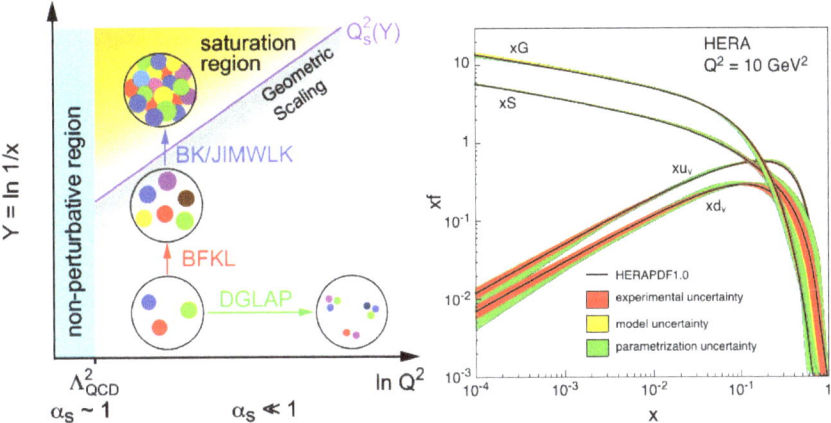

Figure 2. Left: the QCD landscape, the horizontal axis Q^2 represents the resolution of the probe while the y-axis ($\ln(1/x)$) is related to the parton density. **Right**: Parton distribution functions in the proton plotted as functions of Bjorken x; figures from [7].

For large values of Q^2, the coupling constant α_s is small and one expects scattering directly from point-like bare color charges. perturbative Quantum Chromo Dynamics (pQCD) can be then used to reliably predict the hard scattering of partons. For small values of Q^2, in a regime relevant for the description of nucleons and nuclei, one probes longer length scales making QCD non-perturbative and very little is thus calculable. For these small values of Q^2 the content of the nucleon in terms of partons is parameterized using parton distribution functions (PDF) and more recently Generalized Parton Distribution functions (GPD) [8]. Parametrization of PDFs typically requires experimental input (or direct calculations on the lattice). For a given value of Q^2 and decreasing values of x, the density of gluons in the nucleon increases very rapidly (see Figure 2 right). However, for small enough values of x, and large enough values of Q^2 for α_s to be considered small, it is expected that this increase eventually saturates, giving rise to a new regime characterized by weakly-coupled but highly correlated gluon matter called Color Glass Condensate (CGC).

A variety of recent Large Hadron Collider (LHC) results indicate that small systems such as pp and pA exhibit signatures typically expected in larger heavy-ion systems (AA collisions) and resulting from the presence of a QGP. A variety of theories exist which aim at providing explanation to these results some of which include (1) presence of a QGP already in these small systems and (2) universal properties of all nuclei (small and large) in a gluon saturation regime. The first of these explanations requires a careful disentangling of the initial state effects. This is not a trivial task since this is usually achieved using these same small systems as a reference. The second explanation can be tested—with coarse precision and large uncertainties—at current colliders. It is clear that a new generation of results, such as those that will be performed at the Electron Ion Collider (EIC), will be extremely important to help quantify initial state effects with better precision than what is currently achievable. Furthermore, EIC measurements will be pivotal to precisely pin-down the presence of new physics regimes i.e., gluon saturation.

3. The Electron Ion Collider

One of the goals of the lepton–ion (eA) program at an EIC is to unveil the collective behavior of densely packed gluons under conditions where their self-interactions dominate. With its high luminosity and detector acceptance, as well as its span of available collision energies and ion species, the EIC will probe the confined motion as well as the spatial distributions of quarks and gluons inside a nucleus at one tenth of a femtometer resolution. The EIC will be able to detect soft gluons whose energy in the rest frame of the nucleus is less than one tenth of the average binding energy needed to hold the nucleons together to form the nucleus [7]. Thanks to eA collisions with large nuclei, the EIC will reach the saturation regime faster than with ep collisions at similar center of mass energies (cms) energies (Figure 3). This is due to the x and mass number (A) dependence of the saturation scale Q_s, which goes like:

$$Q_s^2(x) \sim A^{1/3}(1/x)^\lambda.$$

The EIC will investigate the onset of saturation, explore its properties and reveal its dynamical behavior. It will also provide a kinematically well defined reference to quantify cold nuclear matter effects. For completeness, it is noted that a similar accelerator proposal (LHeC) with complementary kinematic coverage and physics programme is being evaluated by the European Strategy for Particle Physics [9,10].

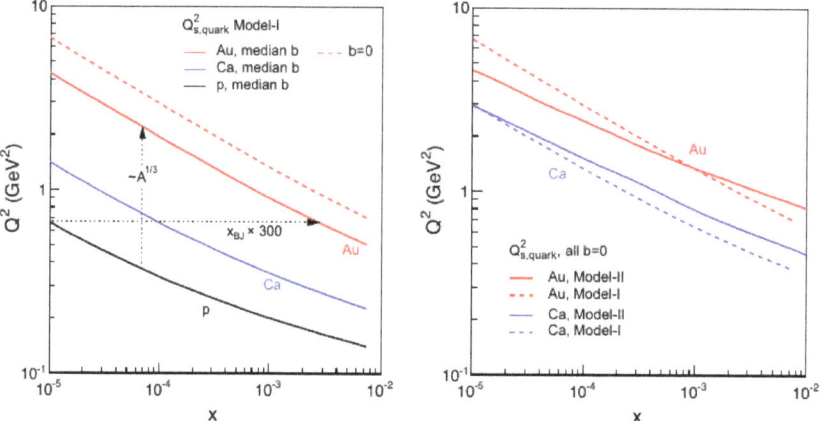

Figure 3. Theoretical expectations for the saturation scale as a function of Bjorken x for the proton along with Ca and Au nuclei. $Q_s^2 \sim 7$ GeV^2 is reached at $x = 10^{-5}$ in e-p collisions at a $\sqrt{s} \sim 1$ TeV while in e–Au collisions, only $\sqrt{s} \sim 60$ GeV is needed to achieve comparable gluon density and the same saturation scale; figure from [7].

The EIC is considered a key component for the future nuclear physics program in the US and as such is among the key recommendations of the Nuclear Science Advisory Committee (NSAC) Long Range Plan from 2015. It has further received a positive and encouraging report from the National Academy of Sciences [11].

3.1. EIC Requirements

- Large luminosity (10^{33}–10^{34} cm^{-2} s^{-1}),
- Center of mass energy (30–140) GeV,
- Hadron and electron beams with high longitudinal spin polarization,
- Ion beams from deuteron to the heaviest stable nuclei,
- Large detector acceptance, in particular for small angle scattered hadrons,
- Optimized high luminosity and high acceptance running modes.

3.2. EIC Designs

The eRHIC design is based on an upgrade to the Relativistic Heavy Ion Collider (RHIC) located at Brookhaven National Laboratory (BNL) in New York:

- New electron injector,
- 5–18 GeV electron energy,
- Energy of heavy ions up to 100 GeV/u,
- $\sqrt{s}$: 20–140 GeV,
- Peak luminosity of $\sim 0.4 \times 10^{34}$ cm^{-2} s^{-1}/A as a base design and 1.0×10^{34} cm^{-2} s^{-1}/A achievable with strong cooling.

The Jefferson Lab EIC (JLEIC) design is based on an upgrade to the Continuous Electron Beam Accelerator Facility (CEBAF) located at the Jefferson Laboratory in Virginia:

- New hadron injector,
- A "figure-8" layout for the booster and collider rings which preserves spin polarization,
- 3–12 GeV electron energy,
- Energy of heavy ions up to 80 GeV/u that could be upgraded to 160 GeV/u,
- $\sqrt{s}$: 20–100 GeV that could be upgraded to 140 GeV,
- Average luminosity per run $\sim 10^{34}$ cm^{-2} s^{-1}/A.

Both designs have science cases by themselves which require a robust integration with detector designs. An ongoing "Generic Detector for an EIC" research and development peer reviewed program is funded by the United States Department of Energy. Thanks to these funds, an active effort exists in which a variety of detector designs and technologies which meet EIC requirements are being explored and tested. Two such examples are cited: the BeAST and JELIC detector R&D efforts. See [12] for a complete list of these programs.

4. Physics at the Energy Frontier: Selection of Recent Results

A short selection of unexpected heavy ion results measured at the LHC is presented which has prompted interpretations. The presence of a mini QGP in small hadronic systems [13,14] has been proposed as an explanation. Other physics mechanisms that do not involve QGP formation have also been proposed including the existence of a gluon saturation regime [15,16], the string percolation model [17] and others. For an interesting review on the subject, see [18]. It is noted that Monte Carlo generators such as PYTHIA have also been used to describe qualitatively some of the data in this document [19].

The selection presented hereafter will highlight the need to better understand small colliding systems if we are to quantify correctly QGP phenomena. The ep and eA collisions of the EIC will undoubtedly contribute to an in-depth understanding of these observations.

4.1. QGP Onset and Strangeness Enhancement

Strangeness enhancement was one of the first proposed signatures of the QGP [20]. The QGP expectation was that strange particle yields would be enhanced with respect to their yield in pp collision. The enhancement would then follow a hierachy based on their strange quark content. This implies that a particle with three strange quarks would be enhanced with respect to a particle with two strange quarks, and even more than a particle with only one strange quark. As predicted, strangeness enhancement was observed in AA collisions at the Super Proton Synchrotron (SPS), RHIC and the LHC (Figure 4) [21].

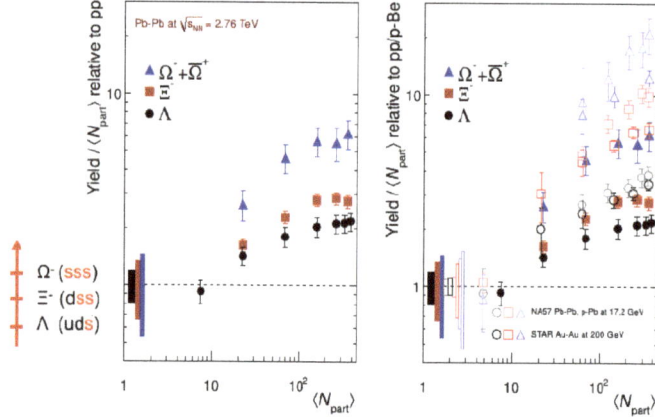

Figure 4. Ratio of strange yields in PbPb collisions with respect to pp collisions as a function of participants. As it is observed in the figure, an enhancement with respect to pp is observed which is larger for Ω (sss) than for Ξ^- (dss) and Λ (uds) [21].

What is unexpected, however, is the observation (Figure 5) that an enhancement of strange particles (K, Λ, Ω) with respect to non strange yields (i.e., π) is also visible in the most violent high multiplicity pp and p-Pb collisions [2].

The mechanisms responsible for the observed enhancement in these small systems might indicate that such system may not be relied upon to discern cold from hot nuclear effects. While more experimental insight is needed to interpret the observed enhancement, it has been proposed that the presence of a strong gluon field leading to the nonlinear regime of gluon saturation [7] may explain these observations.

---

[2] Multiplicity is the number of charged particles in the final state. In pPb and PbPb, this quantity is related to the centrality of the collision.

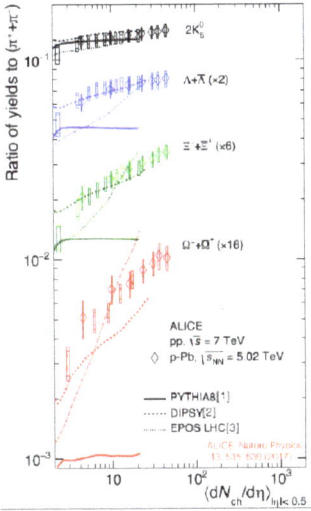

Figure 5. Ratio of strange yields to $\pi^+ + \pi^-$ in pp, p-Pb a a function of average particle multiplicity. A smooth transition is observed as a function of particle multiplicity connecting the small (pp) and larger (pPb) systems [22].

4.2. Heavy Flavor vs. Multiplicity

Heavy flavor probes are ideal to test QGP properties. The contribution of the QCD vacuum condensate to the masses for the three light quark flavors (u, d, s) considerably exceeds the mass generated by the Higgs field. Charm and beauty masses, on the other hand, are not expected to be affected by this QCD vacuum (Figure 6 left), making them ideal probes of the QGP. The mass of the heavy quark itself provides the hard scale for pQCD calculations. This is in contrast to light quarks which often have to rely on the p_T of the final state hadron. In addition, low p_T production of charmonia at forward rapidity (where smaller values of x can be reached) is expected to be sensitive to gluon saturation.

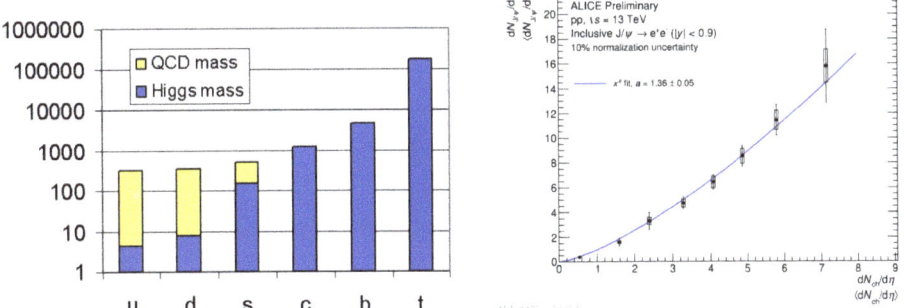

Figure 6. **Left**: Masses of the six quark flavors. The masses generated by electroweak symmetry breaking (current quark masses) are shown in blue; the additional masses of the light quark flavors generated by spontaneous chiral symmetry breaking in QCD (constituent quark masses) are shown in yellow [23]. **Right**: Relative J/ψ production (ordinate) yields as a function of the relative number of charged particles per unit of rapidity (abcissa). The blue line corresponds to a fit of a power law function to the data [24].

Recent results from the ALICE experiment at the LHC show an event activity dependence of inclusive J/ψ and D mesons [25]. The relative charmonium production yield as a function of the per-event relative charged particle multiplicity shows an increase that is faster than linear in pp collisions (Figure 6 right).

Figure 7 (right) shows a similar measurement performed in pPb collisions at negative (Pb-going side), mid and forward rapidity (p-going side). The positive rapidity measurement corresponds to small x values ($\sim 10^{-5}$), a range in which gluon saturation may be present. The observation of similar charged particle multiplicity dependence (Figure 7 left) for both open and hidden charm indicates that hadronization may be of lesser importance.

One plausible physics explanation for the previous results is the existence of a gluon saturation regime as it has been discussed in the introduction of this document. In Figure 8, a CCG [15] calculation which includes gluon saturation effects is compared to ALICE measurements in pp collisions. The calculation describes the data.

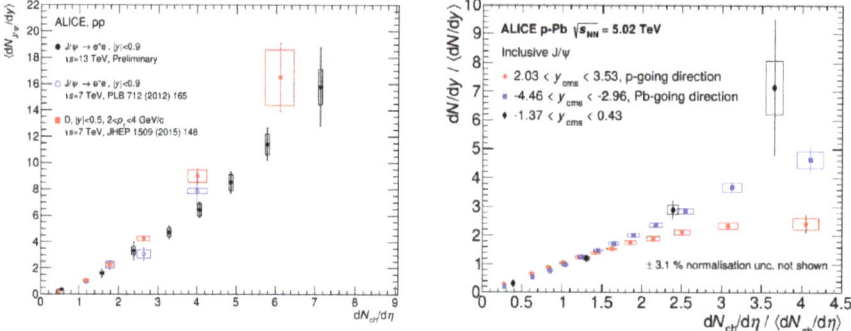

Figure 7. Left: average inclusive J/ψ (closed black and open blue markers), D meson (red closed markers) dependence on charged particle multiplicity in pp collisions and central rapidity. **Right**: inclusive forward and backward rapidity J/ψ's dependence on charged particle multiplicity in p–Pb collisions [25].

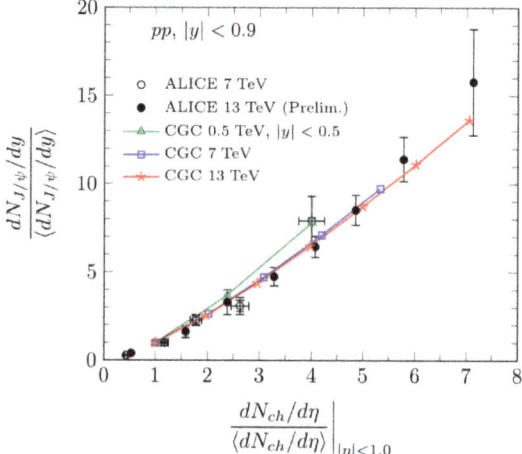

Figure 8. CGC comparisons [15] to recent J/ψ multiplicity results in pp collisions.

4.3. Hydrodynamic Flow

One of the properties of the QGP is that it behaves like a perfect fluid with nearly zero viscosity. This near zero viscosity has been quantified by the correlated momentum anisotropies among the particles produced in the heavy collisions, which result from a common velocity field pattern. This pattern is now identified as collective flow [26]. Among the flow phenomena, two types are highlighted in this document: (1) Radial flow, which typically affects the shape of low p_T spectra, and (2) Elliptic Flow v_2, which is the second coefficient of the Fourier decomposition of particle's momentum azimuthal distributions. This decomposition quantifies the anisotropic particle density which emerges from two nuclei interacting in semi central collisions. A non zero v_2 implies early thermalization of the medium and it is considered a signature of the QGP.

Baryon to meson ratios obtained in PbPb collisions as shown in Figure 9 illustrate the effect of radial flow. Radial flow will push hadrons from low p_T towards intermediate p_T. The effect is expected to be stronger for baryons than for mesons, resulting in a bump in the baryon-to-meson ratio (here proton-to-pions), which depends on the centrality of the collision. Until recently, this was well understood in heavy-ion collisions. What is unexpected, however, is the observation of a similar effect in pp and pPb collisions as shown in Figures 10 and 11. The results would naively imply that thermalization is occurring already in these small systems.

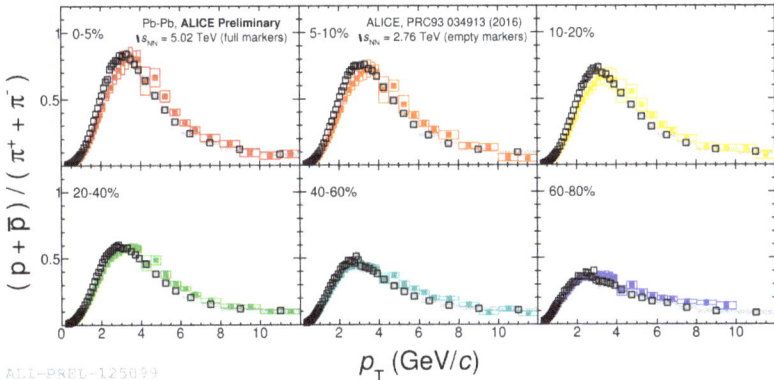

Figure 9. Proton to pion ratio in PbPb collisions as a function of p_T at two $\sqrt{s_{NN}}$ and six centrality classes.

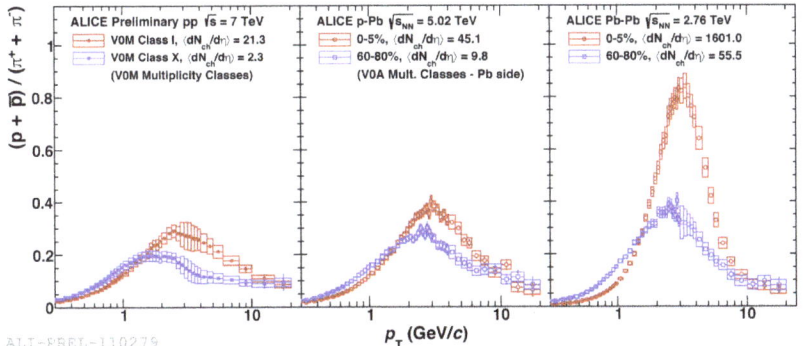

Figure 10. Proton to pion ratio as a function of p_T in pp (**left**), pPb (**middle**) and PbPb (**right**) collisions. The measurements are classified as a function of charged particle multiplicity.

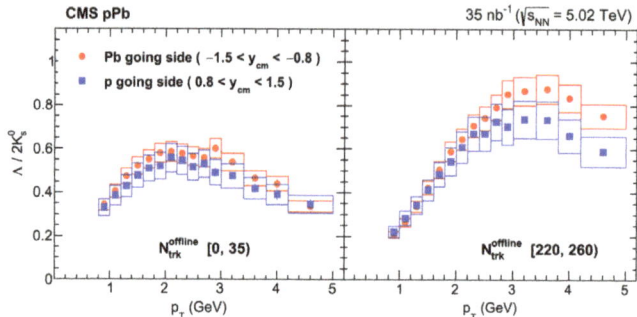

Figure 11. Ratios of p_T spectra, in forward and backward rapidity regions in pPb collisions at $\sqrt{s_{NN}} = 5.02$ TeV. Results are presented for two particle multiplicity ranges; figure and details are taken from [27].

Light meson flow v_2 results in PbPb [28] and reported by the ALICE collaboration are shown in Figure 12 (top figures). At low p_T, as it is the case for many other flow results, the trend is understood as being consistent with a collective expansion within the QGP and has been successfully explained by hydrodynamic models [29]. At intermediate and high p_T constituent quark number scaling takes over (dressed quarks), all mesons fall together and baryons climb above by $\sim 1/3$. What is intriguing on the other hand is that similar signatures are observed in pPb (Figure 12 bottom and Figure 13) and pp collisions (Figure 13 bottom right). Effects that can cause the current observations are either due to initial state effects (saturation), or final state effects (expansion and/or thermal equilibrium). More recently, quantum entanglement has been suggested as a possible explanation [30–32] as well as double parton scattering coupled with the elliptic gluon Wigner distributions (these account for the impact parameter dependence of hard scatterings involving the unintegrated gluon distributions) [33]. This phenomena could be elucidated with a variety of probes at the EIC's lepton–nucleon program, including diffractive measurements of dijet production.

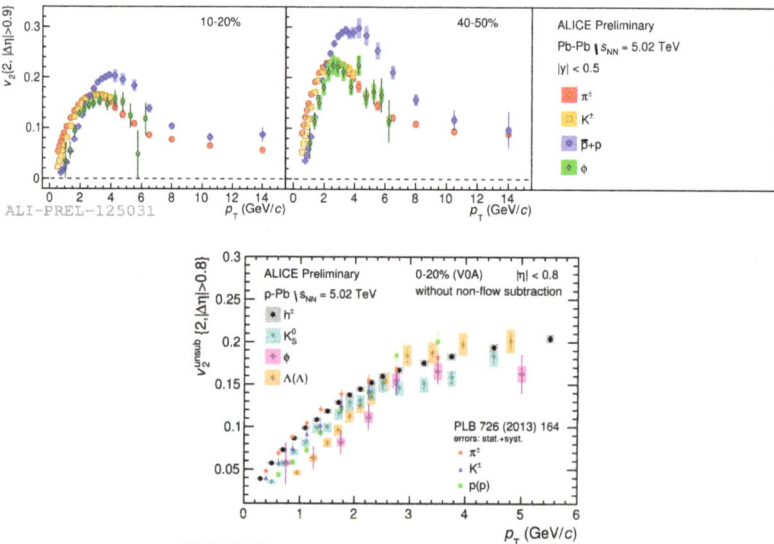

Figure 12. **Top figures**: elliptic flow v_2 in PbPb collisions as a function of p_T in two centrality classes and four particle species [28]; **Bottom figure**: Elliptic flow v_2 in pPb collisions [34].

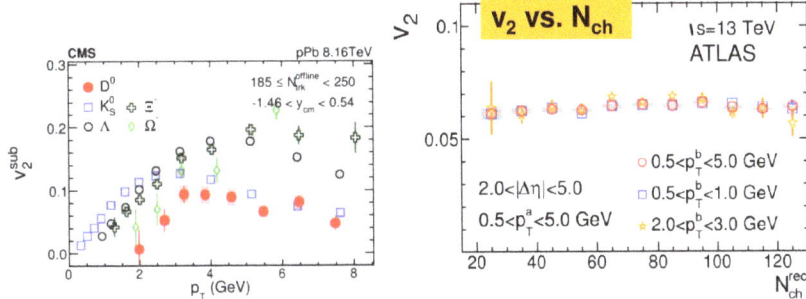

Figure 13. *Left*: v_2 flow as a function of p_T of charm and strange hadrons in high-multiplicity pPb collisions at $\sqrt{s_{NN}} = 8.16$ TeV (CMS Collaboration [35]); *Right*: v_2 as a function of p_T in pp collisions at $\sqrt{s} = 13$ TeV (ATLAS Collaboration [36].)

4.4. Nuclear Modification Factor and Energy Loss in the Medium

The nuclear modification factor R_{AA} is an observable used to quantify the effect of the nuclear medium on particle production. R_{AA} consists of measuring invariant spectra as a function of p_T of particles produced in heavy ion collisions and compared to reference data (pp) at the same energy and scaled by the number of binary collisions. R_{AA} is defined as follows:

$$R_{AA} = \frac{AA}{\text{scaled pp}} = \frac{d^2 N_{AA}/dp_T dy}{<N_{\text{coll}}> d^2 N_{\text{pp}}/dp_T dy}.$$

Values greater than unity would be an indication of production enhancement, while values less than unity will indicate particle suppression in the QGP.

While partons are expected to lose energy when propagating through the dense QGP medium, it is also expected that the amount of energy loss will depend on the parton type and the medium properties. A large number of results, such as those in Figure 14, indicate that the amount of suppression observed in heavy ion collisions is irrelevant of particle mass (or quark content) at high enough p_T.

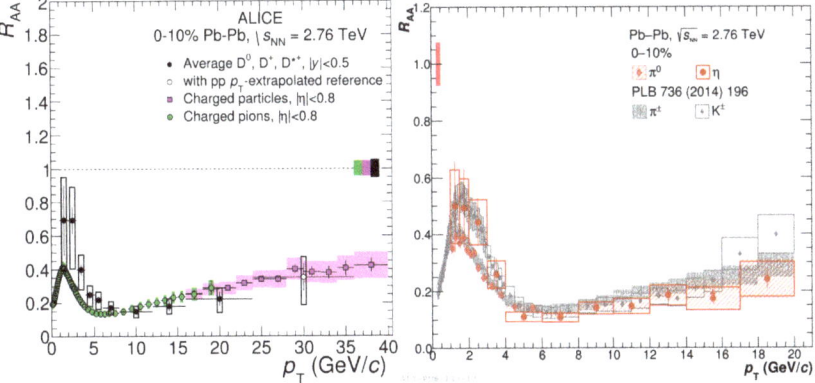

Figure 14. *Left*: Prompt D-meson R_{AA} as a function of p_T compared to the nuclear modification factors of charged pions and charged particles in the 0–10% centrality class [37]. *Right*: R_{AA} of neutral and charged pions, kaons and eta meson [38].

R_{AA} results could largely benefit from independent measurements at the EIC. Measurements such as those illustrated in Figure 15 will study the response of the nuclear medium to a fast moving quark [7,39] and allow proper understanding of hadronization mechanisms.

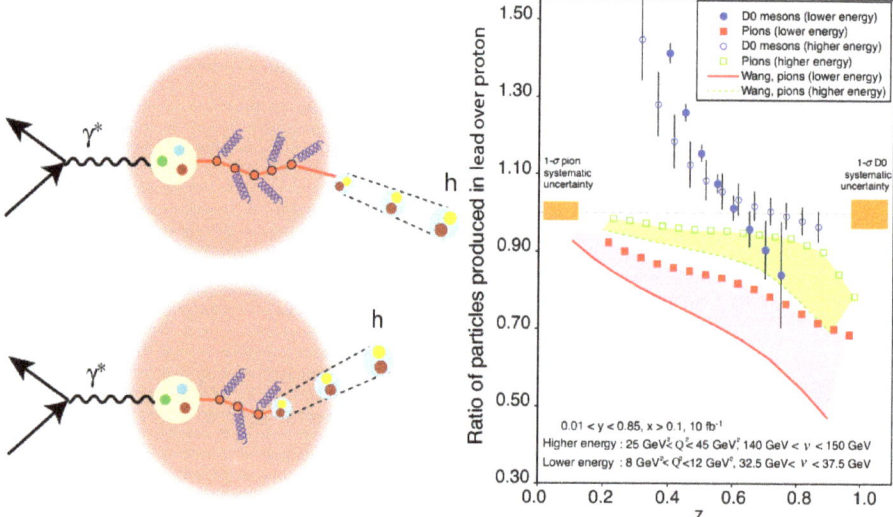

Figure 15. Left: Hadronization schematic illustrating the interaction of a parton moving through cold nuclear matter: the hadron is formed outside (top) or inside (bottom) the nucleus. **Right**: Ratio of semi-inclusive cross section for producing a pion (red) composed of light quarks, and a D^0 meson (blue) composed of heavy quarks in e-Pb collisions to e-d collisions, plotted as function of z, the ratio of the momentum carried by the produced hadron to that of the virtual photon (γ^*), as shown in the plots on the left; figures and descriptions taken from [7].

4.5. Nuclear Parton Distribution Functions

Finally, a careful evaluation of initial state effects such as nuclear modifications of Parton Distribution Functions (nPDFs) is also needed in order to correctly quantify hot nuclear effects present. nPDFs refer to the difference observed between nuclear (bound nucleons) PDFs and free nucleons PDFs (proton, neutron). The nuclear modification of PDFs is due to the interactions between partons from different nucleons. As such, precise measurements of nPDFs are essential in order to understand cold nuclear matter effects that may be convoluted with current heavy ion results. Figure 16 illustrates (in grey) the uncertainty of gluon distributions in the lead nucleus, which is rather large at both low and high x. Measurements that aim at at improving the precision on nPDF are proposed key measurements of the EIC [7].

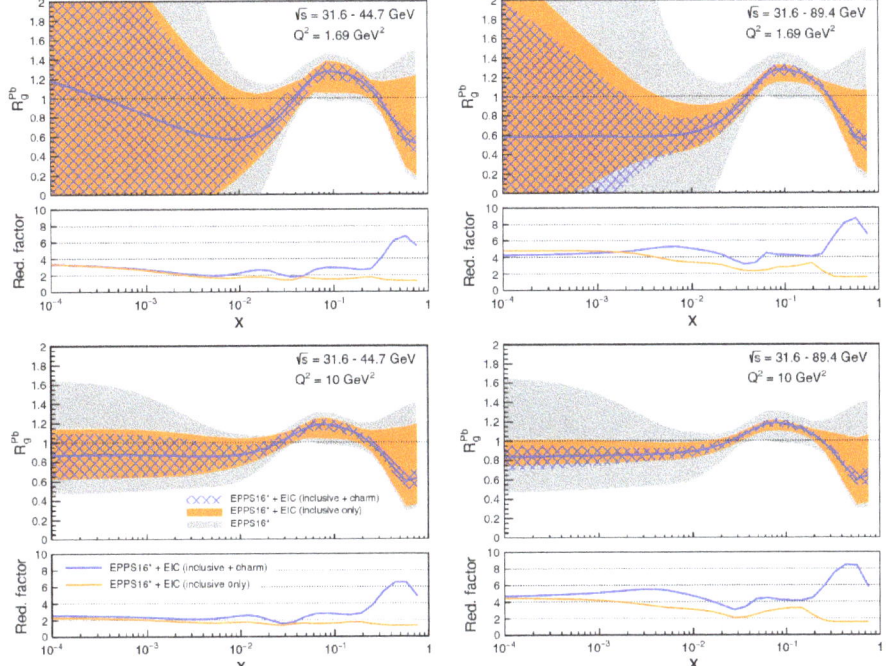

Figure 16. Left: The ratio R_g^{Pb}, from EPPS16∗, of gluon distributions in a lead nucleus relative to the proton, for two different momentum transfers Q^2 possible at the EIC (higher on the bottom figures). Right figures correspond to a larger center of mass energy range considered. The grey band represents the current theoretical uncertainty. The orange (blue hatched) band includes the EIC simulated inclusive (charm quark) reduced cross-section data. The lower panel in each plot shows the reduction factor in the uncertainty with respect to the baseline fit; figures from [40].

5. Conclusions

QCD studies have given us decades of discoveries. Many open questions remain on how the transition from a small system to a dense system takes place: this information is needed to fully understand the properties of the QGP. The current document has given a selection of results that may be better understood and quantified with a new generation of lepton–ion experiments at the EIC.

Funding: This research received no external funding.

Acknowledgments: The author thanks the Center for Frontiers on Nuclear Science for the support that enabled promoting the topics outlined in this document.

Conflicts of Interest: The author declares no conflict of interest.

References

1. Adcox, K. et al. [PHENIX Collaboration]. Suppression of hadrons with large transverse momentum in central Au+Au collisions at $\sqrt{s_{NN}}$ = 130-GeV. *Phys. Rev. Lett.* **2002**, *88*, 022301. [CrossRef] [PubMed]
2. Cao, C.; Elliott, E.; Joseph, J.; Wu, H.; Petricka, J.; Schäfer, T.; Thomas, J.E. Universal Quantum Viscosity in a Unitary Fermi Gas. *Science* **2011**, *331*, 58–61. [CrossRef] [PubMed]
3. Shuryak, E. Physics of Strongly coupled Quark-Gluon Plasma. *Prog. Part. Nucl. Phys.* **2009**, *62*, 48–101. [CrossRef]
4. Adler, C. et al. [STAR collaboration]. Elliptic flow from two and four particle correlations in Au + Au collisions at s(NN)**(1/2) = 130–GeV. *Phys. Rev. C* **2002**, *66*, 034904. [CrossRef]

5. Adamczyk, L.; Adams, J.R.; Adkins, J.K.; Agakishiev, G.; Aggarwal, M.M.; Ahammed, Z.; Ajitanand, N.N.; Alekseev, I.; Anderson, D.M.; Aoyama, R.; et al. Beam Energy Dependence of Jet-Quenching Effects in Au+Au Collisions at $\sqrt{s_{NN}}$ = 7.7, 11.5, 14.5, 19.6, 27, 39, and 62.4 GeV. *Phys. Rev. Lett.* **2018**, *121*, 032301. [CrossRef]
6. Gelis, F.; Iancu, E.; Jalilian-Marian, J.; Venugopalan, R. The Color Glass Condensate. *Ann. Rev. Nucl. Part. Sci.* **2010**, *60*, 463–489. [CrossRef]
7. Accardi, A.; Albacete, J.L.; Anselmino, M.; Armesto, N.; Aschenauer, E.C.; Bacchetta, A.; Boer, D.; Brooks, W.K.; Burton, T.; Chang, N.-B.; et al. Electron Ion Collider: The Next QCD Frontier. *Eur. Phys. J. A* **2016**, *52*, 268. [CrossRef]
8. Ji, X. Generalized parton distributions. *Ann. Rev. Nucl. Part. Sci.* **2004**, *54*, 413–450. [CrossRef]
9. Bruning, O.; Klein, M. *Exploring the Energy Frontier with Deep Inelastic Scattering at the LHC A Contribution to the Update of the European Strategy on Particle Physics*; CERN: Geneva, Switzerland, 2018.
10. Klein, M. Future Deep Inelastic Scattering with the LHeC. In *From My Vast Repertoire ...: Guido Altarelli's Legacy*; Levy, A., Forte, S., Ridolfi, G., Eds.; World Scientific Publishers: Singapore, 2019; pp. 303–347.
11. An Assessment of U.S.-Based Electron-Ion Collider Science. Available online: https://www.nap.edu/catalog/25171/an-assessment-of-us-based-electron-ion-collider-science (accessed on 28 March 2019).
12. EIC Detector Research and Developement. Available online: https://wiki.bnl.gov/conferences/index.php/EIC_R%25D (accessed on 28 March 2019).
13. Zhang, X.; Liao, J. Jet Quenching and Its Azimuthal Anisotropy in AA and possibly High Multiplicity pA and dA Collisions. *arXiv* **2013**, arXiv:1311.5463
14. Park, C.; Shen, C.; Jeon, S.; Gale, C. Rapidity-dependent jet energy loss in small systems with finite-size effects and running coupling. *Nucl. Part. Phys. Proc.* **2017**, *289–290*, 289–292. [CrossRef]
15. Ma, Y.Q.; Tribedy, P.; Venugopalan, R.; Watanabe, K. Event engineering studies for heavy flavor production and hadronization in high multiplicity hadron-hadron and hadron-nucleus collisions. *Phys. Rev. D* **2018**, *98*, 074025. [CrossRef]
16. Mace, M.; Skokov, V.V.; Tribedy, P.; Venugopalan, R. Systematics of azimuthal anisotropy harmonics in proton–nucleus collisions at the LHC from the Color Glass Condensate. *Phys. Lett. B* **2019**, *788*, 161–165. [CrossRef]
17. Braun, M.A.; Dias de Deus, J.; Hirsch, A.S.; Pajares, C.; Scharenberg, R.P.; Srivastava, B.K. De-Confinement and Clustering of Color Sources in Nuclear Collisions. *Phys. Rep.* **2015**, *599*, 1–50. [CrossRef]
18. Dusling, K.; Li, W.; Schenke, B. Novel collective phenomena in high-energy proton–proton and proton–nucleus collisions. *Int. J. Mod. Phys. E* **2016**, *25*, 1630002. [CrossRef]
19. Sjöstrand, T.; Ask, S.; Christiansen, J.R.; Corke, R.; Desai, N.; Ilten, P.; Mrenna, S.; Prestel, S.; Rasmussen, C.O.; Skands, P.Z. An Introduction to PYTHIA 8.2. *Comput. Phys. Commun.* **2015**, *191*, 159–177. [CrossRef]
20. Rafelski, J.; Muller, B. Strangeness Production in the Quark—Gluon Plasma. *Phys. Rev. Lett.* **1982**, *48*, 1066; Erratum in *Phys. Rev. Lett.* **1986**, *56*, 2334. [CrossRef]
21. Abelev, B.B. et al. [ALICE Collaboration]. Multi-strange baryon production at mid-rapidity in Pb-Pb collisions at $\sqrt{s_{NN}}$ = 2.76 TeV. *Phys. Lett.* **2014**, *B728*, 216–227; Erratum in *Phys. Lett. B* **2014**, *734*, 409. [CrossRef]
22. Adam, J. et al. [ALICE Collaboration]. Enhanced production of multi-strange hadrons in high-multiplicity proton-proton collisions. *Nat. Phys.* **2017**, *13*, 535–539.
23. Muller, B. Hadronic signals of deconfinement at RHIC. *Nucl. Phys. A* **2005**, *750*, 84–97. [CrossRef]
24. Weber, S.G. Measurement of J/ψ production as a function of event multiplicity in pp collisions at $\sqrt{s}$ = 13 TeV with ALICE. *Nucl. Phys. A* **2017**, *967*, 333–336. [CrossRef]
25. Adamová, D. et al. [ALICE Collaboration]. J/ψ production as a function of charged-particle pseudorapidity density in p-Pb collisions at $\sqrt{s_{NN}}$ = 5.02 TeV. *Phys. Lett. B* **2018**, *776*, 91–104.
26. Heinz, U.; Snellings, R. Collective flow and viscosity in relativistic heavy-ion collisions. *Ann. Rev. Nucl. Part. Sci.* **2013**, *63*, 123–151. [CrossRef]
27. Khachatryan, V. et al. [The CMS Collaboration]. Multiplicity and rapidity dependence of strange hadron production in pp, pPb, and PbPb collisions at the LHC. *Phys. Lett. B* **2017**, *768*, 103–129. [CrossRef]
28. Noferini, F. ALICE results from Run-1 and Run-2 and perspectives for Run-3 and Run-4. *J. Phys. Conf. Ser.* **2018**, *1014*, 012010. [CrossRef]
29. Hirano, T.; van der Kolk, N.; Bilandzic, A. Hydrodynamics and Flow. *Lect. Notes Phys.* **2010**, *785*, 139–178.

30. Feal, X.; Pajares, C.; Vazquez, R.A. Thermal behavior and entanglement in Pb-Pb and $p-p$ collisions. *Phys. Rev. C* **2019**, *99*, 015205. [CrossRef]
31. Bellwied, R. Quantum entanglement in the initial and final state of relativistic heavy ion collisions. *J. Phys. Conf. Ser.* **2018**, *1070*, 012001. [CrossRef]
32. Baker, O.K.; Kharzeev, D.E. Thermal radiation and entanglement in proton-proton collisions at energies available at the CERN Large Hadron Collider. *Phys. Rev. D* **2018**, *98*, 054007. [CrossRef]
33. Hagiwara, Y.; Hatta, Y.; Xiao, B.W.; Yuan, F. Elliptic flow in small systems due to elliptic gluon distributions? *Phys. Lett. B* **2017**, *771*, 374–378. [CrossRef]
34. Abelev, B.B. et al. [ALICE Collaboration]. Long-range angular correlations of ß, K and p in p-Pb collisions at $\sqrt{s_{NN}}$ = 5.02 TeV. *Phys. Lett. B* **2013**, *726*, 164–177. [CrossRef]
35. Sirunyan, A.M. et al. [CMS Collaboration]. Elliptic flow of charm and strange hadrons in high-multiplicity pPb collisions at $\sqrt{s_{NN}}$ = 8.16 TeV. *Phys. Rev. Lett.* **2018**, *121*, 082301. [CrossRef]
36. Aad, G. et al. [ATLAS Collaboration]. Observation of Long-Range Elliptic Azimuthal Anisotropies in $\sqrt{s}$ =13 and 2.76 TeV pp Collisions with the ATLAS Detector. *Phys. Rev. Lett.* **2016**, *116*, 172301. [CrossRef]
37. Adam, J. et al. [ALICE Collaboration]. Transverse momentum dependence of D-meson production in Pb-Pb collisions at $\sqrt{s_{NN}}$ = 2.76 TeV. *J. High Energy Phys.* **2016**, *2016*, 081. [CrossRef]
38. Acharya, S. et al. [ALICE Collaboration]. Neutral pion and η meson production at mid-rapidity in Pb-Pb collisions at $\sqrt{s_{NN}}$ = 2.76 TeV. *Phys. Rev. C* **2018**, *98*, 044901. [CrossRef]
39. Aschenauer, E.C.; Fazio, S.; Lee, J.H.; Mantysaari, H.; Page, B.S.; Schenke, B.; Ullrich, T.; Venugopalan, R.; Zurita, P. The electron–ion collider: Assessing the energy dependence of key measurements. *Rep. Prog. Phys.* **2019**, *82*, 024301. [CrossRef] [PubMed]
40. Fazio, S. Nuclear Parton Distributions at the future Electron-Ion Collider. *PoS* **2018**, *DIS2017*, 084,

© 2019 by the authors. Licensee MDPI, Basel, Switzerland. This article is an open access article distributed under the terms and conditions of the Creative Commons Attribution (CC BY) license (http://creativecommons.org/licenses/by/4.0/).

Communication

Study of Angular Correlations in Monte Carlo Simulations in Pb-Pb Collisions

Balázs Endre Szigeti [1,2,*] and Monika Varga-Kofarago [2,*]

[1] Department of Atomic Physics, Eötvös Loránd University, Pázmány Péter Sétány 1/A, H-1117 Budapest, Hungary
[2] HAS Wigner Research Centre for Physics, Institute for Particle and Nuclear Physics, Konkoly-Thege Miklós út 29-33, H-1121 Budapest, Hungary
* Correspondence: szigeti.balazs@wigner.mta.hu (B.E.S.); varga-kofarago.monika@wigner.mta.hu (M.V.-K.)

Received: 30 March 2019; Accepted: 24 April 2019; Published: 28 April 2019

Abstract: In two-particle angular correlation measurements, the distribution of charged hadron pairs are evaluated as a function of pseudorapidity ($\Delta\eta$) and azimuthal ($\Delta\varphi$) differences. In these correlations, jets manifest themselves as a near-side peak around $\Delta\eta = 0$, $\Delta\varphi = 0$. These correlations can be used to extract transverse momentum (p_T) and centrality dependence of the shape of the near-side peak in Pb-Pb collision. The shape of the near-side peak is quantified by the variances of the distribution. The variances are evaluated from a fit combining the peak and the background. In this contribution, identified and unidentified angular correlations are shown from Pb-Pb collisions at $\sqrt{s_{NN}}$ = 2.76 TeV from Monte Carlo simulations (AMPT, PYTHIA 8.235/Angantyr). Results show that transport models in AMPT give better results than PYTHIA 8.235/Angantyr when comparing to the experimental results of the ALICE collaboration.

Keywords: angular correlations; Monte Carlo simulations; jet broadening; PYTHIA 8.235, AMPT; Pb-Pb

1. Introduction

In heavy-ion collisions, for processes where the typical momentum transfer is large ($Q >> \Lambda_{QCD}$) partons with high transverse momentum (p_T) are produced. Since the partons carry color charge, they can not exist freely [1] at low energies. For this reason they eventually hadronize into a shower of correlated hadrons called jets. These jets are produced in the early stage of the collisions and propagate through the Quark-Gluon Plasma. During the propagation they interact with the hot and dense medium and lose momentum due to a chain of elementary processes, for instance induced gluon radiation and elastic scattering. These processes are jointly referred to as jet quenching [2]. Experimental studies of these high p_T yields can be used to study the properties of the medium.

In the ALICE experiment [3], unidentified hadron-hadron angular correlations are used to analyze Pb-Pb and pp collision at $\sqrt{s_{NN}}$ = 2.76 TeV in the p_T region of $1 < p_T < 8$ GeV/c. The results show that in central collisions and low p_T cases the jet-peak becomes wider and asymmetric in $\Delta\eta, \Delta\varphi$. Furthermore, a depletion appears around $\Delta\varphi = 0$, $\Delta\eta = 0$. Both the jet broadening and the depletion are seen in AMPT simulations carried out by the ALICE Collaboration [4,5].

In this contribution, identified and unidentified angular correlations are shown from Pb-Pb collisions from Monte Carlo simulations to study which type of particles show similar properties as measured by the ALICE experiment. Moreover, we used Monte Carlo simulations with different physical assumption to study which physical processes are responsible for the observed phenomena.

Universe **2019**, *5*, 97; doi:10.3390/universe5050097   www.mdpi.com/journal/universe

2. Analysis Method

In this work, (identified)hadron-hadron angular correlation measurements are used to compare Monte Carlo simulations with different physical processes. The correlation between particles is measured as a function of pseudorapidity ($\Delta\eta$) and azimuthal ($\Delta\varphi$) differences of trigger and associated particles. In every event, we choose a particle (trigger), then the ($\Delta\eta$) and ($\Delta\varphi$) differences to other particles (associated) are evaluated. This process is then repeated for every particle as trigger from the particular event. Trigger and associated particles are chosen from trigger and associated transverse momentum intervals ($p_{T,trig}$, $p_{T,assoc}$), respectively. The intervals can be disjoint or identical, in the second case only those pairs will be considered, where $p_{T,trig} > p_{T,assoc}$ to avoid double counting. At the experiment the detector acceptance is limited in $|\eta|$, due to this reason $\Delta\eta$ is restricted to $|\Delta\eta| < 2$. The associated yield per-trigger can be expressed in terms of the ratio:

$$\frac{1}{N_{trig}} \frac{d^2 N_{assoc}}{d\Delta\varphi d\Delta\eta} = \frac{\alpha S(\Delta\varphi, \Delta\eta)}{M(\Delta\varphi, \Delta\eta)}, \quad (1)$$

where $S(\Delta\varphi, \Delta\eta)$ is the signal distribution, where the trigger and associated particles are chosen from the same event normalized by the total number of triggers (Figure 1). This is in contrast with $M(\Delta\varphi, \Delta\eta)$ where particles are from separate events scaled by α, the value of $M(\Delta\varphi, \Delta\eta)$ in (0,0). This ratio corrects for pair inefficiencies and detector acceptance effects [5]. In addition to this, the effects from long-lived neutral-particle decays (K_s^0 and Λ) and γ-conversion are removed by cutting on the invariant mass (m_{inv}) of particle pairs. Although, in simulation the cut can be applied directly on the decay products, as particle information is stored during the hadron shower, we would like to maintain consistency with the experimental analysis for better comparison (Figure 2).

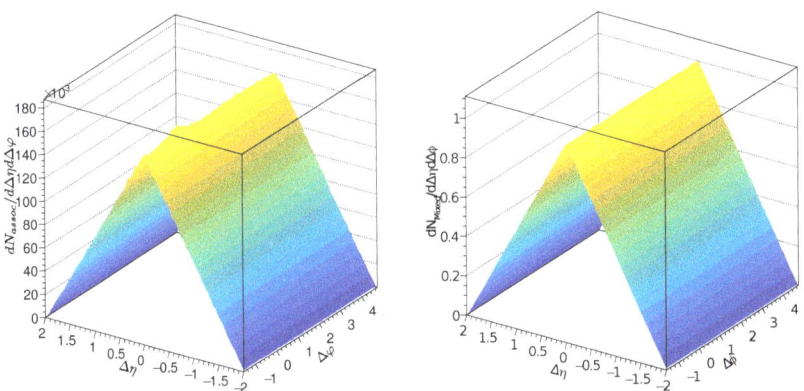

Figure 1. The $S(\Delta\varphi, \Delta\eta)$ (left) and the $M(\Delta\varphi, \Delta\eta)$ (right) distributions in a typical Pb−Pb sample in AMPT at $\sqrt{s_{NN}}$ = 2.76 TeV, where transverse momenta are 1 GeV/c < $p_{T,assoc}$, $p_{T,trigg}$ < 2 GeV/c and centrality is between 10–20%.

To characterize the obtained associated yield per-trigger a combined fit is performed (Figure 2). The shape of the near-side jet peak is parameterized by the variances $\sigma_{\Delta\varphi}$ and $\sigma_{\Delta\eta}$ from the fit by a two-dimensional generalized Gaussian function of the form of Equation (2).

$$G_{\gamma,\omega}(\Delta\varphi,\Delta\eta) = N \frac{\gamma_{\Delta\varphi}\gamma_{\Delta\eta}}{4\omega_{\Delta\varphi}\omega_{\Delta\eta}\Gamma(1/\gamma_{\Delta\varphi})\Gamma(1/\gamma_{\Delta\eta})} exp\left[-\left(\frac{|\Delta\varphi|}{\omega_{\Delta\varphi}}\right)^{\gamma_{\Delta\varphi}} - \left(\frac{|\Delta\eta|}{\omega_{\Delta\eta}}\right)^{\gamma_{\Delta\eta}}\right], \quad (2)$$

where Γ is the gamma function and the ($\omega_{\Delta\varphi}, \omega_{\Delta\eta}, \gamma_{\Delta\varphi}, \gamma_{\Delta\eta}$) parameters characterize the value of the jet-shape. One can note, that the generalized Gaussian distribution equals to a standard Gaussian if $\gamma = 2$, while it is an exponential distribution if $\gamma = 1$. The $\sigma_{\Delta\varphi,\Delta\eta}$ variances are evaluated form the $\omega_{\Delta\varphi,\Delta\eta}$ and $\gamma_{\Delta\varphi,\Delta\eta}$ parameters in the following way.

$$\sigma_{\Delta\varphi;\Delta\eta} = \sqrt{\frac{\omega_{\Delta\varphi;\Delta\eta}^2 \Gamma(3/\gamma_{\Delta\varphi;\Delta\eta})}{\Gamma(1/\gamma_{\Delta\varphi;\Delta\eta})}}. \quad (3)$$

In Pb-Pb collisions, long-range correlations come from collective effects, where one of the essential elements of this is anisotropic flow which can be considered as a background for the jet-peak. The background is characterized by a C_1 constant and the $V_{n\Delta}$ parameters which are Fourier components describing the anisotropic flow.

$$F(\Delta\varphi, \Delta\eta) = C_1 + \sum_{n=2}^{4} 2V_{n\Delta}\cos(n\Delta\varphi) \quad (4)$$

Due to the available simulation statistics the fit was performed separately on the different projections, instead of fitting in two dimensions. Hence, we lose some information about the shape of the jet-peak, but the accuracy of the fit increases.

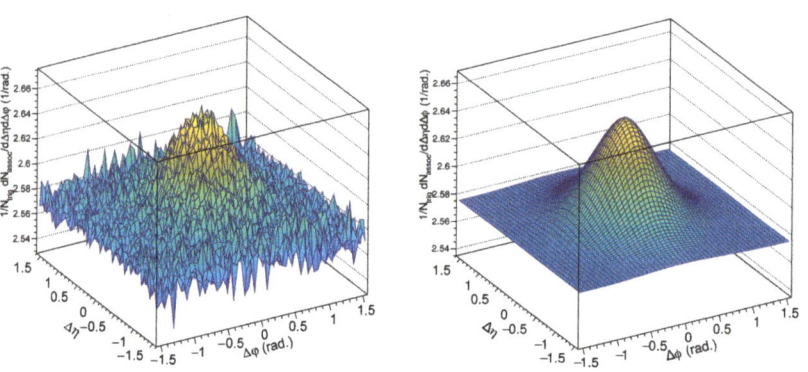

Figure 2. The near-side jet peak (left) and the generalized Gaussian fit (right) in a typical Pb – Pb sample in AMPT at $\sqrt{s_{NN}} = 2.76$ TeV, where transverse momenta are 1 GeV/c < $p_{T,assoc}$, $p_{T,trigg}$ < 2 GeV/c and centrality is between 10–20%.

3. Results

In this work, PYTHIA 8.235/Angantyr [6,7] and AMPT [8] are used to simulate Pb-Pb collisions at $\sqrt{s_{NN}} = 2.76$ TeV. Originally, PYTHIA was developed to simulate pp, p$\bar{\text{p}}$ and e$^+$e$^-$ collision. Recently, a new heavy-ion simulation model has been built on PYTHIA 8.235, called Angantyr. The fomer is based on the Fritiof model [9], and also contains MPI and CR models from PYTHIA [6]. AMPT is based on HIJING and the PYTHIA/JETSET models, and it is directly developed for simulating heavy-ion collisions. In AMPT, the collective effects are described by transport models. In this model, two main settings can be switched on/off: string melting and hadronic rescattering. The PYTHIA 8.235/Angantyr, unlike to the AMPT, does not contain jet-quenching effects.

Unidentified and identified (where trigger particles are kaons or pions) angular correlation are compared at three different centrality bins from PYTHIA 8.235/Angantyr. In the 1 GeV/c < p_T < 2 GeV/c transverse momentum bin, the near-side jet peak was not significant enough for the fit. The obtained variances are depicted in Figure 3. The figure shows that there is a particle species dependence in PYTHIA.

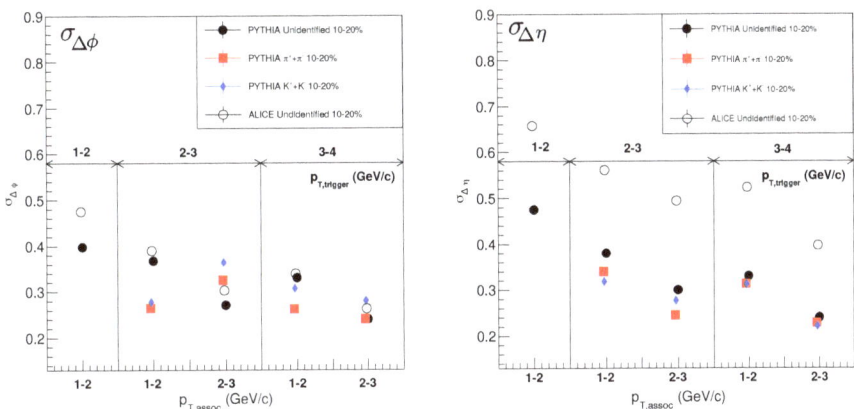

Figure 3. Comparison of the $\sigma_{\Delta\eta}$ and $\sigma_{\Delta\varphi}$ variances from PYTHIA 8.235/Anganty and ALICE measurements from Pb-Pb collisions at $\sqrt{s_{NN}}$ = 2.76 TeV [5].

The PYTHIA 8.235/Angantyr results are compared with the unidentified hadron-hadron correlation from the ALICE experiment. These can be seen in Figure 4. It is clearly visible that PYTHIA gives a better description in the $\Delta\varphi$ direction than in the $\Delta\eta$ direction. However, it does not describe the data well in any of the directions.

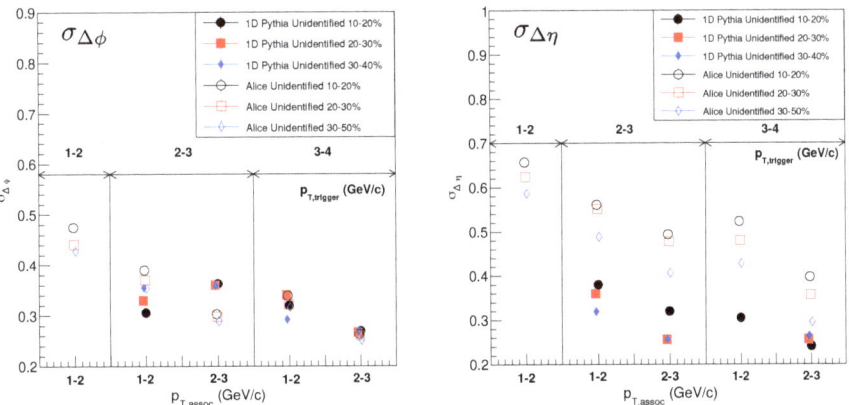

Figure 4. Comprasion of the $\sigma_{\Delta\eta}$ and $\sigma_{\Delta\varphi}$ variances from PYTHIA/Anganty and ALICE measurements from Pb-Pb collisions at $\sqrt{s_{NN}}$ = 2.76 TeV [5].

In the case of AMPT, only the most central bin has been simulated due to the reason that the AMPT simulation time is significantly longer than in the case of PYTHIA 8.235/Angantyr. The obtained $\sigma_{\Delta\varphi}$ and $\sigma_{\Delta\eta}$ variances from AMPT[1] are compared to the ALICE experimental results and to the AMPT results simulated by ALICE collaboration (Figure 5). The figure shows that AMPT gives better description of the widths in $\Delta\varphi$, than in $\Delta\eta$. Furthermore, AMPT gives better trends compared to the experimental results, than PYTHIA 8.235/Angantyr. The results show that there is an asymmetry between the $\Delta\eta$, $\Delta\varphi$ sides in both cases. It is visible that the parameter set used by ALICE to simulate Pb-Pb events overestimates the widths, contrary to our parameter set which underestimates the widths in the $\Delta\eta$ direction. Consequently, there is probably an ideal parameter set where the widths from AMPT fit well in both directions. The difference between the widths from unidentified and identified measurements is not as significant as in PYTHIA 8.235/Angantyr.

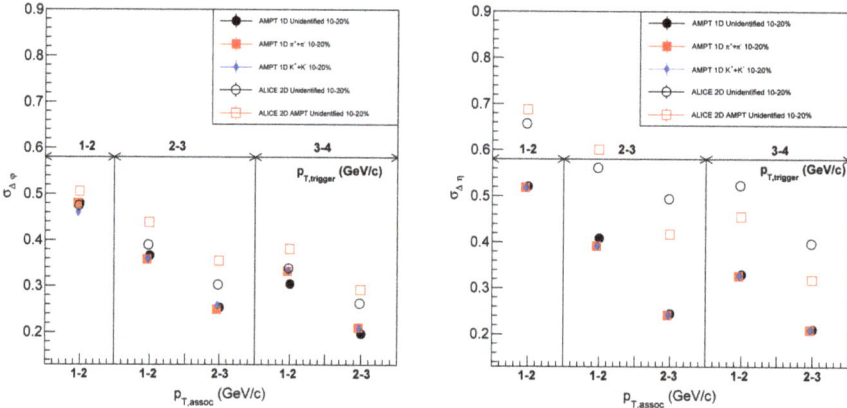

Figure 5. Comparison of the $\sigma_{\Delta\eta}$ and $\sigma_{\Delta\varphi}$ variances from AMPT to the ALICE measurements from Pb-Pb collisions at $\sqrt{s_{NN}} = 2.76$ TeV [5].

4. Conclusions

To summarize, we used unidentified and identified hadron-hadron angular correlations to analyze Monte Carlo simulations. We used a combined fit to characterize the obtained near-side jet peaks. We have seen a hint of particle species dependence in PYTHIA 8.235/Angantyr. As PYTHIA does not describe well the jet-peak widths measured by the ALICE collaboration, we can conclude that MPI and CR models can not describe properly the collective processes seen in Pb-Pb collisions. Furthermore, we found that AMPT gives better results than PYTHIA 8.235/Angantyr when compared to the experimental results. However, we note that it is difficult to find the ideal parameter set in AMPT simulation, due to the reason that the parameter set is large.

Author Contributions: The idea of the measurement is from M.V.-K., and B.E.S. carried out the analysis under M.V.-K.'s supervision.

Funding: This research has been supported by the Hungarian NKFIH/OTKA K 120660 grant.

Conflicts of Interest: The authors declare no conflict of interest. The funders had no role in the design of the study; in the collection, analyses, or interpretation of data; in the writing of the manuscript, or in the decision to publish the results.

---

[1] String melting: on, Hadronic rescattering: off, Lund String fragmentation parameters: 0.30 GeV/c^2, 0.15 GeV/c^2, The p_T cutoff for minijets: 2.0 GeV/c, Quenching flag: off, Shadowing flag: on, Parton screening mass: 2.265 fm^{-1}.

Abbreviations

ALICE   A Large Ion Collider Experiment
QCD   Quantum chromodynamics
AMPT   A Multi-Phase Transport Model
MPI   Multi-Parton Interaction
CR   Color Reconnection

References

1. Gross, D.; Wilczek, F. Ultraviolet Behavior of Non-abelian Gauge Theories. *Phys. Rev. Lett.* **1973**, *30*, 1343. [CrossRef]
2. Cao, S.; Luo, T.; Qin, G.; Wang, X. Heavy and light flavor jet quenching at RHIC and LHC energies. *Phys. Lett. B* **2018**, *777*, 255–259. [CrossRef]
3. The ALICE Collaboration. Performance of the ALICE Experiment at the CERN LHC. *Int. J. Mod. Phys. A* **2014** *29*, 1430044.
4. The ALICE Collaboration. Evolution of the longitudinal and azimuthal structure of the near-side jet peak in Pb-Pb collisions at $\sqrt{s_{NN}} = 2.76$ TeV. *Phys. Rev. C* **2017**, *96*, 034904. [CrossRef]
5. The ALICE Collaboration. Anomalous evolution of the near-side jet peak shape in Pb-Pb collisions at $\sqrt{s_{NN}} = 2.76$ TeV. *Phys. Rev. Lett.* **2017**, *119*, 102301. [CrossRef] [PubMed]
6. Sjöstrand, T.; Ask, S.; Christiansen, J.R.; Corke, R.; Desai, N.; Ilten, P.; Mrenna, S.; Prestel, S.; Rasmussen, C.O.; Skands, P.Z. An Introduction to PYTHIA 8.2. *Comput. Phys. Commun.* **2015**, *191*, 159–177. [CrossRef]
7. Bierlich, C.; Gustafson, G.; Lönnblad, L.; Shah, H. The Angantyr model for Heavy-Ion Collisions in PYTHIA8. *J. High Energy Phys.* **2018**, *134*. [CrossRef]
8. Lin, Z.W.; Ko, C.M.; Li, B.A.; Zhang, B.; Pal, S. A Multi-Phase Transport Model for Relativistic Heavy Ion Collisions. *Phys. Rev. C* **2005**, *72*, 064901. [CrossRef]
9. Andersson, B.; Gustafson, G.; Pi, H. The FRITIOF model for very high energy hadronic collisions. *Z. Phys. C Part. Fields* **1993**, *57*, 485–494. [CrossRef]

© 2019 by the authors. Licensee MDPI, Basel, Switzerland. This article is an open access article distributed under the terms and conditions of the Creative Commons Attribution (CC BY) license (http://creativecommons.org/licenses/by/4.0/).

Article

Bose–Einstein Correlations in pp and pPb Collisions at LHCb [†]

Bartosz Malecki [‡]

Institute of Nuclear Physics Polish Academy of Sciences, PL-31342 Krakow, Poland; Bartosz.Malecki@ifj.edu.pl
† Presented at the 18. Zimanyi School, Budapest 2018.
‡ On behalf of the LHCb Collaboration.

Received: 1 April 2019; Accepted: 22 April 2019; Published: 25 April 2019

Abstract: Bose–Einstein correlations for same-sign charged pions from proton–proton collisions at $\sqrt{s} = 7$ TeV are studied by the Large Hadron Collider beauty (LHCb) experiment. Correlation radii and chaoticity parameters are determined for different regions of charged-particle multiplicity using a double-ratio technique and a Levy parametrization of the correlation function. The correlation radius increases with the charged-particle multiplicity, while the chaoticity parameter decreases, which is consistent with observations from other experiments. A similar study for proton-lead collisions at $\sqrt{s_{NN}} = 5.02$ TeV is proposed. These results can give valuable input for the theoretical models that describe the evolution of the particle source, probing both its potential dependence on pseudorapidity region and differences between proton–proton and proton–lead systems.

Keywords: femtoscopy; small systems; Bose-Einstein correlations; HBT

PACS: 13.87.Ce; 14.40.Aq

1. Introduction

Multi-particle production is a basic process in the field of high energy physics, yet it still lacks a satisfactory description. One of the interesting aspects of this phenomenon is the evolution of the particle source. Intensity interferometry, also known as Hanbury–Brown and Twiss interferometry (HBT) [1] is a very useful tool that can provide information on spatiotemporal structure of this region. This method allows one to observe effects of quantum correlations between same-sign charged hadrons of a given species emitted from a single particle source. Such correlations emerge from the quantum statistics describing the particular particle system. For bosons, Bose–Einstein correlations (BEC) are observed, which are manifested as an enhancement in the production of identical bosons with small relative difference in four-momenta. For fermions, the effect is the opposite and it is called Fermi–Dirac correlations (FDC). Using HBT interferometry, a correlation radius R and a chaoticity parameter λ can be determined. The correlation radius is related to the size of the particle source at freeze-out, while the chaoticity parameter contains information on the coherence of the particle emission.

Comparing results from analyses of the Bose–Einstein correlations in proton–proton (pp) and proton–lead (pPb) collisions is of particular interest. It provides extraordinary input for the development of theoretical models that aim to describe the process of particle production. Some of those models predict contradictory behavior in the size of the particle source in pp and pPb collisions. For example, the hydrodynamical model [2,3] states that the source should have a significantly bigger size in the case of a pPb system, while according to the model based on gluon saturation, this size should be very similar [4,5].

Bose–Einstein correlations were first observed in the field of elementary particle physics in 1959 [6] (Fermi–Dirac correlations slightly later) and were studied in many different collision systems and energies since then. Among others, a number of analyses were conducted for e^+e^- collisions at LEP [7–14] and for pp collisions at LHC by collaborations: ALICE, ATLAS and CMS [15–19]. BEC measurements were also done for heavy ion collisions, e.g., by ALICE [20] and STAR [21]. The pPb system has become of great interest recently, with results from ALICE [22], ATLAS [23] and CMS [24].

Studies of the BEC and FDC effects showed that the correlation parameters depend on a number of factors. It was already indicated by the results from LEP that the correlation radius is related to the mass (species) of the produced hadron—the heavier it is, the smaller the correlation radius. There are several theoretical models that aim to explain this phenomenon [25,26], but the current measurements cannot discard any of them. The correlation effects can be also studied in two or three dimensions—then one can obtain information on the correlation radii in two or three directions, respectively, which gives a more detailed description of the particle source shape. Results from multidimensional analyses of the correlation effects indicate that the source propagates mostly in the beam direction.

Analyses of the BEC effect for different colliding systems and energies showed that R increases with the charged-particle multiplicity (N_{ch}) of an event [27], while λ becomes smaller in most cases. Furthermore, the correlation radius rises approximately linearly with a cube root of the local N_{ch} density in pseudorapidity $<dN_{ch}/d\eta>^{1/3}$. The scaling factor in this relation changes with the type of colliding system [22]. It was also observed that the correlation radius is almost independent of the collision energy, when comparing results for similar N_{ch} values [15,17,19]. Current measurements show that R and λ decrease with the rising average transverse momentum of the particles in a pair (k_T). Many analyses try to determine how the parameters of correlation are related to the factors mentioned and if some scaling behavior can be observed.

The aim of this paper is to present results from the study of the BEC effect for pairs of same-sign charged pions from high-energy pp collisions recorded by the Large Hadron Collider beauty (LHCb) experiment [28]. A one-dimensional analysis is performed and the correlation parameters R and λ are determined for different regions of charged-particle multiplicity. The LHCb detector [29,30] has a unique acceptance ($2.0 < \eta < 5.0$) among other LHC experiments. Thus, the results for the forward direction are the first of their kind and are complementary to observations from other experiments that cover the central acceptance region. It allows one to study the potential dependence of the correlation parameters on pseudorapidity and gives an additional input to understand the process of particle production. A similar study of the BEC effect is ongoing at LHCb for pPb collisions. A direct comparison to the LHCb results on BEC effect in pp collisions will give a strong reference to establish whether the particle production process is different in those two systems. It will also allow one to verify the theoretical models and put additional constraints on their initial parameters (such as the initial transverse size of the colliding system).

2. Materials and Methods

2.1. LHCb Detector

The LHCb detector [29,30] is a single-arm spectrometer (see Figure 1) and covers the pseudorapidity range of $2.0 < \eta < 5.0$, which is a region unique among other experiments at the LHC. The most distinctive features of the LHCb detector are: A very precise system of track and primary vertex (PV) reconstruction, as well as its outstanding particle identification capabilities. The track reconstruction system consists of a vertex detector VELO (which is a silicon-strip detector, surrounding the interaction region), a station of silicon-strip detectors located right before a magnet, the magnet with a bending power of 4 Tm and three stations of silicon-strip and straw drift tubes placed downstream the magnet. The track reconstruction system is capable of determining particles' momenta with uncertainty between 0.5% (for low momenta) and 1% (for momenta at the level of 200 GeV/c). Particle identification is based on information from two Ring Imaging Cherenkov detectors (RICH), a system of calorimeters and muon stations. The RICH detectors are especially important for the BEC analysis, since they allow one to identify species of the charged hadrons.

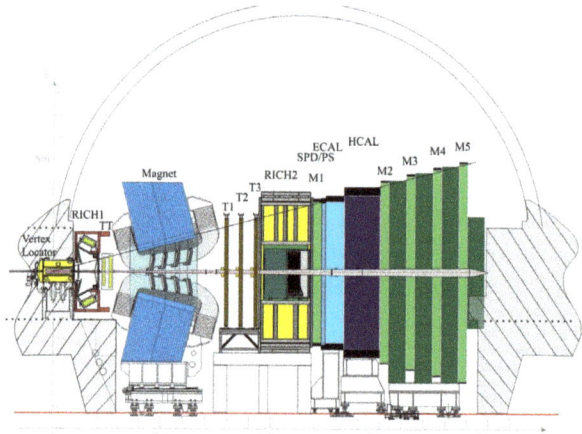

Figure 1. A schematic view of the LHCb detector in y-z plane. The z-axis is the beam axis and the x-axis is perpendicular to the page surface (figure from http://cds.cern.ch/record/1087860; 25 March 2019).

2.2. Data Sample

The data sample used in this study was collected by the LHCb experiment in 2011 from pp collisions at $\sqrt{s} = 7$ TeV centre-of-mass energy and contains 4×10^7 minimum bias events. A corresponding Monte Carlo (MC) sample of 2×10^7 events was prepared with the LHCb software [31] using PYTHIA 8 [32] as the event generator and GEANT4 [33] for the full detector simulation. Futher data analysis is performed within the ROOT framework [34]. The data is divided into three activity classes based on the VELO track multiplicity distribution (see Figure 2), which is a good approximation of the total N_{ch} in an event. The low activity class contains a fraction of 48% PVs with the lowest multiplicities, the medium one corresponds to 37% PVs with higher multiplicities and the high activity class consists of the 15% PVs with highest multiplicities. The VELO track multiplicity is unfolded to the true N_{ch} values using the simulation. The exact ranges of the charged-particle multiplicity regions for each activity class are showed in the Table 1.

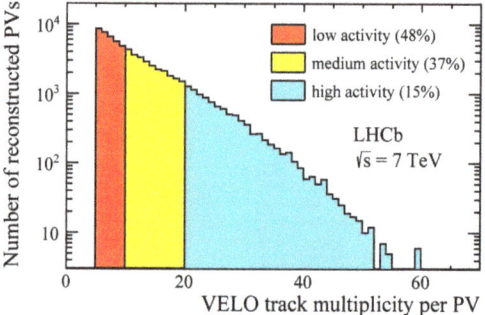

Figure 2. Multiplicity of the reconstructed VELO tracks assigned to a PV for the data. Different colors indicate three activity classes defined as fractions of the full distribution. The Figure is taken from [28].

2.3. Analysis Method

Study of the Bose–Einstein or Fermi–Dirac correlations is usually done using a Lorentz-invariant variable $Q = \sqrt{-(q_1 - q_2)^2}$, which is a measure of proximity in the four-momenta phase space for two particles with four-momenta q_1 and q_2 [35]. Experimentally, the correlation function is defined as:

$$C_2(Q) = N(Q)^{SAME}/N(Q)^{REF},\qquad(1)$$

where $N(Q)^{SAME}$ is the Q distribution of the signal pairs (same-sign charged pions originating from a single PV). The distribution $N(Q)^{REF}$ comes from the reference sample, which should contain all phenomena that are present in the signal pairs, except for the BEC effect. That includes, e.g., long-range correlations (emerging due to energy/momentum conservation law) or Coulomb interactions between the charged pions in their final states. There are several ways of constructing the reference sample, but none of them meets all the above requirements perfectly. In this analysis, the reference pairs are created by mixing particles from different events, where by definition the BEC effect is not present. However, in this way also other kinds of correlations are removed from the sample, such as the long-range ones.

In order to correct for the imperfections in construction of the reference sample, the so-called double-ratio $r_d(Q)$ is introduced. It is a ratio of the correlation functions obtained for data and simulation. In both cases, correlation function is created in exactly the same way, but the BEC effect is switched off in the simulation. Due to that, structures that originate from phenomena properly simulated in the MC sample, are removed from the initial correlation function for data and in an ideal case only the pure BEC signal should be visible. In this way, structures related to, e.g., the long-range correlations can be eliminated from the correlation function. The double-ratio is to a large extent insensitive to effects due to efficiency, detector occupancy and acceptance, as well as the choice of selection criteria.

Coulomb interactions between charged particles in their final states is one of the phenomena that are not present in the simulation. This effect can influence the shape of Q distributions. In the case of same-sign charged particles, a repulsive interaction leads to a decrease in the correlation function for small Q values. For particles with an opposite charge this effect is reversed. The correlation function for data is corrected for the Coulomb interaction effects by applying a Gamov penetration factor [36].

The correlation function is usually parametrized as a Fourier transform of the static source density distribution [37], $C_2(Q) = N(1 + \lambda \exp(-|RQ|^{\alpha_L}))$, where R is the correlation radius, λ denotes the chaoticity parameter and N is a normalization factor. Parameter α_L is a Levy index of stability and corresponds to the assumed source density distribution. In this analysis, an exponential density distribution of a static particle source ($\alpha_L = 1$) is used, which leads to:

$$C_2(Q) = N(1 + \lambda \exp(-RQ)) \times (1 + \delta Q).\qquad(2)$$

The factor $(1 + \delta Q)$ is not related to the BEC effect itself, but it's added to account for the long-range correlations that are manifested for the larger values of Q.

3. Results

Fits to the double-ratio distributions obtained for the three N_{ch} regions are performed using the parametrization (2). An example of the fit for the middle activity class is shown in the Figure 3. A clear enhancement due to the BEC effect is seen for the values of Q approaching 0 GeV. The fit results are summarized in the Table 1. The systematic uncertainty (about 10% in each activity class) is dominated by the MC generator tunings and pile-up effects.

Table 1. Results of fits using parametrization (2) to the double-ratio for three different activity classes and the corresponding N_{ch} bins. Statistical and systematic certainties are given separately (in this order). The Table is taken from [28].

Activity	N_{ch}	R (fm)	λ	δ (GeV^{-1})
Low	[8, 18]	$1.01 \pm 0.01 \pm 0.10$	$0.72 \pm 0.01 \pm 0.05$	$0.089 \pm 0.002 \pm 0.044$
Medium	[19, 35]	$1.48 \pm 0.02 \pm 0.17$	$0.63 \pm 0.01 \pm 0.05$	$0.049 \pm 0.001 \pm 0.009$
High	[36, 96]	$1.80 \pm 0.03 \pm 0.16$	$0.57 \pm 0.01 \pm 0.03$	$0.026 \pm 0.001 \pm 0.010$

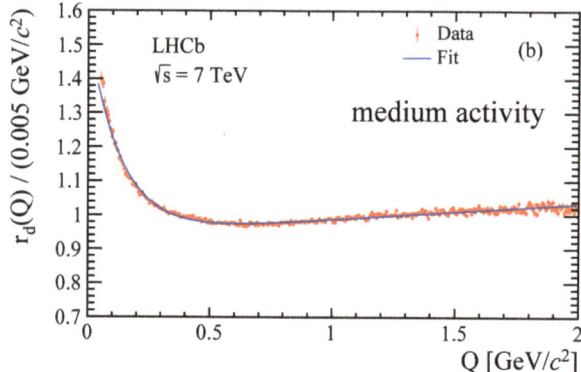

Figure 3. Results of the fit to double-ratio for same-sign charged pions from the middle activity class. The red points represent data with the statistical uncertainties, while the blue line denotes the fit using parametrization (2). The Figure is taken from [28].

The BEC parameters as a function of the activity classes are shown in the Figure 4. It is observed that R increases with the charged-particle multiplicity, while λ decreases, which is consistent with other observations from LEP and LHC. Using the simulation, the LHCb charged-particle multiplicity regions are extrapolated to the corresponding ones in the ATLAS acceptance and a comparison with the ATLAS results [17] is made. It is found that the BEC parameters in the forward direction are slightly lower than those at the central acceptance region. It could indicate that the correlation parameters depend on pseudorapidity. This effect will be studied in more detail by other analyses at LHCb, e.g., in the ongoing research for pPb collisions.

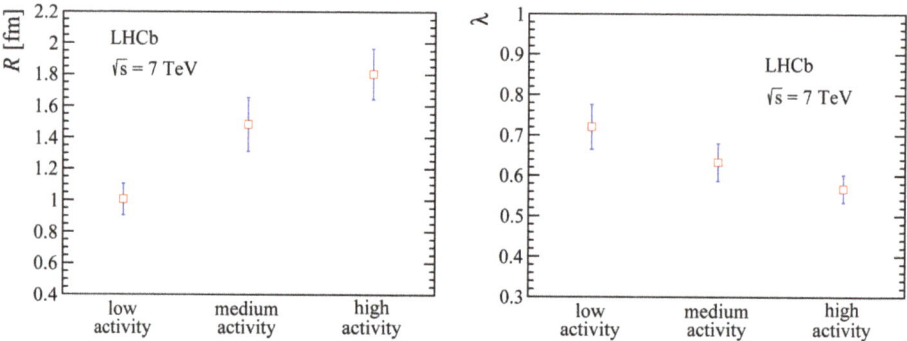

Figure 4. (**left**) Correlation radius R and (**right**) chaoticity parameter λ as a function of activity. Error bars indicate the sum in quadrature of the statistical and systematic uncertainties. The points are placed at the centres of the activity bins. The Figures are taken from [28].

4. Discussion

Bose–Einstein correlations for same-sign charged pions in pp collisions at $\sqrt{s} = 7$ TeV are studied by the LHCb experiment. The correlation radius R and the chaoticity parameter λ are determined for three different regions of the charged-particle multiplicity. It is observed that R increases with N_{ch}, while λ decreases, which is consistent with the previous observations from other experiments at, e.g., LEP and LHC. This measurement is the first of its kind in the forward region and shows the potential of LHCb in similar analyses. The unique acceptance of the LHCb detector among other LHC experiments allows one to obtain results that are complementary to the studies in the central region of pseudorapidity. The LHCb measurement is compared with ATLAS [17] and the correlation parameters in the forward region seem to be slightly lower than those in the central acceptance region. This behavior will be studied in the future analyses at LHCb.

One of the advancing analyses at LHCb is the study of BEC correlations for same-sign charged pions in pPb collisions at $\sqrt{s_{NN}} = 5.02$ TeV centre-of-mass energy per nucleon. This measurement is carried out in different regions of both charged-particle multiplicity and average transverse momentum of the particles in a pair. This will allow one to compare the results to the other LHC experiments and study the potential dependence of the correlation parameters on pseudorapidity. Furthermore, a direct comparison between the LHCb results for both pp and pPb collisions will give insight into the differences in evolution of those two systems. This will provide additional constraints on the parameters of theoretical models that aim to describe the complex process of multi-particle production (such as the initial transverse size of the colliding system).

The usual approach in the BEC studies is to assume a static source of particles, which leads to the Levy-type parametrizations of the correlation function such as (2). One of the alternative strategies is to use a τ-model [38], which accounts for the time evolution of the particle source. This model was proved to be successful in the description of the data in case of e^+e^- collisions [12], however it has not been used for the pPb system yet. Apart from the interesting comparisons between central/forward acceptance regions and pp/pPb collisions, testing the τ-model is one of the goals in the LHCb analysis of the BEC effect in pPb system.

There are many other interesting future directions of the BEC/FDC studies at LHCb, such as 3-body correlations (ongoing), a three-dimensional analysis for same-sign charged pions, or the BEC effect in lead–lead collisions. One of the most intriguing possibilities is to perform this kind of measurement for D mesons, which would be a completely new result in terms of the hadron species. Preliminary studies suggest that such an analysis should be possible with the LHCb data.

Funding: This research was funded by Narodowe Centrum Nauki grant number 2013/11/B/ST2/03829 and 2018/29/N/ST2/01641.

Acknowledgments: This study was partially performed using the PL-GRID infrastructure.

Conflicts of Interest: The authors declare no conflict of interest. The funders had no role in the design of the study; in the collection, analyses, or interpretation of data; in the writing of the manuscript, or in the decision to publish the results.

References

1. Brown, R.H.; Twiss, R.Q. LXXIV. A new type of interferometer for use in radio astronomy. *Lond. Edinb. Dublin Philos. Mag. J. Sci.* **1954**, *45*, 663–682. [CrossRef]
2. Shapoval, V.M.; Braun-Munzinger, P.; Karpenko, I.A.; Sinyukov, Y.M. Femtoscopic scales in p+p and p+Pb collisions in view of the uncertainty principle. *Phys. Lett. B* **2013**, *725*, 139–147. [CrossRef]
3. Bożek, P.; Broniowski, W. Size of the emission source and collectivity in ultra-relativistic p–Pb collisions. *Phys. Lett. B* **2013**, *720*, 250–253. [CrossRef]
4. Bzdak, A.; Schenke, B.; Tribedy, P.; Venugopalan, R. Initial-state geometry and the role of hydrodynamics in proton-proton, proton-nucleus, and deuteron-nucleus collisions. *Phys. Rev. C* **2013**, *87*, 064906. [CrossRef]
5. Dusling, K.; Venugopalan, R. Comparison of the color glass condensate to dihadron correlations in proton-proton and proton-nucleus collisions. *Phys. Rev. D* **2013**, *87*, 094034. [CrossRef]
6. Goldhaber, G.; Goldhaber, S.; Fowler, W.B.; Hoang, T.F.; Kalogeropoulos, T.E.; Powell, W.M. Pion-Pion Correlations in Antiproton Annihilation Events. *Phys. Rev. Lett.* **1959**, *3*, 181–183. [CrossRef]
7. ALEPH Collaboration. Fermi–Dirac correlations in Λ pairs in hadronic Z decays. *Phys. Lett. B* **2000**, *475*, 395–406. [CrossRef]
8. ALEPH Collaboration. Two-dimensional analysis of Bose-Einstein correlations in hadronic Z decays at LEP. *Eur. Phys. J. C* **2004**, *36*, 147–159. [CrossRef]
9. DELPHI Collaboration. Bose-Einstein correlations in the hadronic decays of the Z0. *Phys. Lett. B* **1992**, *286*, 201–210. [CrossRef]

10. DELPHI Collaboration. Kaon interference in the hadronic decays of the Z0. *Phys. Lett. B* **1996**, *379*, 330–340. [CrossRef]
11. L3 Collaboration. Measurement of an elongation of the pion source in Z decays. *Phys. Lett. B* **1999**, *458*, 517–528. [CrossRef]
12. L3 Collaboration. Test of the τ-model of Bose–Einstein correlations and reconstruction of the source function in hadronic Z-boson decay at LEP. *Eur. Phys. J. C* **2011**, *71*, 1648. [CrossRef]
13. OPAL Collaboration. Transverse and longitudinal Bose-Einstein correlations in hadronic Z^0 decays. *Eur. Phys. J. C* **2000**, *16*, 423–433. [CrossRef]
14. OPAL Collaboration. Bose-Einstein correlations in $K^{\pm}K^{\pm}$ pairs from Z^0 decays into two hadronic jets. *Eur. Phys. J. C* **2001**, *21*, 23–32. [CrossRef]
15. ALICE Collaboration. Femtoscopy of pp collisions at $\sqrt{s} = 0.9$ and 7 TeV at the LHC with two-pion Bose-Einstein correlations. *Phys. Rev. D* **2011**, *84*, 112004. [CrossRef]
16. ALICE Collaboration. Ks0Ks0 correlations in pp collisions at $\sqrt{s} = 7$ TeV from the LHC ALICE experiment. *Phys. Lett. B* **2012**, *717*, 151–161. [CrossRef]
17. ATLAS Collaboration. Two-particle Bose–Einstein correlations in pp collisions at $\sqrt{s} = 0.9$ and 7 TeV measured with the ATLAS detector. *Eur. Phys. J. C* **2015**, *75*, 466. [CrossRef]
18. CMS Collaboration. First Measurement of Bose-Einstein Correlations in Proton-Proton Collisions at $\sqrt{s} = 0.9$ and 2.36 TeV at the LHC. *Phys. Rev. Lett.* **2010**, *105*, 032001. [CrossRef]
19. CMS collaboration. Measurement of Bose-Einstein correlations in pp collisions at $\sqrt{s} = 0.9$ and 7 TeV. *J. High Energy Phys.* **2011**, *2011*, 29. [CrossRef]
20. ALICE Collaboration. Two-pion Bose-Einstein correlations in central Pb-Pb collisions at $\sqrt{s_{NN}} = 2.76$ TeV. *Phys. Lett. B* **2011**, *696*, 328–337. [CrossRef]
21. STAR Collaboration. Pion interferometry in Au + Au and Cu + Cu collisions at $\sqrt{s_{NN}} = 62.4$ and 200 GeV. *Phys. Rev. C* **2009**, *80*, 024905. [CrossRef]
22. ALICE Collaboration. Two-pion femtoscopy in p-Pb collisions at $\sqrt{s_{NN}} = 5.02$ TeV. *Phys. Rev. C* **2015**, *91*, 034906. [CrossRef]
23. ATLAS Collaboration. Femtoscopy with identified charged pions in proton-lead collisions at $\sqrt{s_{NN}} = 5.02$ TeV with ATLAS. *Phys. Rev. C* **2017**, *96*, 064908. [CrossRef]
24. CMS Collaboration. Bose-Einstein correlations in pp, pPb, and PbPb collisions at $\sqrt{s_{NN}} = 0.9 - 7$ TeV. *Phys. Rev. C* **2018**, *97*, 064912. [CrossRef]
25. Alexander, G.; Cohen, I.; Levin, E. The dependence of the emission size on the hadron mass. *Phys. Lett. B* **1999**, *452*, 159–166. [CrossRef]
26. Bialas, A.; Kucharczyk, M.; Palka, H.; Zalewski, K. Mass dependence of HBT correlations in e^+e^- annihilation. *Phys. Rev. D* **2000**, *62*, 114007. [CrossRef]
27. AFS Collaboration. Bose-Einstein correlations in $\alpha\alpha$, pp and pp interactions. *Phys. Lett. B* **1983**, *129*, 269–272. [CrossRef]
28. LHCb Collaboration. Bose-Einstein correlations of same-sign charged pions in the forward region in pp collisions at $\sqrt{s} = 7$TeV. *J. High Energy Phys.* **2017**, *2017*, 25. [CrossRef]
29. LHCb Collaboration. The LHCb Detector at the LHC. *J. Instrum.* **2008**, *3*, S08005. [CrossRef]
30. LHCb Collaboration. LHCb detector performance. *Int. J. Mod. Phys. A* **2015**, *30*, 1530022. [CrossRef]
31. Belyaev, I.; Brambach, T.; Brook, N.H.; Gauvin, N.; Corti, G.; Harrison, K.; Harrison, P.F.; He, J.; Ilten, P.H.; Jones, C.R.; et al. Handling of the generation of primary events in Gauss, the LHCb simulation framework. *J. Phys. Conf. Ser.* **2011**, *331*, 032047. [CrossRef]
32. Sjöstrand, T.; Mrenna, S.; Skands, P. A brief introduction to PYTHIA 8.1. *Comput. Phys. Commun.* **2008**, *178*, 852–867. [CrossRef]
33. GEANT4 Collaboration. Geant4—A simulation toolkit. *Nucl. Instrum. Methods Phys. Res. Sect. A: Accel. Spectrom. Detect. Assoc. Equip.* **2003**, *506*, 250–303. [CrossRef]
34. Brun, R.; Rademakers, F. ROOT—An object oriented data analysis framework. *Nucl. Instrum. Methods Phys. Res. Sect. A: Accel. Spectrom. Detect. Assoc. Equip.* **1997**, *389*, 81–86. [CrossRef]
35. Alexander, G. Bose Einstein and Fermi Dirac interferometry in particle physics. *Rep. Prog. Phys.* **2003**, *66*, 481–522. [CrossRef]
36. Pratt, S. Coherence and Coulomb effects on pion interferometry. *Phys. Rev. D* **1986**, *33*, 72–79. [CrossRef]

37. Csörgő, T.; Hegyi, S.; Zajc, W.A. Bose-Einstein correlations for Lévy stable source distributions. *Eur. Phys. J. C* **2004**, *36*, 67–78. [CrossRef]
38. Csörgő, T.; Zimányi, J. Pion interferometry for strongly correlated spacetime and momentum space. *Nucl. Phys. A* **1990**, *517*, 588–598. [CrossRef]

 © 2019 by the authors. Licensee MDPI, Basel, Switzerland. This article is an open access article distributed under the terms and conditions of the Creative Commons Attribution (CC BY) license (http://creativecommons.org/licenses/by/4.0/).

Article

Characterization of Highly Irradiated ALPIDE Silicon Sensors

Valentina Raskina *,† and Filip Křížek *,†

Nuclear Physics Institute of the Czech Academy of Sciences, CZ-25068 Řež, Czech Republic
* Correspondence: raskival@fjfi.cvut.cz (V.R.); Filip.Krizek@cern.ch (F.K.)
† These authors contributed equally to this work.

Received: 20 March 2019; Accepted: 10 April 2019; Published: 14 April 2019

Abstract: The ALICE (A Large Ion Collider Experiment) experiment at CERN will upgrade its Inner Tracking System (ITS) detector. The new ITS will consist of seven coaxial cylindrical layers of ALPIDE silicon sensors which are based on Monolithic Active Pixel Sensor (MAPS) technology. We have studied the radiation hardness of ALPIDE sensors using a 30 MeV proton beam provided by the cyclotron U-120M of the Nuclear Physics Institute of the Czech Academy of Sciences in Řež. In this paper, these long-term measurements will be described. After being irradiated up to the total ionization dose 2.7 Mrad and non-ionizing energy loss 2.7×10^{13} 1 MeV $n_{eq} \cdot cm^{-2}$, ALPIDE sensors fulfill ITS upgrade project technical design requirements in terms of detection efficiency and fake-hit rate.

Keywords: ALICE ITS upgrade; ALPIDE; MAPS; silicon pixel; radiation hardness; cyclotron

1. Introduction

ALICE (A Large Ion Collider Experiment) [1] is a high-energy physics detector at the CERN Large Hadron Collider (LHC). It is designed to study strongly interacting matter in the regime of high-energy densities and temperatures which occur when ultra-relativistic heavy nuclei collide. Under these conditions quarks and gluons escape their confinement in hadrons and form Quark–Gluon Plasma (QGP) [2].

In 2019–2020, the LHC machine will be upgraded. This so-called second long shutdown will be followed by the Run 3 and Run 4 data-taking periods in which ALICE aims to perform detailed measurements of QGP properties using low transverse-momentum (p_T) open-heavy-flavor hadrons, quarkonia, light vector mesons, and low-mass di-leptons [3]. Since these channels have a very small signal-background ratio, large statistics with un-triggered running is needed. ALICE plans to read minimum bias events in a continuous readout mode. Better vertexing and tracking efficiency at low p_T are needed, which requires significant upgrades of ALICE sub-detectors. Another motivation for the ALICE upgrade is the expected increase of delivered luminosity by a factor of 100 in Run 3 and Run 4.

2. Alice Inner Tracking System Upgrade

The ALICE upgrade program includes many sub-projects. This paper is related to the upgrade of the ALICE Inner Tracking System, ITS [4]. This detector is essential for tracking and vertex reconstruction. The main goals of the ITS upgrade are: (i) to improve impact-parameter resolution of reconstructed tracks (by a factor of 5 in the longitudinal direction and by a factor of 3 in the transverse direction), (ii) to improve tracking efficiency and p_T resolution for charged tracks with p_T less than 1 GeV/c, (iii) to increase the readout rate, and (iv) to allow fast insertion and removal of the detector during the end of year technical stops. These goals will be achieved by shifting the first detector layer closer to the beam line from the current 39 mm to 23 mm, by reducing the pixel size, and by reducing the material budget X/X_0 per layer from 1.14% to 0.3% for the three innermost layers and to about 1% for the outer layers. When compared to the current ITS [1], which has 6 cylindrical layers of silicon detectors based on 3 different technologies (pixel, drift, and strip), the new ITS will have 7 layers of silicon pixel sensors called ALPIDEs, see Figure 1. The three innermost layers form the Inner Barrel and the four outer layers form the Outer Barrel. The area covered by ALPIDEs will be about 10 m^2. In total, there will be about 24,000 sensors [5]. The upgraded ITS aims to read out data up to a rate of 100 kHz in Pb–Pb collisions and 1 MHz in pp collisions.

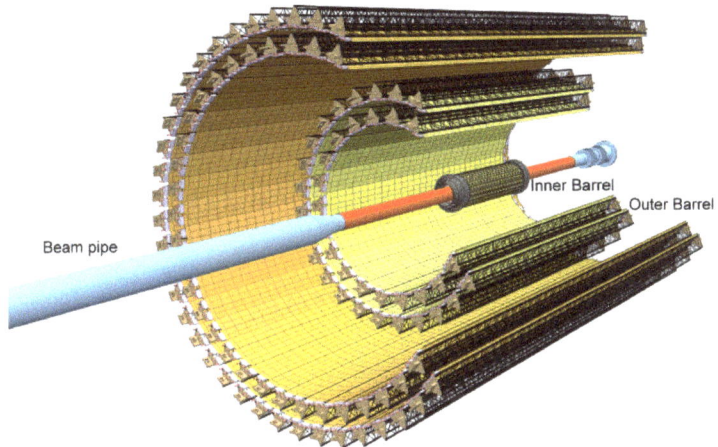

Figure 1. ALICE (A Large Ion Collider Experiment) Inner Tracking System (ITS) after upgrade, taken from [4].

3. The Alpide Sensor

The basic unit of the new ITS is ALPIDE [5–7], which stands for ALice PIxel DEtector. The sensor has a size of 1.5 × 3 cm. It is divided into 512 rows and 1024 columns of pixels with a pitch of 29.24 × 26.88 μm, which makes it possible to achieve a resolution better than 5 μm. The ALPIDE is a Monolithic Active Pixel Sensor (MAPS) which is based on the 180 nm CMOS technology of TowerJazz [8]. This technology uses up to 6 metal layers which in combination with small structure size enables to implement high density CMOS digital circuitry with low power consumption. The average power density across the sensor surface is less than 40 mW cm^{-2} [5,7]. Another important feature is the implementation of a high-resistivity (>1 kΩ) epitaxial layer and a deep p-well, see Figure 2. The deep p-well layer prevents the collection of charge carriers by the n-well of pmos transistors, therefore both nmos and pmos transistors can be implemented in the active pixel area. The thickness of the epitaxial layer is 25 μm. The depletion volume can be increased by applying a moderate reverse bias voltage to the substrate. This also lowers the capacitance of the collection diode and results in a higher collection efficiency [7].

Each ALPIDE pixel contains a sensitive volume, as well as the front-end electronics. The electronics manage charge collection, signal amplification, and discrimination, and write binary hit information to an event buffer. There are several 8-bit DACs (Digital to Analog Converters) on the periphery of the chip, which regulate voltages and currents in the front-end circuits of pixels. The most relevant DACs for the work reported in this paper are voltage V_{CASN} and current I_{THR}, which control charge threshold [6].

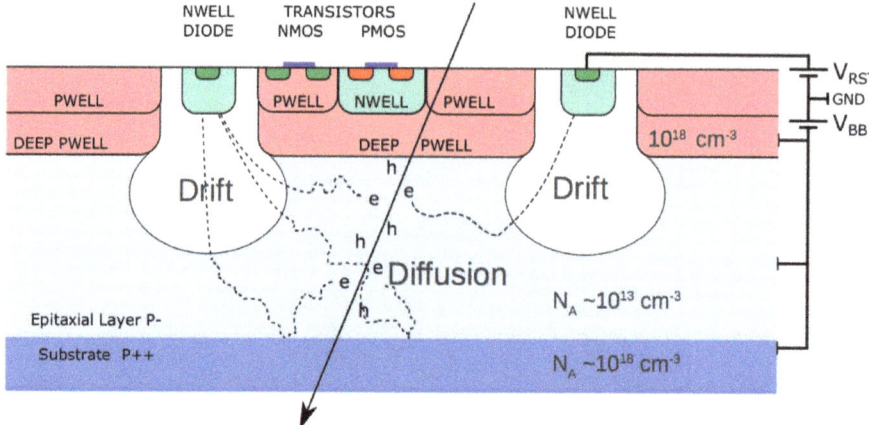

Figure 2. Cross section of an ALPIDE pixel. A charged particle crosses the sensitive volume (high-resistivity epitaxial layer between the substrate and the layer with CMOS transistors) and generates free charge carriers that diffuse across the epitaxial layer until they reach the drift region of a n-well diode, where they are being collected, taken from [4].

Table 1 shows the required parameters of ALPIDE from the technical design report [4] and the achieved performance. The table suggests that the performance of the sensor satisfies all requirements for the Inner and Outer Barrel. After Run 3 and Run 4, the expected total ionizing dose (TID) accumulated by a chip in the Inner Barrel will be 270 krad and the non-ionizing energy loss (NIEL) is expected to be 1.3×10^{12} 1MeV $n_{eq} \cdot cm^{-2}$. However, we tested the ALPIDE chip under up to 10 times larger radiation loads. The corresponding study is reported in this paper.

Table 1. Requirements for a sensor in the Inner and Outer Inner Tracking System (ITS) Barrel and the achieved ALPIDE performance. Taken from [4].

	Inner Barrel	Outer Barrel	ALPIDE Performance
Thickness [μm]	50	100	OK
Spatial resolution [μm]	5	10	~5
Chip dimension [mm]	15 × 30	15 × 30	OK
Power density [mW/cm^2]	<300	<100	<40
Event-time resolution [μs]	<30	<30	~2
Detection efficiency [%]	>99	>99	OK
Fake-hit rate [event^{-1}· pixel^{-1}]	<10^{-6}	<10^{-6}	<10^{-10}
NIEL radiation tolerance [1 MeV $n_{eq} \cdot cm^{-2}$]	1.7×10^{13}	3×10^{10}	OK
TID radiation tolerance [krad]	2700	100	OK

4. Radiation Hardness Tests at the Nuclear Physics Institute

Radiation hardness of ALPIDE sensors was tested using a 30 MeV proton beam provided by the U-120M cyclotron at the Nuclear Physics Institute of the Czech Academy of Sciences (NPI CAS) [9]. For this purpose we use an experimental setup which is shown in Figure 3. The extracted proton beam from the cyclotron has an energy of 34.8 MeV with an RMS of 0.3 MeV and passes through a beam line which is terminated with an energy degrader unit. This unit contains 5 aluminum plates with different thickness which can be inserted into the beam using a remotely controlled pneumatic system. The first aluminum plate is 8 mm thick and serves as a beam stop. Thickness of the second plate is 0.56 mm. This plate is used during the ALPIDE irradiation to make the beam profile wider such that the ALPIDE sensor is irradiated more uniformly across its surface. The other 3 plates were not used in the experiment. The setup with the ALPIDE sensor is placed 130 cm from the end of the beam line. The setup is mounted on a remotely controllable, movable stage and consists of an ionization chamber, another aluminum beam stop plate which can shield the irradiated ALPIDE sensor, and a passive aluminum shielding that protects a readout board connected with the ALPIDE sensor.

The ionization chamber Farmer 30010 from PTW-Freiburg [10] is used to monitor the proton flux and is read out using a UNIDOS E Universal Dosemeter. The chamber has a sensitive volume of 0.6 cm^3 filled with air and provides a linear response to the incoming proton flux up to about 10^9 proton· cm^{-2}· s^{-1} [9]. The vertical distance between the center of the ionization chamber and the center of the ALPIDE sensor is measured using a laser tracker.

Based on GEANT4 simulation, it was estimated that the 0.56 mm thick energy degrader plate and the air decrease the beam energy from 35 MeV to 30 MeV before the beam hits the ALPIDE. The beam intensity profile is measured by moving the whole setup through the beam center along the horizontal and vertical direction. In both cases, the beam profile can be well described by a Gaussian with a standard deviation of about 22 mm. The dose absorbed in the irradiated sample is estimated based on the formula

$$\text{TID}[\text{krad}] = 1.602 \times 10^{-8} \times S[\text{MeV} \cdot \text{cm}^2 \cdot \text{mg}^{-1}] \times F[\text{cm}^{-2}], \quad (1)$$

where S is the linear energy transfer and F is the proton fluence [11]. The non-ionizing enegy loss induced by the 30 MeV proton beam is calculated as

$$\text{NIEL}[1\,\text{MeV}\,n_{eq} \cdot \text{cm}^{-2}] = 2.346 \times F[\text{cm}^{-2}], \quad (2)$$

where the factor 2.346 is a tabled coefficient taken from [12].

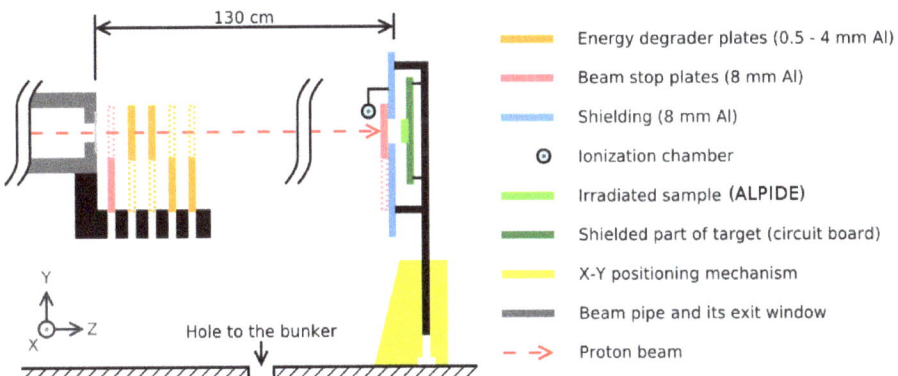

Figure 3. Sketch of the beam route from the beamline exit window to the irradiated sample through energy degrader plates, taken from [9].

Irradiation campaigns of each ALPIDE sensor took place every month since September 2016. ALPIDE was operated with -3 V reverse substrate bias. Sensors were irradiated with proton fluxes of the order of 10^8 proton· cm^{-2}· s^{-1}. The ALPIDE irradiation was carried out as follows. The beam was interrupted periodically using the first beam stop plate. When the beam stop was out of the beam, the sensor was irradiated and analogue and digital currents consumed by the ALPIDE chip were monitored, see Figure 4. When the beam stop interrupted the beam, threshold and DAC scans were performed. The dependence of the absorbed total ionization dose and proton fluence on time for different irradiation campaigns for one chip is shown in Figure 5. The increasing trend in the curves corresponds to the irradiation and the flat trend corresponds to the period when the ALPIDE was not irradiated. The first irradiation campaign that took place in September 2016 was the longest one and the accumulated dose was the highest. In the rest of the campaigns, the chip got about 100 krad, which corresponds to about one third of the total absorbed dose expected during Run 3 and Run 4. After each irradiation campaign, the chip was left to anneal at room temperature.

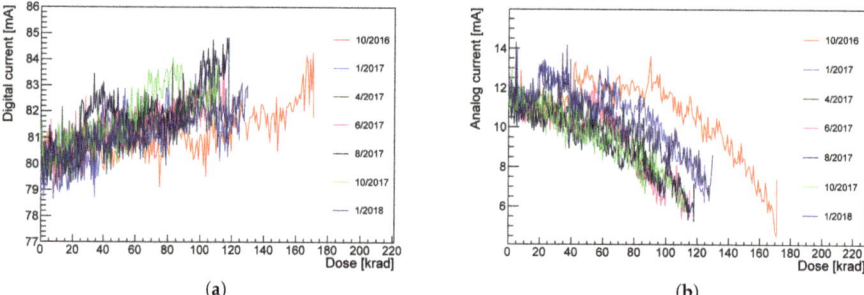

Figure 4. Digital (**a**) and analog (**b**) support currents versus total ionizing dose measured for different irradiation campaigns. The dates of the campaigns are listed in the legends.

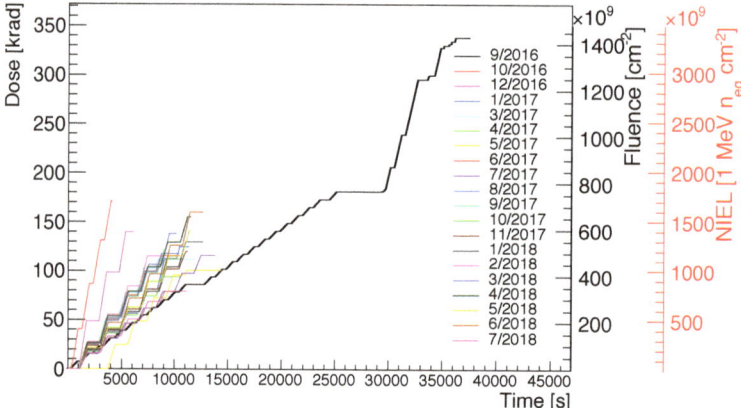

Figure 5. The total ionization dose, the accumulated proton fluence, and the non-ionizing energy loss (NIEL) for different irradiation campaigns. The legend gives the date of the performed campaigns.

In each pixel, charge threshold is measured by injection of a given charge from an injection capacitance to the pixel analog front end. The injection is repeated 50 times and the threshold is defined as a charge which is registered by a pixel with a 50% probability. In the case of ALPIDE sensors, the charge threshold depends mainly on two DACs: I_{THR} which determines the shape of the pulse and V_{CASN} which regulates the baseline voltage [6]. An example of the firing probability of a pixel versus the injected charge is shown in Figure 6. The dependence is fitted by the so-called S-curve

$$S(Q) = \frac{1}{2}\left(50 + 50 \times \text{erf}\left(\frac{Q - Q_{THR}}{\sqrt{2}\sigma}\right)\right), \tag{3}$$

where Q is injected charge and Q_{THR} is threshold. The formula assumes that the temporal noise, which smears the threshold value, has a Gaussian character. The threshold distribution obtained from 10% of pixels for the default settings of the DACs for a non-irradiated chip is shown in Figure 7.

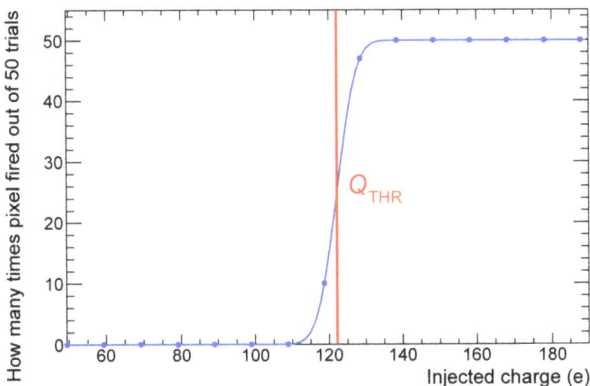

Figure 6. Probability of charge registration in an ALPIDE pixel. The data are fitted with the S-curve (3).

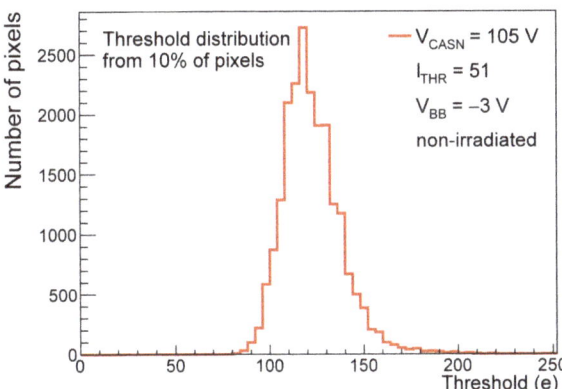

Figure 7. Distribution of charge thresholds from 10% of pixels of a non-irradiated chip.

Figure 8 shows the mean threshold as a function of the accumulated dose for different irradiation campaigns. Initially until October 2017, the ALPIDE was operated using the default DAC settings for -3 V reverse substrate back bias. In this period, the mean threshold was decreasing with the accumulated dose without any sign of annealing. In October 2017, the DAC settings of the chip were changed to increase the threshold and to suppress the noise. Since then we observe that the chip anneals after each campaign.

After obtaining the total ionizing dose of 2700 krad and the NIEL of 2.7×10^{13} 1 MeV $n_{eq} \cdot cm^{-2}$, the chip was characterized at the CERN Proton Synchrotron. There the ALPIDE was tested using a 6 GeV/c pion beam. The sensor was installed in a telescope which consisted of 7 planes of ALPIDE sensors. The tested ALPIDE (device under test, DUT) formed the middle plane of the telescope, the other ALPIDEs served as reference planes for pion track reconstruction. The EUTelescope software [13] was used to estimate the detection efficiency, which was obtained by comparing an extrapolated hit position in the DUT calculated from tracking planes with the actual measured hit position in the DUT.

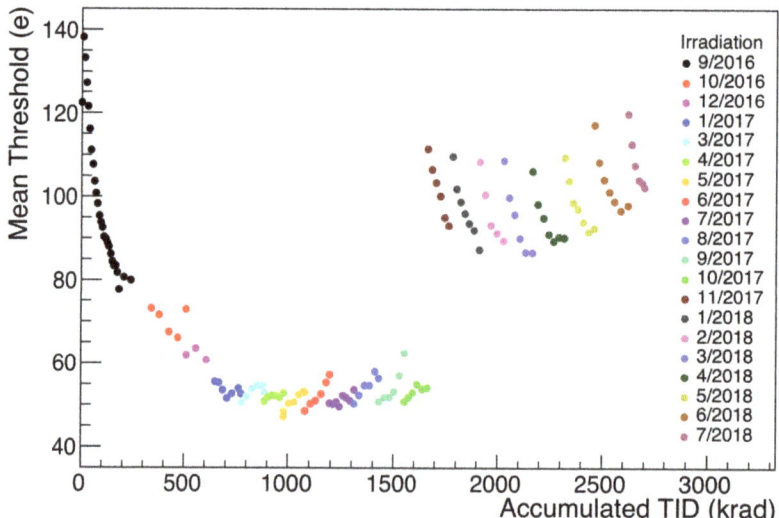

Figure 8. Mean threshold versus accumulated total ionizing dose. The dates of the performed campaigns are quoted in the legend.

Figure 9 shows the detection efficiency and the fake-hit rate as a function of the mean charge threshold for the irradiated chip and for a non-irradiated reference sensor. The red dash-dotted line corresponds to the project limit on fake-hit rate which is 10^{-6}/pixel/event and the black dash-dotted line gives the limit on detection efficiency which should be higher than 99%. As it is seen from the figure, the irradiated sensor still fulfills the requirements of the upgrade project in terms of detection efficiency and the fake-hit rate in the threshold range $\approx$ 150–200 electrons.

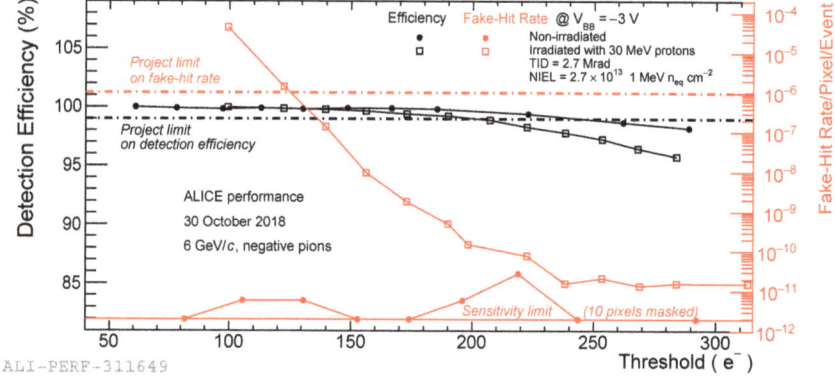

Figure 9. The detection efficiency and the fake-hit rate of the irradiated and non-irradiated ALPIDE sensors versus threshold charge for -3 V reverse substrate bias voltage. The detection efficiency was obtained using a 6 GeV/c pion beam at CERN PS.

5. Conclusions

After reaching ten times the radiation load expected for an ALPIDE sensor in the Inner Barrel of ALICE ITS in Run 3 and Run 4, the sensor is still operational and fulfills project goal requirements. We observed that irradiation with 30 MeV protons caused a steady drop of the mean charge threshold when the chip was operated with the nominal DAC settings for -3 V reverse substrated bias voltage. The initial threshold level could, however, be recovered by retuning DAC settings, which leads to suppression of noisy pixels.

Author Contributions: Conceptualization, F.K.; methodology, F.K.; software, V.R. and F.K.; validation, V.R. and F.K.; formal analysis, V.R. and F.K.; investigation, V.R. and F.K.; resources, F.K.; data curation, F.K.; writing—original draft preparation, V.R.; writing—review and editing, F.K.; visualization, V.R.; supervision, F.K.; project administration, F.K.; funding acquisition, F.K.

Funding: This research was funded by the Ministry of Education, Youth and Sports of the Czech Republic grant number LTT17018.

Conflicts of Interest: The authors declare no conflict of interest.

References

1. Aamodt, K.; Quintana, A.A.; Achenbach, R.; Acounis, S.; Adamová, D.; Adler, C.; Aggarwal, M.; Agnese, F.; Rinella, G.A.; Ahammed, Z.; et al. The ALICE experiment at the CERN LHC. *JINST* **2008**, *3*, S08002.
2. Rafelski, J. *Melting Hadrons, Boiling Quarks*; Springer: Cham, Switerland, 2016 .
3. Abelev, B.; et al. [Alice Collaboration]. Upgrade of the ALICE Experiment: Letter Of Intent. *J. Phys. G* **2014**, *41*, 087001. [CrossRef]
4. Abelev, B.; Real, J.S.; Margotti, A.; Contreras, J.G.; Karavicheva, T.; Arsene, I.C.; Ramello, L.; Gulkanyan, H.; Planinic, M.; Symons, T.J.; et al. Technical Design Report for the Upgrade of the ALICE Inner Tracking System. *J. Phys. G* **2014**, *41*, 087002. [CrossRef]
5. Martinengo, P. The new Inner Tracking System of the ALICE experiment. *Nucl. Instrum. Methods Phys. Res.* **2017**, *967*, 900–903. [CrossRef]
6. Aglieri, G.; Cavicchiolia, C.; Chalmet, P.L.; Chanlekc, N.; Collud, A.; Giubilatoe, P.; Hillemannsa, H.; Juniquea, A.; Keila, M.; Kimg, D. Monolithic active pixel sensor development for the upgrade of the ALICE inner tracking system. *J. Instrum.* **2013**, *8*, C12041. [CrossRef]
7. Mager, M.; et al. [ALICE Collaboration]. ALPIDE, the Monolithic Active Pixel Sensor for the ALICE ITS upgrade. *Nucl. Instrum. Methods Phys. Res.* **2016**, *824*, 434–438. [CrossRef]
8. TowerJazz. Available online: http://www.jazzsemi.com (accessed on 26 January 2019).
9. Křížek, F.; Ferencei, J.; Matlocha, T.; Pospíšil, J.; Príbeli, P.; Raskina, V.; Isakov, A.; Štursa, J.; Vaňát, T.; Vysoká, K. Irradiation setup at the U-120M cyclotron facility. *Nucl. Instrum. Methods Phys. Res.* **2018**, *894*, 87–95. [CrossRef]
10. PTW Freiburg. Available online: http://www.ptw.de/ (accessed on 28 January 2019).
11. The European Space Sciences Committee (ESCC). Single event effects tests method and guidelines. In *ESCC Basic, Specification No. 25100*; European Space Agency: Madrid, Spain, 2016.
12. Displacement Damage in Silicon. Available online: https://rd50.web.cern.ch/rd50/NIEL/default.html (accessed on 28 January 2019).
13. EUTelescope Documentation and Source Code. Available online: http://eutelescope.web.cern.ch (accessed on 1 February 2019).

© 2019 by the authors. Licensee MDPI, Basel, Switzerland. This article is an open access article distributed under the terms and conditions of the Creative Commons Attribution (CC BY) license (http://creativecommons.org/licenses/by/4.0/).

Review

Higgs and BSM Studies at the LHC

Dezső Horváth [1,2,†]

[1] Wigner Research Centre for Physics, H-1121 Budapest, Hungary; horvath.dezso@wigner.mta.hu or dezso.horvath@cern.ch
[2] Institute of Nuclear Research (Atomki), H-4026 Debrecen, Hungary
[†] On behalf of the CMS Collaboration.

Received: 8 March 2019; Accepted: 24 June 2019; Published: 2 July 2019

Abstract: The discovery and study of the Higgs boson at the Large Hadron Collider of CERN has proven the validity of the Brout–Englert–Higgs mechanism of mass creation in the standard model via spontaneous symmetry breaking. The new results obtained by the ATLAS and CMS Collaborations at the LHC show that all measured cross-sections agree within uncertainties with the predictions of the theory. However, the standard model has obvious difficulties (nonzero neutrino masses, hierarchy problem, existence of dark matter, non-existence of antimatter galaxies, etc.), which point towards more possible violated symmetries. We first summarize the present status of the studies of the Higgs boson, including the latest results at 13 TeV p-p collision energy, then enlist some of the problems with possible solutions and the experimental situation regarding them.

Keywords: LHC; ATLAS; CMS; standard model; Higgs boson; symmetry breaking; supersymmetry; dark matter

1. Introduction

The theory behind particle physics, called for historic reasons *the standard model* (SM), is based on local *gauge symmetries* In certain cases, broken symmetries were introduced in order to explain the fundamental experimental observations, the most important ones being parity violation and the Brout–Englert–Higgs (BEH) mechanism of spontaneous symmetry breaking. Left-handed currents could take care of parity violation observed in weak interactions, and using the BEH mechanism, the standard model could account for the masses of the elementary fermions and bosons. In spite of the excellent agreement between the experimental data and the predictions and fitting of the standard model, several mysteries stayed unsolved in particle physics. Numerous extensions for the standard model were proposed to solve those, the most popular one being supersymmetry, a broken fermion-boson symmetry. It helps to interpret the dark matter of cosmology within particle physics and also to solve the hierarchy problem, the quadratic divergence of the calculated mass of the Higgs boson in the standard model. Other problems unsolved in the framework of the standard model are the lack of antimatter in the Universe and baryogenesis, as well as neutrino oscillations. In this review, we shall try to summarize these concepts and the corresponding experimental evidence in high-energy physics.

2. Fundamental Particles in the Standard Model

According to the standard model, the world consists of two kinds of particles, fermions and bosons, different by their spins, intrinsic angular momenta, measured in units of $\hbar$, the reduced Planck constant. The fermions in general have half-integer spins: $S = \frac{1}{2}, \frac{3}{2}, \ldots$, whereas the spins of the bosons are integer: $S = 0, 1, 2, \ldots$ The elementary (fundamental) fermions of the standard model are the leptons and the quarks of three families (see Table 1) with $S = \frac{1}{2}$; those are also called matter particles, as our matter consists of the fermions of the first family. The elementary bosons have integer spins; those mediating the three interactions, the photon, the eight gluons, and the three weak bosons have $S = 1$, whereas the Higgs boson is a scalar particle with zero spin. The LEP experiments have shown by measuring the decay width of the Z boson that only those three families exist with light enough neutrinos to allow for the $Z \to \nu\bar{\nu}$ decay process.

The term *elementary* means that those particles are point-like and structureless, with no excited state. The fundamental fermions have three families, consisting of a pair of quarks and a pair of leptons each. All fermions have antiparticles of opposite charges, but similar other properties. The leptons can propagate freely, but the quarks are confined in composite particles, the hadrons. They are either composite fermions, bound states of three quarks, the baryons (like the proton and neutron), and three antiquarks, antibaryons (like the antiproton), or bosons composed of a quark and an antiquark, the mesons (like the pion). The quarks in the hadrons are bound together by the strong interaction, and because of the two possible compositions, its three-state source is called *color charge*; in analogy with human sight, they are called the colorless states.

Table 1. Leptons and quarks, the three families of basic fermions. T_3 is the third component of the weak isospin; index L stands for the left polarization of the weak isospin doublets; and the apostrophe indicates the mixed quark states.

	Family 1	Family 2	Family 3	Charge	T_3
Leptons	$\begin{pmatrix} \nu_e \\ e \end{pmatrix}_L$	$\begin{pmatrix} \nu_\mu \\ \mu \end{pmatrix}_L$	$\begin{pmatrix} \nu_\tau \\ \tau \end{pmatrix}_L$	0 -1	$+\frac{1}{2}$ $-\frac{1}{2}$
Quarks	$\begin{pmatrix} u \\ d' \end{pmatrix}_L$	$\begin{pmatrix} c \\ s' \end{pmatrix}_L$	$\begin{pmatrix} t \\ b' \end{pmatrix}_L$	$+\frac{2}{3}$ $-\frac{1}{3}$	$+\frac{1}{2}$ $-\frac{1}{2}$

3. Interactions in the Standard Model

The standard model includes three basic interactions derived from local gauge invariances. A fully invariant color-SU(3) accounts for the strong interaction from local SU(3), but the other two are united in a local U(1)⊗SU(2) gauge invariance as an electroweak interaction partially broken by the *spontaneous symmetry breaking* BEH mechanism. The strong interaction originates from the *color charge* of the quarks with its three colors, and it is mediated by eight massless *gluons* carrying both color and anti-color. The weak interaction has three heavy weak bosons as mediators, $W^\pm$ and Z^0, whereas that of of the electromagnetism is the massless and neutral γ photon. Due to Heisenberg's uncertainty principle, these bosons are virtual when they mediate the interactions. A virtual photon as mediator has a finite mass when transferring energy and momentum between two charged particles, and the heavy W boson is many orders of magnitude lighter when mediating the nuclear beta decay. However, they are real particles; they can be also emitted and observed experimentally: everybody can see the visible photons, and even the weak bosons and the gluons can be observed and studied when emitted in high-energy collisions and decay to leptons or hadron showers, jets.

A fermion enters into an interaction if it possesses the corresponding charge. The weak interaction operates on all basic fermions as all of them carry the weak isospin, the electromagnetic one on those having electric charges or magnetic moments, and the strong interaction on the colored fermions, i.e., on the quarks. The photon does not carry charges, but the W boson is electrically charged. The gluons carry the color charge, and so, they have self-interaction as well. In order to ensure a better agreement between theory and experiment, one also needs the existence of a scalar boson, a particle with its charges and spin zero. In the standard model, that scalar particle is the Higgs boson.

The standard model was very thoroughly tested in high-energy experiments, mostly at LEP, the Large Electron-Positron collider of CERN (see Figure 1). All elementary particles of the model were identified and studied experimentally, and much experimental evidence was collected for the three-state color and fractional electric charges of the quarks, as well as for the properties of the gluons and of the heavy weak bosons. Lately, the high-energy experiments concentrated on searching for the only missing particle, the Higgs boson, acquiring more and more precise new data. The latter was achieved mainly at the electron-positron colliders and all results are in reasonable agreement with the predictions of the standard model.

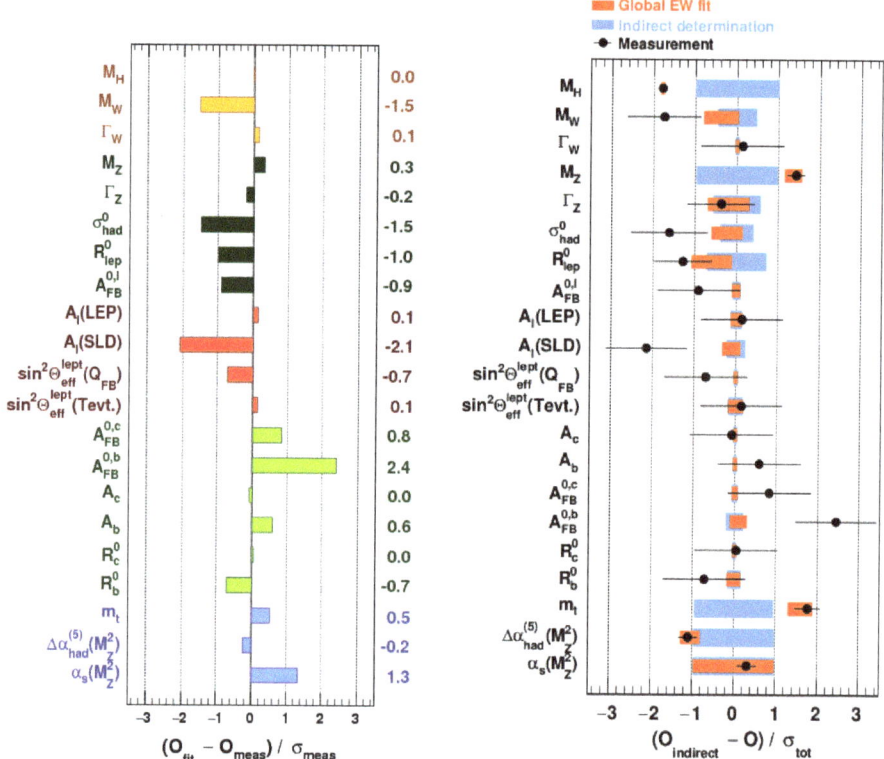

Figure 1. Recent results of testing the standard model [1]. (**Left**) Relative deviations of the measured values for selected model parameters from the fitted ones. (**Right**) Similar fitting performed with the given model parameter left floating, showing the sensitivity of the model to its parameter values.

4. Testing the Standard Model

Figure 1 summarizes the present situation of testing the standard model by comparing measured and calculated values for key parameters [1] using the collected experimental information. On the left plot, relative deviations are shown between the fitted and measured parameter values: that is, a global electroweak fit. On the right side, besides the global fit (in red), another fit is plotted (in blue) with the measured value of the given parameter not included in the fit; that shows the degree of determination of the given parameter by the rest of the measurements, i.e., the sensitivity of the model to the values of the parameters. We can conclude that all parameters agree within their experimental uncertainties with the theoretical predictions. This picture became complete with the measurement of the Higgs boson mass at 125 GeV/c^2. It is interesting that the standard model calculations are not too sensitive to its mass value, the fit without including the Higgs measurement would predict a somewhat lighter Higgs boson (Figure 1, right) with a quite large uncertainty.

Spontaneous Symmetry Breaking

One can derive the three basic interactions from local gauge symmetries, but they cannot produce masses for the elementary fermions and the weak bosons (the photon and the gluons are massless). The model must violate some of those symmetries in order to create masses. Symmetry violation plays a key role in the standard model. Several theoreticians published *spontaneous symmetry breaking* in 1964 independently [2–5], and it is now called the *BEH mechanism* after R. Brout, F. Englert, and P.W. Higgs, although for a long time, it was quoted as the *Higgs mechanism*.

The BEH mechanism assumes that the electroweak vacuum is filled by a potential that spontaneously breaks its perfect symmetry. This can be illustrated by a Mexican hat (Figure 2): its axial symmetry is not violated by a ball on its top, but the ball should roll down, thereby spontaneously breaking the symmetry. The axial symmetry of the potential allows us the choose such a coordinate system where the position of the minimum is determined by one real parameter only (this corresponds to the ball laying on the imaginary axis in Figure 2), that is the vacuum expectation value of the BEH field, determined by the Fermi constant to be $v \approx 246$ GeV.

The BEH potential has a complex doublet field, adding four degrees of freedom to those of the standard model: three of them are used to produce the masses for the weak bosons, and the fourth one makes a scalar particle, the Higgs boson, which is needed by the theory. Thus, the scalar boson appears as a by-product of the mass-creation mechanism. The masses of the weak bosons can actually be calculated by the standard model using the value of the weak coupling (called Fermi constant at low energies) deduced from the beta-decay measurements. Moreover, the BEH mechanism allows us to add fermion mass terms to the electroweak Lagrangian with the fermion masses as free parameters.

We have to emphasize that the masses of the macroscopic objects in our world are predominantly due to the energy content of the proton and the neutron; the quark and electron masses from the BEH mechanism add only a few percent to them.

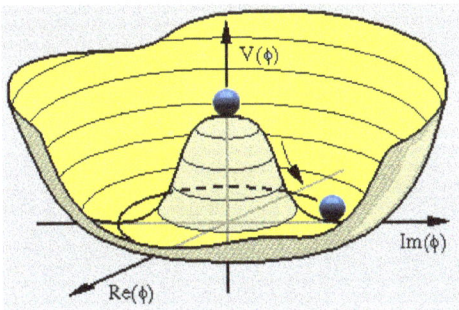

Figure 2. Explanation of the Brout–Englert–Higgs (BEH) mechanism. The axial symmetry of the Mexican hat is not violated by a ball placed on its top, but it will be spontaneously broken when the ball rolls down in the valley. Because of the axial symmetry of the potential, we can rotate the coordinate system so that the ball will be at the point $Im(\Phi) = 0$.

5. Search for the Higgs Boson

All of the fundamental particles of the standard model, the fermions of Table 1, and the bosons mediating the interactions were observed and studied by the high-energy experiments well before the launch of the LHC, but the Higgs boson was not yet. Before the observation of the Higgs boson, the BEH mechanism had no direct experimental proof, and several alternative theories existed (although they had to create a scalar particle, usually as a $t\bar{t}$ bound state). Fitting the standard model has limited, but did not determine the mass of the Higgs boson. This is well demonstrated by the right plot of Figure 1: the Higgs boson mass is not that tightly determined by the rest of the standard model measurements. Thus, the particle physics community invested an incredible amount of effort in searching for the Higgs boson.

5.1. Hunting for the Higgs Boson

In the accelerator experiments, the Higgs boson was searched for at ever-increasing energies by ever-increasing collaborations, usually in the following steps:

- Compose a complete SM background using Monte Carlo simulation including all types of possible events normalized to their cross sections.
- Simulate Higgs-boson signals of all possible production and decay processes with all possible Higgs-boson masses.
- Put all these through your own detector simulation to get events analogous to the measured ones.
- Calculate the number of events expected for signal and background at various conditions at the actually collected experimental luminosity.
- Check whether the expected background predicted by the standard model (SM) agrees with the experimental data yield.
- Optimize the event selection using the simulated events: reduce the B background, and enhance the S signal via maximizing a function of merit, e.g., $N_S/\sqrt{N_B}$ or $N_S/\sqrt{N_S+N_B}$, or using the approximate formula of Cowan et al. [6].

Once we are happy with the simulations and the event selection, we must choose a test statistic. That could be any kind of probability variable characteristic of the given phenomenon: probabilities for having background only, signal, or combinations. A popular one is the Q likelihood ratio of signal + background over background: $Q = \mathcal{L}_{s+b}/\mathcal{L}_b$. Other frequently used ones are the probabilities of *not* having the expected signal above the expected background and the collected data, like:

- CL_b, the signal confidence level assuming background only, i.e., the complete absence of the signal, or
- the so-called *p-value*: the probability that a random fluctuation of the measured background could give the observed excess.

5.2. Exclusion at LEP

The Higgs boson of the standard model is a scalar particle: most of its quantum numbers are zero, although the BEH field is a weak isospin doublet. The main measurable property of the Higgs boson is mass, and all production and decay cross sections depend on its mass. Analyzing all the experimental data before 2012 pointed towards a Higgs-boson mass $m_H \approx 100\,\mathrm{GeV}/c^2$ ($80 \leq m_H \leq 160\,\mathrm{GeV}/c^2$ with a 95% confidence level).

In accelerator experiments, a new theoretical hypothesis is considered to be excluded if its signal does not appear at a $\geq$95% confidence level and observed if its appearance exceeds five uncertainties, i.e., its signal $> 5\sigma$ over the background, where σ is the total experimental uncertainty according to the best honest guess of the experimentalist including both statistics and the possible systematic deviations. Some of the statistical methods used in the search and discovery of the Higgs boson have been reviewed by the author in [7].

CERN used the last two years of the Large Electron-Positron collider (LEP) to search for the Higgs boson and collected more data (luminosity) at energies above the Z boson mass than in the previous decade altogether.

According to the standard model at LEP energies, the dominant production process is $e^-e^+ \to ZH$, and the dominant Higgs boson decay is to the heaviest available fermions, a pair of b quarks [8] (Figure 3). At LEP, the only difference among the various search channels was the different decay process of the accompanying Z boson.

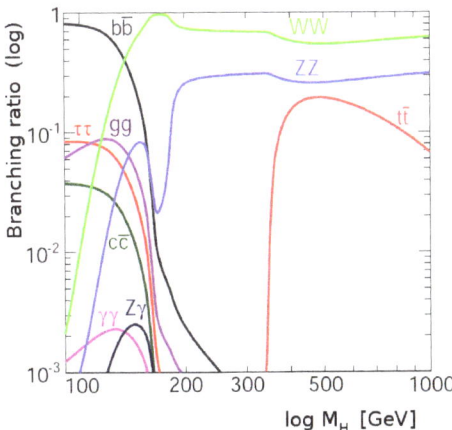

Figure 3. The branching fractions of the various decay channels of the Higgs boson as predicted by the standard model [8]. Both axes are logarithmic. For $m_H \leq 120\,\text{GeV}/c^2$, the H→ $b\bar{b}$ decay channel dominates. Note how small the probabilities of the reactions H→$\gamma\gamma$ and H → ZZ →4 charged leptons are, leading to the discovery of the Higgs boson at the LHC.

The detectors of high-energy experiments are built of cylindrical layers encircled by huge magnets including as much as possible the detector parts. This structure includes a vertex detector right around the interaction point, a tracking system of multiwire chambers or semiconductor detectors of light material for measuring the tracks of the charged particles curved in the magnetic field. The next stage is the electromagnetic calorimeter, something heavy absorbing all electrons and photons, encircled by an even heavier hadron calorimeter, absorbing the pions, protons, neutrons, etc., and outside of an all muon system, following and measuring the high-energy muons leaving the detector. The neutrinos, produced by weak interaction, escape the whole system with no interaction, and they are identified by a momentum imbalance (called *missing momentum*). LEP had four such large experiments located in the four collision points of the collider ring: ALEPH, DELPHI, L3, and OPAL (the present author was in OPAL).

Statistics made a bad joke at LEP: the ALEPH experiment saw a quite significant signal of a Higgs-like boson of a mass of $115\,\text{GeV}/c^2$, whereas the other LEP experiments did not observe any such signal. Finally, a common limit was determined, $114.4\,\text{GeV}/c^2$, below which the Higgs boson was excluded [9] by a 95% confidence level. Many physicists wanted CERN to let LEP work for another year, but the simulations were not very promising for the discovery of a $115\,\text{GeV}/c^2$ Higgs boson, and LEP had to be dismounted and removed from its tunnel to make space for the LHC.

5.3. Search for the Higgs Boson at the LHC

Designing the LHC and its experiments started very early, even before the start of LEP; thus, the construction of its detectors took two decades of hard work by thousands of physicists and engineers before the LHC was launched in 2009. Moreover, during its first two years, the LHC was mostly under development; the actual intensive data taking started in 2011 only.

Similarly to the former LEP ring, the Large Hadron Collider has four beam-beam interaction points with a major experiment (and in some cases, smaller ones associated with them) in each one. The two largest LHC experiments, ATLAS and CMS, have been designed with the main goal of finding the Higgs boson. These collaborations are really large. According to the official statistics, in 2016, CMS had 5250 participants (including 1916 students and 1274 engineers and technicians) from 198 institutions of 45 countries. The largest participant country was the USA, then Italy, Germany, and

Russia. ATLAS has even a somewhat larger size than CMS both in the volume of the detector and in the number of participants.

The various formation processes of the SM Higgs boson in p-p collisions at the LHC are shown in Figure 4. Gluon fusion is by far the dominant reaction, but vector boson fusion is also quite important [8]. From previous experimental data, the general fitting of the standard model predicted a light Higgs boson, with a mass slightly below $100\,\mathrm{GeV}/c^2$. As LEP excluded the Higgs boson below $114\,\mathrm{GeV}/c^2$, ATLAS and CMS had to be prepared for detecting the Higgs boson in the most difficult region, $m_H \approx 120\,\mathrm{GeV}/c^2$, with several competing decay channels (Figure 3). Of course, if the Higgs boson were much lighter or heavier, it would have been observed earlier at LEP or at the Tevatron.

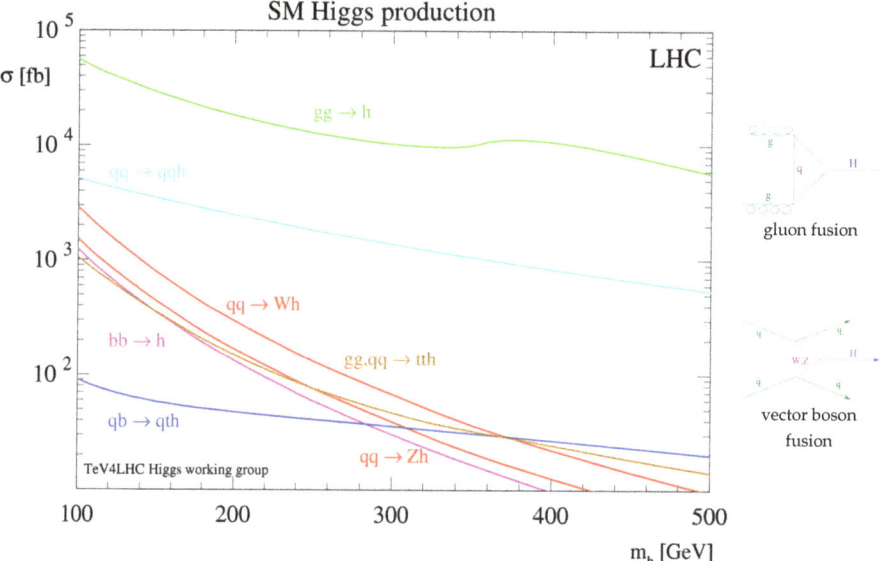

Figure 4. Predictions of the standard model for the production of the Higgs boson in proton-proton collisions [8]. The two most important formation channels are gluon fusion and vector boson fusion, and they were used to discover the Higgs boson. Later, all possible channels were observed and studied.

5.4. Observation at the LHC

The most promising reactions to observe a light SM Higgs boson at the LHC was theoretically predicted to be its decay either to two hard photons or decay to two Z bosons (one of them virtual, of course, as there would not be enough energy to make two $91\,\mathrm{GeV}/c^2$ bosons) and each Z decaying further to a pair of charged leptons: these two reactions could be visible above the high hadron background (see, e.g., Figure 5). In spite of the very low decay branching fractions (BF), these channels promised the best signal/background (S/B) ratios: for $H \to ZZ^* \to \ell^+\ell^-\ell^+\ell^-$ ($\ell = e, \mu$), the BF is low, 1.24×10^{-4}, but the S/B ratio > 1, whereas for $H \to \gamma\gamma$, the BR is much higher, 2.27×10^{-3}, although S/B < 1. Both large LHC experiments, CMS and ATLAS designed their electromagnetic calorimeters with this in mind. The CMS one consists of 75,848 $PbWO_4$ single crystal scintillators, whereas the electromagnetic calorimeter ATLAS is a sampling one based on liquid argon shower detectors (see Table 2).

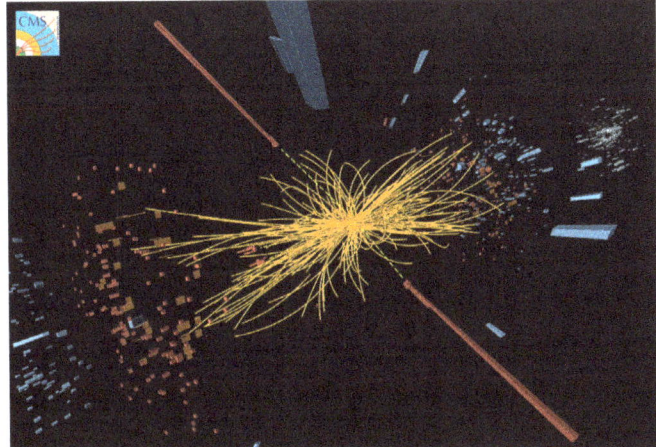

Figure 5. A CMS event: H → γγ candidate. The size of the orange boxes is proportional to the energies deposited in the electromagnetic calorimeter by the neutral photons of invisible tracks. Note the huge hadron background from proton-proton collisions.

In 2012, ATLAS and CMS announced [10,11] that at the LHC, at the collision energies of 7 and 8 TeV, in the two decay channels H → γγ and H → ZZ → $\ell^+\ell^-\ell^+\ell^-$, a new boson was observed with a mass of $m \approx 125\,\text{GeV}/c^2$ by both experiments at a statistical significance of 5σ each (i.e., that much higher than the total σ uncertainties of the measurements). The properties of the observed new particle corresponded to those of the Higgs boson as predicted by the standard model.

5.5. The Two Large Experiments

A combination of ATLAS and CMS data based on 7 and 8 TeV collisions gave the mass estimation: $m_\text{H} = 125.09 \pm 0.24\,\text{GeV}$, including both statistical and systematic uncertainties [12]. This result was made even more convincing by the fact that the construction of ATLAS and CMS was very different. Some of these features are overviewed in Table 2; note, e.g., that CMS has about one tenth of the volume and twice the weight of ATLAS.

Table 2. Design and construction of the detectors of ATLAS and CMS. Very different measuring systems provide very similar data for the Higgs boson. TRD: transition radiation detector, E-m: electromagnetic, LAr: liquid Argon, cal.-m.: calorimeter, scint.: scintillator, h-cal.: hadron calorimeter.

	ATLAS	CMS
Magnet	toroidal + *small(?)* 2 T solenoid	large 3.8 T solenoid
Tracker	semiconductor + TRD	semiconductor
E-m. calorimeter	LAr with steel and Pb	PbWO$_4$ scintillator
Hadron cal.-m.	steel + scint. tiles	brass + scint. tiles
Far forward h-cal.	LAr with Cu and W	steel with quartz Cherenkov
Muon detector	chambers (4 types)	chambers (3 types)
Size	⌀25 m × 46 m (23,000 m^3)	⌀15 m × 21.6 m (3800 m^3)
Trigger	3-level	2-level
Weight, tons	7000 t	14,000 t
Participating scientists	3000	2300

5.6. Mass Measurements

Since the original discovery, an order of magnitude more information was collected in the last three years at a higher collision energy, 13 TeV. As examples of the new measurements, let us show mass spectra measured by ATLAS and CMS at 13 TeV for H → 4ℓ ($\ell = e^\pm, \mu^\pm$) (Figure 6) and for

H → $\gamma\gamma$ (Figure 7). The corresponding mass values for the Higgs boson are listed in Table 3, where the first uncertainties are statistical and the second ones are systematic. Note the remarkable agreement between channels and experiments. As all results were statistically limited and only the 2016 data were used, significant improvement is expected when all 13 TeV data collected in 2016–2018 will be analyzed. One can see in Figure 1 that these values also agree with the standard model calculations very well.

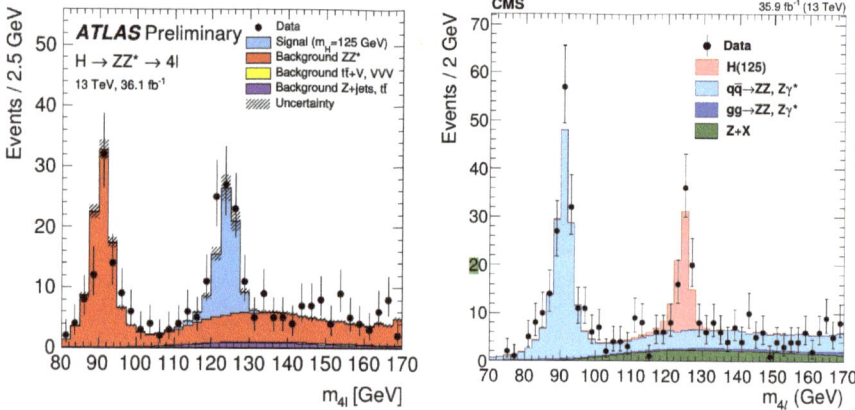

Figure 6. New measurements of the mass of the Higgs boson by the ATLAS [13] and CMS [14] Collaborations at the LHC at 13 TeV p-p collision energy using the H → 4ℓ ($\ell = e^{\pm}, \mu^{\pm}$) decay channel.

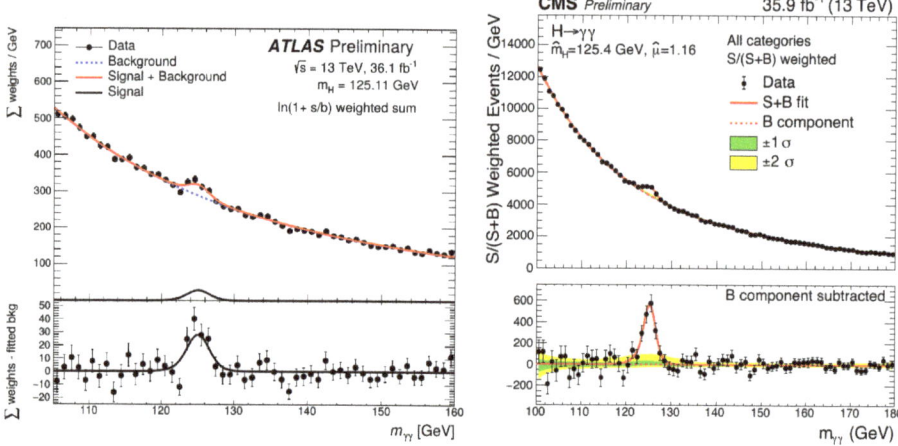

Figure 7. New measurements of the mass of the Higgs boson by the ATLAS [13] and CMS [14] Collaborations at the LHC at 13 TeV p-p collision energy using the H → $\gamma\gamma$ decay channel.

Table 3. New mass values for the Higgs boson measured by ATLAS [13] and CMS [14] at 13 TeV p-p collision energy. The first uncertainty is statistical, the second one systematic. Note the remarkable agreement between channels and experiments.

Experiment	Decay Channel	Mass (GeV/c^2)	Reference
ATLAS	H → 4ℓ	124.79 ± 0.36 ± 0.05	[13]
CMS	H → 4ℓ	125.26 ± 0.20 ± 0.08	[14]
ATLAS	H → $\gamma\gamma$	124.93 ± 0.21 ± 0.34	[13]
CMS	H → $\gamma\gamma$	125.4 ± 0.2 ± 0.2	[15]

5.7. Higgs Couplings

Although much of the new data are still in analysis, all significant production and decay channels of the Higgs boson were identified and studied. The results (Figure 8) show good agreement between the two experiments and also with the predictions for a single Higgs boson with a mass of $125\,\text{GeV}/c^2$. The decay widths were measured for all production and decay channels, and they agreed with those predicted by the standard model.

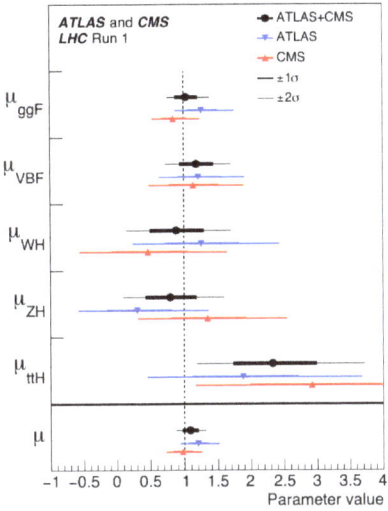 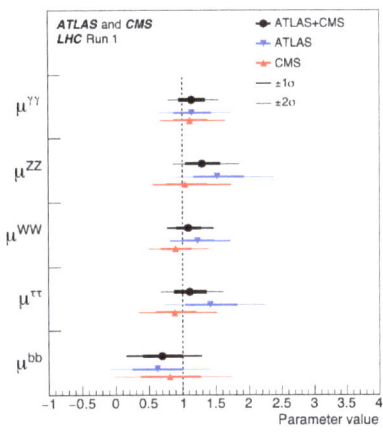

Figure 8. Production and decay rates of the Higgs boson and constraints on its couplings as compared to their predictions by the standard model from a combined ATLAS and CMS analysis of the LHC p-p collision data at $\sqrt{s} = 7$ and $8\,\text{TeV}$ [16].

5.8. Other Decay Channels

Using the 13 TeV data enabled ATLAS and CMS to observe and study the Higgs boson using decay channels other than the most favored four-lepton and $\gamma\gamma$ ones. The third most significant channel is $\text{H} \rightarrow \text{W}^+\text{W}^-$. When in 2012, its contribution at 7 and 8 TeV (Run 1) was added to the four-lepton and $\gamma\gamma$ measurements, it changed the 5σ observational significance for both experiments, increasing it for ATLAS to 6.1σ and decreasing it for CMS to 4.9σ. The analysis of all Run 1 data gave relative signal strengths $\mu = \text{expt/theory} = 1.22^{+0.23}_{-0.21}$ for ATLAS and $\mu = 0.90^{+0.23}_{-0.21}$ for CMS, but the average agreed [16] with the standard model: $\mu_{av} = 1.09^{+0.18}_{-0.16}$.

Another important channel is $\text{H} \rightarrow \tau^+\tau^-$. In Run 1, ATLAS and CMS together managed to observe it with a 5.5σ significance [16]. In Run 2, by a simultaneous analysis of all data obtained at 7, 8, and 13 TeV, CMS reached a 5.9σ significance [17].

In principle, the decay to a pair of b quarks has the highest cross section among the decay channels of the Higgs boson, but it is very difficult to distinguish among all the hadronic activity in the detectors. CMS has managed to observe its signal with a significance of 5.6σ via using all production channels: associated with a vector boson (VH), gluon fusion (gg), vector boson fusion (VBF), and associated with a top pair (ttH); and data collected at 7, 8, and 13 TeV collision energies. The measured signal strength, $\mu = 1.04^{+0.20}_{-0.19}$, again agreed with the SM prediction [18].

According to the standard model, the Higgs boson couples to other particles via their masses. An important test is to check its coupling to the top quark, the heaviest fermion. An analysis of all decay channels at the three LHC energies including the 2016 data gave CMS an observation (Figure 9) of the $\bar{t}tH$ associated production at a 5.2σ significance [19].

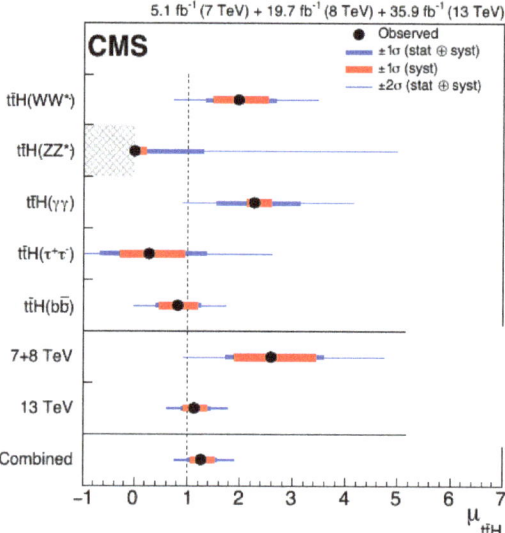

Figure 9. Observation [19] of the production of Higgs bosons associated with a top quark pair: signal strengths normalized to the standard model predictions measured for various decay modes.

6. Vacuum Stability

The observed light mass of the Higgs boson seems to be very exciting for theoreticians, and there was even a special workshop [20] devoted to discuss this mass value in 2013. The reason is that $M_H = 125$ GeV is at the borderline of the stability of the electroweak vacuum on the plane of top mass against Higgs mass as another fine-tuning aspect of the standard model (Figure 10). The vacuum is considered to be stable if it has a single minimum, but it could also be metastable with more than one minima [21].

If our vacuum is metastable and we are not in the deepest minimum, then the universe exists in a *false vacuum*, and the world can shift into the deeper one by a quantum mechanical tunneling effect. In some sense, it could be the end of our world as we know it, as in the new BEH minimum, all the particles would have different masses, and also the strength of the weak interaction would change. Of course, this assumes that the standard model is valid up to extreme high energies. However, Turner and Wilczek stated in 1982 [22] that even if our vacuum is metastable, its expected lifetime could be much longer than the age of the Universe, hence it should appear stable to us.

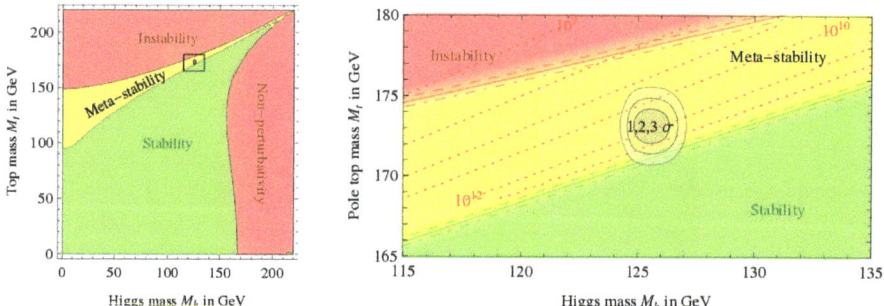

Figure 10. Mass of the top quark against that of the Higgs boson [21]. Note how narrow is the yellow region of metastability around the measured values.

7. Problems of the Standard Model

Thus, we most likely observed the standard model Higgs boson at the LHC. This is, of course, a great success of particle physics as it has proven the BEH mechanism of mass production. However, the fact that the SM seems to describe all high-energy experimental data perfectly is in some sense a problem: it has theoretical difficulties that need new physics to resolve (Figure 11).

- The standard model cannot interpret gravity as a gauge interaction similar to the other three interactions.
- It cannot account for the dark matter as a particle and cannot explain dark energy or the lack of antimatter in the Universe.
- It has an ad hoc symmetry-breaking mechanism to produce masses for the elementary fermions, but cannot explain the nonzero masses of the neutrinos and their oscillation.
- The fractional quantum numbers of the quarks contradict the quantization of the electric charge.
- The coupling constants of the three gauge interactions converge at high energies, but do not meet (Figure 12).
- There is a magic number three for colors, charges, and fermion families with no relation to each other.
- Naturalness (hierarchy) problem: The mass of the Higgs boson quadratically diverges due to radiative corrections.

The hierarchy problem would be solved if the fundamental fermions and bosons existed in pairs with the same properties, just different (one half less) spins, because the largest terms of the corrections (mainly that due to the top quark) could cancel each other.

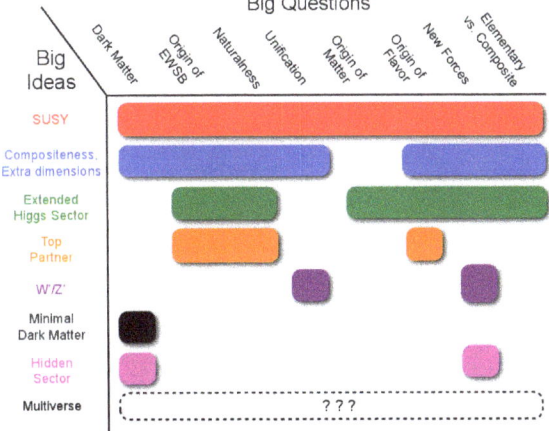

Figure 11. Various problems of the standard model and some theoretical proposals trying to solve them [23]. EWSB: electroweak symmetry breaking.

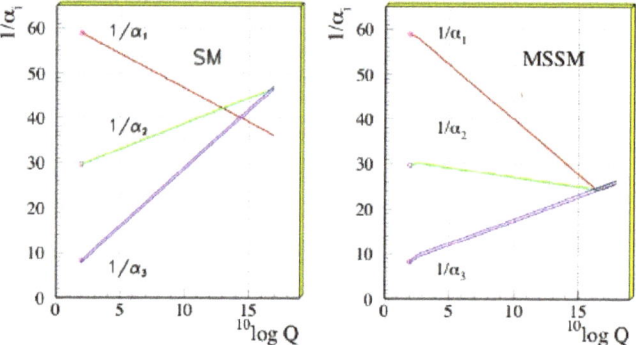

Figure 12. The couplings of the three gauge interactions in the standard model (**left**) converge at high energy, but not to the same point, whereas introducing the supersymmetric particles (**right**) helps to unite the interactions. MSSM: minimal supersymmetric standard model.

8. Supersymmetry

Many extensions have been proposed for the standard model, solving some of the problems above [23]. However, only one of them seems to be able to handle most of those (see Figure 11), *supersymmetry* with the short name *SUSY*. SUSY is originally proposed to solve the hierarchy problem, so it assumes that fermions and bosons exist in identical pairs, with just their spins different by $\frac{1}{2}$. Thus the gauge bosons of the SM should have spin-half fermions and the SM fermions scalar bosons as partners. If this picture has any reality, SUSY is obviously broken at low energy, as we do not observe those partner particles: if they exist at all they must be much heavier. There are also many-many alternative extensions of the standard model proposed to solve these problems and for checking them the experiments search for deviations from SM predictions.

8.1. R Parity and LSP

SUSY's quantum number is called R parity, $R = (-1)^{3B-L+2S}$, where B is the baryon charge, L is the lepton charge, and S is the spin of the particle. It is easy to check that $R = +1$ for all SM particles and $R = -1$ for the SUSY partners. Supersymmetry can unify at high energy the gauge coupling constants (Figure 12). Moreover, it allows including gravity and offers a good candidate for dark matter as the lightest neutral supersymmetric particle (LSP). In the case if SUSY is valid, R parity should be conserved, as we do not see such exotic decays, possibly attributed to supersymmetry. Conservation of the R parity would prevent the LSP from decaying. After the Big Bang, SUSY particles could also be produced in particle-antiparticle pairs, and at the end of the decay cascade, the LSP could just stay on and constitute the dark matter of the universe.

LSP production should make it possible to observe SUSY reactions: supersymmetric particles could be produced in pairs in high-energy collisions and decay via emitting ordinary and SUSY particles. At the end of the decay chain, the lightest one should escape detection, leaving behind a great portion of undetected momentum *missing energy* producing energetically unbalanced events. Unfortunately, leptonic weak decays produce neutrinos, resulting in similar events with missing energy, making a considerable SM background in the search for SUSY phenomena.

In order to eliminate the hierarchy problem, the numbers of ordinary and supersymmetric particle kinds should be equal, thus the left- and right-polarized fermions of the SM should have different SUSY partner bosons, with different masses. Supersymmetry in its minimal form needs two doublet BEH fields, i.e., eight symmetry-breaking fields, and after creating the three weak boson masses, five Higgs bosons will be left, three neutral and two charged ones. As we see already a 125 GeV Higgs boson, we must assume that one of the neutral minimal supersymmetric standard model (MSSM) Higgs bosons (probably the lighter one) corresponds to that one. The other neutral ones could be very heavy, but there must be two charged ones, as well. This clearly shows the importance of the Higgs sector in testing the validity of SUSY.

Due to the rather complicated nature of the proposed SUSY-breaking mechanisms, the MSSM adds 105 new particle masses and coupling constants to the original 19 parameters of the standard model. Thus, SUSY has much too many variables, impossible to test directly. There were several extremely simplified versions of the MSSM (with four or five free parameters), but those were refuted by the earliest runs of the LHC.

8.2. Search for SUSY Phenomena

Recently, the experiments gave up testing definite SUSY models with fixed parameters and rather tried to find deviations from the predictions of the standard model by analyzing events of simple topologies. For that, great amounts of very precise data are needed. Many such topologies are considered by the LHC experiments, but so far, none uncovered any difference from the SM predictions. Nonetheless, these studies are quite useful for the theoreticians, as they help to restrict the parameter space available for the various versions of MSSM. In most of the cases, the available experimental information provides mass limits for SUSY particles in various possible model scenarios from a few hundred GeV up to several TeV's.

Thus, the LHC experiments devote much effort to find or exclude SUSY signatures. ATLAS and CMS studied hundreds of possible supersymmetric scenarios and excluded most of the hypothetical SUSY partner particles with masses below 1 TeV. For solving the hierarchy problem, one needs lighter SUSY particles: even in the case of broken supersymmetry, the mass of the partner of the top quark (the scalar top) cannot be orders of magnitude heavier than the t quark itself, as that would break the elimination of the huge corrections to the mass of the Higgs boson. Generally, the scalar t quark is assumed to be the lightest scalar quark, and the SUSY partner of the tau lepton is assumed to be the lightest among the scalar leptons (reversed particle hierarchy).

9. Unsolved Problems

As shown in Figure 11, in addition to supersymmetry, many different theoretical extensions have been proposed in order to solve the problems of the standard model. All high-energy experiments try to search for such signatures, but so far, no convincing positive result was obtained. Of all extensions, only those models can be tested, of course, that make measurable predictions. All of the hypothetical new particles that were searched for experimentally were excluded at the masses that could be generated at the energies available at the accelerators, and thus, they have lower limits set on their masses, usually above 1 TeV.

9.1. Neutrino Oscillation

Particle oscillation appears when two interactions have different eigenstates in relation to a particle and those eigenstates have a very small mass difference (like in the case of the neutral kaons). Neutrino oscillations, the conversion of one neutrino flavor into another one, were experimentally observed among all three neutrinos and therefore established that neutrinos do have masses. These masses are extremely small, that is why they contribute very little to the measurable quantities in high-energy physics. Nevertheless, neutrino oscillation contradicts the standard model as that assumes massless neutrinos. One cannot add neutrino mass terms to the interaction Lagrangian similarly to those of the quarks and charged leptons. Massive neutrinos should have right-polarized particles and left-polarized antiparticles, which then cannot partake in the SU(2) (charged current) interaction, the only interaction available for neutrinos according to the standard model. Furthermore, for describing the neutrino oscillation, the neutrinos must have two interactions with different eigenstates, but for them, no other interaction is known. Thus, we have another symmetry breaking of unknown origin leading out of the SM world.

9.2. Antimatter

Another big mystery is the apparent lack of antimatter in our universe. According to the theory of the Big Bang, in the beginning, all energy in the Universe had to be in form of radiation, which upon expansion and cooling formed particle–antiparticle pairs, and we do not know where the antimatter went. Explanations were offered based on inflation, the fast expansion of the Universe in the very short first moments of its existence. Nevertheless, there are considerable efforts to check whether or not particle and antiparticle have really identical properties apart from the signs of their charges. At the Antiproton Decelerator of CERN, many such experiments study this question by measuring the charge, mass, and magnetic moment of the antiproton and also the spectral and gravitational properties of the antihydrogen atom, the atomic bound state of an antiproton and a positron. Thus far, none of the measurements uncovered any difference between proton and antiproton within the relative precision of 10^{-12}.

10. Conclusions

The standard model is based on gauge symmetries, but several of them are broken. The P spatial mirror symmetry, i.e., parity conservation, is maximally violated, and the CP (charge + parity reflection) and T (time reversal) symmetries are also slightly broken. Nevertheless, CPT invariance and its consequence, the matter-antimatter symmetry, seems to be fully valid. The BEH mechanism of spontaneous symmetry breaking to produce masses for the elementary particles is well established, as the Higgs boson was observed and its properties agreed very well with the predictions of the standard model. Thus, the standard model seems to be both theoretically and experimentally very well confirmed.

However, in spite of its tremendous success in reproducing experimental data, the standard model has several shortcomings, mostly theoretical problems: divergent Higgs boson mass, unknown dark matter, neutrino mass and oscillations, etc. Various extensions of the standard model were proposed to solve them, all based on breaking some (hypothetical) symmetry, but (1) none of them solves all the problems, although supersymmetry seems to take care of many of them, while (2) their predicted new particles and phenomena have not been observed yet. All particle physics experiments at present working at low and high energies are making tremendous efforts for clarifying these questions, but first of all uncover possible deviations in measurable data from the standard model.

Funding: This work was supported by the Hungarian National Research, Development and Innovation Office via Contracts K-124850 and K-128786.

Conflicts of Interest: The authors declare no conflicts of interest.

References

1. Haller, J.; Hoecker, A.; Kogler, R.; Mönig, K.; Peiffer, T.; Stelzer, J. Update of the global electroweak fit and constraints on two-Higgs-doublet models. *Eur. Phys. J. C* **2018**, *78*, 675. [CrossRef]
2. Higgs, P.W. Broken symmetries, massless particles and gauge fields. *Phys. Lett.* **1964**, *12*, 132–133. [CrossRef]
3. Higgs, P.W. Broken symmetries and the masses of gauge bosons. *Phys. Rev. Lett.* **1964**, *13*, 508. [CrossRef]
4. Englert, F.; Brout, R. Broken symmetry and the mass of gauge vector mesons. *Phys. Rev. Lett.* **1964**, *13*, 321. [CrossRef]
5. Guralnik, G.S.; Hagen, C.R.; Kibble, T.W.B. Global Conservation Laws and Massless Particles. *Phys. Rev. Lett.* **1964**, *13*, 585. [CrossRef]
6. Cowan, G.; Cranmer, K.; Gross, E.; Vitells, O. Asymptotic formulae for likelihood-based tests of new physics. *Eur. Phys. J. C* **2011**, *71*, 1554. [CrossRef]
7. Horváth, D. Twenty years of searching for the Higgs boson: Exclusion at LEP, discovery at LHC. *Mod. Phys. Lett. A* **2014**, *29*, 1430004. [CrossRef]
8. De Florian, D.; LHC Higgs Cross Section Working Group. Handbook of LHC Higgs cross-sections: 4. Deciphering the nature of the Higgs sector. *arXiv* **2016**, arXiv:1610.07922.

9. Barate, R.; ALEPH and DELPHI and L3 and OPAL Collaborations and LEP Working Group for Higgs Boson Searches. Search for the standard model Higgs boson at LEP. *Phys. Lett. B* **2003**, *565*, 61.
10. Aad, G.; ATLAS Collaboration. Observation of a new particle in the search for the Standard Model Higgs boson with the ATLAS detector at the LHC. *Phys. Lett. B* **2012**, *716*, 1–29. [CrossRef]
11. Chatrchyan, S.; CMS Collaboration. Observation of a new boson at a mass of 125 GeV with the CMS experiment at the LHC. *Phys. Lett. B* **2012**, *716*, 30. [CrossRef]
12. Aad, G.; ATLAS and CMS Collaboration. Combined measurement of the Higgs boson mass in pp collisions at $\sqrt{s} = 7$ and 8 TeV with the ATLAS and CMS experiments. *Phys. Rev. Lett.* **2015**, *114*, 191803. [CrossRef] [PubMed]
13. Aaboud, M.; ATLAS Collaboration. Measurement of the Higgs boson mass in the H $\to ZZ^* \to 4\ell$ and H $\to \gamma\gamma$ channels with $\sqrt{s} = 13$ TeV pp collisions using the ATLAS detector. *Phys. Lett. B* **2018**, *784*, 345. [CrossRef]
14. Sirunyan, A.M.; CMS Collaboration. Measurements of properties of the Higgs boson decaying into the four-lepton final state in pp collisions at $\sqrt{s} = 13$ TeV. *J. High Energy Phys.* **2017**, *1711*, 47. [CrossRef]
15. Sirunyan, A.M.; CMS Collaboration. Measurements of Higgs boson properties in the diphoton decay channel in proton-proton collisions at $\sqrt{s} = 13$ TeV. *J. High Energy Phys.* **2018**, *1811*, 185. [CrossRef]
16. Aad, G.; ATLAS and CMS Collaborations. Measurements of the Higgs boson production and decay rates and constraints on its couplings from a combined ATLAS and CMS analysis of the LHC pp collision data at $\sqrt{s} = 7$ and 8 TeV. *J. High Energy Phys.* **2016**, *1608*, 45. [CrossRef]
17. Sirunyan, A.M.; CMS Collaboration. Observation of the Higgs boson decay to a pair of τ leptons with the CMS detector. *Phys. Lett. B* **2018**, *779*, 283. [CrossRef]
18. Sirunyan, A.M.; CMS Collaboration. Observation of Higgs boson decay to bottom quarks. *Phys. Rev. Lett.* **2018**, *121*, 121801. [CrossRef]
19. Sirunyan, A.M.; CMS Collaboration. Observation of t$\bar{t}$H production. *Phys. Rev. Lett.* **2018**, *120*, 231801. [CrossRef]
20. International Workshop entitled *Why M_H = 126 GeV?*, Madrid, Spain, 25–27 September 2013, unpublished. Available online: http://workshops.ift.uam-csic.es/WMH126/index.html (accessed on 11 March 2019).
21. Degrassi, G.; di Vita, S.; Elias-Miro, J.; Espinosa, J.R.; Giudice, G.F.; Isidori, G.; Strumia, A. Higgs mass and vacuum stability in the Standard Model at NNLO. *J. High Energy Phys.* **2012**, *1208*, 98. [CrossRef]
22. Turner, M.S.; Wilczek, F. Might our vacuum be metastable? *Nature* **1982**, *298*, 633. [CrossRef]
23. Gershtein, Y.; Luty, M.; Narain, M.; Wang, L.-T.; Whiteson, D.; Agashe, K.; Apanasevich, L.; Artoni, G.; Avetisyan, A.; Baer, H.; et al. Working group report: New particles, forces, and dimensions. *arXiv* **2013**, arXiv:1311.0299.

© 2019 by the author. Licensee MDPI, Basel, Switzerland. This article is an open access article distributed under the terms and conditions of the Creative Commons Attribution (CC BY) license (http://creativecommons.org/licenses/by/4.0/).

Review

Latest Results from RHIC + Progress on Determining $\hat{q}L$ in RHI Collisions Using Di-Hadron Correlations

Michael J. Tannenbaum

Physics Department, Brookhaven National Laboratory, Upton, NY 11973-5000, USA; mjt@bnl.gov

Received: 15 April 2019; Accepted: 24 May 2019; Published: 5 June 2019

Abstract: Results from Relativistic Heavy Ion Collider Physics in 2018 and plans for the future at Brookhaven National Laboratory are presented.

Keywords: RHIC qhat dihadron correlations

1. Introduction

The Relativistic Heavy Ion Collider (RHIC) at Brookhaven National Laboratory (BNL) is one of the two remaining operating hadron colliders in the world, and the first and only polarized p+p collider. BNL is located in the center of the roughly 200 km long maximum 40 km wide island (named Long Island), and appears on the map as the white circle which is the berm containing the Relativistic Heavy Ion Collider (RHIC). BNL is 100 km from New York City in a region which nurtures science with Columbia University and the Bronx High School of Science indicated (Figure 1). Perhaps more convincing is the list of the many Nobel Prize winners from New York City High School graduates (Figure 2) which does not yet include one of this years Nobel Prize winners in Physics, Arthur Ashkin who graduated from James Madison High school in 1940 and Columbia U. in 1947.

Figure 1. NASA infra-red photo of Long Island and the New York Metro Region from space. RHIC is the white circle to the left of the word BNL. Manhattan Island in New York City, ~100 km west of BNL, is also clearly visible on the left side of the photo, with Columbia U. and Bronx Science High School indicated.

Number of laureates by secondary school	Class	Name of laureate	Award and year		University
8 The Bronx High School of Science, Bronx, New York City, NY	1947	Leon N. Cooper[1]	Physics	1972	Brown University
	1950	Sheldon Glashow[2]	Physics	1979	Columbia University
	1950	Steven Weinberg[3]	Physics	1979	Cornell University
	1949	Melvin Schwartz[3]	Physics	1988	Columbia University
	1950	Russell Hulse[14]	Physics	1993	Princeton University
	1966	H. David Politzer[3]	Physics	2004	California Institute of Technology
	1941	Roy Glauber[15]	Physics	2005	Harvard University
	1959	Robert Lefkowitz[4]	Chemistry	2012	Columbia University
4 James Madison High School, Brooklyn, New York City, NY	1939	Stanley Cohen[15]	Medicine	1986	Vanderbilt University
	1940	Robert Solow[16]	Economics	1987	Massachusetts Institute of Technology
	1943	Martin Lewis Perl[3]	Physics	1995	University of Michigan
	1947	Gary Becker[18]	Economics	1992	University of Chicago
4 Stuyvesant High School, Manhattan, New York City, NY	1941	Joshua Lederberg[19][20]	Medicine	1958	Rockefeller University
	1954	Roald Hoffmann[20][21]	Chemistry	1981	Cornell University
	1944	Robert Fogel[20][22]	Economics	1993	Cornell University
	1963	Richard Axel[20][23]	Medicine	2004	Columbia University
3 Abraham Lincoln High School, Brooklyn, New York City, NY	1933	Arthur Kornberg[31]	Medicine	1959	Stanford University
	1943	Paul Berg[31]	Chemistry	1980	Stanford University
	1933	Jerome Karle[31][32]	Chemistry	1985	City College of New York
3 Far Rockaway High School, Queens, New York City, NY	1935	Richard Feynman[3][34]	Physics	1965	California Institute of Technology
	1940	Burton Richter[3]	Physics	1976	Stanford University
	1942	Baruch Blumberg[34]	Medicine	1976	University of Pennsylvania
3 Townsend Harris High School, Queens, New York City, NY originally Manhattan, New York City, NY	1933	Herbert A. Hauptman[45]	Chemistry	1985	City College of New York
	1933	Julian Schwinger[45]	Physics	1965	Harvard University
	1936	Kenneth Arrow[45]	Economics	1972	City College of New York
2 Brooklyn Technical High School, Brooklyn, New York City, NY	1951	Arno Penzias	Physics	1978	City College of New York
	1922	George Wald	Biology	1987	Harvard University
2 Erasmus Hall High School, Brooklyn, New York City, NY	1919	Barbara McClintock[52]	Medicine or Physiology	1983	Cold Spring Harbor Laboratory
	1944	Eric Kandel[53]	Medicine or Physiology	2000	Columbia University
2 Hastings High School (New York)	1951	Edmund S. Phelps	Economics	2006	Columbia University
Hastings High School (New York)	1962	Robert C. Merton	Economics	1997	MIT Sloan School of Management
2 Martin Van Buren High School, Queens, New York	1967	Frank Wilczek[3]	Physics	2004	University of Chicago Princeton University
	1967	Alvin Roth[58]	Economics	2012	Columbia University Stanford University
2 Walton High School, Bronx, New York City, NY	1941	Rosalyn Sussman Yalow[45]	Medicine and Physiology	1977	Hunter College
	1933	Gertrude B. Elion[45]	Medicine and Physiology	1988	Duke University
1 Manual Training HS, Brooklyn NY	1916	Isidor Isaac Rabi	Physics	1944	Columbia University
1 DeWitt Clinton HS, Bronx, NY	1931	Robert Hofstadter	Physics	1961	Stanford University
1 James Monroe High School, Bronx NY	1939	Leon Max Lederman	Physics	1988	Columbia University
1 New Trier High School, Winnetka, Illinois	1938	Jack Steinberger[3]	Physics	1988	Columbia University
1 Regis High School, Manhattan, New York City, NY	1957	John O'Keefe	Medicine	2014	City College of New York McGill University

Figure 2. From Wikipedia (edited), Physicists in blue and Roald Hoffman a classmate of mine from Columbia.

There also have been many discoveries and Nobel Prizes at BNL (Figure 3).

In particular, Leon Lederman, who made many discoveries at BNL (Figure 4), died this year (2018) at the age of 96. Leon was the most creative and productive high energy physics experimentalist of his generation as well as the physicist with the best jokes. He was also my PhD thesis Professor. For more details, see https://physicstoday.scitation.org/do/10.1063/PT.6.4.20181010a/full/.

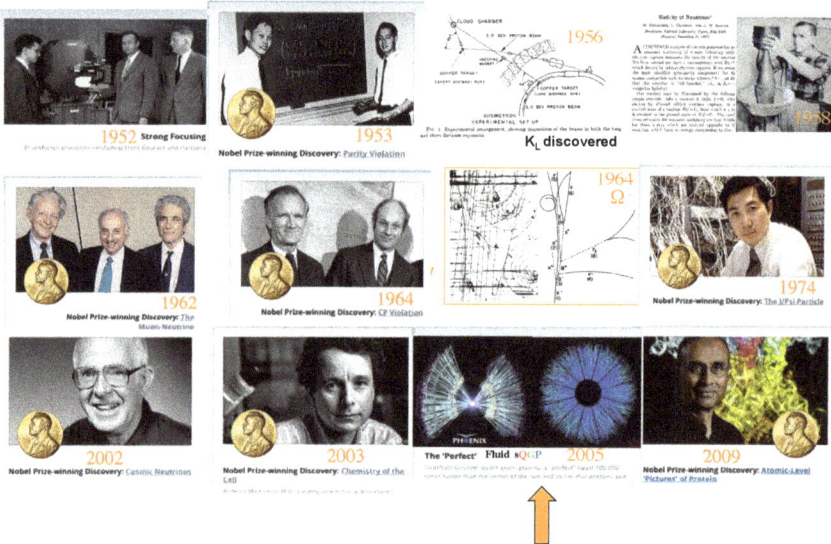

Figure 3. Selected Discoveries and Nobel Prizes at BNL, arrow points to QGP discovery.

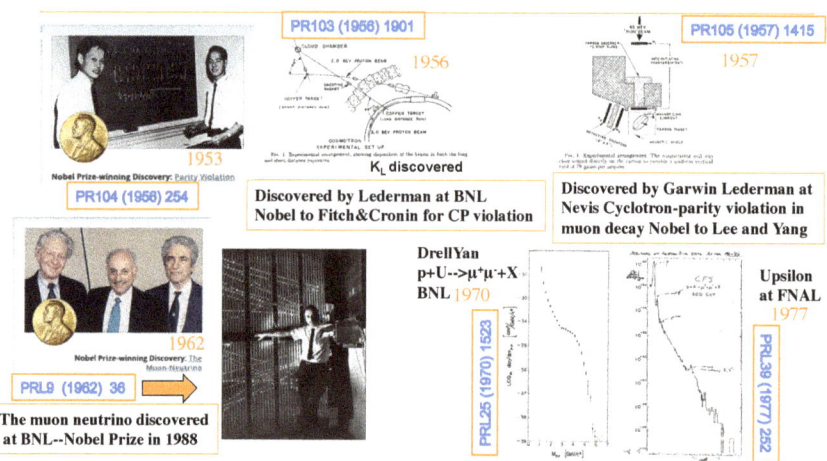

Figure 4. Discoveries by Leon Lederman and close associates at Columbia University.

2. Why RHIC Was Built: To Discover the Quark Gluon Plasma (QGP)

Figure 5 shows central collision particle production in the PHENIX and STAR detectors, which were the major detectors at RHIC.

At the startup of RHIC in the year 2000, there were two smaller more special purpose detectors PHOBOS and BRAHMS, as shown in Figure 6, which finished data taking in 2005.

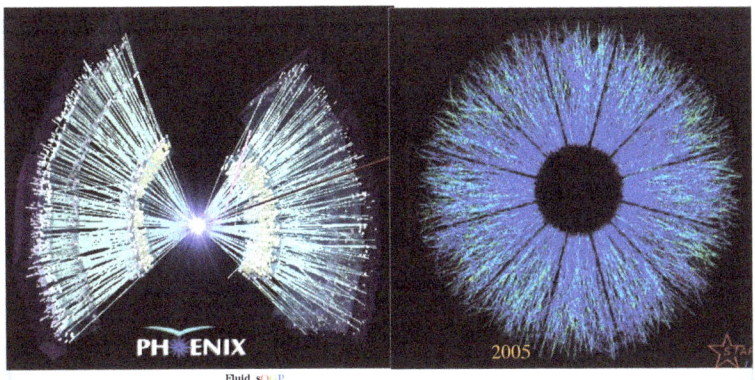

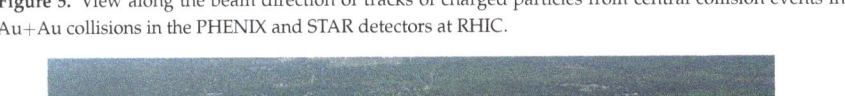

Figure 5. View along the beam direction of tracks of charged particles from central collision events in Au+Au collisions in the PHENIX and STAR detectors at RHIC.

Figure 6. View of RHIC location from the air. The positions of the four original detectors, PHENIX, STAR PHOBOS and BRAHMS are indicated as well as the AGS (with three Nobel Prizes shown in Figure 3).

2.1. The First Major RHIC Experiments

The two major experiments at RHIC were STAR (Figure 7), which is still operating, and PHENIX (Figure 8), which finished data taking at the end of the 2016 run.

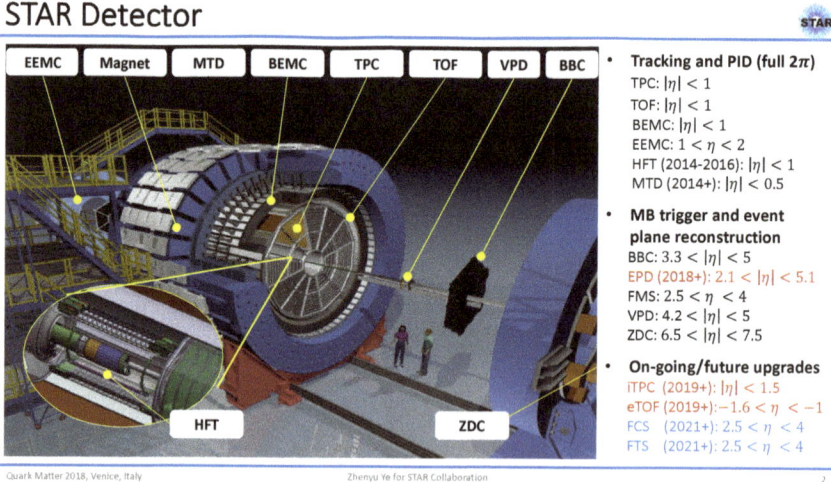

Figure 7. STAR is based on a normal conductor solenoid with Time Projection Chamber for tracking, an EM Calorimeter, Vertex detector and μ detector behind the thick iron yoke.

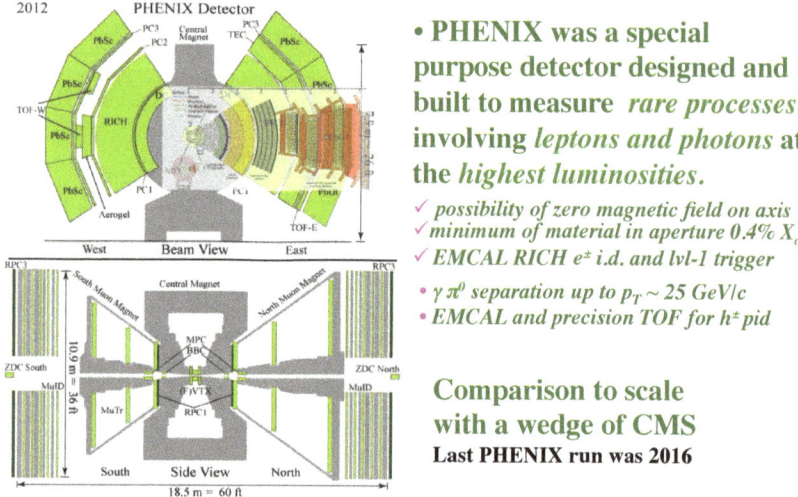

Figure 8. As indicated on the figure, PHENIX is a special purpose detector for electrons and photons but also measures charged hadrons and notably $\pi^0 \to \gamma + \gamma$ at mid-rapidity and muons in the forward direction.

2.2. The New Major RHIC Experiment sPHENIX

sPHENIX is a major improvement over PHENIX with a superconducting thin coil solenoid which was surplus from the BABAR experiment at SLAC and is now working at BNL and has reached its full field (Figure 9).

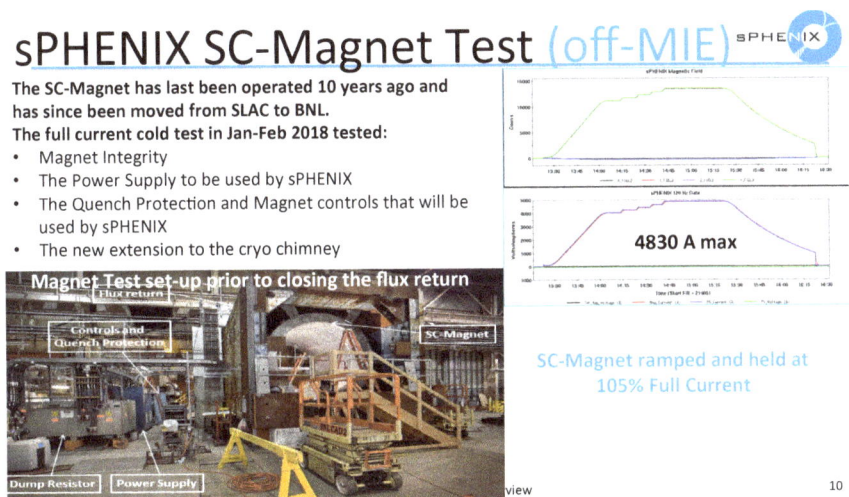

Figure 9. BABAR superconducting solenoid now in operation at BNL.

The design of the sPHENIX experiment is moving along well (Figure 10) with a notable addition of a hadron calorimeter based on the iron return yoke of the solenoid.

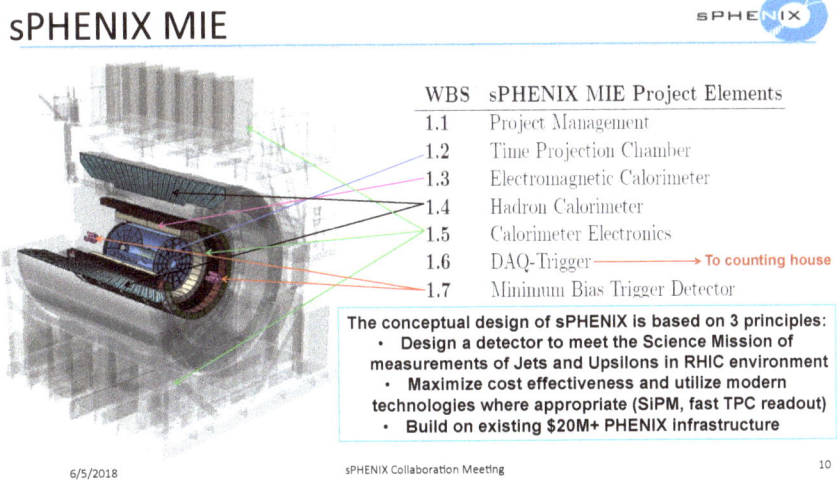

Figure 10. Conceptual design of sPHENIX with major features illustrated.

sPHENIX has been approved by the U. S. Department of Energy (DoE) as a Major Item of Equipment (MIE) with the schedule of critical decisions shown in Figure 11a, and the planned multi-year RHIC runs indicated in Figure 11b. The present sPHENIX collaboration and its evolution is shown in Figure 12.

Critical Decision Level 1 MIE Schedule

Milestone	Schedule Date
CD-0, Approve Mission Need	9/27/2016
CD-1/3A, Approve Alternative Selection and Cost Range. Long Lead Procurements	Q4 FY 2018
CD-2/3, Approve Performance Baseline	Q4 FY 2019
CD-4, Approve Project Completion	Q1 FY 2023

a)

Multi-year run plan for sPHENIX

Year	Species	Energy [GeV]	Phys. Wks	Rec. Lum.	Samp. Lum.	Samp. Lum. All-Z
Year-1	Au+Au	200	16.0	7 nb^{-1}	8.7 nb^{-1}	34 nb^{-1}
Year-2	p+p	200	11.5	—	48 pb^{-1}	267 pb^{-1}
Year-2	p+Au	200	11.5	—	0.33 pb^{-1}	1.46 pb^{-1}
Year-3	Au+Au	200	23.5	14 nb^{-1}	26 nb^{-1}	88 nb^{-1}
Year-4	p+p	200	23.5	—	149 pb^{-1}	783 pb^{-1}
Year-5	Au+Au	200	23.5	14 nb^{-1}	48 nb^{-1}	92 nb^{-1}

b)

Figure 11. (**a**) DoE Critical Decision Schedule; and (**b**) multi-year run plan for sPHENIX.

Figure 12. List of the sPHENIX collaboration members in June 2018 together with photos showing the evolution since December 2015. Dave Morrison (BNL) and Gunther Roland (MIT) are spokespersons.

2.3. Following RHIC in U.S. Nuclear Physics: the Electron Ion Collider (EIC)

Statement by Brookhaven Lab, Jefferson Lab, and the Electron-Ion Collider Users Community on National Academy of Sciences Electron-Ion Collider (EIC) Report

July 24, 2018

On July 24, 2018, a National Academy of Sciences (NAS) committee issued a report of its findings and conclusions related to the science case for a future U.S.-based Electron-Ion Collider (EIC) and the opportunities it would offer the worldwide nuclear physics community.

The committee's report—commissioned by the U.S. Department of Energy (DOE)—comes after 14 months of deliberation and meetings held across the U.S. to gather input from the nuclear science community. The report's conclusions include the following:

- The committee concludes that the science questions regarding the building blocks of matter are compelling and that an EIC is essential to answering these questions.
- The answers to these fundamental questions about the nature of the atoms will also have implications for particle physics and astrophysics and possibly other fields.
- Because an EIC will require significant advances and innovations in accelerator technologies, the impact of constructing an EIC will affect all accelerator-based sciences.
- In summary, the committee concludes that an EIC is timely and has the support of the nuclear science community. The science that it will achieve is unique and world leading and will ensure global U.S. leadership in nuclear science as well as in the accelerator science and technology of colliders.

The first BNL EIC design in 2014 is shown in Figure 13. The 2018 JLab and BNL EIC designs are shown in Figures 14 and 15.

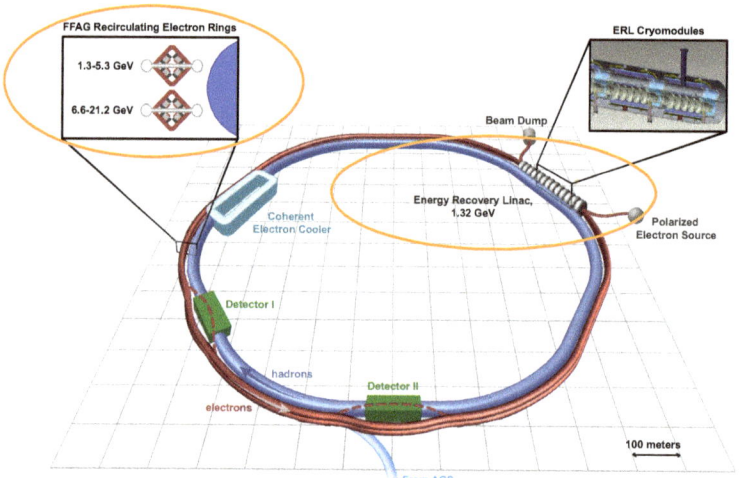

Figure 13. The 2014 cost estimate: BNL $755.9M; Temple NSAC subcommittee cost estimate $1.5B.

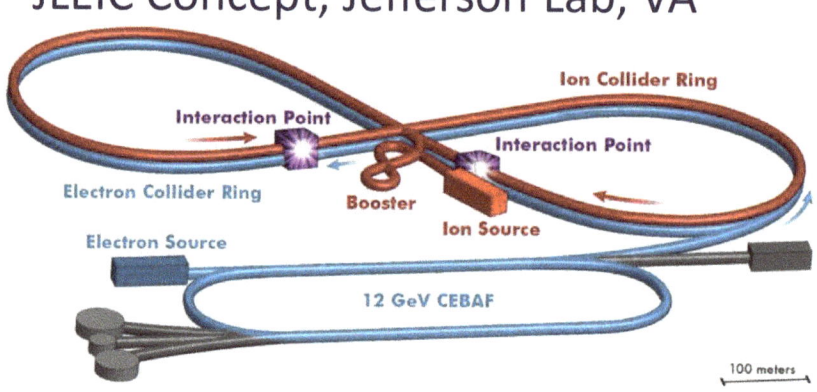

Figure 14. JLab EIC Concept. Temple committee cost estimate also $1.5B but no new accelerator technology required.

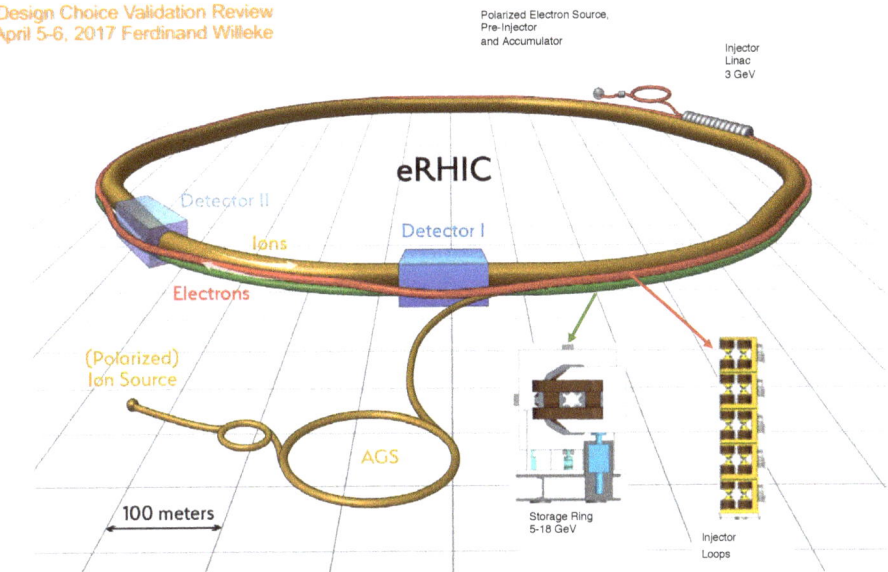

Figure 15. BNL eRHIC design progress 2017. Temple committee cost estimate $1.5B.

The two new designs of the JLab (JLEIC) and BNL (eRHIC) both satisfy the Temple committee cost estimate of $1.5B, but R&D of the novel first BNL design is not idle.

Research and Development (R&D) for an Improved Less Expensive BNL Machine Is Ongoing

BNL and Cornell are in the process of experiments studying an energy recovery linac ERL (Figure 16a). Figure 16b is the main Linac cryo module made from superconducting RF cavities. Figure 16c is a return loop made from fixed-field alternating-gradient (FFAG) optics made with permanent Halbach magnets to contain four beam energies in a single 70 mm-wide beam pipe, designed and prototyped at Brookhaven National Laboratory (BNL).

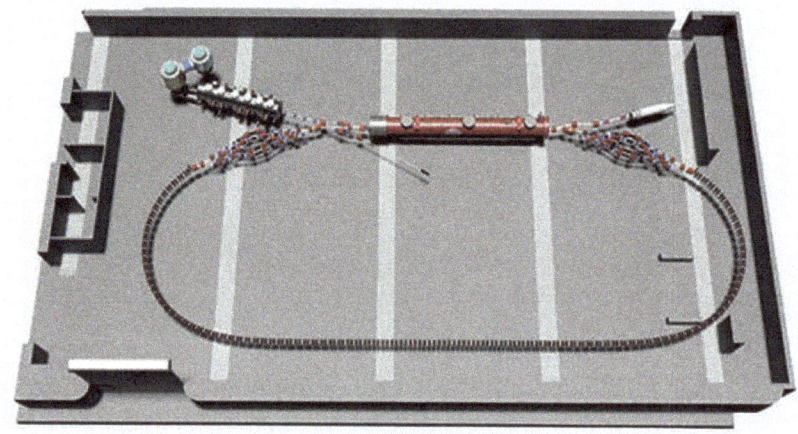

Figure 16. (**a**) CBETA (Cornell-Brookhaven Energy Recovery Linac (ERL)); (**b**) Main Linac cryo module; and (**c**) FFAG permanent loop return loop.

3. RHIC Future Run Plan (Figure 17) and and the Present RHIC Run in 2018 (Figure 18)

3.1. 2018 RHIC Run Is $_{40}Zr^{96} + {}_{40}Zr^{96}$ and $_{44}Ru^{96} + {}_{44}Ru^{96}$, Why?

To determine whether the separation of charges in the flow, v_2, of π^+ and π^- shown in Figure 19 is due to a new phenomenon called the Chiral Magnetic Effect (Figure 20a), the 2018 measurements are made with collisions of Zr+Zr and Ru+Ru, which have the same number of nucleons but different electric charges (Figure 20b). If the effect is larger in Ru+Ru with stronger charge and magnetic field compared to Zr+Zr with the same number of nucleons, it would indicate that the charge asymmetry is a magnetic effect, possibly the Chiral Magnetic Effect.

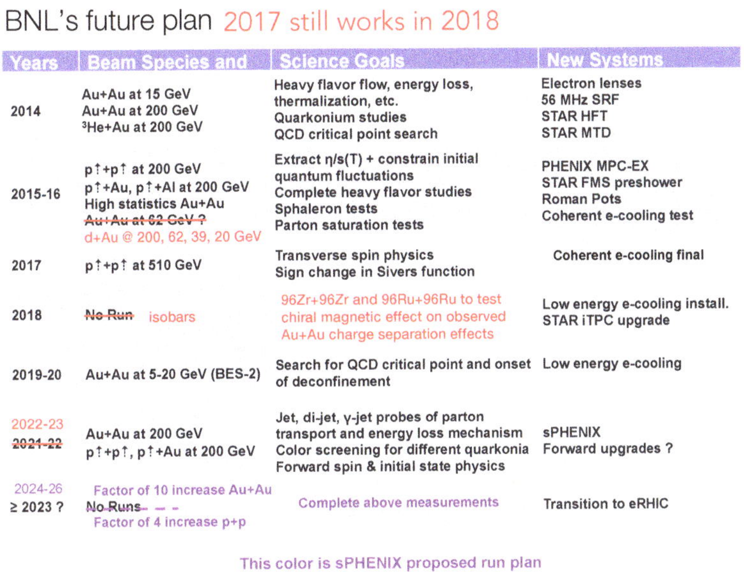

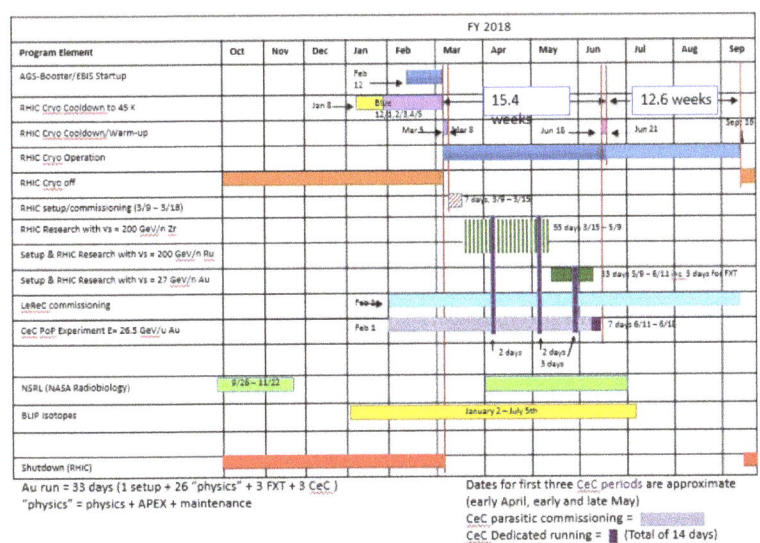

Figure 17. RHIC run plan 2014–2023 (2026?).

Figure 18. The 2018 RHIC run schedule.

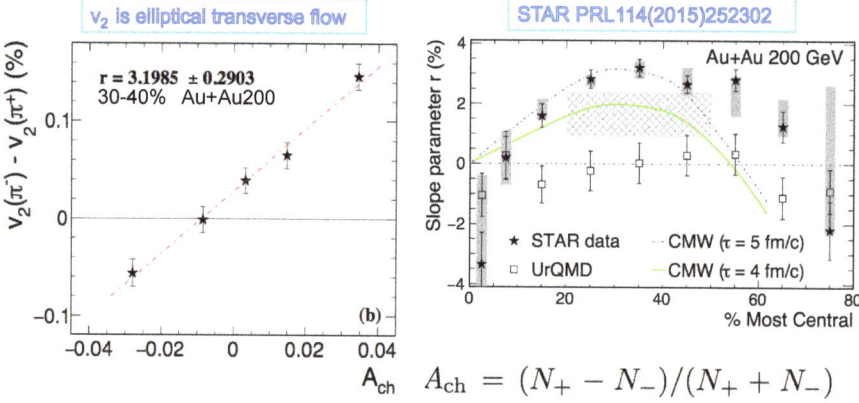

Figure 19. From Article in the BNL news 8 June 2015.

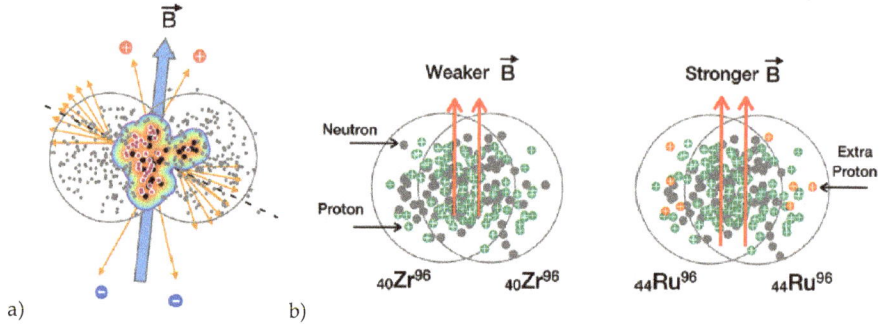

Figure 20. (**a**) Schematic of A+A collision; and (**b**) sketch of the stronger magnetic (B) field in Ru+Ru.

3.2. Vorticity: An Application of Particle Physics to the QGP

It was observed at FERMILAB [1] that forward Λ were polarized in p+Be collisions, where the proton in the $\Lambda \to p + \pi^-$ decay is emitted along the spin direction of the Λ. In the A+A collision (Figure 21a), the forward going beam fragments are deflected outwards so that the event plane and the angular momentum $\hat{J}_{sys}$ of the QGP formed can be determined. STAR claims that the Λ polarization, $\overline{\mathcal{P}}_\Lambda$, is parallel to the angular momentum $\hat{J}_{sys}$ of the QGP everywhere so that the vorticity $\omega = k_B T(\overline{\mathcal{P}}_\Lambda + \overline{\mathcal{P}}_{\overline{\Lambda}})/\hbar$ can be calculated, a good exercise for the reader to see if you can get the $\omega \sim 10^{22}/s$ which is 10^5 times larger than any other fluid [2]. Another interesting thing to note is that the largest vorticity is at

$\sqrt{s_{NN}} = 7.6 - 19$ GeV where the CERN fixed target experiments measure. Does this mean that their fluid (with minimal if any QGP) is also perfect?

STAR team receives secretary's achievement award for vorticity in 2018 (Figure 22).

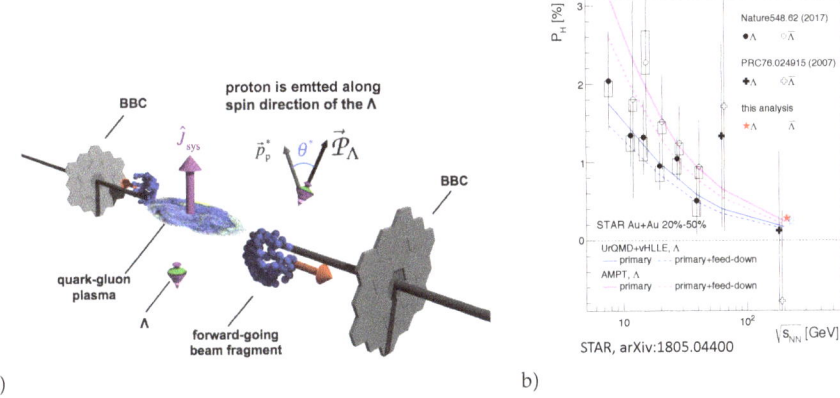

Figure 21. (**a**) Schematic of STAR vorticity detection; and (**b**) polarization $P_H = \overline{\mathcal{P}}_\Lambda$ or $\overline{\mathcal{P}}_{\overline{\Lambda}}$ vs. $\sqrt{s_{NN}}$ [3].

Figure 22. STAR receives an award for vorticity in 2018.

4. The Search for the Quark Gluon Plasma at RHIC

High energy nucleus–nucleus collisions provide the means of creating nuclear matter in conditions of extreme temperature and density, the Quark Gluon Plasma QGP (Figure 23). At large energy or baryon density, a phase transition is expected from a state of nucleons containing confined quarks and gluons to a state of "deconfined" (from their individual nucleons) quarks and gluons covering a volume that is many units of the confinement length.

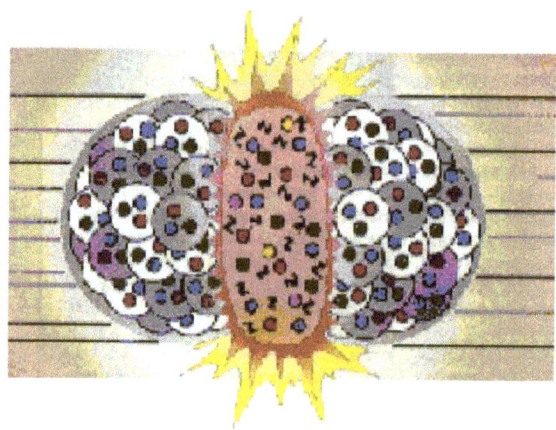

Figure 23. Sketch of nucleus–nucleus collision producing a QGP.

4.1. Anisotropic (Elliptical) Transverse Flow—An Interesting Complication in all A+A Collisions (Figure 24)

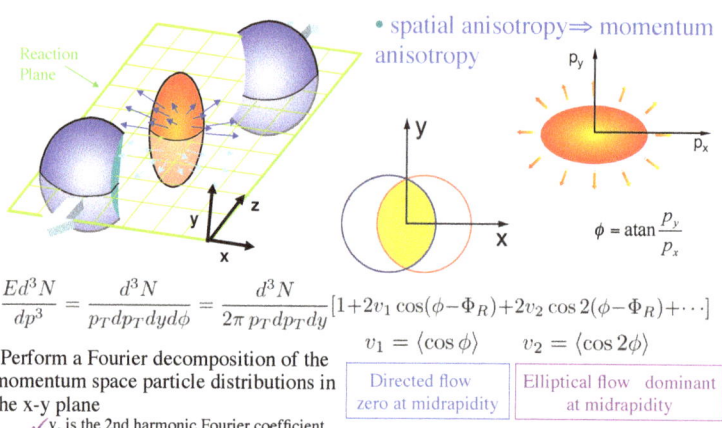

Figure 24. Sketch and definitions of elliptical flow, v_2.

Figure 25 shows that Elliptical flow (v_2) exists in all A+A collisions measured. At very low $\sqrt{s_{NN}}$, the main effect is from nuclei bouncing off each other and breaking into fragments. The negative v_2 at larger $\sqrt{s_{NN}}$ is produced by the effective "squeeze-out" (in the y direction) of the produced particles by slow moving minimally Lorentz-contracted spectators, which block the particles emitted in the reaction plane. With increasing $\sqrt{s_{NN}}$, the spectators move faster and become more contracted so the blocking stops and positive v_2 returns.

4.2. Flow Also Exists in Small Systems and Is Sensitive to the Initial Geometry

Figure 26 shows that flow exists in small p+Au, d+Au, ^{3}He+Au systems with preliminary sensitivity of v_3 to the initial geometry. Figure 27 (Top) shows that v_2 is about the same in all three systems but v_3 is much larger in ^{3}He+Au, clearly indicating the sensitivity of flow to the initial geometry of the collision. Figure 27 (Bottom) shows that there is mass ordering in the flow which is strong evidence for hydrodynamics in these small systems. The solid red and dashed blue lines represent hydrodynamic predictions. These hydrodynamical models, which include the formation of a short-lived QGP droplet, provide the best simultaneous description of the measurements, strong evidence for the QGP in small systems.

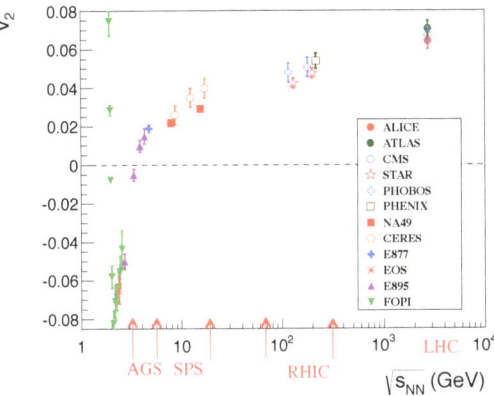

Figure 25. Values of elliptical flow (v_2) as a function of $\sqrt{s_{NN}}$ from all A+A collision measurements.

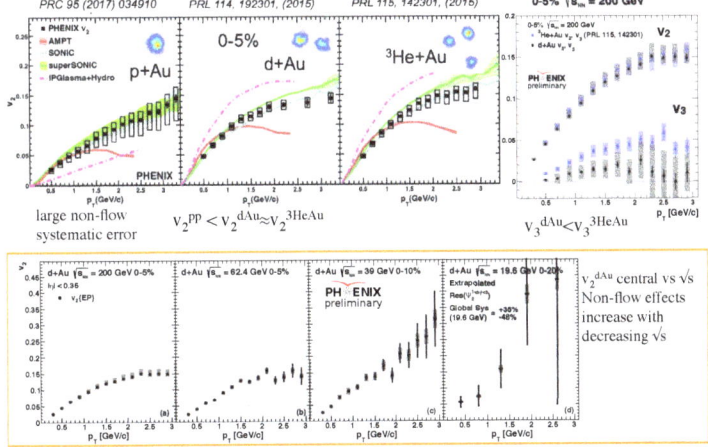

Figure 26. (**Top**) Published PHENIX v_2 measurements in p+Au, and 0-5% central d+Au and ^{3}He+Au collisions at $\sqrt{s_{NN}}$ =200 GeV, with preliminary v_2 and v_3 for the d+Au and ^{3}He+Au compared on the right. (**Bottom**) PHENIX preliminary v_2 in d+Au collisions as a function of $\sqrt{s_{NN}}$ with the centrality indicated illustrating that non-flow effects increase with decreasing $\sqrt{s_{NN}}$.

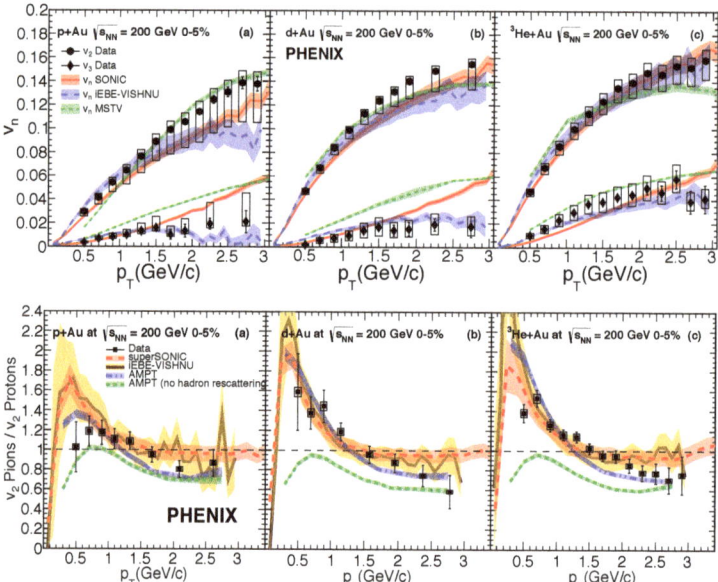

Figure 27. (**Top**) v_2 and v_3 in in 0–5% central (a) p+Au, (b) d+Au, (c) ^{3}He+Au collisions at $\sqrt{s_{NN}}$ = 200 GeV [4]. (**Bottom**) v_2 Pions/v_2 Protons in 0–5% central (a) p+Au, (b) d+Au, (c) ^{3}He+Au collisions at $\sqrt{s_{NN}}$ = 200 GeV [5].

4.2.1. It Takes Two Color Strings for Collectivity—Nagle, J.; et al. [6]

This is an answer to the interesting question of the minimal conditions for collectivity in small systems.

For the case of e$^+$e$^-$ collisions in Figure 28 utilizing the AAMPT framework and a single color string, the results indicate only a modest number of parton–parton scatterings and no observable collectivity signal.

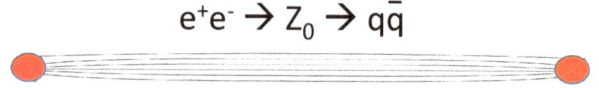

Figure 28. A fundamental point about QCD and the string tension between the q and $\bar{q}$.

However, a simple extension to two color strings (Figure 29), which represent a simplified geometry in p+p collisions, predicts finite long-range two-particle correlations (known as the ridge) and a strong v_2 with respect to the initial parton geometry.

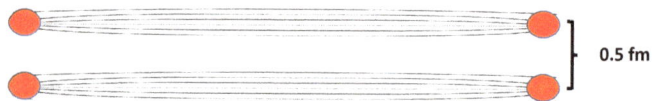

Figure 29. Additional special case—two Strings.

4.2.2. A Fundamental Point about QCD and the String Tension

Unlike an electric or magnetic field between two sources which spreads over all space, in QCD as proposed by Kogut and Susskind [7] the color flux lines connecting two quarks or a $q - \bar{q}$ pair as in Figure 28 are constrained in a thin tube-like region because of the three-gluon coupling. Furthermore, if the field contained a constant amount of color-field energy stored per unit length, this would provide a linearly rising confining potential between the $q - q$ or $q - \bar{q}$ pair.

This led to the Cornell string-like confining potential [8], which combined the Coulomb $1/r$ dependence at short distances from vector-gluon exchange with QCD coupling constant $\alpha_s(Q^2)$, and a linearly rising string-like potential, with string-tension σ,

$$V(r) = -\frac{\alpha_s}{r} + \sigma r \tag{1}$$

which provided confinement at large distances (Equation (1)). Particles are produced by the string breaking (fragmentation).

4.3. The Latest Discovery Claims "Flow" in Small Systems Is From the QGP. How Did We Find the QGP in the First Place?

4.3.1. J/ψ Suppression, 1986

In 1986, T. Matsui and H. Satz [9] said that due to the Debye screening of the color potential in a QGP, charmonium production would be suppressed since the c-$\bar{c}$ could not bind. With increasing temperature, T, in analogy to increasing Q^2, the strong coupling constant $\alpha_s(T)$ becomes smaller, reducing the binding energy, and the string tension, $\sigma(T)$, becomes smaller, increasing the confining radius, effectively screening the potential [10]

$$V(r) = -\frac{4}{3}\frac{\alpha_s}{r} + \sigma r \rightarrow -\frac{4}{3}\frac{\alpha_s}{r}e^{-\mu_D r} + \sigma \frac{(1 - e^{-\mu_D r})}{\mu_D} \tag{2}$$

where $\mu_D = \mu_D(T) = 1/r_D$ is the Debye screening mass. For $r < 1/\mu_D$, a quark feels the full color charge, but, for $r > 1/\mu_D$, the quark is free of the potential and the string tension, effectively deconfined. The properties of the QGP cannot be calculated in QCD perturbation theory but only in Lattice QCD Calculations [11].

J/ψ suppression eventually didn't work because the free c and $\bar{c}$ quarks recombined to make J/ψ's [12]. See Alice publication [13].

4.3.2. Jet Quenching by Coherent LPM Radiative Energy Loss of a Parton in the QGP, 1997

In 1997, Baier, Dokshitzer, Mueller, Peigne, Schiff and Zakharov (BDMPSZ) [14] said that the energy loss from coherent Landau–Pomeranchuk–Migdal (LPM) radiation for hard-scattered partons exiting the QGP would result in an attenuation of the jet energy and a broadening of the jets (Figure 30).

As a parton from hard-scattering in the A+B collision exits through the medium, it can radiate a gluon; and both continue traversing the medium. It is important to understand that "Only the gluons radiated outside the cone defining the jet contribute to the energy loss". In the angular ordering of QCD [15], the angular cone of any further emission will be restricted to be less than that of the previous emission and will end the energy loss once inside the jet cone. This does not work in the QGP so no energy loss occurs only when all gluons emitted by a parton are inside the jet cone. In addition to other issues, this means that defining the jet cone is a big issue—so watch out for so-called trimming.

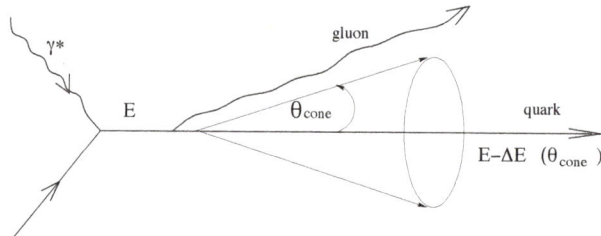

Figure 30. Jet Cone of an outgoing parton with energy E [14].

4.4. BDMPSZ: The Cone, the Energy Loss, Azimuthal Broadening, Is the QGP Signature

The energy loss of the outgoing parton, $-dE/dx$, per unit length (x) of a medium with total length L, is proportional to the total four-momentum transfer-squared, $q^2(L)$, and takes the form:

$$\frac{-dE}{dx} \simeq \alpha_s \langle q^2(L) \rangle = \alpha_s \mu^2 L/\lambda_{\rm mfp} = \alpha_s \hat{q} L$$

where μ, is the mean momentum transfer per collision, and the transport coefficient $\hat{q} = \mu^2/\lambda_{\rm mfp}$ is the four-momentum-transfer-squared to the medium per mean free path, $\lambda_{\rm mfp}$.

Additionally, the accumulated momentum-squared, $\langle p^2_{\perp W} \rangle$ transverse to a parton traversing a length L in the medium is well approximated by

$$\left\langle p^2_{\perp W} \right\rangle \approx \langle q^2(L) \rangle = \hat{q} L \quad .$$

5. Jet Quenching at RHIC, the Discovery of the QGP

The energy loss of an outgoing parton with color charged fully exposed in a medium with a large density of similarly exposed color charges (i.e., a QGP) from Landau–Pomeranchuk–Migdal (LPM) coherent radiation of gluons was predicted in QCD by BDMPSZ [14].

Hard scattered partons (Figure 31a) lose energy going through the medium so that there are fewer partons or jet fragments at a given p_T. The ratio of the measured semi-inclusive yield of, for example, pions in a given A+A centrality class divided by the semi-inclusive yield in a p+p collision times the number of A+A collisions $\langle N_{\rm coll} \rangle$ in the centrality-class is given by the nuclear modification factor, R_{AA} (Figure 31b), which equals 1 for no energy loss.

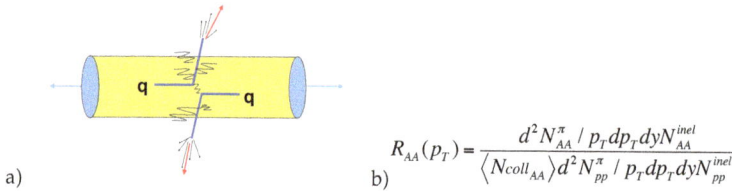

Figure 31. (**a**) Hard quark–quark scattering in an A+A collision with the scattered quarks passing through the medium formed in the collision; and (**b**) nuclear modification factor $R_{AA}(p_T)$.

PHENIX discovered jet quenching of hadrons at RHIC in 2001 [16] (Figure 32). Pions at large $p_T > 2$ GeV/c are suppressed in Au+Au at $\sqrt{s_{NN}}$ =130 GeV compared to the enhancement found at the CERN SpS at $\sqrt{s_{NN}}$ =17 GeV. This is the first regular publication from a RHIC experiment to reach 1000 citations.

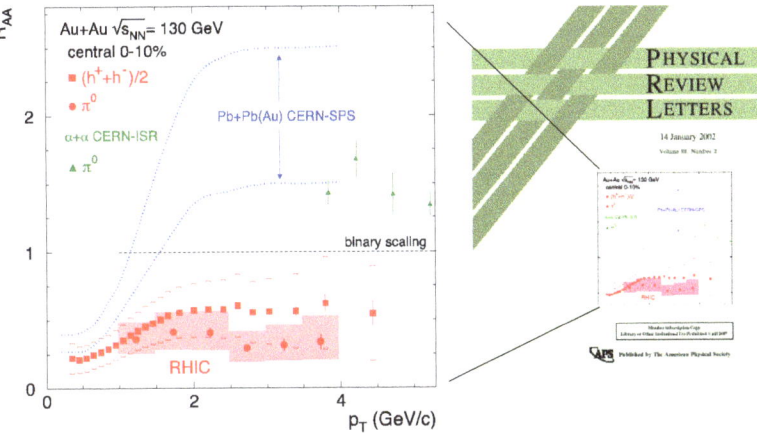

Figure 32. (**left**) Hadron suppression R_{AA} in Au+Au at $\sqrt{s_{NN}}$ = 130 GeV by PHENIX at RHIC compared to enhancement at $\sqrt{s_{NN}}$ = 17 GeV in Pb+Pb at the CERN SpS; and (**right**) plot is from the cover of PRL [16].

5.1. Status of R_{AA} in Au+Au at $\sqrt{s_{NN}}$ = 200 GeV

Figure 33 shows the suppression of all identified hadrons, as well as $e^{\pm}$ from c and b quark decay, with $p_T > 2$ GeV/c measured by PHENIX until 2013. One exception is the enhancement of protons for $2 < p_T < 4$ GeV/c, which are then suppressed at larger p_T. Particle Identification is crucial for these measurements since all particles behave differently. The only particle that shows no-suppression is the direct single γ (from the QCD reaction $g + q \rightarrow \gamma + q$) which shows that the medium produced at RHIC is the strongly interacting QGP since γ rays only interact electromagnetically.

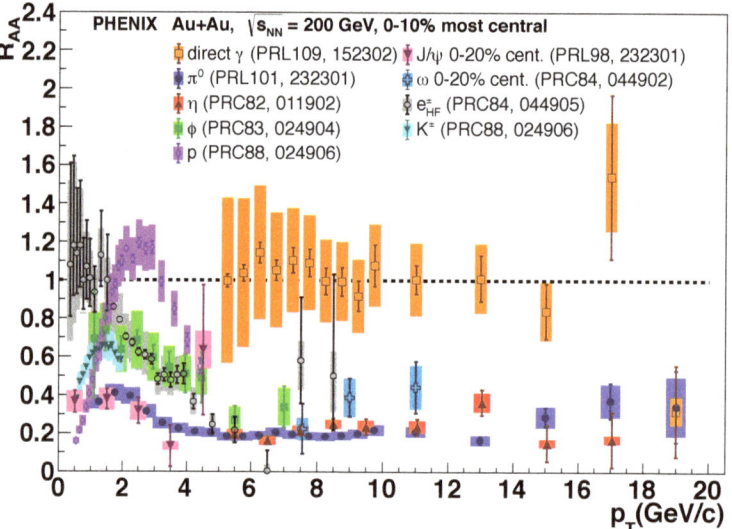

Figure 33. Published PHENIX measurements of R_{AA} with references.

5.2. Recent Measurements to Test the Second BDMPSZ Prediction

(1) The energy loss of the outgoing parton, $-dE/dx$, per unit length (x) of a medium with total length L, is proportional to the total four-momentum transfer-squared, $q^2(L)$, and takes the form:

$$\frac{-dE}{dx} \simeq \alpha_s \langle q^2(L)\rangle = \alpha_s \mu^2 L/\lambda_{\rm mfp} = \alpha_s \hat{q} L$$

where μ, is the mean momentum transfer per collision, and the transport coefficient $\hat{q} = \mu^2/\lambda_{\rm mfp}$ is the four-momentum-transfer-squared to the medium per mean free path, $\lambda_{\rm mfp}$.

(2) Additionally, the accumulated momentum-squared, $\langle p^2_{\perp W}\rangle$ transverse to a parton traversing a length L in the medium is well approximated by

$$\langle p^2_{\perp W}\rangle \approx \langle q^2(L)\rangle = \hat{q} L \qquad \langle \hat{q} L \rangle = \langle k_T^2\rangle_{AA} - \langle k_T'^2\rangle_{pp}. \qquad (3)$$

Although only the component of $\langle p^2_{\perp W}\rangle \perp$ to the scattering plane affects k_T (Figure 34), the azimuthal broadening of the di-jet is caused by the random sum of the azimuthal components $\langle p^2_{\perp W}\rangle/2$ from each outgoing di-jet or $\langle p^2_{\perp W}\rangle = \hat{q} L$.

From the values of R_{AA} observed at RHIC (after 12 years), the JET Collaboration [17] has found that $\hat{q} = 1.2 \pm 0.3$ GeV^2/fm at RHIC, 1.9 ± 0.6 at LHC at an initial time $\tau_0 = 0.6$ fm/c; however, nobody has yet measured the azimuthal broadening predicted. Before proceeding, one has to know the meaning of k_T defined by Feynman, Field and Fox [18] as the transverse momentum of a parton in a nucleon (Figure 34).

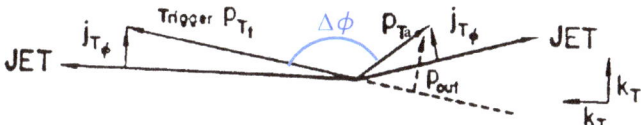

Figure 34. Sketch of a di-jet looking down the beam axis. The k_T from the two jets add randomly and are shown with one k_T perpendicular to the scattering plane, which makes the jets acoplanar in azimuth, and the other k_T parallel to the trigger jet, which makes the jets unequal in energy. in addition, $x_E = p_{Ta} \cos(\pi - \Delta\phi)/p_{Tt}$. The formula for calculating k_T from di-hadron correlations is given in Ref. [19].

5.2.1. The Key New Idea of $\langle k'^2_T \rangle_{pp}$ Instead of $\langle k^2_T \rangle_{pp}$ in Equation (3)

The di-hadron correlations of p_{Ta} with p_{Tt} (Figure 34) are measured in p+p and Au+Au collisions. The parent jets in the original Au+Au collision as measured in p+p will both lose energy passing through the medium but the azimuthal angle between the jets should not change unless the medium induces multiple scattering from $\hat{q}$. Thus, the calculation of k'_T from the di-hadron p+p measurement to compare with Au+Au measurements with the same di-hadron p_{Tt} and p_{Ta} must use the value of $\hat{x}_h$ and $\langle z_t \rangle$ of the parent jets in the A+A collision. The variables are $x_h \equiv p_{Ta}/p_{Tt}$, $\hat{x}_h \equiv \hat{p}_{Ta}/\hat{p}_{Tt}$, $\langle z_t \rangle \equiv p_{Tt}/\hat{p}_{Tt}$, where, e.g., p_{Tt} is the trigger particle transverse momentum and $\hat{p}_{Tt}$ means the trigger jet transverse momentum.

The same values of $\hat{x}_h$ and $\langle z_t \rangle$ in Au+Au and p+p give the cool result [20]:

$$\langle \hat{q}L \rangle = \left[\frac{\hat{x}_h}{\langle z_t \rangle}\right]^2 \left[\frac{\langle p^2_{out} \rangle_{AA} - \langle p^2_{out} \rangle_{pp}}{x^2_h}\right] \quad (4)$$

For di-jet measurements, the formula is even simpler:

(i) $x_h \equiv \hat{x}_h$ because the trigger and away "particles" are the jets; (ii) $\langle z_t \rangle \equiv 1$ because the trigger "particle" is the entire jet not a fragment of the jet; and (iii) $\langle p^2_{out} \rangle = \hat{p}^2_{Ta} \sin^2(\pi - \Delta\phi)$. This reduces the formula for di-jets to:

$$\langle \hat{q}L \rangle = \left[\langle p^2_{out} \rangle_{AA} - \langle p^2_{out} \rangle_{pp}\right] = \hat{p}^2_{Ta}\left[\langle \sin^2(\pi - \Delta\phi) \rangle_{AA} - \langle \sin^2(\pi - \Delta\phi) \rangle_{pp}\right] \quad (5)$$

5.2.2. A Test of Equation (5) for $\langle \hat{q}L \rangle$

Al Mueller et al. [21] gave a prediction for the azimuthal broadening of di-jet angular correlations for 35 GeV jets at RHIC (Figure 35).

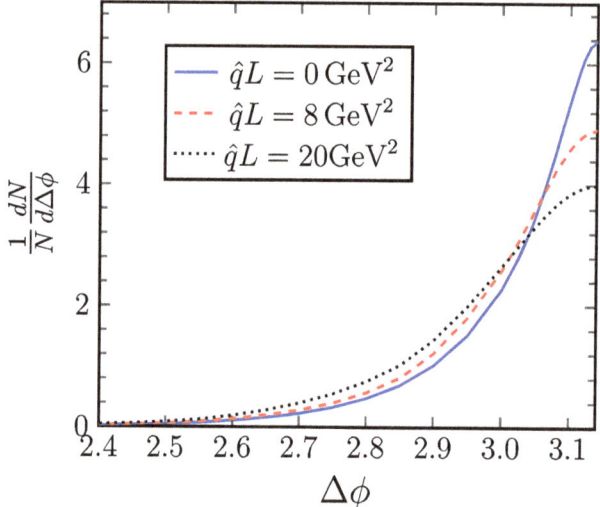

Figure 35. Prediction of folded away azimuthal width of 35 GeV/c Jets at RHIC for several values of $\hat{q}L$.

To check my Equation (5), I measured the half width at half maximum (HWHM), which equals 1.175σ for a Gaussian, for each curve in Figure 35, and calculated $(\sigma \times 35)^2$ to get $\langle p_{out}^2 \rangle$ for each $\hat{q}L$, and used Equation (5) to get 9.6 GeV^2 and 21.5 GeV^2, respectively, for the 8 GeV^2 and 20 GeV^2 plots. This is an excellent result considering that I had to measure the HWHMs in Figure 35 with a pencil and ruler.

5.2.3. How to Calculate $\hat{q}L$ with Equation (4) from Di-Hadron Measurements

The determination of the required quantities is well known to older PHENIXians who have read Ref. [19] or my book [22] as outlined below:

(A) $\langle z_t \rangle$ is calculated from the Bjorken parent–child relation and "trigger bias" [23] (cf. Ref. [24]).

(B) The energy loss of the trigger jet from p+p to Au+Au can be measured by the shift in the p_T spectra [25].

(C) $\hat{x}_h$, the ratio of the away-jet to the trigger jet transverse momenta can be measured by the away particle p_{Ta} distribution for a given trigger particle p_{Tt} taking $x_E = x_h \cos \Delta \phi \approx x_h = p_{Ta}/p_{Tt}$ [19]:

$$\left. \frac{dP_\pi}{dx_E} \right|_{p_{Tt}} = N(n-1)\frac{1}{\hat{x}_h} \frac{1}{(1+\frac{x_E}{\hat{x}_h})^n} . \qquad (6)$$

5.2.4. Example: $\hat{x}_h$ from Fits to the PHENIX Data from Ref. [26]

The fits in Figure 36 work very well, with excellent χ^2/dof. However, it is important to notice that the dashed curve in Au+Au does not fit the data as well as the solid red curve which is the sum of Equation (6) with free parameters + a second term with the form of Equation (6) but with the $\hat{x}_h$ fixed at the p+p value. It is also important to note that the solid red curve between the highest Au+Au data points is notably parallel to the p+p curve. A possible explanation is that, in this region, which is at a fraction $\approx 1\%$ of the dP/dx_E distribution, the highest p_{Ta} fragments are from jets that do not lose energy in the QGP.

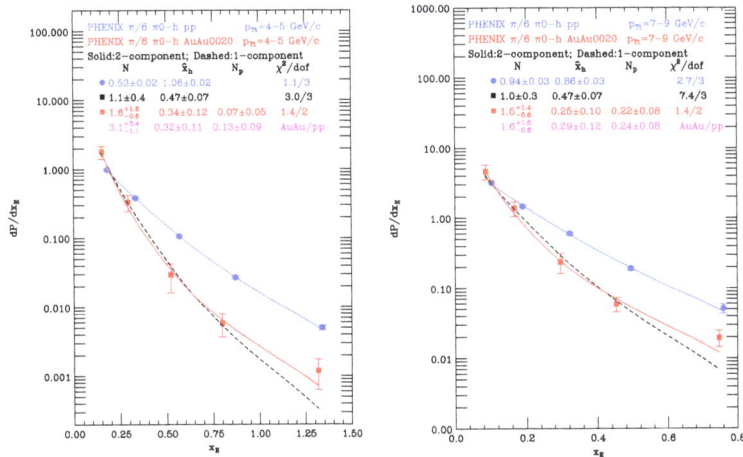

Figure 36. It to x_E distributions for $\pi^0 - h$ correlation in p+p and Au+Au 0–20% central collisions using Equation (6) with the results indicated: (**left**) $4 < p_{Tt} < 5$ GeV/c; and (**right**) $7 < p_{Tt} < 9$ GeV/c.

5.2.5. Results from STAR $\pi^0 - h$ and $\gamma - h$ Correlations [27]

Figure 37 is a table of results of my published calculation [20] of $\langle \hat{q}L \rangle$ from the STAR data. The errors on the STAR $\langle \hat{q}L \rangle$ here (with the *) are much larger than stated in my published calculation because I made a trivial mistake, which is corrected here. In addition, the new values of $\langle \hat{q}L \rangle$ reflect that Equation (4) defines $\langle \hat{q}L \rangle$ not $\langle \hat{q}L \rangle /2$.

STAR PLB771						
$\sqrt{s_{NN}} = 200$	$\langle p_{Tt} \rangle$	$\langle p_{Ta} \rangle$	$\langle z_t \rangle$	$\hat{x}_h$	$\langle p_{out}^2 \rangle$	$\sqrt{\langle k_T^2 \rangle}$
Reaction	GeV/c	GeV/c		GeV/c		GeV/c
p+p	14.71	1.72	0.80 ± 0.05	0.84 ± 0.04	0.263 ± 0.113	2.34 ± 0.34
p+p	14.71	3.75	0.80 ± 0.05	0.84 ± 0.04	0.576 ± 0.167	2.51 ± 0.31
Au+Au 00-12%	14.71	1.72	0.80 ± 0.05	0.36 ± 0.05	0.547 ± 0.163	2.28 ± 0.35
Au+Au 00-12%	14.71	3.75	0.80 ± 0.05	0.36 ± 0.05	0.851 ± 0.203	1.42 ± 0.22
p+p comp	14.71	1.72	0.80 ± 0.05	0.36 ± 0.05	0.263 ± 0.113	1.006 ± 0.18
p+p comp	14.71	3.75	0.80 ± 0.05	0.36 ± 0.05	0.576 ± 0.167	1.076 ± 0.18
					$\langle \hat{q}L \rangle$ GeV^2	
Au+Au 00-12%	14.71	1.72			$4.21 \pm 3.24^*$	
Au+Au 00-12%	14.71	3.75			$0.86 \pm 0.87^*$	

Figure 37. $\hat{q}L$ result table for STAR $\pi^0 - h$, $12 < p_{Tt} < 20$ GeV/c 0-20% centrality.

5.3. Some $\langle \hat{q}L \rangle$ Results from PHENIX [26]

The away widths from PHENIX $\pi^0 - h$ correlations [26] are shown in Figure 38 with the calculated $\hat{q}L$ values for $\pi^0 - h$ $\sqrt{s_{NN}} = 200$ GeV, 20–60% centrality, $5 < p_{Tt} < 7$ GeV/c shown in Figure 39 and $7 < p_{Tt} < 9$ GeV/c in Figure 40.

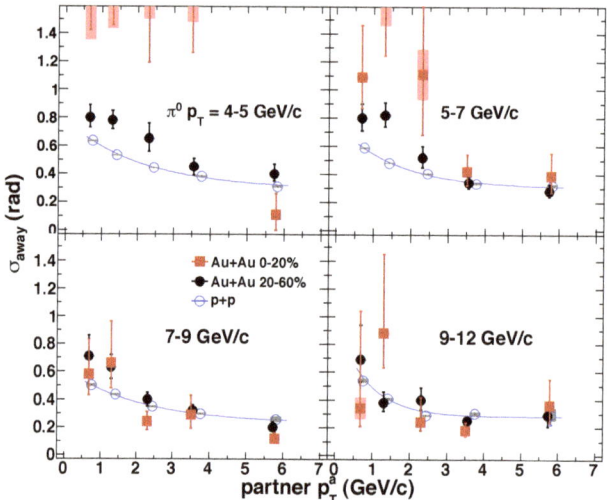

Figure 38. Away widths from $\pi^0 - h$ correlations as function of partner p_T, i.e., p_{Ta}, in Au+Au 0–20% and 20–60% and p+p collisions at $\sqrt{s_{NN}}$ = 200 GeV for four ranges of trigger p_{Tt} indicated [26].

PHENIX PRL104						
$\sqrt{s_{NN}} = 200$	$\langle p_{Tt} \rangle$	$\langle p_{Ta} \rangle$	$\langle z_t \rangle$	$\hat{x}_h$	$\langle p_{out}^2 \rangle$	$\sqrt{\langle k_T^2 \rangle}$
Reaction	GeV/c	GeV/c		GeV/c		GeV/c
p+p	5.78	1.42	0.60 ± 0.06	0.96 ± 0.02	0.434 ± 0.010	3.13 ± 0.37
p+p	5.78	2.44	0.60 ± 0.06	0.96 ± 0.02	0.934 ± 0.031	3.18 ± 0.34
p+p	5.78	3.76	0.60 ± 0.06	0.96 ± 0.02	1.523 ± 0.061	2.74 ± 0.29
p+p	5.78	5.82	0.60 ± 0.06	0.96 ± 0.02	3.339 ± 0.351	2.73 ± 0.32
Au+Au 20-60%	5.78	1.30	0.62 ± 0.06	0.69 ± 0.05	0.867 ± 0.116	4.04 ± 0.61
Au+Au 20-60%	5.78	2.31	0.62 ± 0.06	0.69 ± 0.05	1.291 ± 0.308	2.88 ± 0.54
Au+Au 20-60%	5.78	3.55	0.62 ± 0.06	0.69 ± 0.05	1.370 ± 0.249	1.90 ± 0.32
Au+Au 20-60%	5.78	5.73	0.62 ± 0.06	0.69 ± 0.05	2.562 ± 0.620	1.66 ± 0.31
p+p comp	5.78	1.30	0.62 ± 0.06	0.69 ± 0.05	0.434 ± 0.010	2.39 ± 0.32
p+p comp	5.78	2.31	0.62 ± 0.06	0.69 ± 0.05	0.934 ± 0.031	2.34 ± 0.29
p+p comp	5.78	3.55	0.62 ± 0.06	0.69 ± 0.05	1.522 ± 0.061	2.03 ± 0.25
p+p comp	5.783	5.73	0.62 ± 0.06	0.69 ± 0.05	3.339 ± 0.351	1.93 ± 0.26
					$\langle \hat{q}L \rangle$ .01	$\langle \hat{q}L \rangle$ GeV^2
Au+Au 20-60%	5.78	1.30			6.9 ± 3.6	10.6 ± 3.8
Au+Au 20-60%	5.78	2.31			2.3 ± 2.1	2.8 ± 2.4
Au+Au 20-60%	5.78	3.55			0.35 ± 0.93	−0.5 ± 0.9
Au+Au 20-60%	5.78	5.73			−0.75 ± 1.0	−1.0 ± 0.9

Figure 39. $\hat{q}L$ result table for PHENIX $\pi^0 - h$, $5 < p_{Tt} < 7$ GeV/c 20–60% centrality.

PHENIX PRL104						
$\sqrt{s_{NN}}=200$	$\langle p_{Tt}\rangle$	$\langle p_{Ta}\rangle$	$\langle z_t\rangle$	$\hat{x}_h$	$\langle p^2_{out}\rangle$	$\sqrt{\langle k_T^2\rangle}$
Reaction	GeV/c	GeV/c		GeV/c		GeV/c
p+p	7.83	1.42	0.64 ± 0.06	0.86 ± 0.03	0.360 ± 0.017	2.98 ± 0.41
p+p	7.83	2.44	0.64 ± 0.06	0.86 ± 0.03	0.694 ± 0.048	2.99 ± 0.34
p+p	7.83	3.76	0.64 ± 0.06	0.86 ± 0.03	1.213 ± 0.109	2.76 ± 0.32
p+p	7.83	5.82	0.64 ± 0.06	0.86 ± 0.03	2.177 ± 0.424	2.48 ± 0.38
Au+Au 20-60%	7.83	1.30	0.66 ± 0.06	0.62 ± 0.04	0.548 ± 0.107	3.35 ± 0.64
Au+Au 20-60%	7.83	2.31	0.66 ± 0.06	0.62 ± 0.04	0.803 ± 0.177	2.45 ± 0.46
Au+Au 20-60%	7.83	3.55	0.66 ± 0.06	0.62 ± 0.04	1.237 ± 0.232	2.08 ± 0.34
Au+Au 20-60%	7.83	5.73	0.66 ± 0.06	0.62 ± 0.04	1.300 ± 0.350	1.29 ± 0.27
p+p comp	7.83	1.30	0.66 ± 0.06	0.62 ± 0.04	0.360 ± 0.017	2.28 ± 0.33
p+p comp	7.83	2.31	0.66 ± 0.06	0.62 ± 0.04	0.694 ± 0.048	2.22 ± 0.28
p+p comp	7.83	3.55	0.66 ± 0.06	0.62 ± 0.04	1.213 ± 0.109	2.05 ± 0.26
p+p comp	7.83	5.73	0.66 ± 0.06	0.62 ± 0.04	2.177 ± 0.424	1.76 ± 0.28
					$\langle\hat{q}L\rangle$ .01	$\langle\hat{q}L\rangle$ GeV^2
Au+Au 20-60%	7.83	1.30			9.3 ± 6.3	6.0 ± 3.7
Au+Au 20-60%	7.83	2.31			2.4 ± 2.2	1.1 ± 1.9
Au+Au 20-60%	7.83	3.55			1.0 ± 1.2	0.11 ± 1.1
Au+Au 20-60%	7.83	5.73			−1.2 ± 1.0	−1.4 ± 1.0

Figure 40. $\hat{q}L$ result table for PHENIX π^0-h, $7<p_{Tt}<9$ GeV/c 20–60% centrality.

5.4. Conclusions

It appears that the method works and gives consistent results for all the $\hat{q}L$ calculations shown (Figures 37, 39 and 40). In the lowest $p_{Ta}\sim 1.5$ GeV/c bin, the results are all consistent with the JET collaboration [17] result, $\hat{q}=1.2\pm 0.3$ GeV^2/fm or $\hat{q}L=8.4\pm 2.1$ GeV^2 for $L=7$ fm, the radius of an Au nucleus. However, for $p_{Ta}>2.0$ GeV/c, all the results are consistent with $\hat{q}L=0$. Personally, I think that this is where the first gluon emitted in the medium was inside the jet cone, so that all further emissions were also inside the jet cone due to the angular ordering of QCD so that there is no evident suppression; or that jets with fragments with $p_T\geq 3$ GeV/c, which are distributed narrowly about the jet axis, are not strongly affected by the medium [28]. I think that this also agrees with the observation in Figure 36 that two or three orders of magnitude down in the $x_E=p_{Ta}/p_{Tt}$ distributions the A+A best fit is parallel to the p+p measurement, which means that these A+A fragments are from jets that have not lost energy. This is consistent with all the $I_{AA}=x_E^{AA}/x_E^{pp}=(p_{Ta}^{AA}/p_{Ta}^{pp})|_{p_{Tt}}$ distributions ever measured (e.g., Figures 41 and 42), which decrease with increasing p_{Ta} until $p_{Ta}\approx 3$ GeV/c and then remain constant because the A+A and p+p distributions are parallel due to no jet energy loss for fragments in this range.

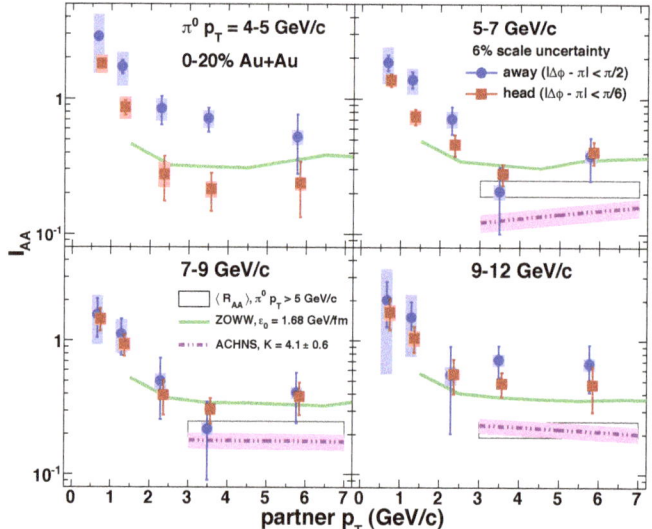

Figure 41. PHENIX I_{AA} distribution [26].

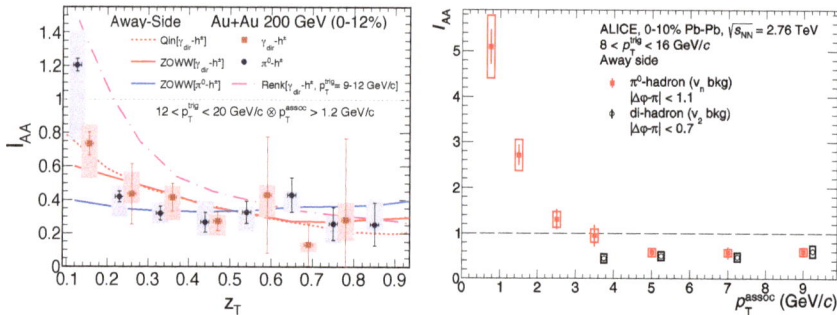

Figure 42. (**left**) STAR I_{AA} distribution [27]; and (**right**) ALICE I_{AA} distribution [29].

Funding: The research was supported by U. S. Department of Energy, DE-SC0012704.

Conflicts of Interest: The author declares no conflict of interest.

References

1. Bunce, G.; et al. [WMR Collaboration]. Λ^0 Hyperon Polarization in Inclusive Production by 300-GeV Protons on Beryllium. *Phys. Rev. Lett.* **1976**, *36*, 1113–1116. [CrossRef]
2. Adamczyk, L.; et al. [STAR Collaboration]. Global Λ hyperon polarization in nuclear collisions. *Nature* **2017**, *548*, 62–65.
3. Adam, J.; et al. [STAR Collaboration]. Global polarization of Λ hyperons in Au+Au collisions at $\sqrt{s_{NN}}$ = 200 GeV. *Phys. Rev. C* **2018**, *98*, 014910. [CrossRef]

4. Aidala, C.; et al. [PHENIX Collaboration]. Creation of quark-gluon plasma droplets with three distinct geometries. *Nat. Phy.* **2019**, *15*, 214–220.
5. Adare, A.; et al. [PHENIX Collaboration]. Measurements of mass-dependent azimuthal anisotropy in central p+Au, d+Au and ^{3}He+Au collisions at $\sqrt{s_{NN}}$ = 200 GeV. *Phys. Rev. C* **2018**, *97*, 064904. [CrossRef]
6. Nagle, J.L.; Belmont, R.; Hill, K.; Orjuela Koop, J.; Perepelitsa, D.V.; Yin, P.; Lin, Z.-W.; McGlinchey, D. Minimal conditions for collectivity in e^+e^- and $p + p$ collisions. *Phys. Rev. C* **2018**, *97*, 024909. [CrossRef]
7. Kogut, J.; Susskind, L. Vacuum polarization and the absence of free quarks in four dimensions. *Phys. Rev. D* **1974**, *9*, 3501–3512. [CrossRef]
8. Eichten, E.; Gottfried, K.; Kinoshita, T.; Kogut, J.; Lane, K.D.; Yan, T.-M. Spectrum of Charmed Quark-Antiquark Bound States. *Phys. Rev. Lett.* **1975**, *34*, 369–372. [CrossRef]
9. Matsui, T.; Satz, H. J/ψ Suppression by Quark-Gluon Plasma Formation. *Phys. Lett. B* **1987**, *178*, 416–422. [CrossRef]
10. Satz, H. Colour deconfinement in nuclear collisions. *Rep. Prog. Phys.* **2000**, *63*, 1511–1574. [CrossRef]
11. Soltz, R.A.; DeTar, C.; Karsch. F.; Mukherjee, S.; Vranas, P. Lattice QCD Thermodynamics with Physical Quark Masses. *Ann. Rev. Nucl. Part. Sci.* **2015**, *65*, 379. [CrossRef]
12. Braun-Munzinger, P.; Stachel, J. (Non) thermal aspects of charmonium production and a new look at J/ψ suppression. *Phys. Lett. B* **2000**, *490*, 196. [CrossRef]
13. Abelev, B.; et al. [ALICE Collaboration]. J/ψ Suppression at Forward Rapidity in Pb-Pb Collisions at $\sqrt{s_{NN}}$ = 2.76 TeV. *Phys. Rev. Lett.* **2012**, *109*, 072301. [CrossRef] [PubMed]
14. Baier, R.; Schiff, D.; Zakharov, B.G. Energy Loss in Perturbative QCD. *Ann. Rev. Nucl. Part. Sci.* **2000**, *50*, 37–69. [CrossRef]
15. Mueller, A.H. On the multiplicity of hadrons in QCD jets. *Phys. Lett. B* **1981**, *104*, 161–164. [CrossRef]
16. Adcox, K.; et al. [PHENIX Collaboration]. Suppression of Hadrons with Large Transverse Momentum in Central Au+Au Collisions at $\sqrt{s_{NN}}$ =130 GeV. *Phys. Rev. Lett.* **2002**, *88*, 022301. [CrossRef] [PubMed]
17. Burke, K.M.; et al. [JET Collaboration]. Extracting the jet transport coefficient from jet quenching in high-energy heavy-ion collisions. *Phys. Rev. C* **2014**, *90*, 014909. [CrossRef]
18. Feynman, R.P.; Field, R.D.; Fox, G.C. Correlations among particles and jets produced with large transverse momenta. *Nucl. Phys. B* **1977**, *128*, 1–65. [CrossRef]
19. Adler, S.S.; et al. [PHENIX Collaboration]. Jet properties from dihadron correlations in p+p collisions at $\sqrt{s}$ = 200 GeV. *Phys. Rev. D* **2006**, *74*, 072002. [CrossRef]
20. Tannenbaum, M. J. Measurement of $\hat{q}$ in Relativistic Heavy Ion Collisions using di-hadron correlations. *Phys. Lett. B* **2017**, *771*, 553–557. [CrossRef]
21. Mueller, A.H.; Wu, B.; Xiao, B.-W.; Yuan, F. Probing transverse momentum broadening in heavy ion collisions. *Phys. Lett. B* **2016**, *763*, 208–212. [CrossRef]
22. Rak, J.; Tannenbaum, M.J. *High p_T Physics in the Heavy Ion Era*; Cambridge University Press: Cambridge, UK, 2013; pp. 1–387.
23. Jacob, M.; Landshoff, P.V. Large Transverse Momentum and Jet Studies. *Phys. Repts.* **1978**, *48*, 285–350. [CrossRef]
24. Adare, A.; et al. [PHENIX Collaboration]. Double-helicity dependence of jet properties from dihadrons in longitudinally polarized p+p collisions at $\sqrt{s_{NN}}$ = 200 GeV. *Phys. Rev. D* **2010**, *81*, 012002. [CrossRef]
25. Adare, A.; et al. [PHENIX Collaboration]. Neutral pion production with respect to centrality and reaction plane in Au+Au collisions at $\sqrt{s_{NN}}$ = 200 GeV. *Phys. Rev. C* **2013**, *87*, 034911. [CrossRef]
26. Adare, A.; et al. [PHENIX Collaboration]. Transition in Yield and Azimuthal Shape Modification in Dihadron Correlations in Relativistic Heavy Ion Collisions. *Phys. Rev. Lett.* **2010**, *104*, 252301. [CrossRef] [PubMed]
27. Adamczyk, L.; et al. [STAR Collaboration]. Jet-like correlations with direct-photon and neutral-pion triggers at $\sqrt{s_{NN}}$ = 200 GeV. *Phys. Lett. B* **2016**, *760*, 689–696.[CrossRef]

28. Methar-Tani, Y.; Milhano, J.G.; Tywoniuk, K. Jets in Heavy-Ion Collisions. *Int. J. Mod. Phys. A* **2013**, *28*, 1340013. [CrossRef]
29. Adam, J.; et al. [ALICE Collaboration]. Jet-like correlations with neutral pion triggers in pp and central Pb-Pb collisions at 2.76 TeV. *Phys. Lett. B* **2016**, *763*, 238–250. [CrossRef]

© 2019 by the author. Licensee MDPI, Basel, Switzerland. This article is an open access article distributed under the terms and conditions of the Creative Commons Attribution (CC BY) license (http://creativecommons.org/licenses/by/4.0/).

MDPI
St. Alban-Anlage 66
4052 Basel
Switzerland
www.mdpi.com

Universe Editorial Office
E-mail: universe@mdpi.com
www.mdpi.com/journal/universe

Disclaimer/Publisher's Note: The statements, opinions and data contained in all publications are solely those of the individual author(s) and contributor(s) and not of MDPI and/or the editor(s). MDPI and/or the editor(s) disclaim responsibility for any injury to people or property resulting from any ideas, methods, instructions or products referred to in the content.